Lectures on Differential Geometry I

Will J. Merry

Lectures on Differential Geometry I

Differentiable Manifolds

Springer

Will J. Merry *(deceased)*
Departement Mathematik
ETH Zürich
Zürich, Switzerland

ISBN 978-3-032-03732-9 ISBN 978-3-032-03733-6 (eBook)
https://doi.org/10.1007/978-3-032-03733-6

Mathematics Subject Classification: 53-01

Cover illustration: Plot of the 2-dimensional manifold z = cos(x)sin (y) © Martin Merry 2025

This Springer imprint is published by the registered company Springer Nature Switzerland AG
The registered company address is: Gewerbestrasse 11, 6330 Cham, Switzerland

Will J. Merry
1984–2022

Foreword

I first met the student Will Merry at a summer school at the Korea Institute for Advanced Study in Seoul in 2010. At that time, Will was a brilliant PhD student in Cambridge, working under the supervision of Gabriel Paternain. A year later, I served as an external referee for his PhD defence. We became friends, and later I had the pleasure of working with the researcher Will Merry. Since 2015 an assistant professor of mathematics at ETH Zürich, he had become a deeply original and creative mathematician, and collaborating with him was an enriching experience. Now that he is no longer with us, I am discovering the teacher Will Merry. It is both a beautiful revelation and a confirmation of the extraordinary qualities I had already come to appreciate in the student and the researcher.

The book you are holding in your hands—or reading on a screen—originates from a two-semester course in differential geometry that Will taught twice at ETH, in 2018/2019 and 2020/2021. His PhD student Ana Žegarac told Will's father, Martin Merry, about the existence of these lecture notes and helped him recover them from an Overleaf account. With the precious assistance of Urs Fuchs, Martin Merry edited the manuscript and shaped it into this two-volume textbook.

The book assumes very few prerequisites: a solid foundation in analysis (in one and several variables) and in linear algebra is essential, but no prior knowledge of surfaces in Euclidean space, general topology, or algebraic topology is required. Yet, it sets itself the ambitious goal of introducing readers to a broad spectrum of the language and foundational results of modern differential geometry: abstract manifolds first and foremost, but also Lie groups, homogeneous spaces, fibre bundles, de Rham cohomology, connections and curvature, and Riemannian geometry.

With its fresh and informal style, the book clearly benefits from the short time span between Will the student and Will the teacher. The teacher has not forgotten the difficulties of the student and addresses many natural questions that arise when encountering this subject for the first time: Why is it useful to view a tangent vector as a derivation? Why is a connection on a vector bundle sometimes regarded as a distribution of linear subspaces, sometimes as a way to differentiate certain objects, sometimes as a means of transporting vectors along curves?

The first volume begins with the definition of abstract manifolds and the foundations of calculus on manifolds. It includes a digression on Lie groups and homogeneous spaces, which not only provide systematic ways of constructing new manifolds from old ones but also supply the right language for discussing symmetry in geometry. Vector bundles are introduced alongside more general fibre bundles, which play an important role in the second volume. The first volume concludes with a study of differential forms, Stokes' theorem, and de Rham cohomology. Remaining faithful to the idea of avoiding prerequisites from algebraic topology, Will explains de Rham's theorem by first introducing a model for singular homology based on smooth singular cubes.

The first part of the second volume is devoted to connections on vector bundles, their holonomy, and their curvature. After a discussion of characteristic classes, the theory of connections on principal bundles is developed. The Ambrose–Singer holonomy theorem, already presented in the context of vector bundles, is proven in this more general setting. The remainder of Volume 2 focuses on the Levi-Civita connection and the foundations of Riemannian geometry: the Riemann curvature tensor and its derived notions of curvature, geodesics, length, and the metric structure of a Riemannian manifold.

The first volume is suitable for a one-semester introduction to differential geometry, and the second for a more advanced follow-up course. Chapters are short and roughly correspond to 90-minute lectures. Many chapters include Bonus Material, featuring topics that are not essential for subsequent reading—such as technical proofs or deeper insights.

The theory and examples are further developed through 18 Problem Sheets, which may be used in tutorials. Problems which are either harder or require more background knowledge are marked with a ♣ symbol. Solutions to the problems, written by Will's students, are provided at the end of each volume. There are also Bonus Problems—either harder exercises or open-ended questions meant to encourage reflection. They all deserve a ♣, and their solutions are not included.

The book is also well suited for self-study and can serve as a hands-on introduction to more advanced texts. I remember trying, as a student, to read the classic two-volume treatise *Foundations of Differential Geometry* by Shoshichi Kobayashi and Katsumi Nomizu. I always failed, and eventually absorbed the material in a more haphazard way while doing research. I believe that having had Will's book as a guide would have spared me those frustrations.

Ruhr-Universität Bochum
Bochum, Germany
May 20, 2025

Alberto Abbondandolo

Preface

Will Merry died tragically young in 2022. In addition to his mathematical research he was a very capable and enthusiastic teacher of mathematics. His goal as a maths lecturer was not only to present the material in a comprehensible and enjoyable form but also to make the student see why the material was important, and how it related to other branches of mathematics. This is something he excelled at: after his death one of his students said:

"There is a famous allegory which says most mathematicians are like frogs, who can only see and appreciate what is close to them. A few of them, however, are like birds that fly above all else and find the beauty of the overall landscape. Dr. Merry had the gift to make all his students feel like one of these birds for an hour, so that they were free to admire all of mathematics up to the far horizon."

This book and its companion book (*Differential Geometry II*) [1] are based on a two semester course on Differential Geometry that Will gave at ETH Zürich in the academic year 2020–2021. The two books—like the course—aim to give the reader a complete introduction to the language of differential geometry. Like his lectures, the books also cover material in a range of adjacent fields to provide context for the material.

While the material in these two volumes is based on a two semester lecture course, the material can be presented in a number of ways to fit other course lengths—for example, if a shorter first semester course is required, then the material on Lie groups and homogeneous spaces could be delayed to a second semester.

The diagram on the next page shows the dependencies between the different parts of the course, to help readers who want to omit some material, or follow a different path through the books. Solid arrows indicate dependencies: to understand the material in a given block the reader needs to understand the material in all the blocks pointing to this block with solid arrows. The dashed arrows show blocks that it is natural to go to after finishing a block, but which could be considered in a different sequence.

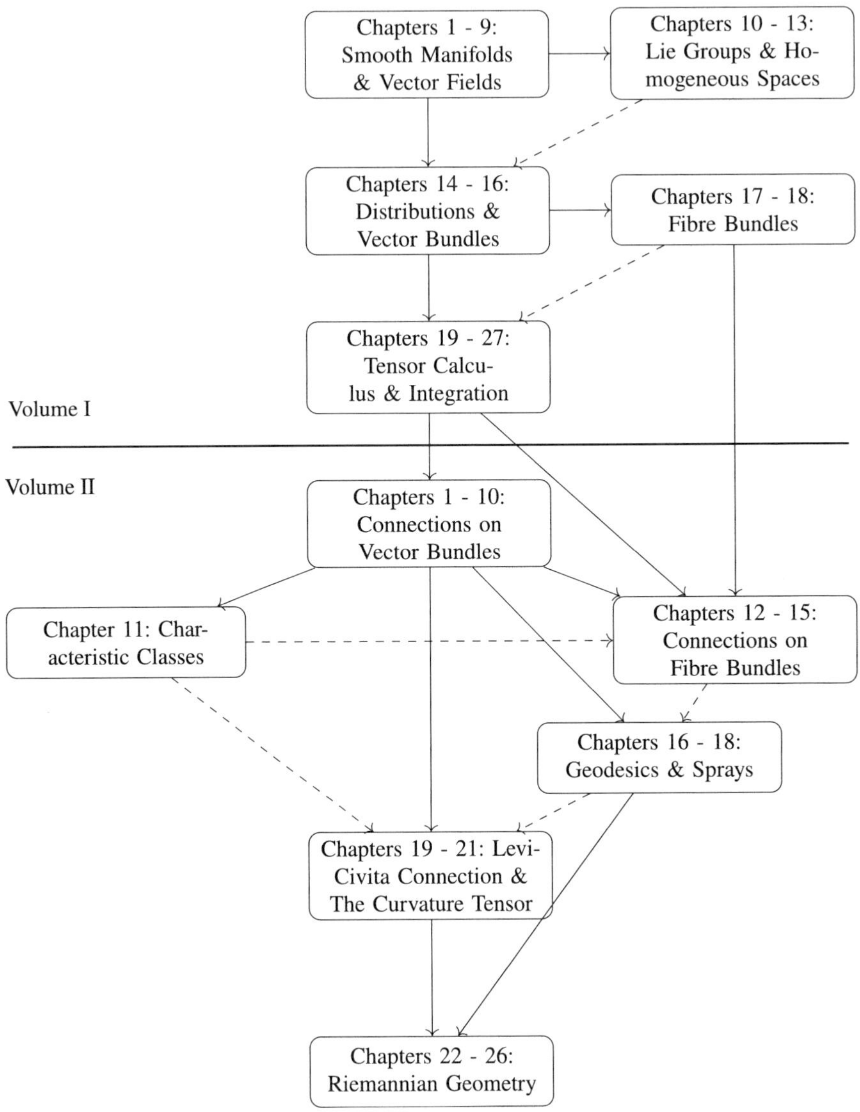

Reading Guide and Dependencies

Many chapters end with a bonus section which expands on the core material. These bonus sections contain material which may

- Be only tangentially related to the central material of the book
- Be rather technical or difficult
- Be just a sketch

- Require more background knowledge (of, e.g., algebraic topology, functional analysis, etc.) than the rest of the book assumes

Such material is described succinctly as *unexaminable*.

Competing Interests The author has no competing interests to declare that are relevant to the content of this manuscript.

Acknowledgements

I am grateful to the many people who have helped make this book a reality. First and foremost I thank Dr Urs Fuchs, whose detailed review of the content and patient explanations of things I needed to change have been invaluable. I thank Professor Alberto Abbondandolo for his initial review of the manuscript, his Foreword, the Reading Guide, and *his* patience in explaining things to me. I thank Dr Remi Lodh at Springer for his support in getting these notes published. And I thank Will's friends and colleagues for their help and contributions, in particular Paul Biran and Alessio Pellegrini, as well as Mads Bisgaard, Giada Franz, Francesco Palmurella, Alessandro Pigati, Alexandre Puttick, and Jagna Wisniewska. And I especially thank Ana Žegarac.

Contents

Chapter 1
Smooth Manifolds

Let us begin with a short history lesson on how we learned to identify (continuously) differentiable functions.

(i) **(High school)** A function $f\colon \mathbb{R} \to \mathbb{R}$ is continuous if its graph doesn't have any jumps. The derivative $f'(t)$ at a point t is the slope of the graph of $f(t) = s$ at the point t.

(ii) **(First class in Analysis)** The (ε, δ) definition of continuity. A function $f\colon \mathbb{R} \to \mathbb{R}$ is differentiable at the point t if the limit

$$\lim_{s\to 0} \frac{f(t+s) - f(t)}{s}$$

exists. This limit is denoted by $f'(t)$. The function f is continuously differentiable if $t \mapsto f'(t)$ is itself a continuous function.

(iii) **(Second class in Analysis)** Now you learned how to handle functions with more than one variable. Suppose $f\colon \mathbb{R}^m \to \mathbb{R}^n$ is a continuous function. Then f is differentiable at $p \in \mathbb{R}^m$ if there exists a *linear* map $\ell\colon \mathbb{R}^m \to \mathbb{R}^n$ (that is, an $n \times m$ matrix) such that

$$\lim_{\|\xi\|\to 0} \frac{\|f(p+\xi) - f(p) - \ell\xi\|}{\|\xi\|} = 0. \tag{1.1}$$

We denote ℓ by $Df(p)$. It is the matrix of partial derivatives of $f = (f^1, \dots, f^n)$ at the point $p = (u^1, \dots, u^m)$:

$$Df(p) = \begin{pmatrix} \frac{\partial f^1}{\partial u^1}(p) & \cdots & \frac{\partial f^1}{\partial u^m}(p) \\ \vdots & \ddots & \vdots \\ \frac{\partial f^n}{\partial u^1}(p) & \cdots & \frac{\partial f^n}{\partial u^m}(p) \end{pmatrix}$$

W. J. Merry, *Lectures on Differential Geometry I*,
https://doi.org/10.1007/978-3-032-03733-6_1

Of course, this reduces to the same definition as before if $m = n = 1$, since a 1×1 matrix is just a number, and in this case $Df(p)$ is simply multiplication by the number $f'(p)$. As before, the function f is continuously differentiable if $p \mapsto Df(p)$ is a continuous function (this is now a function $\mathbb{R}^m \longrightarrow \{n \times m \text{ matrices}\} \cong \mathbb{R}^{mn}$).

(iv) **(First class in topology)** Suppose now X and Y are topological spaces and $f: X \longrightarrow Y$ is a function. You learned that f is continuous if $f^{-1}(U)$ is an open set in X for every open set U in Y. If X and Y are metric spaces then this reduces to the old (ε, δ) definition of continuity. But how does one define differentiability in this setting? Equation (1.1) *does not make sense* any more, since in an arbitrary topological space one cannot simply "add" points, and there is no such thing as a "linear" map!

Here endeth the history lesson.

TL;DR:

- It's easy to differentiate functions on Euclidean spaces.
- Most topological spaces are not Euclidean spaces.
- Bummer.

Indeed, this is a real shame. Measuring the rate at which things change—that is, differentiating them—is absolutely crucial to all applications of mathematics (and is arguably the single most important concept in theoretical physics). However most "real life" systems are *not* defined on open sets in Euclidean spaces (the whole point of your topology course was to introduce classes of spaces appropriate for such models).

This is where differential geometry comes in. Our first aim is to define a special type of topological space, called a **smooth manifold**, on which it *is* possible to make sense of differentiating a continuous function. The definition of a smooth manifold will:

- Include open sets in Euclidean spaces as a special case.
- Be sufficiently general so that the topological spaces that occur in "real life" systems (in theoretical physics, economics, computer science, robotics, genetics, cooking etc) are smooth manifolds.

So let's get started.

In fact, we will define smooth manifolds in two stages. We will first define a **topological manifold**, which is a topological space that locally resembles Euclidean space. We will then endow a topological manifold with an additional piece of data called a **smooth structure**. The smooth structure is what will allow us to actually go ahead and differentiate things. A topological manifold equipped with a smooth structure is then called a **smooth manifold**.

We first recall a few concepts from point-set topology.

DEFINITION 1.1. A topological space X is said to be **metrisable** if there exists a metric on X which induces the given topology.

Thus a metrisable topological space is a topological space which is homeomorphic to a metric space. Non-metrisable topological spaces[1] crop up quite frequently in functional analysis and algebraic topology. In geometry, however, such spaces are abominations, and we will exclude them right from the start.

Definition 1.2. A metrisable topological space X is said to be **separable** if there exists a countable dense subset.

It is easy to come up with examples of non-separable metrisable spaces—for example, an uncountable disjoint union of metrisable spaces.

Definition 1.3. A topological space X is said to be **locally Euclidean of dimension** m if every point has a neighbourhood[2] which is homeomorphic to $\mathbb{R}^m$.

Thus a topological space is locally Euclidean of dimension m if locally it "looks" like the Euclidean space $\mathbb{R}^m$. We are now ready for the first key definition of the course.

Definition 1.4. A **topological manifold of dimension** m is a separable metrisable topological space which is locally Euclidean of dimension m.

As already alluded to, the most important part of the definition of a topological manifold is the locally Euclidean part. Metrisability and separability are included solely to rule out pathologies. In general the phrase "topological manifold" means a topological manifold of some unspecified dimension m.

Convention. We will typically use the letters M, N, and L to denote manifolds. Unless specified otherwise, the dimension of a manifold should be assumed to be the corresponding lowercase letter. Thus M has dimension m, and N has dimension n, and so on.

> The bonus section at the end of this chapter contains additional background information on the point-set topological properties of manifolds.

Remark 1.5. Suppose $m \neq n$ are two non-negative integers. Is it possible for a (non-empty) topological space to be locally Euclidean of dimension m *and* locally Euclidean of dimension n? Equivalently, is $\mathbb{R}^m$ homeomorphic to $\mathbb{R}^n$ for $m \neq n$? The answer to this is "no", but this is surprisingly difficult to prove. This result is called the *Invariance of Domain Theorem*, and was first proved by Brouwer in 1912. The easiest proof uses tools from algebraic topology.

Examples 1.6. Here are some examples.

(i) $\mathbb{R}^m$ is trivially a topological manifold of dimension m. More generally, any m-dimensional vector space[3] is a topological manifold of dimension m.

---

[1] Exercise: Find examples thereof.

[2] We use the convention that a neighbourhood of a point is an *open* set containing that point.

[3] We use the convention that all vector spaces are real and finite dimensional, unless specified otherwise.

(ii) The *open* ball

$$B^m := \{p \in \mathbb{R}^m \mid \|p\| < 1\}$$

is a topological manifold of dimension m.[4] More generally, every non-empty open subset of a topological manifold of dimension m is also a topological manifold of dimension m.

(iii) A **non-example**: For $m > 0$ the *closed* unit ball

$$D^m := \{p \in \mathbb{R}^m \mid \|p\| \leq 1\}$$

is *not* a topological manifold of dimension m. (It is an illustrative exercise to try and work out why.) In fact, D^m is an example of a more general concept of a **manifold with boundary** that we will come back to later in Chapter 24.

We will see more interesting examples later in this chapter.

Let us now get back to the point of view discussed at the beginning of the chapter: we are trying to develop a class of topological spaces in which it is possible to differentiate functions. One might naively believe that the locally Euclidean condition built into the definition of a topological manifold is enough. Indeed, to check whether a function $f: \mathbb{R}^m \to \mathbb{R}^n$ is differentiable at a point $p \in \mathbb{R}^m$, we need only examine f in a small neighbourhood of p: this is clear from (1.1). Thus if we are given a continuous map between two topological manifolds, we can locally view it as a continuous map between two Euclidean spaces, and thus we could conceivably say our original map is differentiable if this local map is. But herein lies a problem: a topological manifold is only locally *homeomorphic* to Euclidean space, and a different choice of homeomorphism might affect whether the local map is differentiable or not.

The solution to this is to introduce more structure. Before doing so, let us recall the chain rule for continuously differentiable functions between Euclidean spaces. We will give two different versions: one for the total differential $Df(p)$ (the matrix) and one for the partial derivatives $\frac{\partial f^i}{\partial u^j}$.

PROPOSITION 1.7 (THE CHAIN RULE). *Let $\mathcal{O} \subset \mathbb{R}^m$, $\Omega \subset \mathbb{R}^n$ be open sets. Let $f: \mathcal{O} \to \mathbb{R}^n$ and $g: \Omega \to \mathbb{R}^l$ be continuously differentiable functions satisfying $f(\mathcal{O}) \subset \Omega$.*

(i) *The function $g \circ f$ is also continuously differentiable, and its derivative at the point p is given by*

$$D(g \circ f)(p) = Dg(f(p)) \circ Df(p).$$

(ii) *Write $(u^1, \ldots, u^m)$ for the coordinates on $\mathbb{R}^m$ and $(v^1, \ldots, v^n)$ for the coordinates on $\mathbb{R}^n$, and write $f = (f^1, \ldots, f^n)$ and $g = (g^1, \ldots, g^l)$. Then the partial derivatives of $g \circ f$ are given by*

$$\frac{\partial (g^i \circ f)}{\partial u^j}(p) = \sum_{k=1}^{n} \frac{\partial g^i}{\partial v^k}(f(p)) \frac{\partial f^k}{\partial u^j}(p), \qquad \text{for all } 1 \leq i \leq l, 1 \leq j \leq m.$$

We now define higher order derivatives.

[4] Exercise: prove this!

DEFINITION 1.8. Let $O \subset \mathbb{R}^m$ and $\Omega \subset \mathbb{R}^n$ be non-empty open sets and suppose $f\colon O \to \Omega$ is a differentiable map. We say that f is of class C^k if each partial derivative $\frac{\partial f^i}{\partial u^j}$ is an $(k-1)$-times continuously differentiable function. We say that f is a **smooth** or of class C^∞ if f is of class C^k for every $k \geq 1$. If f is both smooth and bijective and the inverse function is also smooth then we say that f is a **diffeomorphism**.

It follows from part (ii) of Proposition 1.7 that the composition of smooth functions defined on open sets in Euclidean spaces is again a smooth function.

REMARK 1.9. If f is a diffeomorphism then necessarily $m = n$. This follows immediately from part (i) of Proposition 1.7, which tells us that if f is a diffeomorphism then $Df(p)$ is an invertible matrix. (Its inverse is given by $D(f^{-1})(f(p))$.) An $n \times m$ matrix can only be invertible if $m = n$. Thus in particular $\mathbb{R}^m$ cannot be *diffeomorphic* to $\mathbb{R}^n$ for $m \neq n$ (compare to Remark 1.5).

With these preliminaries in hand, let us get started on the definition of a smooth manifold.

DEFINITION 1.10. Let M be a topological manifold of dimension m. A **smooth atlas** on M is a collection

$$\mathcal{X} = \{x_a\colon U_a \to O_a \mid a \in A\}$$

where $\{U_a \mid a \in A\}$ is an open cover of M, each O_a is an open set in $\mathbb{R}^m$, and each $x_a\colon U_a \to O_a$ is a homeomorphism such that the following *compatibility condition* is satisfied: Suppose $a, b \in A$ are such that $U_a \cap U_b \neq \emptyset$. Then the composition

$$x_b \circ x_a^{-1}\colon x_a(U_a \cap U_b) \to x_b(U_a \cap U_b)$$

should be a diffeomorphism. This makes sense, since both $x_a(U_a \cap U_b)$ and $x_b(U_a \cap U_b)$ are open subsets of $\mathbb{R}^m$. We call the maps x_a the **charts** of the atlas $\mathcal{X}$, and the compositions $x_b \circ x_a^{-1}$ the **transition functions** of the atlas.

CONVENTION. We typically denote points in manifolds by the letters p and q, and charts on manifolds by the letters x and y. The phrase "let (U, x) be a chart about p" is short for: let $x\colon U \to O$ be a chart on M with $p \in U$.

We say that two smooth atlases $\mathcal{X}$ and $\mathcal{Y}$ are **equivalent** if their union is also a smooth atlas, that is, if given any chart x of $\mathcal{X}$ and any chart y of $\mathcal{Y}$ such that the domains of x and y intersect, the composition $y \circ x^{-1}$ is also a diffeomorphism. It is immediate that this notion defines an equivalence relation on the set of smooth atlases on a given topological manifold.

DEFINITION 1.11. A **smooth structure** on a topological manifold is an equivalence class of smooth atlases.

REMARK 1.12. Given an equivalence class of smooth atlases, there is a unique *maximal* smooth atlas in that class (simply take the union of all the atlases in the given equivalence class). Thus there is a one-to-one correspondence between smooth structures and maximal smooth atlases. Since dealing with equivalence relations can be tedious, it is usually more convenient to regard a smooth structure as a maximal smooth atlas, and we will do so without further comment.

We now finally arrive at the main definition of this first chapter.

DEFINITION 1.13. A **smooth manifold of dimension** m is a pair $(M, \mathfrak{X})$ where M is a topological manifold of dimension m and $\mathfrak{X}$ is a smooth structure on M.

Since a smooth atlas is contained in a unique maximal smooth atlas, it is sufficient when defining a smooth manifold to specify a smooth atlas on the underlying topological manifold. Whenever possible we will omit the $\mathfrak{X}$ from the notation and just write M. For smooth manifolds the fact that the dimension is well-defined is much easier than for topological manifolds (we only need Remark 1.9, which does not require any algebraic topology).

EXAMPLE 1.14. The standard smooth structure on $\mathbb{R}^m$ is the one containing the smooth atlas consisting of exactly one chart: the identity map $\mathrm{id}\colon \mathbb{R}^m \to \mathbb{R}^m$. The reason for the word "standard" will become clear by the end of the chapter. More generally, if V is any m-dimensional real vector space, then the standard smooth structure on V is the one induced by the smooth atlas consisting of a single chart $\ell\colon V \to \mathbb{R}^m$, where ℓ is some linear isomorphism.[5]

Just as with topological manifolds, an open subset of a smooth manifold is also a smooth manifold:

LEMMA 1.15. *Let M be a smooth manifold of dimension m and let $W \subset M$ be a non-empty open set. Then W naturally inherits the structure of a smooth manifold of dimension m.*

Proof. We have already remarked in part (i) of Example 1.6 that W is a topological manifold of dimension m. Let $\mathfrak{X} = \{x_a\colon U_a \to O_a \mid a \in A\}$ be a smooth atlas on M. Then

$$\left\{x_a|_{W\cap U_a}\colon W \cap U_a \to x_a(W \cap U_a) \subset O_a \mid a \in A\right\}$$

is a smooth atlas for W. ∎

Thus any open subset of a vector space is a smooth manifold. Let us now consider a slightly less trivial example. Recall we denote by S^m the unit sphere:

$$S^m := \left\{p \in \mathbb{R}^{m+1} \mid \|p\| = 1\right\}.$$

PROPOSITION 1.16. *The sphere S^m is a compact[6] smooth manifold of dimension m.*

Proof. We give S^m the subspace topology from $\mathbb{R}^{m+1}$. Then S^m is certainly a separable and metrisable. We will directly exhibit a smooth atlas on S^m (thus proving at the same time that S^m is a topological manifold). Let $p_N = (0, \dots, 0, 1)$ denote the "north pole" and let $p_S := (0, \dots,, 0, -1)$ denote the "south pole". Let $U_N = S^m \setminus \{p_N\}$ and $U_S := S^m \setminus \{p_S\}$. Then $\{U_N, U_S\}$ is an open cover of $S^{\cdot}$. Define charts

[5] Exercise: Why is this independent of the choice of ℓ?

[6] See Definition 1.28 for the definition of compact.

$$x_N \colon U_N \to \mathbb{R}^m, \qquad x_N(u^1, \dots, u^{m+1}) := \frac{1}{1 - u^{m+1}}(u^1, \dots, u^m)$$

and

$$x_S \colon U_S \to \mathbb{R}^m, \qquad x_S(u^1, \dots, u^{m+1}) := \frac{1}{1 + u^{m+1}}(u^1, \dots, u^m).$$

The maps x_N and x_S are stereographic projection from the north and south pole respectively. Both the transition maps

$$x_N \circ x_S^{-1} \colon \mathbb{R}^m \setminus \{0\} \to \mathbb{R}^m \setminus \{0\},$$
$$x_S \circ x_N^{-1} \colon \mathbb{R}^m \setminus \{0\} \to \mathbb{R}^m \setminus \{0\}$$

are given by

$$(u^1, \dots, u^m) \mapsto \frac{1}{\sum_{i=1}^m (u^i)^2}(u^1, \dots, u^m)$$

which is obviously a diffeomorphism. Thus we have defined a smooth atlas on S^m. We refer to this smooth structure as the standard smooth structure on S^m. ■

All we really needed to do in the previous proof was check differentiability of the transition function $x_N \circ x_S^{-1}$. This is because (as a subset of $\mathbb{R}^{m+1}$), S^m already carried a nice topology. Sometimes however we will want to build a smooth manifold "from scratch". For this, the next result is very useful.

Proposition 1.17 (Constructing smooth manifolds). *Let M be a set. Suppose we are given a collection $\{U_a \mid a \in A\}$ of subsets of M together with bijections $x_a \colon U_a \to O_a$, where O_a is an open subset of $\mathbb{R}^m$. Assume in addition that:*

(i) *For any $a, b \in A$, $x_a(U_a \cap U_b)$ is open in $\mathbb{R}^m$.*
(ii) *If $U_a \cap U_b \neq \emptyset$, the map $x_b \circ x_a^{-1} \colon x_a(U_a \cap U_b) \to x_b(U_a \cap U_b)$ is a diffeomorphism.*
(iii) *Countably many of the U_a cover M.*
(iv) *If $p \neq q$ are points in M then either there exists a such that p and q both belong to U_a, or there exists a, b with $U_a \cap U_b = \emptyset$ such that $p \in U_a$ and $q \in U_b$.*

Then M has a unique smooth manifold structure for which the collection $\{x_a \colon U_a \to O_a \mid a \in A\}$ is a smooth atlas.

The proof is essentially trivial: we simply took the definition of a smooth manifold and inserted it into the hypotheses.

Proof. Define a topology on M by declaring all the x_a to be homeomorphisms. That this is well-defined topology follows from the fact that the x_a are bijections, together with (i) and (ii). The locally Euclidean property is then immediate. Properties (iii) and (iv) guarantee this topology is metrisable and separable, thus turning M into a topological manifold. Finally the fact that $\{x_a \colon U_a \to O_a \mid a \in A\}$ is a smooth atlas on M is clear from (ii). ■

Remark 1.18. Historically, a manifold M (smooth or topological) was called *open* if M was non-compact and *closed* if M was compact. This however is bad terminology for two reasons:

(i) Thought of as an abstract topological space, every manifold is both open and closed! (This is true of any topological space.)
(ii) If however our given manifold M is a subspace of a larger space N, then it does make sense to ask whether M is open or closed in the subspace topology of N. For example, the unit ball B^m is open in $\mathbb{R}^m$ and the unit sphere S^m is closed in $\mathbb{R}^{m+1}$. Historically, all manifolds were thought of as subspaces—actually *submanifolds*[7]—of some Euclidean space $\mathbb{R}^m$, and in fact any manifold can be embedded inside Euclidean space.[8] However even then the terminology "open" and "closed" does not make sense! For instance, if we identify $\mathbb{R}^2$ with the set of points in $\mathbb{R}^3$ whose last coordinate is zero then $\mathbb{R}^2$ is closed as a subspace of $\mathbb{R}^3$, but $\mathbb{R}^2$ is not compact as a manifold.

Thus throughout this course, we will only use the words "open" and "closed" in their topological context (i.e. to speak of open sets and closed sets). If we wish to indicate a given manifold is compact, we will use the rather more logical terminology "compact manifold".

The only caveat to this is that when we define (both smooth and topological) manifolds with boundary later on (Chapter 21), we will need to differentiate between the terms "compact manifold with boundary" and "compact manifold without boundary". Indeed, as we have already mentioned, the closed unit ball D^m is an example of a compact smooth manifold with boundary.

On Problem Sheet A at the end of this chapter there are many more examples (and non-examples) of smooth manifolds for you to play with. Going back to the general theory, we have now achieved the goal we set out at the beginning of the chapter: to come up with an appropriate class of topological spaces for which it makes sense to say whether a map is differentiable or not.

Definition 1.19. Let $\varphi \colon M \to N$ be a continuous map between two smooth manifolds. We say that φ is of class C^k if for every point $p \in M$, if (U, x) is any chart on M with $p \in U$ and (V, y) is any chart on N with $\varphi(U) \subset V$, the composition

$$y \circ \varphi \circ x^{-1} \colon x(U) \to y(V)$$

is of class C^k. If φ is of class C^k for all k then we say φ is **smooth** (or of class C^∞). If φ is smooth and bijective and the inverse function $N \to M$ is also smooth then φ is said to be a **diffeomorphism**.

It follows from the definition of smooth atlases that it does not matter which charts we use to check differentiability (i.e. we could replace "any chart" with "every chart" above).[9]

Examples 1.20.

---

[7] We will define submanifolds precisely in Chapter 5.

[8] In the smooth case, this is known as the *Whitney Embedding Theorem*, which we will prove in Chapter 7.

[9] See Problem A.1.

(i) If $(M, \mathcal{X})$ is a smooth manifold and $x : U \to O$ belongs to $\mathcal{X}$, then if we think of U and O as smooth manifolds in their own right (using Lemma 1.15 and Example 1.14) then x is a diffeomorphism.
(ii) Similarly if $W \subset M$ is any open set (endowed with the smooth structure from Lemma 1.15) then the inclusion map $\iota : W \hookrightarrow M$ is a smooth map.

The next result also follows immediately from the chain rule in Euclidean spaces (Proposition 1.7).

PROPOSITION 1.21. *Let M, N and L be smooth manifolds, and suppose $\varphi : M \to N$ and $\psi : N \to L$ are smooth maps. Then $\psi \circ \varphi : M \to L$ is smooth.*

Proof. Let $p \in M$. Let (U, x) be a chart on M containing p, let (V, y) be a chart on N containing $\varphi(p)$, and let (W, z) be a chart on L containing $\psi(\varphi(p))$. We want to show that the composition $z \circ (\psi \circ \varphi) \circ x^{-1}$ is smooth where defined. But

$$z \circ (\psi \circ \varphi) \circ x^{-1} = \left(z \circ \psi \circ y^{-1}\right) \circ \left(y \circ \varphi \circ x^{-1}\right),$$

and by assumption each of the two bracketed terms on the right-hand side is a smooth map. Since the composition of smooth maps (defined on open sets in Euclidean space) is smooth, the left-hand side is also smooth. ∎

REMARK 1.22. Consider the following curiosity. We have defined what it means for a continuous map between two smooth manifolds to be differentiable (Definition 1.19), but we have not defined what the derivative $D\varphi(p)$ is yet! This is somehow backwards—in normal calculus one first defines the derivative $Df(p)$ and then says the map is differentiable if the derivative $Df(p)$ always exists. In fact, the definition of the derivative of a map between two smooth manifolds is a little tricky, and this is what we will do in the next three chapters.

A smooth structure is defined as an equivalence class of smooth atlases. We can take this one step further and look at equivalence classes of smooth structures.

DEFINITION 1.23. We say that two smooth structures $\mathcal{X}_1$ and $\mathcal{X}_2$ on a given topological manifold M belong to the same **diffeomorphism class** if there exists a diffeomorphism $(M, \mathcal{X}_1) \to (M, \mathcal{X}_2)$. This is clearly another equivalence relation. We write $\mathcal{S}(M)$ for the set of diffeomorphic classes of smooth structures on M.

EXAMPLE 1.24. As an example to show that smooth structures and diffeomorphism classes really are different concepts, take $M = \mathbb{R}$. Let $\mathcal{X}$ denote the maximal smooth atlas containing the chart $t \mapsto t^3$. On Problem Sheet A you will check that this is *not* the same smooth structure as the standard one described in Example 1.14. However, there is an obvious diffeomorphism between the two smooth structures (namely, $t \mapsto t^3$). Thus they belong to the same diffeomorphism class.

REMARK 1.25. Does every topological manifold admit a smooth structure (i.e. can every topological manifold be turned into a smooth manifold)? Can a topological manifold admit more than one diffeomorphism class? These questions are typically very hard to solve (and there are many open problems). Here are some interesting facts, all of which are way too hard to prove in this course.

(i) If M is a topological manifold of dimension 0,1,2 or 3 then $\mathcal{S}(M)$ consists of exactly one element.
(ii) In higher dimensions, there may be more than one diffeomorphism class. For example, $\mathcal{S}(S^7)$ has exactly 28 elements (15 if one ignores orientations), and there are more than sixteen million different elements in $\mathcal{S}(S^{31})$! On the other hand. $\mathcal{S}(S^{61})$ consists of exactly one element, but for any odd number $m \geq 63$, one has $\#\mathcal{S}(S^m) \geq 2$.
(iii) For any $m \neq 4$, $\mathbb{R}^m$ admits a unique diffeomorphism class. However $\mathcal{S}(\mathbb{R}^4)$ has infinitely many elements. In general the most "wild" phenomena occur in dimension 4.
(iv) There exist topological manifolds that do not admit *any* smooth structures at all: $\mathcal{S}(M) = \emptyset$.

Bonus Material for Chapter 1

Defining topological manifolds as separable metrisable spaces that are locally Euclidean has the advantage of being concise, but in practice it can be hard to check. In this bonus section we recall some additional material from point-set topology, and explore alternative ways to define topological manifolds.

Definition 1.26. Let X be a topological space. We say that X is **Hausdorff** if for every pair $p \neq q$ of points in X, there are open subsets $U, V \subset X$ such that $p \in U$, $q \in V$ and $U \cap V = \emptyset$.

Any metrisable space is Hausdorff.

Definition 1.27. A topological space X is said to be **connected** if it is not the disjoint union of non-empty open sets. A topological space X is said to be **path connected** if for any two points $p, q \in X$ there exists a continuous map $\gamma\colon [0,1] \to X$ such that $\gamma(0) = p$ and $\gamma(1) = q$.

A path connected space is connected, but the converse need not hold.

In general any topological space can be decomposed into its **connected components** (resp. **path components**), where the connected component (resp. path component) containing a given point x is the union of all the connected (resp. path connected) sets containing x.

Recall that an **open cover** of a topological space X is a collection $\{U_a \mid a \in A\}$ of open subsets of X, where A is some index set, such that $X = \bigcup_{a \in A} U_a$. If the index set A is a finite set, we say that the open cover is a **finite cover**. A **subcover** of an open cover $\{U_a \mid a \in A\}$ consists of a subset $A' \subset A$ such that the collection $\{U_a \mid a \in A'\}$ is still an open cover.

Definition 1.28. Let X be a topological space. We say that X is **compact** if every open cover has a finite subcover.

Compact spaces are typically the most "useful" class of topological spaces, in the sense that many powerful theorems only hold for compact spaces. Unfortunately, since manifolds include Euclidean spaces as a special case, they are certainly not always compact.[10]

We therefore introduce a weaker condition, which requires two more preliminary definitions about covers. Suppose $\{U_a \mid a \in A\}$ is an open cover. A **refinement** is another open cover $\{V_b \mid b \in B\}$ with the property that for every $b \in B$ there exists $a \in A$ such that $V_b \subset U_a$. Next, an open cover $\{U_a \mid a \in A\}$ of X is said to be **locally finite** if for every $x \in X$ there exists a neighbourhood W of x such that the set $\{a \in A \mid U_a \cap W \neq \emptyset\}$ is a finite set.

Definition 1.29. A topological space X is said to be **paracompact** if every open cover has a locally finite refinement.

Thus compact spaces are obviously paracompact, but the latter is more general. For instance, $\mathbb{R}^m$ is paracompact, but as we have just observed, not compact. In fact, the following result holds.

Theorem 1.30 *Every metrisable space is paracompact.*

If X is a topological space then a **basis** for the topology on X is a set $\mathcal{B}$ of open sets of X with the property that every open set in X is a union of sets in $\mathcal{B}$.

Definition 1.31. A topological space is said to be **second countable** if it admits a countable basis.[11]

The following proposition gives two alternative characterisations of topological manifolds, which often are easier to verify.

Proposition 1.32. *Let M be a locally Euclidean topological space. The following are equivalent*:

(i) *M is a topological manifold.*
(ii) *M is Hausdorff, paracompact, and has at most countably many connected components.*
(iii) *M is Hausdorff and second countable.*

Here are some more point-set topology definitions.

Definition 1.33. A topological space is said to be **Lindelöf** if every open cover has a *countable* subcover.[12]

Any locally compact paracompact space with at most countably many components is Lindelöf.

---

[10] A subset $K \subset \mathbb{R}^m$ is compact if and only if it is closed and bounded—this is the Heine-Borel Theorem [7].

[11] A metrisable space is second countable if and only if it is separable (consider balls of rational radii).

[12] Clearly compact $\Rightarrow$ Lindelöf.

Definition 1.34. A topological space X is **normal** if given any two closed disjoint subsets K_1, K_2 of X there are open sets U_1, U_2 of X such that $K_i \subset U_i$ for $i = 1, 2$ and $U_1 \cap U_2 = \emptyset$.

Every paracompact Hausdorff space is normal.[13]

Definition 1.35. A topological space X is said to be **locally compact** if for every point $p \in X$ there exists a compact set K and a neighbourhood U of x such that $U \subset K$.

Clearly any compact space is locally compact.
If the topological space is Hausdorff, locally compact is equivalent to asking that every point has a neighbourhood with compact closure.

Definition 1.36. A topological space X is **locally path connected** if for every point $p \in X$ and every neighbourhood U of p, there exists a path connected neighbourhood V of p with $V \subset U$.[14]

Topological manifolds enjoy all of these properties.

Proposition 1.37. *Topological manifolds are normal Lindelöf spaces which are both locally compact and locally path connected. Moreover a topological manifold is connected if and only if it is path connected.*

We conclude this chapter with another somewhat esoteric remark about infinite-dimensional manifolds. This is for *interest only*—we will not use infinite-dimensional manifolds in this course.

Definition 1.38. Fix a Banach space E. We say that a topological space X is **locally modelled on** E if every point in X has a neighbourhood which is homeomorphic to an open set in E.

Definition 1.39. A **topological Banach manifold** is a separable metrisable topological space which is locally modelled on some Banach space E.[15]

A smooth Banach manifold is defined similarly—here we use the fact that differentiating functions on Banach spaces works in exactly the same way as differentiating functions on Euclidean spaces.

You should compare this to how you initially learned linear algebra. To begin with all vector spaces were finite-dimensional and linear operators were just matrices. Then two years later they told you that actually things could be infinite-dimensional. All the theorems you knew and loved from linear algebra continued to hold (provided a few more assumptions were made), only the proofs were much harder and it was no longer called "linear algebra", it was called "functional analysis". The same is true in differential geometry—infinite-dimensional differential geometry is sometimes referred to as "global analysis".

---

[13] If $\{x\}$ is closed for all x in X then normal $\Rightarrow$ Hausdorff.

[14] For a locally path connected space, the path components and the connected components coincide.

[15] Since any Euclidean space is a Banach space, any topological manifold is also a topological Banach manifold.

Example 1.40. As a concrete example of an infinite-dimensional manifold, let M and N be two finite-dimensional smooth manifolds, and let $0 \le k < \infty$. Then the space $C^k(M, N)$ of maps from M to N of class C^k is an infinite-dimensional Banach manifold.

A constant thorn in the side of global analysists is the fact that the space $C^\infty(M, N)$ of smooth maps from M to N is *not* a Banach manifold.[16]

[16] This is because $C^\infty(\mathbb{R}^m, \mathbb{R}^n)$ is not a Banach space.

Problem Sheet A

A (♣) indicates the problem is non-examinable. This does not necessarily imply it is difficult.

Problem A.1 Let $\varphi\colon M \to N$ be a continuous map between two smooth manifolds. Prove that the following two statements are equivalent:

(i) For every point $p \in M$, if (U, x) is **any** chart on M with $p \in U$ and (V, y) is **any** chart on N with $\varphi(U) \subseteq V$, the composition

$$y \circ \varphi \circ x^{-1}\colon x(U) \to y(V)$$

is of class C^k.

(ii) For every point $p \in M$, **there exists** a chart (U, x) on M with $p \in U$ and a chart (V, y) on N with $\varphi(U) \subseteq V$ such that the composition

$$y \circ \varphi \circ x^{-1}\colon x(U) \to y(V)$$

is of class C^k.

Problem A.2 Prove that the set $\mathrm{GL}(m)$ of invertible $m \times m$ matrices is a smooth manifold of dimension m^2.

Problem A.3 Let M and N be two smooth manifolds of dimension m and n respectively. Prove that $M \times N$ is a smooth manifold of dimension $m + n$. Deduce that the m-dimensional torus:

$$T^m := \underbrace{S^1 \times \cdots \times S^1}_{m \text{ times}} \subset \mathbb{R}^{2m}.$$

is a compact smooth manifold of dimension m.

Problem A.4 Let $\mathbb{R}P^m$ denote m-dimensional real projective space, i.e. the space of lines through the origin in $\mathbb{R}^{m+1}$. Prove that $\mathbb{R}P^m$ is a compact smooth manifold of dimension m.

Problem A.5 Let $G(k, m)$ denote the set of k-dimensional linear subspaces of $\mathbb{R}^m$. We call $G(k, m)$ a Grassmannian manifold. Prove that $G(k, m)$ is a compact smooth manifold and compute its dimension.

Problem A.6 Let X denote the union of the x-axis and the y-axis in $\mathbb{R}^2$. Prove that X is *not* a topological manifold.

Problem A.7 Let Y denote the "pinched 2-dimensional torus", as shown in Figure A.1. Prove that Y is *not* a topological manifold.

Problem A.8 Show that the smooth atlas on $\mathbb{R}$ consisting of the single chart $x\colon \mathbb{R} \to \mathbb{R}$ given by $x(t) = t^3$ defines a smooth structure that is *different* to the "standard" smooth structure (the latter is the smooth structure containing the identity map as a chart). Prove however that both the smooth structures belong to the same diffeomorphism class.

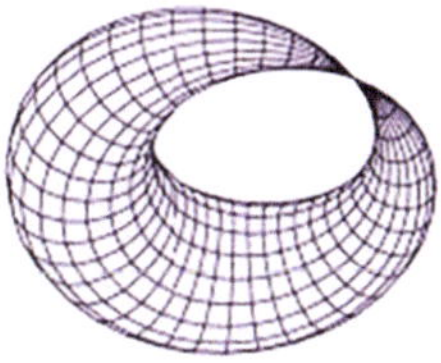

Fig. A.1: The pinched torus

Bonus Problem(s) for Sheet A

These problem(s) are hard, and are included for enthusiasts only. Solutions will not be provided.

(♣)PROBLEM A.9 Let M be a compact connected topological manifold of dimension one. Prove that M is homeomorphic to S^1.

Chapter 2
Tangent Spaces

The goal of the next few chapters is to associate to an m-dimensional smooth manifold an m-dimensional vector space, denoted by T_pM, to each point $p \in M$. We call T_pM the **tangent space** to M at p. Although it *won't* be immediate from the definition why, the tangent space is what you would naturally "guess" it would be. See Figure 2.1 for the case of S^2 (which should be thought of as sitting inside $\mathbb{R}^3$).

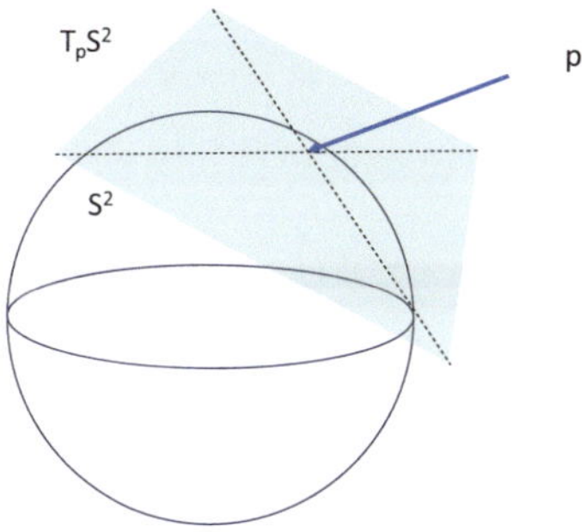

Figure 2.1: The tangent space to S^2 at a point p

We will use this construction to define the *derivative* of a smooth map $\varphi: M \to N$: this will be a *linear* map $D\varphi(p): T_pM \to T_{\varphi(p)}N$ for each $p \in M$. In Chapter 5 we will "glue" the vector spaces together to form one larger space called the **tangent bundle** of M. This will be a smooth manifold of twice the dimension of M. A smooth map $\varphi: M \to N$ will then induce a smooth map $D\varphi: TM \to TN$. In Chapter 6 we will look at submanifolds—it will not be until then that we can rigorously prove that the tangent space we define in this chapter really is the actual "tangent space" as in Figure 2.1 (cf. Example 6.16).

W. J. Merry, *Lectures on Differential Geometry I*,
https://doi.org/10.1007/978-3-032-03733-6_2

Definition 2.1. A **smooth function** on a manifold is a smooth map $f: M \to \mathbb{R}$ in the sense of Definition 1.19, where $\mathbb{R}$ is given the standard smooth structure from Example 1.14. Thus f is a smooth function if for any chart $x: U \to O$ on M, the composition $f \circ x^{-1}: O \to \mathbb{R}$ is a smooth function (in the normal sense).

Convention. We typically use the symbols φ, ψ to denote smooth maps from one manifold to another, and f, g for smooth maps from a manifold to a Euclidean space.

We denote by $C^\infty(M)$ the space of smooth functions. If $W \subset M$ is an open set, we define $C^\infty(W)$ to be the space of smooth functions that are only defined on W (where W is thought of as a smooth manifold in the sense of Lemma 1.15). The space $C^\infty(M)$ is an **unital algebra**[1] (and thus in particular a ring and a vector space), under the operations

$$(f+g)(p) := f(p) + g(p), \qquad (fg)(p) := f(p)g(p),$$

and $(cf)(p) := cf(p)$ for $c \in \mathbb{R}$.

We will always write $\circ$ to denote composition, while juxtaposition indicates the pointwise product. See Definition 19.18 if you are unfamiliar with algebras.

Before going any further, let us go back to $\mathbb{R}^m$ and introduce some more notation. To begin with, this will feel somewhat redundant, but we will see in the next chapter that it makes the various formulae easier to understand. Slightly abusively, we denote by $u^i: \mathbb{R}^m \to \mathbb{R}$ the function

$$(u^1, \dots, u^m) \mapsto u^i. \tag{2.1}$$

Let e_i denote the ith standard basis vector in $\mathbb{R}^m$, so that

$$u^i(e_j) = \delta^i_j, \tag{2.2}$$

where δ^j_i is the *Kronecker delta* defined by

$$\delta^j_i = \begin{cases} 1, & i = j, \\ 0, & i \neq j. \end{cases}$$

Now suppose $f: O \subset \mathbb{R}^m \to \mathbb{R}^n$ is a smooth map defined on an open subset O of $\mathbb{R}^m$. We will start by defining the derivative of f, Df.

If $p \in O$ and $\xi \in \mathbb{R}^m$ then the vector $Df(p)\xi$ can be thought of as the partial derivative of f in the direction ξ:

$$Df(p)\xi = \lim_{t \to 0} \frac{f(p + t\xi) - f(p)}{t}.$$

Definition 2.2. We abbreviate $Df(p)e_j$ by $D_j f(p)$:

$$D_j f(p) = Df(p)e_j = \lim_{t \to 0} \frac{f(p + te_j) - f(p)}{t}.$$

[1] That is, a vector space where you can also multiply two elements together, and where the multiplication has a unit.

Let us summarise the various different ways we can write the derivative:

Let $f : O \subset \mathbb{R}^m \to \mathbb{R}^n$ be a smooth map, and let $p \in O$. Then:

- $Df(p)$ is a $n \times m$ matrix.
- $D_j f(p)$ is an element of $\mathbb{R}^n$. It is the jth column of the matrix $Df(p)$.
- $D(u^i \circ f)(p)$ is a linear map from $\mathbb{R}^m$ to $\mathbb{R}$. One can think of it as the ith row of the matrix $Df(p)$.
- $D_j(u^i \circ f)(p)$ is a number. It is the (i, j)th entry of the matrix $Df(p)$.

In more familiar notation

$$D_j(u^i \circ f)(p) = \frac{\partial f^i}{\partial u^j}(p). \tag{2.3}$$

In general we will prefer the slightly more cumbersome expression on the left-hand side of (2.3). This is because in the next chapter the symbol $\frac{\partial}{\partial x^i}$ will take on a special meaning, cf. Example 3.6.

Remark 2.3. In our new notation, part (ii) of the chain rule in Euclidean spaces (Proposition 1.7) reads:

$$D_j(u^i \circ g \circ f)(p) = \sum_{k=1}^{n} D_k(u^i \circ g)(f(p))D_j(u^k \circ f)(p).$$

Going back to manifolds, we can use the (u^i) to give examples of smooth functions.

Example 2.4. If $x : U \to O$ is a chart on M, for each $i = 1, \ldots, m$ the function $u^i \circ x$ is a smooth function on U.

This type of smooth function is especially important, so it gets its own special name.

Definition 2.5. If $p \in M$ and x is a chart defined on a neighbourhood of p then we write

$$x^i := u^i \circ x.$$

We call the functions (x^i) the **coordinates** of the chart x, and we say that the (x^i) are **local coordinates about** p.

We use the convention that the local coordinates of a chart are always written with the same letter: thus if y is another chart then $y^i := u^i \circ y$. Since the local coordinates uniquely determine the chart, this convention also works backwards. Thus the phrase "let (z^i) be local coordinates about p" is shorthand for: let z be a chart on M containing p, and set $z^i := u^i \circ z$.

Remark 2.6. Consider $\mathbb{R}^m$ as a smooth manifold with the single chart $\mathrm{id} : \mathbb{R}^m \to \mathbb{R}^m$ (cf. Example 1.14). Then the local coordinates of id are simply the (u^i).

Let us say a few words on the philosophy behind the notation, which may help you in chapters to come.

Differential Geometry is essentially a way of formalising calculus so that it makes sense on smooth manifolds. The formalism is designed to make things "look" as similar as possible to calculus on Euclidean space. This means that if you are ever stuck when trying to compute something (for instance, a derivative), you can just "pretend" that everything is actually defined on Euclidean space, and then simply follow the normal rules of multivariable calculus. Magically, it just works!

DEFINITION 2.7. Let M be a smooth manifold and let $p \in M$. Let U and V be two neighbourhoods of p, and suppose $f \in C^\infty(U)$ and $g \in C^\infty(V)$. We say that f and g have the same **germ** at p if there exists a smaller neighbourhood $W \subset U \cap V$ of p such that

$$f|_W \equiv g|_W.$$

One can think of this as follows: define an equivalence relation on the set of smooth functions defined on a neighbourhood of p by saying that $(U, f) \sim (V, g)$ if there exists a neighbourhood $W \subset U \cap V$ such that $f|_W \equiv g|_W$. A germ is then an equivalence class under this relation. We denote the germ by $\underline{f}$ and we let $\mathcal{F}_p M$ denote the set of germs at p.

In fact, $\mathcal{F}_p M$ is another algebra.[2] We can add germs together: if $\underline{f}$ and $\underline{g}$ are two germs with representatives (U, f) and (V, g) respectively, then $\underline{f} + \underline{g}$ is the germ represented by $(U \cap V, f + g)$. Similarly $\underline{f}\,\underline{g}$ is the germ represented by $(U \cap V, fg)$, and for a real number c, $c\underline{f}$ is the germ represented by (U, cf). We denote by $\underline{c}$ the germ of any function which is constant and equal to c in a neighbourhood of p. The map $\mathbb{R} \to \mathcal{F}_p M$ given by $c \mapsto \underline{c}$ is then an inclusion of algebras.

A germ at p has a well-defined value at p (although nowhere else), and this gives us the map

$$\mathrm{eval}_p : \mathcal{F}_p M \to \mathbb{R}, \qquad \mathrm{eval}_p(\underline{f}) := f(p), \tag{2.4}$$

where (U, f) is any representative of $\underline{f}$.

The next example motivates the approach we will take to tangent vectors.

EXAMPLE 2.8. Let $f : O \to \mathbb{R}$ be smooth, where $O \subset \mathbb{R}^m$ is open. Let $p \in O$ and $\xi \in \mathbb{R}^m$. The usual interpretation of the derivative is that the matrix $Df(p)$ eats the vector ξ to produce a real number $Df(p)\xi$. However we could flip this on its head and think of (p, ξ) as being fixed, and instead let f vary. To this end, let us denote by

$$\xi_p : C^\infty(O) \to \mathbb{R}, \qquad \xi_p(f) := Df(p)\xi.$$

It follows from (1.1) that differentiability is a local property, in the sense that the value of $\xi_p(f)$ depends only on the germ of f at p.[3] Thus we can think of ξ_p as defining a

[2] And in fact is a local ring; see Lemma 2.15.

[3] The formal definition of a "local property" will come in Chapter 16.

linear map

$$\xi_p : \mathcal{F}_p O \to \mathbb{R}, \qquad \xi_p(\underline{f}) := Df(p)\xi$$

(here we are thinking of O as a smooth manifold). In fact, the map $\xi_p : \mathcal{F}_p O \to \mathbb{R}$ is not just any linear map, it is also a *derivation* in the sense that

$$\xi_p(\underline{f}\,\underline{g}) = \mathrm{eval}_p(\underline{f})\xi_p(\underline{g}) + \mathrm{eval}_p(\underline{g})\xi_p(\underline{f}).$$

Indeed, this is just a fancy way of expressing the *Leibniz rule*:

$$D(fg)(p)\xi = f(p)Dg(p)\xi + g(p)Df(p)\xi.$$

Following Example 2.8, we define a tangent vector as a derivation on the space of germs.

DEFINITION 2.9. Let M be a smooth manifold and let $p \in M$. A **tangent vector at** p is a linear map

$$\xi : \mathcal{F}_p M \to \mathbb{R}$$

which satisfies the *derivation property*:

$$\xi(\underline{f}\,\underline{g}) = \mathrm{eval}_p(\underline{f})\xi(\underline{g}) + \mathrm{eval}_p(\underline{g})\xi(\underline{f}).$$

CONVENTION. We typically denote tangent vectors with the symbols ξ, ζ.

Since a tangent vector is a linear map from the vector space $\mathcal{F}_p M$ to $\mathbb{R}$, the set of tangent vectors is itself a vector space. We denote this by $T_p M$ - this is the tangent space of M at p.

The next observation follows directly from the definition:

LEMMA 2.10. *Let M be a smooth manifold and let $W \subset M$ be a non-empty open set. Regard W as a smooth manifold in its own right, as in Lemma* 1.15. *Then for any $p \in W$ there is a canonical identification $T_p M = T_p W$.*

Proof. Since W is open it there is a canonical isomorphism $\mathcal{F}_p M \cong \mathcal{F}_p W$. ∎

Here is an easy lemma about derivations.

LEMMA 2.11. *Suppose $\xi : \mathcal{F}_p M \to \mathbb{R}$ is a tangent vector at p and $\underline{c} \in \mathcal{F}_p M$ is a constant germ. Then $\xi(\underline{c}) = 0$.*

Proof. Since $\underline{c} = c\,\underline{1}$ we have $\xi(\underline{c}) = c\,\xi(\underline{1})$ by linearity. But by the derivation property:

$$\xi(\underline{1}) = \xi(\underline{1}\,\underline{1}) = 2\ \mathrm{eval}_p(\underline{1})\xi(\underline{1}) = 2\,\xi(\underline{1})$$

and thus $\xi(\underline{1}) = 0$. Thus also $\xi(\underline{c}) = 0$. ∎

In the special case where $O \subset \mathbb{R}^m$ is an open set, Example 2.8 showed that every vector $\xi \in \mathbb{R}^m$ defines an element of $\xi_p \in T_p O$ (in the sense of Definition 2.9). In fact, these are *all* the elements of $T_p O$, although this requires a bit of work to see. More generally, one has:

Theorem 2.12 *Let M be a smooth manifold of dimension m and let $p \in M$. Then the vector space $T_p M$ has dimension m.*

Theorem 2.12 is not immediate. Indeed, from Definition 2.9 it is not remotely clear why $T_p M$ should even be finite-dimensional! We will prove Theorem 2.12 in the next chapter by explicitly finding a basis of $T_p M$.

Bonus Material for Chapter 2

In this bonus section we explore two further properties of the algebras $C^\infty(M)$ and $\mathcal{F}_p M$.

Lemma 2.13. *Let M be a manifold of dimension $m > 0$, and let $W \subset M$ be a non-empty open set. Then as a real vector space, $C^\infty(W)$ is always infinite-dimensional.*

Proof. Let $f \in C^\infty(W)$ be any smooth function which is not constant on some connected component of W. Then $f(W)$ is an infinite subset of $\mathbb{R}$ (since it contains an interval).[4]

Consider now the vector space $\mathbb{R}[t]$ of all polynomials. This is an infinite-dimensional vector space—a basis is the set of monomials $\{t^k \mid k \geq 0\}$. Any polynomial $P(t)$ is completely determined by its values on an infinite set, and thus if $P \in \mathbb{R}[t]$ then P is completely determined by its values on $f(W)$. Therefore

$$\{P \circ f \mid P \in \mathbb{R}[t]\} \subset C^\infty(W)$$

is an infinite-dimensional subspace of $C^\infty(W)$. ■

Now for some abstract ring theory. This material is not remotely relevant to the course, so ignore it if the terms are not familiar.

Definition 2.14. A ring is said to be a **local** ring if it contains a unique maximal left ideal.

Lemma 2.15. *The ring $\mathcal{F}_p M$ is a local ring.*

Proof. The map $\mathrm{eval}_p \colon \mathcal{F}_p M \to \mathbb{R}$ from (2.4) is clearly a ring homomorphism. Thus the kernel of eval_p is an ideal in the ring $\mathcal{F}_p M$. Since the map eval_p is surjective (as $\mathrm{eval}_p(\underline{c}) = c$), this is actually a maximal ideal. In fact, it is the unique maximal ideal, since if $\mathrm{eval}(\underline{f}) \neq 0$ then $\underline{f}$ is invertible in $\mathcal{F}_p M$. Indeed, if (U, f) is a representative of $\underline{f}$ then there exists $V \subset U$ such that f is never zero on V. Thus there is a well-defined function $g := 1/f \colon V \to \mathbb{R}$, and $\underline{g}$ is then an inverse to $\underline{f}$. This completes the proof. ■

[4] If you are worried why such a function exists, use Lemma 3.2 from the next chapter.

Chapter 3
Partitions of Unity

We begin this chapter by reformulating the definition of a tangent vector in a slightly more convenient way. Since germs are defined via equivalence classes, they are often tedious to work with, and we would like to dispense with them.

DEFINITION 3.1. Let M be a smooth manifold, let $p \in M$, and let W be any neighbourhood of p (for instance W could be all of M). A **derivation on $C^\infty(W)$ at** p is a linear map $\zeta \colon C^\infty(W) \to \mathbb{R}$ which satisfies the *derivation property*

$$\zeta(fg) = f(p)\zeta(g) + g(p)\zeta(f).$$

If $\xi \in T_pM$ then ξ naturally defines a derivation ζ of $C^\infty(W)$ for any open W containing p by setting

$$\zeta(f) := \xi(\underline{f}). \tag{3.1}$$

In fact, the converse is also true, as we will prove in Proposition 3.3 below. First we need a preliminary lemma. To state it, recall that for a smooth function $f \colon M \to \mathbb{R}$, we denote by $\operatorname{supp}(f)$ the **support** of f, defined by:

$$\operatorname{supp}(f) := \overline{\{p \in M \mid f(p) \neq 0\}}.$$

LEMMA 3.2 (BUMP FUNCTIONS). *Let M be a smooth manifold and let $K \subset U$ be subsets, where K is closed and U is open. Then there exists a smooth function $\chi \colon M \to \mathbb{R}$ such that*:

(i) $0 \le \chi(p) \le 1$ *for all* $p \in M$,
(ii) $\operatorname{supp}(\chi) \subset U$,
(iii) $\chi(p) = 1$ *for all* $p \in K$.

A function χ satisfying the three conditions of Lemma 3.2 is called a **bump function**. The proof of Lemma 3.2 will be carried out at the end of this chapter, when we discuss **partitions of unity**. It is *not* obvious—as we will see this is the main reason we imposed

W. J. Merry, *Lectures on Differential Geometry I*,
https://doi.org/10.1007/978-3-032-03733-6_3

the additional point-set topological conditions (metrisable and separable) on top of the locally Euclidean property.

PROPOSITION 3.3. *Let M be a smooth manifold, let $p \in M$, and let W be any neighbourhood of p. Then there is a linear isomorphism between T_pM and the space of derivations of $C^\infty(W)$ at p.*

Proof. Let W be a neighbourhood of p. We prove the result in three steps.

1. Let $\zeta\colon C^\infty(W) \to \mathbb{R}$ be a derivation at p. Suppose $f \in C^\infty(W)$ is identically zero on a neighbourhood $V \subset W$ of p. We claim that $\zeta(f) = 0$. For this, choose a bump function $\chi\colon M \to \mathbb{R}$ such that $\chi(p) = 1$ and $\operatorname{supp}(\chi) \subset V$ (by applying Lemma 3.2 with $K = \{x\}$ and "U" equal to V). Let $g = \chi f$, thought of as a function $W \to \mathbb{R}$. Then g is identically zero, and hence $\zeta(g) = 0$ by linearity. But by the derivation property

$$\begin{aligned}\zeta(g) &= \zeta(\chi f)\\ &= \chi(p)\zeta(f) + f(p)\zeta(\chi)\\ &= \zeta(f)\end{aligned}$$

since $\chi(p) = 1$ and $f(p) = 0$. Thus $\zeta(f) = 0$.

2. Suppose now $\underline{f} \in \mathcal{F}_pM$. We claim that we can always find a representative for $\underline{f}$ with domain W, i.e. a smooth function $g\colon W \to \mathbb{R}$ such that $\underline{g} = \underline{f}$. For this, let (V, f) be any representative of $\underline{f}$. By shrinking V if necessary, we may assume that $V \subset W$. Now choose a smaller neighbourhood U of p with $\overline{U} \subset V \subset W$.[1] Our goal now is to *extend* f to a smooth function g defined on W such that $g|_U = f$. For this, we apply Lemma 3.2 again, this time with $K = \overline{U}$ and "U" equal to V. Now consider the smooth function

$$g\colon W \to \mathbb{R}, \qquad g(p) := \begin{cases}\chi(p)f(p), & x \in V,\\ 0, & x \in W \setminus V.\end{cases}$$

Since $g|_U = f$, we certainly have $\underline{g} = \underline{f}$.

3. We now complete the proof. Let $\zeta\colon C^\infty(W) \to \mathbb{R}$ be a derivation at p. We define a linear map $\xi\colon \mathcal{F}_pM \to \mathbb{R}$ by setting

$$\xi(\underline{f}) := \zeta(f), \qquad \text{where } (W, f) \text{ is any representative of } \underline{f}.$$

That such a representative (W, f) exists was the content of Step 2, and the fact that ξ is well-defined follows from Step 1. Indeed, if (W, h) was another representative of $\underline{f}$ then by assumption there exists a smaller neighbourhood V of p such that $f|_V \equiv h|_V$. Then by linearity $\zeta(f) - \zeta(h) = \zeta(f - h)$ and $\zeta(f - h) = 0$ by Step 1. Finally, it is clear that ξ is a derivation. This association $\zeta \mapsto \xi$ obviously inverts (3.1), and thus this completes the proof. ■

Thanks to Proposition 3.3, we will from now always regard a tangent vector ξ as a derivation of $C^\infty(W)$ at p for any open W containing p. We emphasise that Proposi-

[1] The existence of such a set U follows from the fact M is metrisable.

tion 3.3 implies that it doesn't matter which W we choose. Typically we take W either to be the domain of a chart, or the whole manifold M. The next statement is a reformulation of Lemma 2.11 (or alternatively Step 1 of Proposition 3.3).

COROLLARY 3.4. *Let M be a smooth manifold, let $p \in M$, and let $f \in C^\infty(W)$ for some open W containing p. If f is constant in a neighbourhood of p then $\xi(f) = 0$ for all $\xi \in T_pM$.*

REMARK 3.5. Given Proposition 3.3, you may wonder why we didn't immediately define T_pM as the vector space of derivations of $C^\infty(M)$ at p. There are (at least) four reasons:

(i) Using germs better encapsulates the fact that differentiation is a local property.
(ii) An advantage of the germ approach was that Lemma 2.10 is tautological; if we had defined T_pM directly as derivations of $C^\infty(M)$ at p then Lemma 2.10 would have required proof — namely, one would have had to directly show that derivations of $C^\infty(M)$ at p are isomorphic to derivations of $C^\infty(W)$ at p. This is essentially the statement of Proposition 3.3, and thus we wouldn't have saved any time by avoiding germs.
(iii) In certain other geometric categories, the analogue of Lemma 3.2 is false. For instance, there is an analogous theory of **analytic manifolds**, which are defined in exactly the same way as smooth manifolds, except the word "smooth" should be replaced with "real-analytic" everywhere (thus an analytic manifold has a real-analytic atlas, and maps between real-analytic manifolds are required to be real-analytic, etc). We will not study analytic manifolds in this course, although they are very important in certain fields. In the real-analytic category Lemma 3.2 is false: there do *not* exist real-analytic bump functions.[2] Thus for analytic manifolds, Proposition 3.3 is false, and one is forced to work with germs to define the tangent space.
(iv) Later in this book (Chapter 17) we will discuss **sheaves**, and germs are a motivating example for the construction of the **stalk** of a sheaf.

Let us give now give a concrete example of a tangent vector.

EXAMPLE 3.6. Let M be a smooth manifold of dimension m, and let (U, x) be a chart on M with local coordinates (x^i). Let p be any point in U. Define a derivation on $C^\infty(U)$ at p by:

$$\left.\frac{\partial}{\partial x^i}\right|_p : C^\infty(U) \to \mathbb{R}, \qquad \left.\frac{\partial}{\partial x^i}\right|_p (f) := D_i(f \circ x^{-1})(x(p)),$$

where the right-hand side uses the convention from Definition 2.2. We will shortly prove that the collection $\left.\frac{\partial}{\partial x^i}\right|_p$ for $i = 1, \dots, m$ form a basis of T_pM, thus establishing Theorem 2.12.

We will need the following easy lemma from multivariable calculus. Recall an open set $O \subset \mathbb{R}^m$ such that $0 \in O$ is said to be **star-shaped** if given any $p \in O$, the line segment from 0 to p is also contained in O.

[2] Exercise: Why?

LEMMA 3.7. *Let $O \subset \mathbb{R}^m$ be a star-shaped open set. Suppose $h\colon O \to \mathbb{R}$ is a smooth function. Then there exist m smooth functions $g_i\colon O \to \mathbb{R}$ for $i = 1, \ldots, m$ such that $g_i(0) = D_i h(0)$ and such that*

$$h = h(0) + \sum_{i=1}^{m} u^i\, g_i,$$

where u^i is as in (2.1).

Proof. Fix $q = (a^1, \ldots, a^m) \in O$ and consider the line segment $\gamma(t) = tq$. Set $\delta := h \circ \gamma\colon [0,1] \to \mathbb{R}$. Then by the chain rule

$$\delta'(t) = \sum_{i=1}^{m} a^i\, D_i h(tq).$$

Thus

$$h(q) - h(0) = \delta(1) - \delta(0) = \int_0^1 \delta'(t)\, dt = \sum_{i=1}^{m} a^i \int_0^1 D_i h(tq)\, dt.$$

Since $a^i = u^i(q)$ by definition, the claim follows with

$$g_i(q) := \int_0^1 D_i h(tq)\, dt.$$

■

Theorem 2.12 from the last chapter follows immediately from the next statement.

PROPOSITION 3.8. *Let M be a smooth manifold of dimension m. Let $x\colon U \to O$ be a chart on M, and fix $p \in U$. Then any tangent vector $\xi \in T_pM$ can be uniquely written as a linear combination*

$$\xi = \sum_{i=1}^{m} a^i \left.\frac{\partial}{\partial x^i}\right|_p.$$

In fact, $a^i = \xi(x^i)$. Thus $\left\{ \left.\frac{\partial}{\partial x^i}\right|_p \mid i = 1, \ldots, m \right\}$ is a basis of T_pM.

Proof. We may assume without loss of generality that $x(p) = 0$ and that O is star-shaped. Let $f \in C^\infty(U)$ and apply Lemma 3.7 with $h := f \circ x^{-1}$. We obtain $f = f(p) + \sum_{i=1}^m x^i\,(g_i \circ x)$, where

$$g_i(0) = D_i(f \circ x^{-1})(0) = \left.\frac{\partial}{\partial x^i}\right|_p (f).$$

Thus for any derivation ξ, one has

$$\xi(f) = \underbrace{\xi(f(p))}_{=0} + \sum_{i=1}^{m} \left(\xi(x^i) g_i(0) + \underbrace{x^i(p)}_{=0} \, \xi(g_i \circ x) \right)$$
$$= \sum_{i=1}^{m} \xi(x^i) \, \frac{\partial}{\partial x^i}\bigg|_p (f),$$

where we used Corollary 3.4 and the assumption that $x(p) = 0$.

This shows that $\left\{ \frac{\partial}{\partial x^i}\big|_p \mid i = 1, \ldots, m \right\}$ spans $T_p M$. It remains to prove linear independence. For this we note that:

$$\begin{aligned} \frac{\partial}{\partial x^i}\bigg|_p (x^j) &= \frac{\partial}{\partial x^i}\bigg|_p (u^j \circ x) \\ &= D_i(u^j \circ x \circ x^{-1})(x(p)) \\ &= D_i u^j(x(p)) \\ &= \delta_i^j, \end{aligned} \tag{3.2}$$

where we used the fact that $Du^j = u^j$ as u^j is a linear function, together with (2.2). Thus if $\zeta := \sum_{i=1}^m b^i \frac{\partial}{\partial x^i}\big|_p = 0$ then feeding x^j to ζ gives $b^j = 0$. This shows linear independence, and thus completes the proof. ■

Remark 3.9. Suppose x and y are two charts about p, with corresponding coordinate systems (x^i) and (y^i). Taking $\xi = \frac{\partial}{\partial y^j}\big|_p$ in Proposition 3.8 tells us that

$$\frac{\partial}{\partial y^j}\bigg|_p = \sum_{i=1}^{m} \frac{\partial}{\partial y^j}\bigg|_p (x^i) \, \frac{\partial}{\partial x^i}\bigg|_p .$$

But unravelling the definitions,

$$\begin{aligned} \frac{\partial}{\partial y^j}\bigg|_p (x^i) &= D_j(x^i \circ y^{-1})(y(p)) \\ &= D_j(u^i \circ x \circ y^{-1})(y(p)), \end{aligned}$$

which is just the (i, j)th entry of the matrix $D(x \circ y^{-1})(y(p))$. Thus we have shown:

The transition matrix from the basis

$$\left\{ \frac{\partial}{\partial y^i}\bigg|_p \mid i = 1, \ldots, m \right\} \quad \text{to the basis} \quad \left\{ \frac{\partial}{\partial x^i}\bigg|_p \mid i = 1, \ldots, m \right\}$$

is given by the matrix $D(x \circ y^{-1})(y(p))$.

We conclude this chapter by introducing partitions of unity, and using these to prove Lemma 3.2.

DEFINITION 3.10. Let M be a smooth manifold. A **partition of unity** is a collection $\{\kappa_a \mid a \in A\}$ of smooth functions $\kappa_a : M \to \mathbb{R}$ such that:

(i) $0 \le \kappa_a(p) \le 1$ for all $p \in M$ and $a \in A$.
(ii) The collection $\{\operatorname{supp}(\kappa_a) \mid a \in A\}$ is locally finite, i.e. any $p \in M$ has a neighbourhood that intersects at most finitely many of $\operatorname{supp}(\kappa_a)$.
(iii) For all $p \in M$ one has

$$\sum_{a \in A} \kappa_a(p) = 1$$

(note by (ii) this sum only has finitely many non-zero terms for every p).

We say that a partition of unity $\{\kappa_a \mid a \in A\}$ is **subordinate** to an open cover $\{U_a \mid a \in A\}$ if $\operatorname{supp}(\kappa_a) \subset U_a$ for each $a \in A$.

Theorem 3.11 (Partitions of Unity) *Let M be a smooth manifold. For any open cover of M, there exists a partition of unity subordinate to that cover.*

The proof of Theorem 3.11 is carried out in the bonus section below. Lemma 3.2 is an easy consequence of Theorem 3.11:

Proof of Lemma 3.2. Consider the open cover $\{U, M \setminus K\}$ of M. By Theorem 3.11 there exists a partition of unity $\{\kappa_U, \kappa_{M \setminus K}\}$. The function $\chi := \kappa_U$ has the properties we desire. ∎

Bonus Material for Chapter 3

In this bonus section we carry out the proof of Theorem 3.11. In fact, we will first establish a special case of Lemma 3.2 where the smaller set K is *compact* (instead of merely closed).

LEMMA 3.12 (BUMP FUNCTIONS, THE COMPACT CASE). *Let M be a smooth manifold and let $K \subset U$ be subsets, where K is compact and U is open. Then there exists a smooth function $\chi : M \to \mathbb{R}$ such that:*

(i) $0 \le \chi(p) \le 1$ *for all* $p \in M$,
(ii) $\operatorname{supp}(\chi) \subset U$,
(iii) $\chi(p) = 1$ *for all* $p \in K$.

Proof. We prove the result in four steps.

1. We first prove that for any pair of real numbers $r < R$ there exists a smooth function $f : \mathbb{R} \to [0, 1]$ such that $f(t) = 1$ for $t \le r$, $f(t) = 0$ for all $t \ge R$, and $0 < f(t) < 1$ for all $t \in (r, R)$. For this, consider the function

$$h\colon \mathbb{R} \to \mathbb{R}, \qquad h(t) := \begin{cases} e^{-1/t}, & t > 0, \\ 0, & t \leq 0. \end{cases}$$

A somewhat tedious computation[3] shows that h is smooth. Our desired function f is then given by

$$f(t) := \frac{h(R-t)}{h(R-t) + h(t-r)}.$$

One can easily check this function f has the desired properties.

2. Now let us extend this to $\mathbb{R}^m$. Let $B_r \subset \mathbb{R}^m$ denote the open ball of radius r about the origin (so that $B_1 = B^m$). Then for any $0 < r < R$ there exists a smooth function $g\colon \mathbb{R}^m \to \mathbb{R}$ such that $g(p) = 1$ for all $p \in \overline{B}_r$, $g(p) = 0$ on $\mathbb{R}^m \setminus B_R$, and $0 < g(p) < 1$ for all $p \in B_R \setminus \overline{B}_r$. Indeed, the function $g(p) := f(\|p\|)$, where f is as in the previous step works.

3. Now let M be a smooth manifold, let $p \in M$, and let U be an arbitrary neighbourhood of p. Then we can choose a smaller neighbourhood $V \subset U$ of p with $\overline{V} \subset U$ that has the following property: there exists a smooth function $\chi\colon M \to \mathbb{R}$ such that $\chi(p) = 1$ for all $p \in \overline{V}$, $0 \leq \chi(p) \leq 1$ for all $p \in M$, and $\chi(p) = 0$ for all $p \in M \setminus U$. This follows from the previous step, by choosing an appropriate chart about p.

4. We now complete the proof. For each point $p \in K$, choose neighbourhoods $V_p \subset U$ such that $\overline{V}_p \subset K$ (such sets exist as metrisable spaces are normal, cf. Definition 1.34). Since K is compact, there are finitely many points $p_1, \ldots, p_N$ such that $K \subset \bigcup_{i=1}^N V_{p_i}$. For each i, choose functions $\chi_i\colon M \to \mathbb{R}$ such that $\chi_i(p) = 1$ for all $p \in \overline{V}_i$, $0 \leq \chi_i(p) \leq 1$ for all $p \in M$, and $\chi_i(p) = 0$ for all $p \in M \setminus U$. Now set

$$\chi := 1 - \prod_{i=1}^{N} (1 - \chi_i).$$

One easily checks this χ does the trick. ■

We now prove the following alternative version of Theorem 3.11. This proof assume you are familiar with paracompact space—see Definition 1.29.

Theorem 3.13 *Let M be a smooth manifold. Let $\{U_a \mid a \in A\}$ be an open cover of M. There exists a locally finite refinement $\{V_b \mid b \in B\}$ and a partition of unity $\{\kappa_b \mid b \in B\}$ subordinate to $\{V_b \mid b \in B\}$ with the additional property that* $\operatorname{supp}(\kappa_b)$ *is a compact subset of M for every $b \in B$.*

Of course, the main content of the theorem is the existence of the partition of unity $\{\kappa_b \mid b \in B\}$—the existence of the locally finite refinement $\{V_b \mid b \in B\}$ is just the very definition of paracompactness.

Proof. Paracompactness guarantees us the existence of a locally finite refinement $\{V_b \mid b \in B\}$. In fact, we can do a little better than this: we can find a locally finite refinement $\{V_b \mid b \in B\}$ together with another open cover $\{W_b \mid b \in B\}$ (with the

[3] Exercise: Enjoy.

same index set B) such that $\overline{W_b}$ is compact for each $b \in B$ and such that $\overline{W_b} \subset V_b$. This argument uses the fact that M is also locally compact (Definition 1.35). We won't dwell on the details as they not important to the main theme of the course.

We now apply Lemma 3.12 to each pair $\overline{W_b} \subset V_b$ to obtain a smooth function $\chi_b \colon M \to \mathbb{R}$ such that $0 \le \chi_b(p) \le 1$ for all $p \in M$, $\chi_b|_{\overline{W_b}} \equiv 1$, and $\operatorname{supp}(\chi_b) \subset V_b$ is compact. The desired partition of unity is then given by

$$\kappa_b := \frac{\chi_b}{\sum_{b \in B} \chi_b}.$$

This completes the proof. ∎

We conclude by proving Theorem 3.11.

Proof of Theorem 3.11. Let $\{U_a \mid a \in A\}$ be an arbitrary open cover.

Let $\{V_b \mid b \in B\}$ be a locally finite refinement and let $\{\kappa_b \mid b \in B\}$ be a partition of unity subordinate to $\{V_b \mid b \in B\}$, whose existence is guaranteed by Theorem 3.13. Choose a function $\beta \colon B \to A$ such that $V_b \subset U_{\beta(b)}$ for each $b \in B$. Now define

$$\kappa_a := \sum_{b \in \beta^{-1}(a)} \kappa_b.$$

If $\beta^{-1}(a) = \emptyset$ this should be interpreted as the zero function. Then

$$\begin{aligned} \operatorname{supp}(\kappa_a) &= \overline{\bigcup_{b \in \beta^{-1}(a)} \{x \in M \mid \kappa_b(p) \neq 0\}} \\ &= \bigcup_{b \in \beta^{-1}(a)} \operatorname{supp}(\kappa_b) \subset U_a, \end{aligned}$$

where the second equality used the fact that $\{\operatorname{supp}(\kappa_b) \mid b \in B\}$ is a locally finite.[4] It is immediate that the collection $\{\operatorname{supp}(\kappa_a) \mid a \in A\}$ is locally finite, and thus we conclude that $\{\kappa_a \mid a \in A\}$ is another partition of unity which is subordinate to our original cover $\{U_a \mid a \in A\}$. ∎

[4] Note however that κ_a need not have compact support.

Chapter 4
The Derivative

Let us now finally define the derivative of a smooth map.

Definition 4.1. Let M and N be smooth manifolds, and let $\varphi\colon M \to N$ be a smooth map. Fix $p \in M$ and $\xi \in T_pM$. We define a tangent vector $\zeta \in T_{\varphi(p)}N$ by setting

$$\zeta(f) := \xi(f \circ \varphi), \qquad \forall\, f \in C^\infty(N).$$

It is clear ζ is a linear derivation of $C^\infty(N)$ at $\varphi(p)$, and hence an element of $T_{\varphi(p)}N$. Moreover if we denote ζ by $D\varphi(p)\xi$ then it is immediate that the map $\xi \mapsto D\varphi(p)\xi$ is a linear map. We call this linear map the **derivative of** φ **at** p.

The chain rule becomes essentially tautologous.

Proposition 4.2 (The chain rule on manifolds). *Let M, N and L be smooth manifolds, and suppose $\varphi\colon M \to N$ and $\psi\colon N \to L$ are smooth maps. Then*

$$D(\psi \circ \varphi)(p) = D\psi(\varphi(p)) \circ D\varphi(p).$$

Proof. Take $\xi \in T_pM$ and $f \in C^\infty(L)$. Then

$$\begin{aligned} \big(D(\psi \circ \varphi)(p)\xi\big)(f) &= \xi(f \circ \psi \circ \varphi) \\ &= \big(D\varphi(p)\xi\big)(f \circ \psi) \\ &= D\psi(\varphi(p)) \circ \big(D\varphi(p)\xi\big)(f). \end{aligned}$$

The claim follows. ∎

Remark 4.3. You may wonder why the chain rule is so (suspiciously) easy to prove. After all, the Euclidean version (Proposition 1.7) is quite tricky. Does Proposition 4.2 give a shortcut to proving the Euclidean version? The answer is sadly no: indeed, we already used the Euclidean version at least twice (in Proposition 1.21 and Lemma 3.7),

W. J. Merry, *Lectures on Differential Geometry I*,
https://doi.org/10.1007/978-3-032-03733-6_4

and hence any attempt to "prove" the Euclidean version via Proposition 4.2 would yield a circular argument.

Let us compute the map $D\varphi(p)$ in local coordinates.

LEMMA 4.4. *Let $\varphi\colon M \longrightarrow N$ be a smooth map between two smooth manifolds, where M has dimension m and N has dimension n. Fix $p \in M$, and let (U, x) be a chart on M about p and (V, y) be a chart on N about $\varphi(p)$. Then the matrix of $D\varphi(p)$ with respect to the bases $\left\{\frac{\partial}{\partial x^j}\big|_p \mid j = 1, \dots, m\right\}$ of T_pM and $\left\{\frac{\partial}{\partial y^i}\big|_{\varphi(p)} \mid i = 1, \dots, n\right\}$ of $T_{\varphi(p)}N$ is given by the matrix $D(y \circ \varphi \circ x^{-1})(x(p))$.*

Proof. We compute

$$\begin{aligned} D\varphi(p)\left(\frac{\partial}{\partial x^j}\bigg|_p\right) &= \sum_{i=1}^n D\varphi(p)\left(\frac{\partial}{\partial x^j}\bigg|_p\right)(y^i)\,\frac{\partial}{\partial y^i}\bigg|_{\varphi(p)} \\ &= \sum_{i=1}^n \frac{\partial}{\partial x^j}\bigg|_p (y^i \circ \varphi)\,\frac{\partial}{\partial y^i}\bigg|_{\varphi(p)} \\ &= \sum_{i=1}^n D_j(u^i \circ y \circ \varphi \circ x^{-1})(x(p))\,\frac{\partial}{\partial y^i}\bigg|_{\varphi(p)} \end{aligned}$$

The number $D_j(u^i \circ y \circ \varphi \circ x^{-1})(x(p))$ is the (i, j)th entry of the matrix $D(y \circ \varphi \circ x^{-1})(x(p))$, and thus the proof is complete. ∎

REMARK 4.5. Suppose $f\colon \mathcal{O} \subset \mathbb{R}^m \longrightarrow \mathbb{R}^n$ is smooth. We now have two (!) definitions of the map $Df(p)$. We temporarily write the two maps as $Df(p)^{\mathrm{calc}}$ and $Df(p)^{\mathrm{man}}$. Thus $Df(p)^{\mathrm{calc}}$ is a linear map $\mathbb{R}^m \longrightarrow \mathbb{R}^n$; it is the matrix of partial derivatives, as at the beginning of Chapter 1. Meanwhile $Df(p)^{\mathrm{man}}$ is a linear map $T_p\mathbb{R}^m \longrightarrow T_{f(p)}\mathbb{R}^n$. In fact, these are the "same" map.[1] Indeed, we apply Lemma 4.4 with x and y the respective identity maps. Then the (i, j)th entry of $Df(p)^{\mathrm{man}}$ is given by

$$D_j(u^i \circ f)(p) = \frac{\partial f^i}{\partial u^j},$$

which is also the (i, j)th entry of $Df(p)^{\mathrm{calc}}$. From now on we will drop the "calc" and "man" superscripts, and just call both maps $Df(p)$. It should be clear from the context which is meant.

We now give an entirely different way of defining tangent vectors. This approach is not quite as aesthetically pleasing as using derivations, but it has the advantage that it is easier to compute.

Suppose $\gamma\colon (a, b) \longrightarrow \mathbb{R}^m$ is a smooth map. We usually write the coordinate on $\mathbb{R} = \mathbb{R}^1$ as t instead of u^1, and we denote the derivative of γ at a point t by $\gamma'(t)$. Writing $\gamma = (\gamma^1, \dots, \gamma^m)$, the vector $\gamma'(t)$ is just the row vector $((\gamma^1)'(t), \dots, (\gamma^m)'(t))$. Our aim now is to extend this to manifolds.

[1] A coordinate-free proof of this fact is given in Corollary 4.14 below.

Definition 4.6. A **curve** in a smooth manifold M is a smooth map $\gamma\colon (a,b) \to M$, where we think of (a,b) as a 1-dimensional smooth manifold. Now fix $t \in (a,b)$. There are, a priori, two different ways we could define an element $\dot\gamma(t)$ of $T_{\gamma(t)}M$, which we will call the **velocity vector of γ at time** t.

(i) Firstly, we can define a derivation on $C^\infty(M)$ at $\gamma(t)$ by setting[2]

$$\dot\gamma(t)(f) := (f \circ \gamma)'(t), \qquad f \in C^\infty(M). \tag{4.1}$$

(ii) Secondly, if we think of γ as a smooth map between manifolds then we can define a tangent vector $\dot\gamma(t)$ at $\gamma(t)$ via the derivative $D\gamma(t)$:

$$\dot\gamma(t) := D\gamma(t)\left(\frac{\partial}{\partial t}\Big|_t\right) \in T_{\gamma(t)}M. \tag{4.2}$$

To see that these two definitions agree, let x be a chart defined on a neighbourhood of $\gamma(t)$ with local coordinates (x^i). Let $\gamma^i := x^i \circ \gamma$ so that γ^i is a curve in $\mathbb{R}$. Applying Proposition 3.8 to (4.1), we see that

$$\dot\gamma(t) = \sum_{i=1}^m (\gamma^i)'(t)\,\frac{\partial}{\partial x^i}\Big|_{\gamma(t)}, \tag{4.3}$$

since

$$\dot\gamma(t)(x^i) = (x^i \circ \gamma)'(t) = (\gamma^i)'(t).$$

But similarly by applying Lemma 4.4 to (4.2) we see that this definition also gives the same formula (4.3) for $\gamma'(t)$.

Lemma 4.7. *Let M be a smooth manifold and let $\gamma, \delta\colon (-\varepsilon, \varepsilon) \to M$ be two smooth curves such that $\gamma(0) = \delta(0)$. Then $\dot\gamma(0) = \dot\delta(0)$ as elements of $T_{\gamma(0)}M$ if and only if for some (and hence any) chart (U, x) defined on a neighbourhood of $\gamma(0)$, we have*

$$(x \circ \gamma)'(0) = (x \circ \delta)'(0). \tag{4.4}$$

Proof. The stated condition is equivalent to requiring that $(\gamma^i)'(0) = (\delta^i)'(0)$ for each i, where $\gamma^i = x^i \circ \gamma$ and $\delta^i = x^i \circ \delta$. The claim follows from (4.3), since $\left\{\frac{\partial}{\partial x^i}\big|_{\gamma(0)} \mid i = 1, \dots, m\right\}$ is a basis of $T_{\gamma(0)}M$. ∎

What is less clear is that *every* tangent vector can be written as the velocity vector of a curve.

Proposition 4.8. *Let M be a smooth manifold of dimension m, let $p \in M$ and let $\xi \in T_pM$. There exists a smooth curve $\gamma\colon (-\varepsilon, \varepsilon) \to M$ such that $\gamma(0) = p$ and $\dot\gamma(0) = \xi$.*

---

[2] The use of both a dot and a dash to denote derivatives in (4.1) is deliberate: a dot denotes a tangent vector in a manifold, whereas a dash denotes the normal derivative from calculus. See Definition 4.11 below.

Proof. Choose a chart $x\colon U \to O \subset \mathbb{R}^m$, where O is an open set containing 0 such that $x(p) = 0$. Write

$$\xi = \sum_{i=1}^{m} a^i \left.\frac{\partial}{\partial x^i}\right|_p,$$

where the a^i are real numbers. For sufficiently small $\varepsilon > 0$, the vector $(ta^1, \ldots, ta^m)$ belongs to O for all $|t| < \varepsilon$. This means that if we define

$$\gamma\colon (-\varepsilon, \varepsilon) \to M, \qquad \gamma(t) := x^{-1}(ta^1, ta^2, \ldots, ta^m),$$

then γ is well-defined, smooth, and satisfies $\gamma(0) = p$. Moreover (4.3) shows us that $\dot{\gamma}(0) = \xi$. ■

Remark 4.9. This tells us that we can make the following alternative definition of T_pM: a tangent vector at $p \in M$ is an equivalence class of smooth curves $\gamma\colon (-\varepsilon, \varepsilon) \to M$ such that $\gamma(0) = p$, where $\gamma \sim \delta$ if and only if for some chart x centred about p, (4.4) holds.

Note however that this only works because we have *already* established that T_pM was a vector space with basis $\left\{ \left.\frac{\partial}{\partial x^i}\right|_p \right\}$. If one wanted to *start* with this definition of T_pM, one would need to use Problem B.1 to endow T_pM with a vector space structure.

Let us examine how velocity vectors behave with respect to smooth maps.

Proposition 4.10. *Let $\varphi\colon M \to N$ be a smooth map between two smooth manifolds, and let $\gamma\colon (a, b) \to M$ be a smooth curve in M. Then $\delta := \varphi \circ \gamma$ is a smooth curve in N, and*

$$D\varphi(\gamma(t))\dot{\gamma}(t) = \dot{\delta}(t).$$

Proof. We will give two proofs, one for each of the two (equivalent) definitions (4.1) and (4.2) of $\dot{\gamma}(t)$. Of course these are really the same proof.

(i) *Proof using* (4.1) *as the definition of* $\dot{\gamma}(t)$: Take $f \in C^\infty(N)$. Then by the definition of $D\varphi(p)$ and (4.1)

$$\begin{aligned} D\varphi(\gamma(t))\dot{\gamma}(t)(f) &= \dot{\gamma}(t)(f \circ \varphi) \\ &= (f \circ \varphi \circ \gamma)'(t) \\ &= (f \circ \delta)'(t) \\ &= \dot{\delta}(t)(f) \end{aligned}$$

(ii) *Proof using* (4.2) *as the definition of* $\dot{\gamma}(t)$: For this we simply use the chain rule (Proposition 4.2):

$$\begin{aligned}
D\varphi(\gamma(t))\dot{\gamma}(t) &= D\varphi(\gamma(t)) \circ D\gamma(t)\left(\frac{\partial}{\partial t}\Big|_t\right) \\
&= D(\varphi \circ \gamma)(t)\left(\frac{\partial}{\partial t}\Big|_t\right) \\
&= D\delta(t)\left(\frac{\partial}{\partial t}\Big|_t\right) \\
&= \dot{\delta}(t).
\end{aligned}$$

This completes the proof (twice). ■

We now use our new definition of a tangent vector to prove that the tangent space of a vector space is canonically isomorphic to the vector space itself.

DEFINITION 4.11. Let E be a vector space of dimension m, endowed with its standard smooth structure (cf. Example 1.14). Fix $p \in E$. Define the **dash-to-dot map**

$$\mathcal{J}_p \colon E \to T_pE, \qquad \mathcal{J}_p\xi := \dot{\gamma}(0), \qquad \text{where } \gamma(t) := p + t\xi.$$

Note that $\xi = \gamma'(0)$ (normal derivative), and hence $\mathcal{J}_p$ is the map

$$\mathcal{J}_p \colon \gamma'(0) \mapsto \dot{\gamma}(0),$$

which explains the name "dash to dot".

LEMMA 4.12. *The dash-to-dot map is a canonical isomorphism.*

Proof. The smooth structure on E is determined by taking a chart which is a linear isomorphism $\ell \colon E \to \mathbb{R}^m$, cf. Example 1.14. Let $\ell^i := u^i \circ \ell$ denote the local coordinates of such a chart. The map ℓ determines a basis $\{v^i\}$ of E via the equation $\ell v_i = e_i$. If one writes an arbitrary vector in E in terms of this basis as $\xi = \sum_{i=1}^m a^i v_i$ then $a^i = \ell^i(\xi)$. Now with $\gamma(t) := p + t\xi$ one has

$$\begin{aligned}
\mathcal{J}_p\xi &= \dot{\gamma}(0) \\
&= \sum_{i=1}^m \dot{\gamma}(0)(\ell^i)\frac{\partial}{\partial \ell^i}\Big|_p \\
&= \sum_{i=1}^m (\ell^i \circ \gamma)'(0)\frac{\partial}{\partial \ell^i}\Big|_p \\
&= \sum_{i=1}^m \ell^i(\xi)\frac{\partial}{\partial \ell^i}\Big|_p \\
&= \sum_{i=1}^m a^i\frac{\partial}{\partial \ell^i}\Big|_p .
\end{aligned}$$

This shows that the matrix of $\mathcal{J}_p$ with respect to the basis $\{v_i\}$ of E and $\left\{\frac{\partial}{\partial \ell^i}\Big|_p\right\}$ of T_pE is simply given by identity map, which in particular is an isomorphism. ■

The proof of Lemma 4.12 required us to fix a basis of E (i.e. to choose a chart ℓ) in order to prove that $\mathcal{J}_p$ was an isomorphism. But the definition of $\mathcal{J}_p$ did *not* require us to choose a basis of E. This explains the "canonical" in the statement of Lemma 4.12.

REMARK 4.13. If we go back to our original definition of tangent vectors as derivations, the dash-to-dot map is given by taking directional derivatives:

$$(\mathcal{J}_p\xi)(f) = Df(p)\xi, \qquad f \in C^\infty(E).$$

Compare to Example 2.8.

Remark 4.5 can now be expressed in a coordinate-free manner.

COROLLARY 4.14. *Suppose $f : O \subset \mathbb{R}^m \to \mathbb{R}^n$ is smooth. Then using the notation from Remark 4.5, the following diagram commutes:*[3]

$$\begin{array}{ccc}
\mathbb{R}^m & \xrightarrow{\;Df(p)^{\mathrm{calc}}\;} & \mathbb{R}^n \\
{\scriptstyle \mathcal{J}_p}\downarrow & & \downarrow{\scriptstyle \mathcal{J}_{f(p)}} \\
T_p\mathbb{R}^m & \xrightarrow[\;Df(p)^{\mathrm{man}}\;]{} & T_{f(p)}\mathbb{R}^n
\end{array}$$

We conclude with the following statement, whose proof is deferred to Problem Sheet B.

LEMMA 4.15. *Let E and F be vector spaces and assume that $\ell: E \to F$ is a linear map. Then for any $p \in E$ the following diagram commutes:*

$$\begin{array}{ccc}
E & \xrightarrow{\;\ell\;} & F \\
{\scriptstyle \mathcal{J}_p}\downarrow & & \downarrow{\scriptstyle \mathcal{J}_{\ell(p)}} \\
T_pE & \xrightarrow[\;D\ell(p)\;]{} & T_{\ell(p)}F
\end{array}$$

[3] This means that going clockwise is the same as going anticlockwise.

Chapter 5
The Tangent Bundle

We begin this chapter by defining the cotangent space of a manifold. We then move onto the tangent and cotangent bundles. We conclude by recalling the Euclidean versions of the Inverse and Implicit Function Theorems, and generalising the Inverse Function Theorem to manifolds. In the next chapter we will generalise the Implicit Function Theorem to manifolds.

Definition 5.1. Let M be a smooth manifold of dimension m and let $p \in M$. We denote the dual vector space $(T_pM)^*$ of T_pM by T_p^*M and call it the **cotangent space** of M at p.

Convention. We write a typical element of cotangent space with the symbols λ and η. We sometimes write λ_p to indicate that $\lambda \in T_p^*M$.

The cotangent space T_p^*M is another vector space of dimension m. Since elements of T_pM are linear derivations eating functions, the standard duality construction tells us that we can interpret elements of T_p^*M as functions eating linear derivations.

Example 5.2. Let M be a smooth manifold of dimension m and let $p \in M$. Let U be a neighbourhood of p and let $f \in C^\infty(U)$. Then f defines an element $df_p \in T_p^*M$ by

$$df_p(\xi) := \xi(f), \qquad \xi \in T_pM.$$

One calls df_p the **differential** of f at p.

Remark 5.3. Thus df_p is a linear function $T_pM \to \mathbb{R}$. In contrast, the **derivative**[1] $Df(p)$ is a linear function $T_pM \to T_{f(p)}\mathbb{R}$. The two are related via the dash-to-dot map $\mathcal{J}_{f(p)} \colon \mathbb{R} \to T_{f(p)}\mathbb{R}$ in the sense that the following commutes:

[1] Note: "derivative" and "differential" are two different words!

W. J. Merry, *Lectures on Differential Geometry I*,
https://doi.org/10.1007/978-3-032-03733-6_5

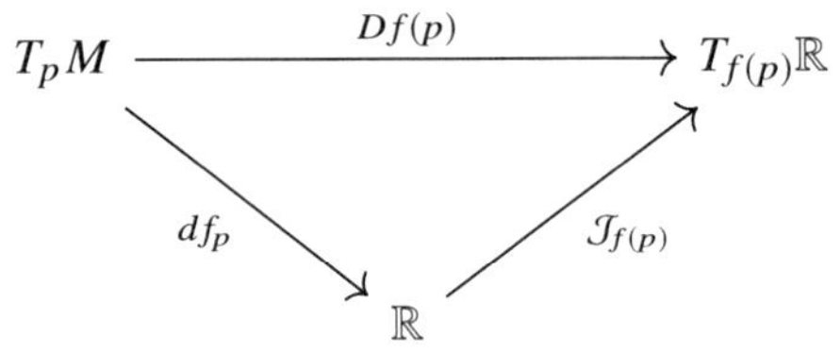

i.e. $Df(p) = \mathcal{J}_p \circ df_p$.

Since by definition $\mathcal{J}_{f(p)}(1) = \frac{\partial}{\partial t}\big|_{f(p)}$, this means that

$$Df(p)\xi = df_p(\xi)\,\frac{\partial}{\partial t}\bigg|_{f(p)}, \qquad \forall\, \xi \in T_pM.$$

Proposition 5.4. *Let M be a smooth manifold of dimension m and let $p \in M$. Let (U, x) be a chart about p, with corresponding local coordinates (x^i). Then $\{dx^i_p\}$ is a basis of T^*_pM.*

Proof. We need only note that $\{dx^i_p\}$ is the dual basis to $\left\{\frac{\partial}{\partial x^i}\big|_p\right\}$ since

$$dx^j_p\left(\frac{\partial}{\partial x^i}\bigg|_p\right) = \frac{\partial}{\partial x^i}\bigg|_p (x^j) = \delta^j_i,$$

by (3.2) in the proof of Proposition 3.8 ■

We now aim to "glue" the vector spaces T_pM together into one big manifold TM.

Definition 5.5. Let M be a smooth manifold. The **tangent bundle**[2] of M is the disjoint union of the tangent spaces:

$$TM = \bigsqcup_{p \in M} T_pM.$$

We adopt a somewhat flexible notation for points in TM. If $\xi \in T_pM$ we write the corresponding point in TM either simply by ξ again, or as a pair (p, ξ). There is a map $\pi\colon TM \to M$ given by $\pi(p, \xi) = p$. We call π the **footpoint** map.

As it stands TM is only a set. Let us now prove it is actually a smooth manifold.

Theorem 5.6 *Let M be a smooth manifold of dimension m. The smooth structure on M naturally induces a smooth structure on TM, making TM into a smooth manifold of dimension $2m$. Moreover the map $\pi\colon TM \to M$ is smooth.*

Proof. Let $\mathcal{X} = \{x_a\colon U_a \to O_a \mid a \in A\}$ be our smooth atlas on M. Write $x^i_a = u^i \circ x_a$ for the local coordinates of x_a. We build a chart $\tilde{x}_a\colon \pi^{-1}(U_a) \to O_a \times \mathbb{R}^m$ by setting

$$\tilde{x}_a(p, \xi) = \left(x_a(p), \sum_{i=1}^m (dx^i_a)_p(\xi)\, e_i\right), \qquad p \in U_a,\ \xi \in T_pM.$$

[2] The tangent bundle is the prototypical example of a more general construction called a **vector bundle** over a smooth manifold. We will take up their study in Chapter 13.

We will prove that if $U_a \cap U_b \neq \emptyset$ then for all $z \in x_a(U_a \cap U_b)$ and $\zeta \in \mathbb{R}^m$, one has:

$$\tilde{x}_b \circ \tilde{x}_a^{-1}(z, \zeta) = \left(x_b \circ x_a^{-1}(z), D(x_b \circ x_a^{-1})(z)\zeta\right), \tag{5.1}$$

From this it follows from Proposition 1.17 that TM is a smooth manifold. To prove (5.1), write $\zeta = \sum_{j=1}^m c^j e_j$ and set $p := x_a^{-1}(z) \in U_a \cap U_b$. Then

$$\tilde{x}_a^{-1}(z, \zeta) = \left(p, \sum_{j=1}^m c^j \frac{\partial}{\partial x_a^j}\Big|_p\right).$$

Fix $1 \leq i \leq m$. We compute

$$\begin{aligned} (dx_b^i)_p\left(\sum_{j=1}^m c^j \frac{\partial}{\partial x_a^j}\Big|_p\right) &= \sum_{j=1}^m c^j (dx_b^i)_p\left(\frac{\partial}{\partial x_a^j}\Big|_p\right) \\ &= \sum_{j=1}^m c^j \frac{\partial}{\partial x_a^j}\Big|_p (x_b^i) \end{aligned}$$

By Remark 3.9, the number $\frac{\partial}{\partial x_a^j}\Big|_p (x_b^i)$ is the (i, j)th entry of the matrix $D(x_b \circ x_a^{-1})(z)$. Thus

$$\sum_{i=1}^m (dx_b^i)_p\left(\sum_{j=1}^m c^j \frac{\partial}{\partial x_a^j}\Big|_p\right) e_i = D(x_b \circ x_a^{-1})(z)\zeta,$$

and (5.1) is proved.

The right-hand side of (5.1) is a diffeomorphism by assumption. Thus

$$\tilde{\mathfrak{X}} = \left\{\tilde{x}_a \colon \pi^{-1}(U_a) \to O_a \times \mathbb{R}^m \mid a \in A\right\}$$

is a smooth atlas on TM. This proves that TM is a smooth manifold of dimension $2m$. To check that π is smooth, we simply observe that if $z \in O_a$ and $\zeta \in \mathbb{R}^m$ then

$$x_a \circ \pi \circ \tilde{x}_a^{-1}(z, \zeta) = z,$$

which is obviously smooth. ∎

We can use the tangent bundle to unify the derivatives $D\varphi(p)$ from Definition 4.1 into a single map.

Definition 5.7. Let $\varphi \colon M \to N$ be a smooth map between two smooth manifolds. Define the **derivative of** φ to be the map

$$D\varphi \colon TM \to TN, \qquad D\varphi(p, \xi) := (\varphi(p), D\varphi(p)\xi).$$

On Problem Sheet C you will prove this map is smooth.

Definition 5.8. Let M be a smooth manifold. The **cotangent bundle** of M is the disjoint union of the cotangent spaces:

$$T^*M = \bigsqcup_{p \in M} T^*_p M.$$

As with TM, points in T^*M will sometimes be denoted by (p, λ), or λ_p, or sometimes just λ. We denote again by $\pi\colon T^*M \to M$ the **footpoint** map $\pi(p, \lambda) = p$.

On Problem Sheet B you will show that T^*M is also naturally a smooth manifold of twice the dimension of M.

We conclude this chapter by discussing the Inverse and Implicit Function Theorems. We state the Euclidean version of the Inverse Function Theorem, and use it to prove the manifold version of the Inverse Function Theorem, and the Euclidean version of the Implicit Function Theorem. In the next chapter we will take this one step further and prove a version of the Implicit Function Theorem for manifolds.

We say that a smooth map $f\colon O \subset \mathbb{R}^m \to \mathbb{R}^n$ has **rank** k at $p \in O$ if the $n \times m$ matrix $Df(p)$ has rank k. We say that f has **maximal rank at** p if the rank of f at p is as large as it can be (which is thus equal to the minimum of m and n). If $m = n$ then f has maximal rank at p if and only if $Df(p)$ is invertible.

Theorem 5.9 (The Inverse Function Theorem) *Let $f\colon O \subset \mathbb{R}^m \to \mathbb{R}^m$ be a smooth map, where O is open. Let $p \in O$ and assume the matrix $Df(p)$ has maximal rank ($= m$). Then there exists a neighbourhood $\Omega \subset O$ of p such that the restriction $f\colon \Omega \to f(\Omega)$ is a diffeomorphism.*

The theorem immediately carries over to manifolds. We say that a smooth map $\varphi\colon M \to N$ has **rank** k at a point p if the linear subspace $D\varphi(p)(T_pM)$ has dimension k inside of $T_{\varphi(p)}N$.

Theorem 5.10 (The Inverse Function Theorem for manifolds) *Let M and N be smooth manifolds of the same dimension m and suppose $\varphi\colon M \to N$ is a smooth map. Let $p \in M$ and assume that φ has maximal rank ($= m$) at p. Then there exists a neighbourhood W of p such that the restriction $\varphi\colon W \to \varphi(W)$ is a diffeomorphism.*

Proof. The assertion is purely local. Choose a chart $x\colon U \to O$ on M at p and a chart $y\colon V \to \Omega$ on N at $\varphi(p)$ such that $\varphi(U) \subset V$. Since x and y are diffeomorphisms (cf. Example 1.20), the derivative of the map

$$y \circ \varphi \circ x^{-1}\colon x(U) \to y(V)$$

has rank m at $x(p)$. Thus by Theorem 5.9 there exists $O_0 \subset O$ such that $y \circ \varphi \circ x^{-1}|_{O_0}$ is a diffeomorphism. Then using once more that x and y are diffeomorphisms, if $W := x^{-1}(O_0)$ then $\varphi|_W\colon W \to \varphi(W)$ is also a diffeomorphism. ■

We now move onto the Implicit Function Theorem. We shall give a quick proof using the Inverse Function Theorem.

Theorem 5.11 (The Implicit Function Theorem) *Let O be a neighbourhood of 0 in $\mathbb{R}^m$ and suppose $f\colon O \to \mathbb{R}^n$ is a smooth map such that $f(0) = 0$, and such that $Df(0)$ has maximal rank.*

(i) **The case** $m \leq n$**:** *Let* $\iota \colon \mathbb{R}^m \to \mathbb{R}^n$ *denote the inclusion*

$$\iota(u^1, \ldots, u^m) := (u^1, \ldots, u^m, 0, \ldots, 0). \tag{5.2}$$

Then there exists a chart y *about* 0 *on* $\mathbb{R}^n$ *such that* $y \circ f = \iota$ *on a neighbourhood of* 0 *in* $\mathbb{R}^m$.

(ii) **The case** $m \geq n$**:** *Let* $\rho \colon \mathbb{R}^m \to \mathbb{R}^n$ *denote the projection*

$$\rho(u^1, \ldots, u^m) := (u^1, \ldots, u^n). \tag{5.3}$$

Then there exists a chart x *about* 0 *in* $\mathbb{R}^m$ *such that* $f \circ x = \rho$ *on a neighbourhood of* 0 *in* $\mathbb{R}^m$.

Proof. We start with (i). The matrix $Df(0)$ has rank m. By rearranging the coordinate functions $f^i = u^i \circ f$ if necessary (this corresponds to composing f with a linear isomorphism $\mathbb{R}^n \to \mathbb{R}^n$, which is a diffeomorphism), we may assume that the $m \times m$ submatrix $\left(\frac{\partial f^i}{\partial u^j}(0)\right)_{1 \leq i,j \leq m}$ is invertible. Now define a map

$$\tilde{f} \colon O \times \mathbb{R}^{n-m} \to \mathbb{R}^n$$

by

$$\tilde{f}(u^1, \ldots, u^n) = f(u^1, \ldots, u^m) + (0, \ldots, 0, u^{m+1}, \ldots, u^n).$$

Then $\tilde{f} \circ \iota = f$ and the derivative $D\tilde{f}(0)$ takes the following form:

$$D\tilde{f}(0) = \begin{pmatrix} \left(\frac{\partial f^i}{\partial u^j}(0)\right)_{1 \leq i,j \leq m} & 0 \\ * & \mathrm{id}_{\mathbb{R}^{n-m}} \end{pmatrix}$$

where $*$ denotes the other entries of $Df(0)$. Thus $\det D\tilde{f}(0) \neq 0$, and consequently $D\tilde{f}(0)$ has rank n. Thus by Theorem 5.9, there exists a neighbourhood $O_0 \subset O \times \mathbb{R}^{n-m}$ of the origin $0 \in \mathbb{R}^n$ such that $\tilde{f} \colon O_0 \to \tilde{f}(O_0)$ is a diffeomorphism. If y denotes the inverse to $\tilde{f}|_{O_0}$ then $y \circ f = y \circ \tilde{f} \circ \iota = \iota$. This proves (i).

The proof of (ii) is very similar. This time we may assume that the submatrix $\left(\frac{\partial f^i}{\partial x^j}(0)\right)_{1 \leq i,j \leq n}$ is invertible, and we define $\tilde{f} \colon O \to \mathbb{R}^m$ by

$$\tilde{f}(u^1, \ldots, u^m) := (f(u^1, \ldots, u^m), u^{n+1}, \ldots, u^m).$$

Then $f = \rho \circ \tilde{f}$ and the derivative $D\tilde{f}(0)$ takes the following form:

$$D\tilde{f}(0) = \begin{pmatrix} \left(\frac{\partial f^i}{\partial x^j}(0)\right)_{1 \leq i,j \leq n} & * \\ 0 & \mathrm{id}_{\mathbb{R}^{m-n}} \end{pmatrix}$$

This is invertible, whence $\tilde{f}$ has a local inverse x, and $f \circ x = \rho \circ \tilde{f} \circ x = \rho$. ∎

Problem Sheet B

A (♣) indicates the problem is non-examinable. This does not necessarily imply it is difficult.

PROBLEM B.1 Let $\{V_a \mid a \in A\}$ be a family of vector spaces indexed by a set A, and let W be a fixed set. Suppose that for each $a \in A$ we are given a bijection $\ell_a : V_a \to W$ such that for any $a, b \in A$, the composition $\ell_b^{-1} \circ \ell_a : V_a \to V_b$ is a linear isomorphism. Prove that there is a unique vector space structure on W such that each ℓ_a is a linear isomorphism.

(♣)PROBLEM B.2 Let M be a smooth manifold of dimension m with maximal smooth atlas $\mathcal{X}$. Given a point $p \in M$, let $\mathcal{X}_p \subset \mathcal{X}$ denote the set of charts (U, x) such that $p \in U$. Define an equivalence relation on $\mathbb{R}^m \times \mathcal{X}_p$ by saying

$$(\xi, x) \sim (\zeta, y) \qquad \Leftrightarrow \qquad D(y \circ x^{-1})(x(p))\xi = \zeta.$$

(i) Prove that this is indeed a well-defined equivalence relation.
(ii) Write $[\xi, x]$ to denote the equivalence class of (ξ, x), and let $\mathcal{T}_p M$ denote the set of equivalence classes. Prove that for every $x \in \mathcal{X}_p$ the map $\ell_x : \mathbb{R}^m \to \mathcal{T}_p M$ given by

$$\ell_x \xi := [\xi, x]$$

is a bijection. Deduce that $\mathcal{T}_p M$ admits a unique vector space structure such that each ℓ_x is a linear isomorphism.
(iii) Let $(U, x) \in \mathcal{X}_p$, and let $\tilde{\ell}_x : \mathbb{R}^m \to T_p M$ denote the linear isomorphism defined by

$$\tilde{\ell}_x e_i = \frac{\partial}{\partial x^i}\bigg|_p .$$

Prove that there exists a linear isomorphism $\kappa_p : \mathcal{T}_p M \to T_p M$ which in addition satisfies

$$\kappa_p \circ \ell_x = \tilde{\ell}_x,$$

for every $(U, x) \in \mathcal{X}_p$. Deduce that $\mathcal{T}_p M$ is another equivalent way to define the tangent space of a manifold.

PROBLEM B.3 Let E and F be vector spaces and assume that $\ell : E \to F$ is a linear map. Prove that for any $p \in E$ the following diagram commutes:

$$\begin{array}{ccc} E & \xrightarrow{\quad \ell \quad} & F \\ \Big\downarrow \mathcal{J}_p & & \Big\downarrow \mathcal{J}_{\ell(p)} \\ T_p E & \xrightarrow[D\ell(p)]{} & T_{\ell(p)} F \end{array}$$

Problem B.4 Let M be a smooth manifold of dimension m. Prove that the cotangent bundle T^*M is naturally a smooth manifold of dimension $2m$.

Bonus Problem(s) for Sheet B

These problem(s) are hard, and are included for enthusiasts only. Solutions will not be provided.

Recall from Proposition 1.32 that a locally Euclidean space M is a topological manifold if and only if it is Hausdorff, paracompact, and has at most countably many connected components. The following three problems show that no two of these conditions imply the third.

(♣)Problem B.5 Consider the subspace $S := \mathbb{R} \times \{-1\} \cup \mathbb{R} \times \{1\} \subseteq \mathbb{R}^2$ together with its subspace topology and define an equivalence relation on S by setting

$$(u_1, v_1) \sim (u_2, v_2) \quad \Leftrightarrow \quad u_1 = u_2 \quad \text{and} \quad u_1, u_2 \neq 0.$$

Equip $M = S/\!\sim$ with the quotient topology. Prove that M is locally Euclidean and paracompact, but not Hausdorff.

(♣)Problem B.6 Consider $\mathbb{R}^2$ as a set and equip it with the topology $\mathcal{T}$ generated by the basis

$$\mathcal{B} = \{U \times \{a\} \mid U \subseteq \mathbb{R} \text{ open}, a \in \mathbb{R}\}.$$

Define

$$x_a \colon \mathbb{R} \times \{a\} \to \mathbb{R}, \qquad x_a(p, a) = p$$

and set $\mathcal{X} = \{x_a \mid a \in \mathbb{R}\}$. Prove that the topological space $(\mathbb{R}^2, \mathcal{T})$ is locally Euclidean, Hausdorff and paracompact, but that it has an uncountable number of connected components.

(♣)Problem B.7 Let $H \subset \mathbb{R}^2$ be the right half plane

$$H := \left\{(u, v) \in \mathbb{R}^2 \mid u > 0\right\},$$

endowed with the subspace topology from $\mathbb{R}^2$. Given $c \in \mathbb{R}$, let $M_c \subset \mathbb{R}^3$ be the set

$$M_c := \{(u, v, c) \mid u \leq 0\},$$

endowed with the subspace topology from $\mathbb{R}^3$. Then set

$$M := H \sqcup \bigsqcup_{c \in \mathbb{R}} M_c,$$

Next, given $a, b, c, \varepsilon \in \mathbb{R}$ with $a < b$ and $\varepsilon > 0$, let

$$U(a, b, c, \varepsilon) := \{(u, v) \in H \mid 0 < u < \varepsilon,\ c + au < v < c + bu\} \subset H,$$

and

$$V(a, b, c, \varepsilon) := \{(u, v, c) \mid -\varepsilon < u \leq 0,\ a < v < b\}.$$

Define a topology on M by declaring that a basis is given by all sets of the following three forms:

(i) open sets in H,
(ii) open sets in $\operatorname{int} M_c$,
(iii) each union $U(a, b, c, \varepsilon) \cup V(a, b, c, \varepsilon)$.

Prove that M is locally Euclidean, connected and Hausdorff but not paracompact.

Chapter 6
Submanifolds

We begin this chapter by proving a version of the Implicit Function Theorem 5.11 for manifolds. We remind the reader that unless stated otherwise, M should always be assumed to have dimension m and N should always be assumed to have dimension n. As in the statement of Theorem 5.11, we must make a case distinction depending as to which of m and n is larger. Unlike in the Euclidean case, however, the two statements are not analogous to each other—as we will see, the case $m \leq n$ is straightforward, but the case $m \geq n$ is rather deeper.

> We first deal with the case where $m \leq n$.

DEFINITIONS 6.1. Let $\varphi \colon M \to N$ be a smooth map.

- We say that φ is an **immersion**[1] if the linear map $D\varphi(p) \colon T_pM \to T_{\varphi(p)}N$ is injective for every $p \in M$.
- If in addition φ itself is injective then we say that φ is an **injective immersion**.
- If in addition φ maps M homeomorphically onto $\varphi(M)$ (where $\varphi(M)$ is endowed with the subspace topology in N) we say that φ is an **embedding**.

REMARK 6.2. If M is compact, then an injective immersion $\varphi \colon M \to N$ is automatically an embedding, as you will prove on Problem Sheet C. However in the non-compact case, this need not be the case (see again Problem Sheet C). An immersion is always *locally* an embedding, as the next result shows.

The next result is the manifold version of part (i) of the Implicit Function Theorem 5.9.

PROPOSITION 6.3. *Suppose $\varphi \colon M \to N$ is an immersion. Then for any $p \in M$, there exists a neighbourhood U of p and a chart $y \colon V \to \Omega$ on N, where V is some neighbourhood of $\varphi(p)$ such that*:

[1] Note an immersion can only exist when $m \leq n$.

W. J. Merry, *Lectures on Differential Geometry I*,
https://doi.org/10.1007/978-3-032-03733-6_6

(i) *One has*:

$$\varphi(U) \cap V = \left\{ q \in V \mid y^{m+1}(q) = \cdots = y^n(q) = 0 \right\}. \tag{6.1}$$

(ii) $\varphi|_U$ *is an embedding.*

Proof. The assertion is again local. Let $\iota\colon \mathbb{R}^m \to \mathbb{R}^n$ denote the inclusion, as in part (i) of the Implicit Function Theorem 5.11. Let x denote a chart on M with $x(p) = 0$ and let z denote a chart on N with $z(\varphi(p)) = 0$. Then $z \circ \varphi \circ x^{-1}$ has maximal rank at 0, and hence by part (i) of the Implicit Function Theorem there exists a chart $\tilde{y}$ on $\mathbb{R}^n$ about 0 and a neighbourhood O of 0 in $\mathbb{R}^m$ such that

$$\tilde{y} \circ z \circ \varphi \circ x^{-1}|_O = \iota|_O.$$

Set $U := x^{-1}(O)$ and set $y := \tilde{y} \circ z$. Then after restricting the domain if necessary, (6.1) holds. To prove the second statement, simply note that $\varphi|_U = y^{-1} \circ \iota \circ x|_U$ is the composition of embeddings, and thus is an embedding. ■

Remark 6.4. If φ is an embedding then the set $\varphi(U)$ from Proposition 6.3 can be written as $\varphi(U) = \varphi(M) \cap W$ for some open set $W \subset N$. (This is just the definition of the subspace topology). Replacing V with $W \cap V$, (6.1) becomes

$$\varphi(M) \cap V = \left\{ q \in V \mid y^{m+1}(q) = \cdots = y^n(q) = 0 \right\}. \tag{6.2}$$

Definition 6.5. Let M and N be manifolds with $M \subset N$ (as sets). We say that M is a **embedded submanifold** of N if the inclusion $M \hookrightarrow N$ is an embedding. If the inclusion is merely an immersion, we say that M is an **immersed submanifold**.[2]

If M is an embedded submanifold of N then Remark 6.4 tells us we can always choose charts on N that satisfy (6.2). We give such a chart a special name:

Definition 6.6. Let M be an embedded submanifold of N. A **slice chart for M in N** is a chart $y\colon V \to \Omega$ on N such that

$$M \cap V = \left\{ q \in V \mid y^{m+1}(q) = \cdots = y^n(q) = 0 \right\}.$$

In fact, the existence of slice charts is an "if and only if" condition, in the sense that we can use slice charts to endow a subset with a smooth structure. The next result makes this more precise.

Proposition 6.7. *Let N be a smooth manifold and suppose $M \subset N$ is a subset with the property that around every point $p \in M$ there exists a chart $y\colon V \to \Omega$ on N with $p \in V$ such that*

$$M \cap V = \left\{ q \in V \mid y^{m+1}(q) = \cdots = y^n(q) = 0 \right\}. \tag{6.3}$$

Then if we endow M with the subspace topology on N, M is a topological manifold of dimension m, and moreover it has a smooth structure that makes it into an embedded submanifold of N.

[2] Note the inclusion is always injective!

Proof. Let $\rho\colon \mathbb{R}^n \to \mathbb{R}^m$ denote the projection[3]

$$\rho(u^1, \dots, u^n) := (u^1, \dots, u^m).$$

Fix $p \in M$ and let $y\colon V \to \Omega$ be a chart as in (6.3). Let $U := M \cap V$ and let $O := \rho(y(U))$, and set $x := \rho \circ y|_U$. If M is given the subspace topology then x is a homeomorphism. If we do this at every point $p \in M$, we end up with a collection of maps for which the hypotheses of Proposition 1.17 are satisfied. Thus M is a smooth manifold of dimension m. Moreover the topology on M that Proposition 1.17 provides coincides with the subspace topology, since the maps x were already homeomorphisms in the subspace topology. Finally if $i\colon M \hookrightarrow N$ denotes the inclusion then with y, x as above, one has $y \circ i \circ x^{-1} = \iota$, where ι was defined in (5.2). Since ι is smooth, so is i. ■

Remark 6.8. If $\varphi\colon M \to N$ is an injective immersion then M is diffeomorphic to an immersed submanifold of N — namely, $M \cong \varphi(M)$. The same is true for embeddings.

We now move to the case where $m \geq n$.

Definitions 6.9. Let $\varphi\colon M \to N$ be smooth. A point $p \in M$ is said to be a **regular point** of φ if φ has rank n at p. (Note this can only happen when $m \geq n$.) A point $p \in M$ is called a **critical point** if it is not a regular point. Similarly a point $q \in N$ is called a **regular value** if every point in $\varphi^{-1}(q)$ is a regular point. A point $q \in N$ is called a **critical value** if it is not a regular value.[4]

We now state the manifold version of part (ii) of the Implicit Function Theorem 5.11. Despite the fact that this is only "half" of the Euclidean Implicit Function Theorem, this result is usually called "the" Implicit Function Theorem (and Proposition 6.3 doesn't get a name).

Theorem 6.10 (The Implicit Function Theorem for manifolds) *Let $\varphi\colon M \to N$ be a smooth map and suppose $q \in N$ is a regular value of φ such that $L := \varphi^{-1}(q)$ is not empty. Then L is a topological manifold of dimension $m - n$. Moreover there exists a smooth structure on L which makes L into a smooth embedded submanifold of M.*

Unlike Proposition 6.3 this is a much deeper result, as the assertion is *not* local—there is no reason why L should be contained in the domain of a chart on M. This proof is deferred to the bonus section at the end of the chapter, since it is rather fiddly.

Definition 6.11. A smooth map $\varphi\colon M \to N$ is called a **submersion** if every point of M is a regular point of φ, i.e. if $D\varphi(p)$ is surjective for every $p \in M$.

Thus if φ is a submersion then by the Implicit Function Theorem 6.10, every point $p \in M$ belongs to the $(m - n)$-dimensional embedded submanifold $\varphi^{-1}(\varphi(p))$.

---

[3] **Warning**: We are still in the case $m \leq n$. Thus m and n have switched roles compared to the projection from (5.3).

[4] If $q \in N \setminus \varphi(M)$ then q is vacuously a regular value of φ.

DEFINITION 6.12. Let $\varphi\colon M \to N$ be a smooth map. Fix $p \in M$. We say that φ admits **smooth local sections**[5] if for every $p \in M$ there exists a neighbourhood U of p and a neighbourhood V of $\varphi(p)$, together with a smooth map $\psi\colon V \to U$ such that

$$\varphi \circ \psi = \mathrm{id}, \qquad \text{on } V.$$

PROPOSITION 6.13. *Let $\varphi\colon M \to N$ be a submersion. Then φ is an open map which admits smooth local sections.*

Proof. Let $p \in M$. From the Implicit Function Theorem 5.11, we may choose a chart $x\colon U \to O$ on M about p and a chart $y\colon V \to \Omega$ on N about $\varphi(p)$ such that $y \circ \varphi \circ x^{-1}$ is of the form

$$(u^1, \dots, u^m) \mapsto (u^1, \dots, u^n). \tag{6.4}$$

By shrinking the domains if necessary we may assume

$$O = O_1 \times O_2 \subset \mathbb{R}^n \times \mathbb{R}^{m-n}$$

and that $\Omega = O_1$. Define

$$\psi\colon V \to U, \qquad \psi(q) := x^{-1}\big(y(q), \pi_2(x(p))\big)$$

Then ψ is smooth and by (6.4) one has $\varphi \circ \psi = \mathrm{id}$ on V. The fact that φ is an open map is clear from the representation (6.4), since a small open cube $\{|u^i| < \varepsilon \mid i = 1, \dots, m\}$ is mapped onto the small open cube $\{|u^i| < \varepsilon, \mid i = 1, \dots, n\}$. ■

COROLLARY 6.14. *Let $\varphi\colon M \to N$ be a surjective submersion. Then φ is a quotient map.*[6]

We now identify the tangent space of a submanifold produced via the Implicit Function Theorem.

PROPOSITION 6.15. *Let $\varphi\colon M \to N$ be a smooth map and let $q \in N$ be a regular value of φ such that $L := \varphi^{-1}(q) \neq \emptyset$. Let $i\colon L \hookrightarrow M$ denote the inclusion. Then for all $p \in L$, one has*

$$Di(p)(T_pL) = \ker D\varphi(p).$$

Proof. By assumption both sides are linear subspaces of T_pM of dimension $m - n$, so it suffices to show that $Di(p)(T_pL) \subset \ker D\varphi(p)$. For this take $f \in C^\infty(N)$ and $\xi \in T_pL$. Then by the chain rule (Proposition 4.2), one has

$$\begin{aligned} D\varphi(p) \circ \big(Di(p)\xi\big)(f) &= \big(D(\varphi \circ \iota)(p)\xi\big)(f) \\ &= \xi(f \circ \varphi \circ i). \end{aligned}$$

But $f \circ \varphi \circ i \in C^\infty(L)$ is the constant function $p \mapsto f(q)$ and hence by Corollary 3.4 one has $\xi(f \circ \varphi \circ i) = 0$. The result follows. ■

[5] The motivation behind the name "section" will come in Chapter 15.

[6] A surjective continuous map $f\colon X \to Y$ is a **quotient map** if for $V \subset Y$, one has V open $\Leftrightarrow f^{-1}(V)$ open.

Proposition 6.15 finally allows us to recover the "intuitive" definition of the tangent space for S^2 given in Figure 2.1 at the beginning of Chapter 2.

EXAMPLE 6.16. Let $f\colon \mathbb{R}^{m+1} \to \mathbb{R}$ be the map $f(p) = \|p\|^2 - 1$. It is straightforward to check that f is smooth and that the only critical point of f is $0 \in \mathbb{R}^{m+1}$. Thus $0 \in \mathbb{R}$ is a regular value of f, and so by the Implicit Function Theorem 6.10, $S^m = f^{-1}(0)$ is a smooth manifold of dimension m. This is the same smooth structure as the one given in Proposition 1.16. If we denote by $i\colon S^m \to \mathbb{R}^{m+1}$ the inclusion then (as you will check on Problem Sheet C), one has

$$Di(p)(T_pS^m) = \mathcal{J}_p(p^\perp), \tag{6.5}$$

where $\mathcal{J}_p : \mathbb{R}^{m+1} \to T_p\mathbb{R}^{m+1}$ is the dash-to-dot map from Definition 4.11, and

$$p^\perp := \left\{q \in \mathbb{R}^{m+1} \mid \langle p, q\rangle = 0\right\},$$

for $\langle\cdot,\cdot\rangle$ the standard Euclidean dot product. Now a moment's thought shows that (6.5) implies that the tangent space to S^m at a point p is the hyperplane tangent to S^m at p, as Figure 2.1 claimed.

We now state a version of Sard's Theorem valid for manifolds.

Theorem 6.17 (Sard's Theorem for Manifolds) *Let $\varphi\colon M \to N$ be a smooth map. The set of critical values of φ has measure zero and is nowhere dense. The set of regular values of φ is residual and thus dense in N. In particular, if $m < n$ then every point of M is necessarily a critical point of φ, and hence $N \setminus \varphi(M)$ is dense in N.*

A sketch of the proof of Theorem 6.17 is in the bonus section below.

We next discuss a generalisation of the Implicit Function Theorem in Euclidean spaces (Theorem 5.11), which will occasionally be useful. Suppose O is a neighbourhood of 0 in $\mathbb{R}^m$ and $f\colon O \to \mathbb{R}^n$ is a smooth map. As Theorem 5.11 showed, if we assumed that the rank of f at 0 was maximal (and thus either equal to m or n, depending which was larger), then the rank of f was also maximal for all p near 0 too. Thus having maximal rank is an *open* condition.

If the rank is not maximal, then it can "jump", i.e. if f has (non-maximal) rank k at 0 then for p arbitrarily close to 0 the rank of f at p can be different to k. However if one adds as a hypothesis that the rank of f does *not* jump, then an analogous result to Theorem 5.11 holds. Here is a precise statement:

Theorem 6.18 (The Constant Rank Theorem) *Let O be a neighbourhood of 0 in $\mathbb{R}^m$ and suppose $f\colon O \to \mathbb{R}^n$ is a smooth map such that $f(0) = 0$. Assume that f has constant rank k for all $p \in O$, and let $\kappa\colon \mathbb{R}^m \to \mathbb{R}^n$ denote the map*

$$\kappa(u^1, \dots, u^m) = (u^1, \dots, u^k, 0, \dots, 0). \tag{6.6}$$

There exists chart x about 0 on $\mathbb{R}^m$ and a chart y about 0 on $\mathbb{R}^n$ such that $y \circ f \circ x = \kappa$ on a neighbourhood of 0 on $\mathbb{R}^m$.

The proof is similar to (albeit slightly messier than) Theorem 5.11, and we omit the details. Just as in Proposition 6.3, one can immediately translate this to a local statement about smooth maps between manifolds:

COROLLARY 6.19. *Let $\varphi\colon M \to N$ be a smooth map. Let $p \in M$ and assume there exists a neighbourhood of p such that φ has constant rank k on that neighbourhood. Then there exists a chart x on M about p and a chart y on N about $\varphi(p)$ such that $y \circ \varphi \circ x^{-1} = \kappa$, where κ is as in* (6.6).

Bonus Material for Chapter 6

In this bonus section we prove Theorem 6.10 and sketch the proof of Sard's Theorem 6.17.

Proof of Theorem 6.10. We prove the result in four steps.

1. Let us first fix some notation. Write $\mathbb{R}^m = \mathbb{R}^n \times \mathbb{R}^{m-n}$. Let ρ_1 and ρ_2 denote the two projections $\mathbb{R}^m \to \mathbb{R}^n$ and $\mathbb{R}^{m-n}$ respectively:

$$\rho_1(u^1, \ldots, u^m) := (u^1, \ldots, u^n), \qquad \rho_2(u^1, \ldots, u^m) := (u^{n+1}, \ldots, u^m),$$

and let $\jmath\colon \mathbb{R}^{m-n} \to \mathbb{R}^m$ denote the inclusion onto the last $m - n$ coordinates:

$$\jmath(u^1, \ldots, u^{m-n}) = (0, \ldots, 0, u^1, \ldots, u^{m-n}).$$

Now let $y\colon V \to y(V) \subset \mathbb{R}^n$ denote a chart on N such that $y(q) = 0$. Fix a point $p \in L$ and let $x\colon U \to x(U) \subset \mathbb{R}^m$ denote a chart on M such that $x(p) = 0$. Then $y \circ \varphi \circ x^{-1}$ has maximal rank n at $0 \in \mathbb{R}^m$, and hence by part (ii) of the Implicit Function Theorem 5.11 there exists a chart z on $\mathbb{R}^m$, defined on an open ball O containing the origin such that

$$y \circ \varphi \circ x^{-1} \circ z|_O = \rho_1|_O.$$

Shrinking O if necessary, we may assume $O = \rho_1(O) \times \rho_2(O) \subset \mathbb{R}^n \times \mathbb{R}^{m-n}$. Set $\Omega := \rho_2(O)$. Then

$$y \circ \varphi \circ x^{-1} \circ z \circ \jmath|_\Omega = \rho_1 \circ \jmath|_\Omega \equiv 0.$$

Thus if $\sigma := x^{-1} \circ z \circ \jmath|_\Omega$ then $\sigma(\Omega) \subset L$.

2. In this step we show that the inclusion $L \hookrightarrow M$ is a topological embedding in a neighbourhood of p. For this, it suffices to show that σ maps Ω homeomorphically onto a neighbourhood of p in L in the subspace topology. This means that we must prove:

$$\sigma(\Omega) = L \cap (x^{-1} \circ z)(O). \tag{6.7}$$

It is clear that the left-hand side of (6.7) is contained in the right-hand side. Indeed, we have

$$\begin{aligned}\sigma(\Omega) &= (x^{-1} \circ z \circ \jmath)(\Omega) \\ &= (x^{-1} \circ z)(\mathcal{O} \cap (0 \times \mathbb{R}^{m-n})) \\ &\subset L \cap (x^{-1} \circ z)(\mathcal{O}).\end{aligned}$$

To see the other direction, if $v \in L \cap (x^{-1} \circ z)(\mathcal{O})$ then $v = (x^{-1} \circ z)(u)$ for a unique $u \in \mathcal{O}$, and since

$$\begin{aligned}\rho_1(u) &= \left(y \circ \varphi \circ x^{-1} \circ z\right)(u) \\ &= y \circ \varphi(v) \\ &= 0,\end{aligned}$$

we can write $v = (0, v_0)$ for a unique $v_0 \in \Omega$. Then $v = \sigma(v_0)$. This proves the other inclusion, and hence establishes (6.7).

3. We now show that L is a smooth manifold. Using the notation from above, set $W := \sigma(\Omega)$ and let $w := \sigma^{-1}$. We shall show that $w \colon W \to \Omega$ can serve as a chart on L. More precisely, we claim that the collection of all such charts, as p ranges over L, determines a smooth structure on L. To see this, suppose p_1 was another point in L with corresponding chart $x_1 \colon U_1 \to x_1(U_1) \subset \mathbb{R}^m$. Assume that $U \cap U_1 \neq \emptyset$. Let z_1 denote the corresponding diffeomorphism of $\mathbb{R}^m$, and define σ_1 and w_1 similarly. Then by assumption $x_1 \circ x^{-1}$ is a diffeomorphism where defined, and hence so is $\tau := z_1^{-1} \circ x_1 \circ x^{-1} \circ z$. Moreover from (6.7) we can write $\tau(0, u) = (0, \tau_1(u))$ for τ_1 a diffeomorphism defined on a neighbourhood of 0 in $\mathbb{R}^{m-n}$. Thus

$$w_1 \circ w^{-1} = \jmath^{-1} \circ \tau \circ \jmath = l_1$$

is a diffeomorphism where defined. This shows that we have built a smooth structure on L.

4. To complete the proof, we show that the inclusion $\iota \colon L \hookrightarrow M$ is smooth. For this we note that with x, w and z as above,

$$x \circ \iota \circ w^{-1} = x \circ \iota \circ \sigma = z \circ \jmath,$$

which is smooth. This completes the proof. ■

The Implicit Function Theorem also generalises to constant rank maps.

Theorem 6.20 (Constant Rank Implicit Function Theorem) *Suppose $\varphi \colon M \to N$ has constant rank k. Take $q \in \varphi(M)$ and set $L := \varphi^{-1}(q)$. Then L is a topological manifold of dimension $m - k$. Moreover there exists a smooth structure on L which makes L into a smooth embedded submanifold of M, and if $i \colon L \hookrightarrow M$ denotes the inclusion then for all $p \in L$, one has*

$$Di(p)(T_p L) = \ker D\varphi(p).$$

The proof of Theorem 6.20 proceeds in an analogous fashion to that of Theorem 6.10, only starting with Theorem 6.18 instead.

We conclude by giving a brief sketch of the proof of Sard's Theorem 6.17.[7]

Proof of Sard's Theorem 6.17. The classical version of Sard's Theorem says that if $\mathcal{O} \subset \mathbb{R}^m$ is an open set and $f : \mathcal{O} \to \mathbb{R}^n$ is a smooth map, then the set of critical values of f has measure zero in $\mathbb{R}^n$. Since manifolds are second countable they can be covered by countably many open sets that are diffeomorphic to balls in Euclidean spaces, cf. Proposition 1.32. As the countable union of measure zero sets is also of measure zero, the result follows. ■

Sard's Theorem is the main reason we require that manifolds have at most countably many components (cf. Proposition 1.32)—the theorem is false if this condition is not imposed.

[7] A nice proof can be found in Chapter 3 of Milnor's classic textbook "Topology from the Differentiable Viewpoint" [2].

Chapter 7
The Whitney Theorems

In this chapter we will prove two famous theorems of Whitney. The first states that *every* smooth manifold can be embedded inside Euclidean space.

Recall that a continuous function $f\colon X \to Y$ between two topological spaces is **proper** if the preimage of any compact set in Y is compact in X. If X is compact and Y is Hausdorff then every continuous function is proper.

Theorem 7.1 (The Strong Whitney Embedding Theorem) *Let M be a smooth manifold of dimension m. Then there exists a proper embedding $\varphi\colon M \to \mathbb{R}^{2m}$.*

Theorem 7.1 is a genuinely difficult result. It is much easier to prove that M always embeds in $\mathbb{R}^{2m+1}$ (this is sometimes called the "Weak Whitney Embedding Theorem"). This is still too hard for us, however, so we will prove this only for the special case of compact manifolds. We call this the "Baby Whitney Embedding Theorem".

Theorem 7.2 (The Baby Whitney Embedding Theorem) *Let M be a compact smooth manifold of dimension m. Then there exists a (proper) embedding $\varphi\colon M \to \mathbb{R}^{2m+1}$.*

The "proper" is in parentheses, as this is automatic when M is compact.

Proof. We prove the result in four steps.

1. We begin by showing that M admits an embedding into *some* Euclidean space $\mathbb{R}^n$ (this method will typically produce a very large n). In the next step we will reduce n down to $2m + 1$.

Since M is compact we can find a finite cover $\{V_1, \dots, V_k\}$ of open sets, with the property that there exist charts (U_i, x_i) for $i = 1, \dots, k$ with $\overline{V}_i \subset U_i$. Now let $\chi_i\colon M \to \mathbb{R}$ denote a bump function (whose existence is guaranteed by Lemma 3.2) such that $\chi_i(\overline{V}_i) \equiv 1$, $0 \le \chi_i(p) \le 1$ for all $p \in M$ and $\operatorname{supp}(\chi_i) \subset U_i$. Set $f_i = \chi_i\, x_i$, which we think of as a function from $M \to \mathbb{R}^m$ by extending it to be zero outside of U_i. Then define

$$\varphi\colon M \to \mathbb{R}^{(m+1)k}, \qquad \varphi(x) := \Big(f_1(x), \dots, f_k(x), \chi_1(x), \dots, \chi_k(x)\Big).$$

W. J. Merry, *Lectures on Differential Geometry I*,
https://doi.org/10.1007/978-3-032-03733-6_7

We claim that φ is an injective immersion. Since M is compact, it then follows from Problem C.3 that φ is an embedding. To see that φ is injective, suppose $\varphi(p) = \varphi(q)$. Since the sets $\{V_i\}$ cover M, there is some i such that $p \in V_i$, and hence $\chi_i(p) = 1$. Since $\varphi(p) = \varphi(q)$, we also have $\chi_i(q) = 1$, and thus $q \in \operatorname{supp}(\chi_i) \subset U_i$. Then also

$$x_i(p) = \chi_i(p)x_i(p) = f_i(p) = f_i(q) = \chi_i(q)x_i(q) = x_i(q).$$

But x_i is a diffeomorphism, and hence in particular injective. Thus $p = q$.

Finally, to check φ is an immersion, pick an arbitrary $p \in M$. Then $p \in V_i$ for some i. Since $\chi_i \equiv 1$ on a neighbourhood of p, we have $Df_i(p) = Dx_i(p)$, which is injective. Thus also $D\varphi(p)$ is injective. This completes the proof of the weak version we wished to prove, where we took $n = (m+1)k$.

2. Replacing M by $\varphi(M)$, we now have $M \subset \mathbb{R}^n$. Assume that $n > 2m+1$, otherwise there is nothing to prove. Think of $\mathbb{R}^{n-1}$ as sitting inside $\mathbb{R}^n$ as the hyperplane $\{(u^1, \ldots, u^n) \mid u^n = 0\}$. Given $\xi \in \mathbb{R}^n \setminus \mathbb{R}^{n-1}$, let $\rho_\xi \colon \mathbb{R}^n \to \mathbb{R}^{n-1}$ denote the projection parallel to ξ, that is, the unique linear map with

$$\ker \rho_\xi = \mathbb{R} \cdot \xi.$$

We will look for unit vectors ξ with the property that

$$\rho_\xi|_M \colon M \to \mathbb{R}^{n-1}$$

is an embedding. Using Problem C.3 again, it suffices to show that $\rho_\xi|_M$ is an injective immersion. But what does that mean in this context? In words, saying that $\rho_\xi|_M$ is injective is saying that ξ is not parallel to any secant of M, that is

$$\xi \neq \frac{p-q}{\|p-q\|}, \qquad \forall\, p, q \in M. \tag{7.1}$$

(Throughout this chapter, the norm $\|\cdot\|$ denotes the standard Euclidean norm. This is true both for points in M and for points in $TM \subset T\mathbb{R}^n = \mathbb{R}^{2n}$.)

The kernel of the linear map ρ_ξ is the line through ξ. Since ρ_ξ is linear, its derivative is the same linear map. Thus a tangent vector $\zeta \in T_pM$ lies in the kernel of $D\rho_\xi(p)$ if and only if ζ is parallel to ξ. We therefore see that ρ_ξ is an immersion if

$$\xi \neq \frac{\zeta}{\|\zeta\|}, \qquad \forall\, \zeta \in T_pM,\ \forall\, p \in M. \tag{7.2}$$

3. We will use Sard's Theorem 6.17 to prove a ξ exists such that both (7.1) and (7.2) are satisfied. For (7.1), consider the map

$$\psi \colon (M \times M) \setminus \Delta \to S^{n-1}, \qquad (p,q) \mapsto \frac{p-q}{\|p-q\|}.$$

Here Δ is the *diagonal* inside $M \times M$:

$$\Delta := \{(x,x) \mid x \in M\}.$$

Clearly ξ satisfies (7.1) if and only if ξ is *not* in the image of ψ. Note that $(M\times M)\setminus\Delta$ is an open set of $M\times M$, and thus $(M\times M)\setminus\Delta$ is a manifold of dimension $2m$ by Lemma 1.15 and Problem A.3. The map ψ is visibly smooth. Since $2m < n-1 = \dim S^{n-1}$, by Sard's Theorem 6.17 the image of ψ is nowhere dense in S^{n-1}. Thus in particular, any non-empty open set of S^{n-1} contains a point ξ satisfying (7.1).

Now we consider (7.2). It suffices to check that it holds for all vectors ζ of norm 1. To this end we focus on the *unit tangent bundle*

$$SM := \{(p,\zeta) \in TM \mid \|\zeta\| = 1\}.$$

To see this is a manifold, consider the map $h\colon T\mathbb{R}^n \to \mathbb{R}$ given by $h(p,\zeta) = \|\zeta\|^2$. It is easy to see that 1 is a regular value of $h|_{TM}$ and that $SM = h|_{TM}^{-1}(1)$. Thus by the Implicit Function Theorem 6.10, SM is a manifold of dimension $2m-1$. Moreover since M is compact so is SM. Since $M \subset \mathbb{R}^n$, we have

$$TM \subset M \times \mathbb{R}^n \subset \mathbb{R}^{2n} = T\mathbb{R}^n$$

(See part (i) of Problem D.1 if you are worried about the identification $T\mathbb{R}^n \cong \mathbb{R}^{2n}$.)

Similarly SM is identified with a subset of $M \times S^{n-1}$. Projecting onto the second factor, this gives us a map

$$SM \to M \times S^{n-1} \to S^{n-1}$$

which geometrically takes a unit vector based at a point in M and translates it to a unit vector based at the origin in $\mathbb{R}^n$. Using Sard's Theorem 6.17 again, the image of the composite map $SM \to S^{n-1}$ is nowhere dense. Since SM is compact, it follows that the complement—let us call it W—of the image is a dense open set in S^{n-1}. Thus W meets $S^{n-1} \cap \left(\mathbb{R}^n \setminus \mathbb{R}^{n-1}\right)$ in an non-empty open set W_0. From what we already proved, such a non-empty open set W_0 contains a vector ξ which is not in the image of ψ.

4. We now complete the proof. The choice of ξ found above gives us an embedding $\rho_\xi\colon M \to \mathbb{R}^{n-1}$. If $n-1 = 2m+1$ we are done, if not then $n-1 > 2m+1$, and the same argument again works to provide a new embedding in $\mathbb{R}^{n-2}$. By induction, we eventually obtain our desired embedding $M \to \mathbb{R}^{2m+1}$. ■

Remark 7.3. Extending Theorem 7.2 to cover all smooth manifolds (not just compact ones) is not that much more work.[1] We emphasise though that the stronger result (Theorem 7.1, where $2m+1$ is reduced down to $2m$) is *much* harder.

Theorem 7.1 implies one could equivalently *define* a manifold as an embedded submanifold of Euclidean space.

Definition 7.4 (Alternative Definition of a Manifold). Let $m \leq n$. A subset $M \subset \mathbb{R}^n$ is called a **smooth manifold[2] of dimension** m if each point p in M has a neighbourhood V in $\mathbb{R}^n$ such that $M \cap V$ is diffeomorphic to an open set in $\mathbb{R}^m$.

[1] This appears as bonus problem D.8 on Problem Sheet D.

[2] This is how manifolds are defined in most "baby" courses on differential geometry.

In more detail, this means: for each point $p \in M$ there exists an open set $O \subset \mathbb{R}^m$ and a neighbourhood $V \subset \mathbb{R}^n$ of p, together with an injective smooth map $\sigma\colon O \to \mathbb{R}^n$ of maximal rank m everywhere such that $\sigma(O) = M \cap V$ and $x := \sigma^{-1}|_{M\cap V}\colon M \cap V \to O$ is continuous (where M is given the subspace topology of $\mathbb{R}^n$). One usually calls σ a **parametrisation** of M. The inverse x of σ is then a chart on M in the normal sense. Note that if $m = n$ then this forces M to be an open subset of $\mathbb{R}^n$, and hence if M is compact and non-empty then one necessarily has $m < n$.

REMARK 7.5. Definition 7.4 is superficially much simpler than our original definition (Definition 1.13)—there is no need to first define topological manifolds, or even mention metrisability and separability. The equivalence of the definitions follows from Theorem 7.1 and the existence of slice charts (Definition 6.6). Moreover it is immediate from Definition 7.4 that manifolds are metrisable, since any subset of a metric space inherits a metric that determines its subspace topology.

You might therefore reasonably ask: was there any point in the abstract definition? The answer is of course **"yes"**, as we will now try to explain.

An embedded submanifold of Euclidean space should really be thought of as a pair (M, φ), where M is an (abstract) smooth manifold and φ is a choice of embedding. However it is possible to embed a given manifold in many different ways, and moreover if you can embed M in $\mathbb{R}^n$ then you can also embed M in $\mathbb{R}^k$ for any $k \geq n$. A different choice of embedding can lead to dramatically different geometry (this will be particularly evident when we study *Riemannian geometry* in Volume II, starting in Chapter 10. Thus when proving results about embedded submanifolds, one always needs to ask the question: is this proof really a statement about the manifold itself, or does it depend on the embedding? This can often vastly complicate the proofs. The upshot is that having a more complicated definition leads to simpler proofs, and hence in the long run—since you only need to define things once but there are many theorems to prove!—it is better to work with the abstract definition whenever possible.

Still another reason to prefer the abstract definition is the following: One of the key applications of differential geometry in theoretical physics is Einstein's theory of *General Relativity*. Here one views the universe as 4-dimensional (curved) space-time. In the finite universe model, the spatial part of space-time is taken to be a compact 3-dimensional manifold. Since (by definition) the universe is "everything", it doesn't make any sense *at all* to require the theory to begin by embedding the universe in a larger Euclidean space...

We now aim to prove another theorem, also due to Whitney, called the *Whitney Approximation Theorem*, that allows us replace a continuous map with a smooth one. We begin with the following statement, which says a continuous function from a manifold to a Euclidean space can be approximated arbitrarily well by a smooth one.

PROPOSITION 7.6. *Let M be a smooth manifold and let $h\colon M \to \mathbb{R}^n$ be a continuous function. Given any positive continuous function $\delta\colon M \to \mathbb{R}$, there exists a smooth function $f\colon M \to \mathbb{R}^n$ such that*

$$\|f(p) - h(p)\| < \delta(p), \qquad \forall\, p \in M.$$

Proof. Fix $p \in M$ and let U_p be a neighbourhood of p such that for all $q \in U_p$, one has

$$\delta(q) > \frac{1}{2}\delta(p), \qquad \|h(q) - h(p)\| < \frac{1}{2}\delta(p).$$

Such a neighbourhood exists as h and δ are assumed to be continuous. Then in particular we have that

$$\|h(q) - h(p)\| < \delta(q), \qquad \forall\, q \in U_p.$$

The collection $\{U_p \mid p \in M\}$ is an open cover of M. Let $\{\kappa_p \mid p \in M\}$ be a partition of unity subordinate to this open cover and define

$$f : M \to \mathbb{R}^n, \qquad f(q) := \sum_{p \in M} \kappa_p(q)h(p).$$

Recall that the right-hand side is actually a finite sum at every point, since $\{\text{supp}(\kappa_p)\}$ is locally finite, and hence f is smooth. Moreover since $\sum_p \kappa_p \equiv 1$ and $\text{supp}(\kappa_p) \subset U_p$, one has for any $q \in M$ that

$$\begin{aligned}
\|f(q) - h(q)\| &= \Big\| \sum_{p \in M} \kappa_p(q)h(p) - h(q) \Big\| \\
&= \Big\| \sum_{p \in M} \kappa_p(q)h(p) - \sum_{p \in M} \kappa_p(q)h(q) \Big\| \\
&\le \sum_{p \in M} \kappa_p(q)\|h(q) - h(p)\| \\
&< \sum_{p \in M} \kappa_p(q)\delta(q) = \delta(q).
\end{aligned}$$

This completes the proof. ∎

Our aim now is to improve Proposition 7.6 to the case where the target space is another manifold, not a Euclidean space. The "obvious" tactic (given that we just proved the Whitney Embedding Theorem) is to embed the target manifold in a Euclidean space, and then approximate via the result we just proved. Unfortunately this doesn't quite work, as even though the function f can be chosen to be very close to h, it may still be the case that f "misses" our newly embedded manifold (remember an embedded manifold is not an open subset unless it is of full dimension). Thus we need a way to correct this. We will do this by making use of *tubular neighbourhoods*, which will be defined shortly.

Definition 7.7. Let M be an embedded submanifold of $\mathbb{R}^n$. We define the **normal space to M at p** to be the $(n-m)$-dimensional subspace $\text{Nor}_p M \subset T_p\mathbb{R}^n$ consisting of all vectors that are orthogonal to T_pM with respect to the Euclidean dot product. We define the **normal bundle** of M as the set

$$\operatorname{Nor} M := \left\{(p, \xi) \in T\mathbb{R}^n = \mathbb{R}^n \times \mathbb{R}^n \mid p \in M, \xi \in \operatorname{Nor}_p M\right\}.$$

On Problem Sheet C you are asked to prove that $\operatorname{Nor} M$ is an embedded n-dimensional submanifold of $T\mathbb{R}^n = \mathbb{R}^n \times \mathbb{R}^n$. We define a map

$$T\colon \operatorname{Nor} M \to \mathbb{R}^n, \qquad T(p, \xi) := p + \xi.$$

We emphasise that this only makes sense as M is embedded in $\mathbb{R}^n$. In general one cannot add points together on a manifold! The map T is smooth, since it is the restriction to $\operatorname{Nor} M$ of the addition map $\mathbb{R}^n \times \mathbb{R}^n \to \mathbb{R}^n$. If O_M denotes the **zero section**:[3]

$$O_M := \{(p, 0) \mid p \in M\}.$$

then one has

$$T(O_M) = M.$$

Thus it is reasonable to hope that a small neighbourhood of O_M in $\operatorname{Nor} M$ gets mapped under T to a small neighbourhood of M in $\mathbb{R}^n$. This motivates the following definition.

Definition 7.8. A **tubular neighbourhood of** M is a neighbourhood U of M in $\mathbb{R}^n$ which is the diffeomorphic image under T of an open subset $V \subset \operatorname{Nor} M$ of the form

$$V = \left\{(p, \xi) \in \operatorname{Nor} M \mid \|\xi\| < \varepsilon(p)\right\}, \tag{7.3}$$

where $\varepsilon\colon M \to \mathbb{R}$ is a strictly positive continuous function.

It is a non-trivial fact that such neighbourhoods always exist:

Theorem 7.9 (The Tubular Neighbourhood Theorem) *Every embedded submanifold $M \subset \mathbb{R}^n$ admits a tubular neighbourhood.*

The proof is deferred to the bonus section below.

Definition 7.10. Let $Y \subset X$ be a subspace of a topological space. A **retraction** of X onto Y is a continuous map $r\colon X \to Y$ such that $r|_Y$ is the identity map on Y.

Corollary 7.11. *Let $M \subset \mathbb{R}^n$ be an embedded submanifold, and let U be a tubular neighbourhood of M. There exists a smooth map $r\colon U \to M$ which is both a retraction and a submersion.*

Proof. Let $T\colon V \subset \operatorname{Nor} M \to U$ be our tubular neighbourhood, and write $\pi\colon \operatorname{Nor} M \to M$ for the footpoint map that sends a pair (p, ξ) to p. Define $r\colon U \to M$ by $r := \pi \circ T^{-1}|_U$. Since $T|_V$ is a diffeomorphism and π is clearly a submersion, it follows that r is a submersion. Finally since $T(p, 0) = p$, we see that

$$r(p) = \pi \circ T^{-1}(p) = p,$$

and hence r is a retraction. ■

[3] The explanation for the name "zero section" will come in Chapter 10, when we discuss vector bundles.

Recall that if $h_0, h_1 : X \to Y$ are two continuous maps, we say they are **homotopic** if there exists a continuous map $H : X \times [0,1] \to Y$ such that $H(\cdot, 0) = h_0$ and $H(\cdot, 1) = h_1$. We can now state and prove the Whitney Approximation Theorem. We will use this result later when we discuss the *homotopy invariance of de Rham cohomology* in Chapter 23.

Theorem 7.12 (The Whitney Approximation Theorem) *Let $h : M \to N$ be a continuous map between two smooth manifolds. Then h is homotopic to a smooth map $\varphi : M \to N$.*

Proof. By the Whitney Embedding Theorem 7.1, we may assume that N is a properly embedded submanifold of some Euclidean space $\mathbb{R}^n$. Let U be a tubular neighbourhood of N, and let $r : U \to N$ be a smooth submersive retraction (whose existence is guaranteed by Corollary 7.11). Given $p \in N$, let

$$0 < \varepsilon(p) := \sup \left\{ \varepsilon \le 1 \mid B_\varepsilon(p) \subset U \right\},$$

where $B_\varepsilon(p)$ denotes the ball of radius ε about p in the Euclidean norm. We claim that ε is actually a continuous function. To see this let $p, q \in N$ and first suppose that $\|p - q\| < \varepsilon(p)$. Then for $\delta := \varepsilon(p) - \|p - q\|$, one has by the triangle inequality that $B_\delta(q) \subset B_{\varepsilon(p)}(p)$, and hence $\varepsilon(q) \ge \varepsilon(p) - \|p - q\|$. Thus if $\|p - q\| < \varepsilon(p)$ then

$$\varepsilon(p) - \varepsilon(q) \le \|p - q\|.$$

On the other hand, if $\varepsilon(p) \le \|p - q\|$ then since $\varepsilon(q) > 0$ by definition, one trivially also has

$$\varepsilon(p) - \varepsilon(q) \le \|p - q\|.$$

Reversing the roles of p and q shows that

$$\|\varepsilon(p) - \varepsilon(q)\| \le \|p - q\|,$$

which proves ε is continuous. Now define

$$\delta := \varepsilon \circ h : M \to \mathbb{R}.$$

Then δ is a continuous positive function, and hence by Proposition 7.6, there exists a smooth function $f : M \to \mathbb{R}^n$ such that

$$\|f(p) - h(p)\| < \delta(p), \qquad \forall\, p \in M.$$

Define

$$H : M \times [0,1] \to N, \qquad H(p,t) := r((1-t)h(p) + t f(p)).$$

This is well-defined due to our choice of function δ, which implies that $(1-t)h(p) + t f(p) \in U$ for all $t \in [0,1]$. Since r is the identity on $N \subset U$ and h takes values in N, we see that $H(\cdot, 0) = h$. Moreover if $\varphi := r \circ f$ then φ is smooth and $H(\cdot, 1) = \varphi$. This completes the proof. ■

Bonus Material for Chapter 7

In this bonus section we make some additional remarks about the Strong Whitney Embedding Theorem 7.1, prove the Tubular Neighbourhood Theorem, and finally discuss an improvement to the Whitney Approximation Theorem.

REMARKS 7.13.

(i) The Whitney Embedding Theorem is sharp in the sense that if $m = 2^k$ then $\mathbb{R}P^m$ cannot be embedded in $\mathbb{R}^{2m-1}$. This can be proved using characteristic classes. We will come back to this in Volume II, cf. Volume II Corollary 11.20

(ii) There are various other versions of the Whitney Embedding Theorem. For instance, if M is a compact *orientable* smooth manifold of dimension m (we will define orientability in Chapter 24) then M embeds inside $\mathbb{R}^{2m-1}$. This does not contradict the previous statement, since for m even $\mathbb{R}P^m$ is not orientable.

(iii) In many cases the upper bound can be improved—for instance, in Chapter 1 we saw that S^m embeds into $\mathbb{R}^{m+1}$. Another result (due to Haefliger) is that if M is a compact smooth manifold of dimension m whose homotopy groups $\pi_i M$ vanish for $i \leq k$ then if $2k + 3 \leq m$ one can embed M in $\mathbb{R}^{2m-k}$. In general, if eM denotes the optimal n such that M embeds inside $\mathbb{R}^n$ then computing eM is an open problem for many manifolds M.

Let us now prove the Tubular Neighbourhood Theorem.

Proof of the Tubular Neighbourhood Theorem 7.9. We prove the result in four steps.

1. We will prove that $DT(p, 0)$ is invertible at every point $(p, 0) \in O_M$. Since $T|_{O_M} : O_M \to M$ is obviously a diffeomorphism, one sees that $DT(p, 0)$ maps $T_{(p,0)} O_M \subset T_{(p,0)} \operatorname{Nor} M$ isomorphically onto $T_p M$. Secondly, if we restrict T to the fibre $\operatorname{Norm}_p M$, T just becomes the affine map $\xi \mapsto p + \xi$, and thus $DT(p, 0)$ maps $T_{(p,0)} \operatorname{Norm}_p M$ isomorphically onto $\operatorname{Norm}_p M$ by Lemma 4.15.

Thus by the Inverse Function Theorem 5.10 we see that for each $p \in M$ there exists an $\varepsilon_p > 0$ such that if

$$U(p, \varepsilon_p) := \left\{(q, \xi) \in \operatorname{Nor} M \mid \|p - q\| < \varepsilon_p \text{ and } \|\xi\| < \varepsilon_p\right\}$$

then $T|_{U(p,\varepsilon_p)}$ is a diffeomorphism. To complete the proof we need to show that there is open set of the form (7.3) on which T is a global diffeomorphism.

2. Let $\varepsilon \colon M \to \mathbb{R}$ be the function that assigns to a point $x \in M$ the supremum of all $\varepsilon \leq 1$ such that T is a diffeomorphism on $U(x, \varepsilon)$. Then ε is strictly positive, as $\varepsilon(p) \geq \varepsilon_p$. We now claim that ε is actually a continuous function. This argument is essentially identical to the proof of the Whitney Approximation Theorem, but we give it again anyway. So suppose $p, q \in M$ and suppose that $\|p - q\| < \varepsilon(p)$. Then for $\delta := \varepsilon(p) - \|p - q\|$, one has by the triangle inequality that $U(y, \delta) \subset U(p, \varepsilon(p))$, and hence $\varepsilon(q) \geq \varepsilon(p) - \|p - q\|$. Thus if $\|p - q\| < \varepsilon(p)$ then

$$\varepsilon(p) - \varepsilon(q) \leq \|p - q\|.$$

On the other hand, if $\varepsilon(p) \leq \|p - q\|$ then since $\varepsilon(q) \geq 0$ by definition, one trivially also has

$$\varepsilon(p) - \varepsilon(q) \leq \|p - q\|.$$

Reversing the roles of p and q shows that

$$|\varepsilon(p) - \varepsilon(q)| \leq \|p - q\|,$$

which proves ε is continuous.

3. Set

$$V := \left\{(p, \xi) \in \operatorname{Nor} M \mid \|\xi\| < \frac{1}{2}\varepsilon(p)\right\}.$$

We claim that T is injective on V. Indeed, suppose (p, ξ) and (q, ζ) both belong to V and satisfy

$$p + \xi = T(p, \xi) = T(q, \zeta) = q + \zeta.$$

Without loss of generality, assume $\varepsilon(q) \leq \varepsilon(p)$. Then

$$\begin{aligned} \|p - q| &= \|\xi - \zeta\| \\ &\leq \|\xi\| + \|\zeta\| \\ &\leq \frac{1}{2}\varepsilon(p) + \frac{1}{2}\varepsilon(p) \\ &= \varepsilon(p), \end{aligned}$$

where the first equality used $p+\xi = q+\zeta$. Thus both (p, ξ) and (q, ζ) belong to $U(p, \varepsilon(p))$. But on this set T is injective by construction. Thus $(p, \xi) = (q, \zeta)$ as required.

4. We complete the proof. Set $U := T(V)$. Then U is open as T is a local diffeomorphism. Since $T|_V$ is injective, we see that $T: V \to U$ is smooth bijection, and hence (as T is a local diffeomorphism), also a diffeomorphism. This completes the proof. ■

We conclude this chapter with a couple of additional remarks about the Whitney Approximation Theorem. First, a definition.

Definition 7.14. Suppose M and N are smooth manifolds and $A \subset M$ is an arbitrary set. We say a map $\varphi: A \to N$ is **smooth on** A if it can be locally smoothly extended, i.e. if for every $p \in A$ there exists a neighbourhood U of p in M and a smooth map $\tilde{\varphi}: U \to N$ such that $\tilde{\varphi}|_{U \cap A} = \varphi$.

With a little bit more work, Theorem 7.12 can be improved to give the following statement:

Theorem 7.15 *Suppose $h \colon M \to N$ is a continuous map between two smooth manifolds. Suppose $A \subset M$ is a closed set and $h|_A$ is smooth in the sense of Definition* 7.14. *Then h is homotopic to a smooth map φ such that $\varphi|_A \equiv h|_A$. In fact, h and φ are homotopic via a homotopy H with the property that $H(p,t) = h(p)$ for all $t \in [0,1]$.*

One can also play the same game with smooth homotopies. Two smooth maps $\varphi, \psi \colon M \to N$ are **smoothly homotopic** if there exists a smooth map[4]

$$H : M \times [0,1] \to N$$

such that $H(\cdot, 0) = \varphi$ and $H(\cdot, 1) = \psi$.

Theorem 7.16 (The Homotopy Whitney Approximation Theorem) *If $\varphi, \psi \colon M \to N$ are two smooth maps between smooth manifolds which are homotopic (in the normal sense), then they are also smoothly homotopic. Moreover if the given normal homotopy H from φ to ψ is stationary*[5] *on some closed set A then the approximating smooth homotopy can also be chosen to be stationary on A.*

[4] Note that we are using Definition 7.14 again here to make sense of this.

[5] i.e. $H(p,t) = \varphi(p)$ for all $p \in A$—note this implies $\varphi|_A \equiv \psi|_A$.

Problem Sheet C

A (♣) indicates the problem is non-examinable. This does not necessarily imply it is difficult.

Problem C.1 Let M and N be smooth manifolds. Prove that for all $(p, q) \in M \times N$ there is a canonical isomorphism

$$T_{(p,q)}(M \times N) = T_p M \times T_q N.$$

Problem C.2 Let $\varphi \colon M \to N$ be a smooth map. Prove that $D\varphi \colon TM \to TN$ is also smooth. Prove that if $\varphi \colon M \to N$ is an embedding then so is $D\varphi \colon TM \to TN$.

Problem C.3 Let $\varphi \colon M \to N$ be an injective immersion with M compact. Prove that φ is an embedding. Give an example to show that this need not be true if M is not compact.

Problem C.4 Let $\mathcal{O}$ be an open subset in $\mathbb{R}^m$ and suppose $f : \mathcal{O} \to \mathbb{R}$ is smooth. Define $g : \mathcal{O} \to \mathbb{R}^{m+1}$ by

$$g(x) = (x, f(x)).$$

Prove that g is a smooth embedding, and hence that $g(\mathcal{O})$ is a smooth embedded m-dimensional submanifold of $\mathbb{R}^{m+1}$. (We call $g(\mathcal{O})$ the **graph** of f.)

Problem C.5 Let $i \colon S^m \hookrightarrow \mathbb{R}^{m+1}$ denote the inclusion. Prove that

$$Di(p)[T_p S^m] = \mathcal{J}_p(p^\perp),$$

where $\mathcal{J}_p : \mathbb{R}^{m+1} \to T_p\mathbb{R}^{m+1}$ is the dash-to-dot map and

$$p^\perp := \left\{ q \in \mathbb{R}^{m+1} \mid \langle p, q \rangle = 0 \right\},$$

for $\langle \cdot, \cdot \rangle$ the standard Euclidean dot product.

(♣)Problem C.6 Prove that $\operatorname{Nor} M$ is an embedded n-dimensional submanifold of $T\mathbb{R}^n = \mathbb{R}^{2n}$, where $\operatorname{Nor} M$ is as in Definition 7.7

Problem C.7 Let $\varphi \colon M \to N$ be a smooth map. Assume that the rank of φ is constant on all of M.

(i) Assume that φ is injective. Prove that φ is an immersion.
(ii) Assume that φ is surjective. Prove that φ is a submersion.
(iii) Assume that φ is bijective. Prove that φ is a diffeomorphism.

Bonus Problem(s) for Sheet C

These problem(s) are hard, and are included for enthusiasts only. Solutions will not be provided.

(♣)PROBLEM C.8 Let $\varphi\colon M \to N$ be smooth, and let $L \subset N$ be an embedded submanifold. We say that φ is **transverse** and **regular** at L if

$$D\varphi(p)(T_pM) + T_{\varphi(p)}L = T_{\varphi(p)}N, \qquad \forall\, p \in \varphi^{-1}(L).$$

Assume that $\varphi^{-1}(L) \neq \emptyset$. Prove that if φ is transverse and regular at L then $\varphi^{-1}(L)$ is a smooth embedded submanifold of M of dimension $m - n + l$.

The Implicit Function Theorem 6.10 is the special case where L is a point.

(♣)PROBLEM C.9 Let M be a smooth manifold and let N denote a covering space of M. Prove that N is a topological manifold, and moreover that there exists a unique smooth structure on N such that N is a smooth manifold and the covering projection $\pi\colon N \to M$ is smooth.

Chapter 8
Vector Fields

In this chapter we will define **vector fields**, which are smooth sections of the tangent bundle. We first introduce the following standard notational convention, which will hold for the remainder of the book.[1]

> **The Einstein Summation Convention**. If the same index appears *exactly* twice in any monomial, written once as an *upper* index and once as a *lower* index, then that term is understood to be summed over all possible values of that index.

Here are two examples:

(i) If e_i denotes the standard ith basis vector in $\mathbb{R}^m$, then we write

$$v = a^i e_i \qquad \text{as an abbreviation for} \qquad v = \sum_{i=1}^{m} a^i e_i$$

(ii) If M is an m-dimensional smooth manifold, $p \in M$, and (x^i) are local coordinates about p, then for $\xi \in T_p M$ we write

$$\xi = a^i \frac{\partial}{\partial x^i}\bigg|_p \qquad \text{as an abbreviation for} \qquad \xi = \sum_{i=1}^{m} a^i \frac{\partial}{\partial x^i}\bigg|_p$$

Here $\frac{\partial}{\partial x^i}$ is understood to have i as a lower index, despite the fact that x^i has i as an upper index, because it is on the bottom of a fraction.

[1] We deliberately chose to delay introducing this convention until now, so that you could see how cumbersome proofs with multiple summation signs are (e.g. Theorem 5.6), and thus fully appreciate the new convention!

W. J. Merry, *Lectures on Differential Geometry I*,
https://doi.org/10.1007/978-3-032-03733-6_8

This convention will vastly simplify equations throughout the course. For instance, when we start to talk about tensors, we will have cause to consider quantities which have local expressions such as

$$A = A^{ij}_{kl} \frac{\partial}{\partial x^i} \otimes \frac{\partial}{\partial x^j} \otimes dx^k \otimes dx^l,$$

which is much simpler than writing this abomination

$$A = \sum_{i=1}^{m} \sum_{j=1}^{m} \sum_{k=1}^{m} \sum_{l=1}^{m} A^{ij}_{kl} \frac{\partial}{\partial x^i} \otimes \frac{\partial}{\partial x^j} \otimes dx^k \otimes dx^l.$$

The caveat is that in order for the convention to "work", the choice of whether to write a given quantity as an upper index or a lower index is *not* arbitrary.

Definition 8.1. Let M be a smooth manifold and let $W \subset M$ be a non-empty open set (possibly equal to all of M). A **vector field** X on W is a smooth map $X \colon W \to TM$ (where we regard W as a smooth manifold in its own right) that satisfies the **section property**:

$$\pi(X(p)) = p, \qquad \forall\, p \in W, \tag{8.1}$$

where $\pi \colon TM \to M$ is the footpoint map. We denote by $\mathfrak{X}(W)$ the set of all vector fields on W.

Convention. Vector fields will typically be written with capital letters: X, Y and Z.

Equation 8.1 is equivalent to requiring that $X(p) \in T_pM$ for each $p \in W$. Thus a vector field can be thought of as a smoothly varying choice of tangent vector at each point.

Let us give various equivalent ways of expressing what smooth means in this context. Let $x \colon U \to O$ be a chart on M, and suppose $X \colon U \to TM$ is *any* function satisfying the section property (8.1) (not necessarily smooth). Let $p \in U$. Since $\left\{ \frac{\partial}{\partial x^i}\big|_p \mid i = 1, \ldots, m \right\}$ is a basis of T_pM, we can write[2]

$$X(p) = X^i(p) \frac{\partial}{\partial x^i}\bigg|_p, \tag{8.2}$$

for some real numbers $X^i(p)$. If we do this for every point $p \in U$, we can think of the X^i as defining functions $X^i \colon U \to \mathbb{R}$. In general these functions X^i need not even be continuous, but as we will shortly see, if X is smooth (i.e. a vector field on U) then the X^i are actually smooth functions.

Here is yet another way to think about it. Suppose $f \in C^\infty(U)$, and let as before X denote any map $U \to TM$ satisfying the section property. Then for any given $p \in U$, thinking of $X(p)$ as a derivation of $C^\infty(U)$ at p, we can feed f to $X(p)$ to get a number $X(p)(f)$. This gives us a function $X(f) \colon U \to \mathbb{R}$:

---

[2] Note here we are using the Einstein Summation Convention to omit the $\sum_{i=1}^{m}$.

$$X(f)(p) := X(p)(f), \qquad \forall\, p \in U. \tag{8.3}$$

Once again, if X is just any map satisfying the section property then $X(f)$ will not in general even be continuous. However if X is smooth (i.e. a vector field) then $X(f)$ is smooth.

Proposition 8.2. *Let M be a smooth manifold and let $W \subset M$ be a non-empty open set. Let $X\colon W \to TM$ be any function satisfying the section property* (8.1). *Then the following are equivalent.*

(i) *X is a vector field on W.*
(ii) *If $x\colon U \to O$ is any chart on M with $U \subset W$ then the functions X^i defined in* (8.2) *belong to $C^\infty(U)$.*
(iii) *If $V \subset W$ is any open set (possibly equal to all of W) and $f \in C^\infty(V)$ then the function $X(f)$ defined by* (8.3) *also belongs to $C^\infty(V)$.*

Proof. We begin by proving that (i) $\Leftrightarrow$ (ii).

Let $p \in W$, and let $x\colon U \to O$ be a chart about p. By definition, the function X^i defined in (8.2) is smooth if and only if $X^i \circ x^{-1}\colon O \to \mathbb{R}$ is smooth in the normal sense.

Note that by Proposition 3.8 and the definition of dx^i the function $X^i \circ x^{-1}$ can alternatively be written as

$$X^i \circ x^{-1} = dx^i \circ X \circ x^{-1} \tag{8.4}$$

The right-hand side of this equation should be understood as the function $q \mapsto dx^i_{x^{-1}(q)}\big(X(x^{-1}(q))\big)$ for $q \in O$.

Now let us recall from the proof of Theorem 5.6 that a chart $x\colon U \to O$ on M defines a chart $\tilde{x}\colon \pi^{-1}(U) \to O \times \mathbb{R}^m$ on TM by[3]

$$\tilde{x}(p, \xi) = \big(x(p), dx^i_p(\xi)\, e_i\big), \qquad p \in U,\ \xi \in T_pM.$$

By definition, X is smooth at p if and only if the composition

$$\tilde{x} \circ X \circ x^{-1}\colon O \to O \times \mathbb{R}^m$$

is smooth at $x(p)$. Explicitly this is the map

$$O \to O \times \mathbb{R}^m, \qquad q \mapsto \Big(q, dx^i_{x^{-1}(q)}\big(X(x^{-1}(q))\big)\, e_i\Big). \tag{8.5}$$

Applying (8.4) tells us that (8.5) can equivalently be written as

$$O \to O \times \mathbb{R}^m, \qquad q \mapsto \Big(q, X^i(x^{-1}(q))e_i\Big). \tag{8.6}$$

Thus (8.6) is smooth if and only if $X^i \circ x^{-1}$ is smooth for each $i = 1, \ldots, m$. This proves (i) $\Leftrightarrow$ (ii).

[3] Note how prettier this formula is with the Einstein Summation Convention in effect.

Now let us prove (ii) $\Rightarrow$ (iii). Let $V \subset W$ and let $f \in C^\infty(V)$. Choose a chart $x\colon U \to O$ with $U \subset V$. Then for $p \in U$, we have that

$$X(f)(p) = X^i(p) \left.\frac{\partial}{\partial x^i}\right|_p (f)$$

The function $p \mapsto \left.\frac{\partial}{\partial x^i}\right|_p (f)$ is smooth—this is just the function $p \mapsto D_i(f \circ x^{-1})(x(p))$. By (ii) the X^i are also smooth functions, and hence $X(f)$ is a finite sum of the pointwise product of smooth functions and hence is smooth. Thus $X(f)$ is smooth on U. But since U was arbitrary and smoothness is a local property, it follows that $X(f)$ is smooth on all of V. This proves (iii).

Finally we note that (ii) is a special case of (iii): if $x\colon U \to O$ is a chart about p with local coordinates (x^i), then the function X^i defined in (8.2) is simply the function $X(x^i)$, and thus (iii) implies X^i is smooth. This completes the proof. ∎

Warning: One must be careful with notation here: $X(x^i)$ is a function defined on U. Despite the suggestive notation, however, this is *not* the "composition" $X \circ x^i$. Indeed, the expression $X \circ x^i$ makes no sense at all, since x^i takes values in $\mathbb{R}$, and X cannot eat numbers. Thus

$$X(x^i) \neq X \circ x^i.$$

In contrast, the composition $X \circ x^{-1}$ from the left-hand side of (8.4) really does mean composition. Moreover one cannot "feed" x^{-1} to X to produce a function $X(x^{-1})$, since x^{-1} is not a smooth function on M.

The confusion could be avoided by simply declaring that we only use the notation $X(f)$ in the sense of a vector field eating a function, and only use the notation $X \circ x^{-1}$ to denote actual composition. In practice, however, whilst we will never write $X \circ f$ to denote the function $X(f)$, it is too cumbersome to only use composition notation for expressions such as $X \circ x^{-1}$. See for instance the right-hand sides of (8.5) and (8.6).

Thus in the future whenever you see the notation $X(w)$ for some object w, you should double check exactly what w is, before deciding on how to interpret $X(w)$.

EXAMPLE 8.3. Suppose $x\colon U \to O$ is a chart on M with local coordinates (x^i). Then we can think of $\frac{\partial}{\partial x^i}$ as defining a vector field on U via:

$$\frac{\partial}{\partial x^i}(p) := \left.\frac{\partial}{\partial x^i}\right|_p.$$

It is immediate from Proposition 8.2 that $\frac{\partial}{\partial x^i}$ is smooth.

We now introduce a notational convention that is both totally logical and somewhat confusing at the same time:

Definition 8.4. If $f \in C^\infty(U)$ then we denote the function $\frac{\partial}{\partial x^i}(f)$ from (8.3) with $X = \frac{\partial}{\partial x^i}$ by $\frac{\partial f}{\partial x^i}$. Thus $\frac{\partial f}{\partial x^i}$ is the function

$$\frac{\partial f}{\partial x^i}(p) := \frac{\partial}{\partial x^i}(p)(f) = \frac{\partial}{\partial x^i}\bigg|_p (f) = D_i(f \circ x^{-1})(x(p)).$$

If our given manifold is an open subset of $\mathbb{R}^m$ and x is the identity chart with local coordinates (u^i) then the notation $\frac{\partial f}{\partial u^i}$ is consistent with the "usual" definition of partial derivative.

Let us now continue with the general case, where $W \subset M$ is any non-empty open subset. The space $\mathfrak{X}(W)$ is a real vector space under pointwise addition:

$$(X + Y)(p) := X(p) + Y(p), \qquad (cX)(p) := cX(p)$$

In fact, $\mathfrak{X}(W)$ forms a module over the ring $C^\infty(W)$ by defining

$$(fX)(p) := f(p)X(p), \qquad X \in \mathfrak{X}(W),\ f \in C^\infty(W).$$

In order for this to be well-defined, one needs to know that e.g. $X + Y$ is smooth and fX is smooth. This however is immediate from Proposition 8.2.

Remark 8.5. Pay attention to the ordering. If $X \in \mathfrak{X}(W)$ and $f \in C^\infty(W)$ then $X(f)$ belongs to $C^\infty(W)$ whereas fX belongs to $\mathfrak{X}(W)$!

We now extend Definition 3.1 to derivations that are not based at a point.

Definition 8.6. Let M be a smooth manifold and let $W \subset M$ be a non-empty open set. A **derivation on** $C^\infty(W)$ is a linear map[4]

$$\mathcal{X} \colon C^\infty(W) \to C^\infty(W)$$

satisfying the *derivation property*

$$\mathcal{X}(fg) = f\mathcal{X}(g) + g\mathcal{X}(f), \qquad \forall f, g \in C^\infty(W).$$

Let us temporarily denote by $\mathfrak{X}^{\mathrm{deriv}}(W)$ the set of derivations on W. Observe that $\mathfrak{X}^{\mathrm{deriv}}(W)$ is another module over $C^\infty(W)$. It follows from Proposition 3.3 that any vector field $X \in \mathfrak{X}(W)$ defines a derivation $\mathcal{X} \in \mathfrak{X}^{\mathrm{deriv}}(W)$ via

$$\mathcal{X}(f) := X(f)$$

In fact, the converse is true.

Proposition 8.7. *Let M be a smooth manifold and let $W \subset M$ be a non-empty open set. Then $\mathfrak{X}^{\mathrm{deriv}}(W)$ and $\mathfrak{X}(W)$ are isomorphic as modules over $C^\infty(W)$.*

[4] Note that by "linear map" we always mean an $\mathbb{R}$-linear map, rather than a $C^\infty(W)$-linear map.

Proof. Suppose $\mathcal{X}$ is a derivation on $C^\infty(W)$. Fix $p \in W$. Then $\mathcal{X}$ defines a derivation on $C^\infty(W)$ at p, which we suggestively write as $X(p)$, via the formula

$$X(p)(f) := \mathcal{X}(f)(p), \qquad \forall f \in C^\infty(W).$$

Proposition 3.3 tells us that we can then think of X as defining a map $W \to TM$ via $p \mapsto X(p)$. We claim that X is smooth, and hence defines a vector field on W. For this by part (iii) of Proposition 8.2, we need only check that $X(f)$ is smooth for any $f \in C^\infty(W)$. But by construction $X(f) = \mathcal{X}(f)$, which is smooth by assumption. It is then clear that the map between $\mathfrak{X}^{\mathrm{deriv}}(W)$ and $\mathfrak{X}(W)$ given by $X \to \mathcal{X}$ is $C^\infty(W)$-linear and hence an isomorphism of modules over $C^\infty(W)$. ■

From now on we will identify a vector field $X \in \mathfrak{X}(W)$ with the corresponding derivation $\mathcal{X}$ of $C^\infty(W)$ and write both with Latin letters X. We will also abandon the notation $\mathfrak{X}^{\mathrm{deriv}}(W)$ and just write $\mathfrak{X}(W)$. Our next goal is to turn $\mathfrak{X}(W)$ into an algebra, that is, to have a bilinear operation

$$\mathfrak{X}(W) \times \mathfrak{X}(W) \to \mathfrak{X}(W).$$

The naive guess would be to try composition of derivations:

$$X \circ Y \colon C^\infty(W) \to C^\infty(W), \qquad (X \circ Y)(f) := X(Y(f)).$$

Unfortunately, this is *not* a derivation. Indeed, if we take $f, g \in C^\infty(W)$ and compute:

$$\begin{aligned}(X \circ Y)(fg) &= X\Big(fY(g) + gY(f)\Big) \\ &= \Big(f(X \circ Y)(g) + g(X \circ Y)(f)\Big) + \Big(X(f)Y(g) + X(g)Y(f)\Big)\end{aligned}$$

However, observe that the "error" term $X(f)Y(g) + X(g)Y(f)$ is symmetric in X and Y. This means that if we consider the **commutator**

$$[X, Y] := X \circ Y - Y \circ X$$

then the error term cancels, and thus $[X, Y]$ is a derivation. We have thus justified the following definition.

Definition 8.8. Let $X, Y \in \mathfrak{X}(W)$. Then the commutator $[X, Y] := X \circ Y - Y \circ X$ is another derivation. We call $[X, Y]$ the **Lie bracket** of X and Y.

Remark 8.9. **Warning:** A few authors[5] define the Lie bracket with the *opposite* sign: $[X, Y] := Y \circ X - X \circ Y$. From a "high-level" point of view, this other sign convention is actually the "correct" one, but this requires a little bit of infinite-dimensional Lie group theory to understand, as we will explain at the end of Chapter 13. The convention we are using, namely $[X, Y] := X \circ Y - Y \circ X$, is consistent with the majority of the literature.

[5] Notably, Joel Robbin and Dietmar Salamon use the other sign convention in their wonderful lecture notes [3].

The next proposition gives a formula for $[X, Y]$ in coordinates. The proof is deferred to Problem Sheet D.

PROPOSITION 8.10. *Let (U, x) be a chart on M with local coordinates (x^i) and let $X, Y \in \mathfrak{X}(U)$. Write $X = X^i \frac{\partial}{\partial x^i}$ and $Y = Y^i \frac{\partial}{\partial x^i}$. Then*

$$[X, Y] = \left(X^i \frac{\partial Y^j}{\partial x^i} - Y^i \frac{\partial X^j}{\partial x^i} \right) \frac{\partial}{\partial x^j},$$

where $\frac{\partial Y^j}{\partial x^i}$ and $\frac{\partial X^j}{\partial x^i}$ are the functions from Definition 8.4.

In order to explain the name, we need an algebraic definition.

DEFINITION 8.11. A (real) **Lie algebra** is a vector space $\mathfrak{g}$ endowed with a bilinear operation called the **Lie bracket**

$$\mathfrak{g} \times \mathfrak{g} \to \mathfrak{g}, \qquad (v, w) \mapsto [v, w]$$

which in addition is antisymmetric, $[v, w] = -[w, v]$ and satisfies the **Jacobi identity**

$$[u, [v, w]] + [v, [w, u]] + [w, [u, v]] = 0, \qquad \forall\, u, v, w \in \mathfrak{g}.$$

Thus a Lie algebra is a non-associative algebra. The name "Lie" comes from the Norwegian mathematician Sophus Lie [13]. It is traditional to write Lie algebras using Fraktur symbols $\mathfrak{g}$ and $\mathfrak{h}$. The **dimension** of a Lie algebra $\mathfrak{g}$ is simply the dimension of $\mathfrak{g}$ as a vector space. If $\mathfrak{g}$ is a Lie algebra then a linear subspace $\mathfrak{h} \subset \mathfrak{g}$ is called a **Lie subalgebra** if $[v, w] \in \mathfrak{h}$ for all $v, w \in \mathfrak{h}$.

EXAMPLE 8.12. Here are some examples of Lie algebras:

(i) The cross product $[x, y] := x \times y$ makes $\mathbb{R}^3$ into a 3-dimensional Lie algebra.
(ii) The set $\mathrm{Mat}(n)$ of $n \times n$ matrices is an n^2-dimensional Lie algebra under the normal commutator $[A, B] := AB - BA$.
(iii) If V is any vector space then we can turn V into a (rather boring) Lie algebra by defining $[v, w] := 0$. Such an Lie algebra is called **abelian**.

You will probably not be surprised to learn we have just constructed another example:

Theorem 8.13 *Let M be a smooth manifold and let $W \subset M$ be an open set. Then $\mathfrak{X}(W)$ is a Lie algebra.*

Proof. The only thing left to check is the Jacobi identity. This is Problem D.3 on Problem Sheet D. ∎

REMARK 8.14. As long as $\dim M > 0$ then for any non-empty open subset W, $\mathfrak{X}(W)$ is an infinite-dimensional Lie algebra.

To see this we pick a chart U inside W such that $\overline{U}$ is compact and a subset of W. Denote by $C_c^\infty(U)$ the class of smooth functions on U whose support $\mathrm{supp}(f)$ is a compact subset of U. Since U is infinite, Lemma 3.2 implies that $C_c^\infty(U)$ is infinite-dimensional (the function constructed in Lemma 3.2 is a function in $C^\infty(M)$ with support in U, but the support is automatically compact since $\overline{U}$ is compact).

Note that any $f \in C^\infty(U)$ can be viewed as an element of $C^\infty(W)$ by extending f by 0 in the complement of its support; similarly for any $X \in \mathfrak{X}(W)$ and $f \in C_c^\infty(U)$ the vector field $fX \in \mathfrak{X}(U)$ can be viewed as a vector field on W.

Thus since $\dim M > 0$, we can use the coordinate vector field $\frac{\partial}{\partial x_1} \in \mathfrak{X}(U)$ to define a linear injective map $C_c^\infty(U) \to \mathfrak{X}(W)$, $f \mapsto f\frac{\partial}{\partial x_1}$. Since we saw that $C_c^\infty(U)$ is infinite dimensional, so is $\mathfrak{X}(W)$.

We conclude this chapter by looking at how functions and vector fields can be "pushed forward" with a diffeomorphism.

Definition 8.15. Let $\varphi\colon M \to N$ be a diffeomorphism. We define an algebra homomorphism

$$\varphi_*\colon C^\infty(M) \to C^\infty(N), \qquad f \mapsto \varphi_*(f)$$

where

$$\varphi_*(f) := f \circ \varphi^{-1}.$$

Warning! Many authors use the notation φ_* for the derivative $D\varphi$.

The claim that φ_* is an algebra homomorphism is just the assertion that

$$\varphi_*(f+g) = \varphi_*(f) + \varphi_*(g), \qquad \varphi_*(fg) = \varphi_*(f)\varphi_*(g), \qquad \varphi_*(cf) = c\varphi_*(f)$$

for all $f, g \in C^\infty(M)$ and $c \in \mathbb{R}$, which is immediate from the definitions.

Definition 8.16. Suppose $\varphi\colon M \to N$ is a diffeomorphism and $X \in \mathfrak{X}(M)$. We define the **pushforward** vector field $\varphi_* X \in \mathfrak{X}(N)$ by defining

$$(\varphi_* X)(q) := D\varphi(\varphi^{-1}(q))X(\varphi^{-1}(q)).$$

To check this is well defined, we need to know that (a) $(\varphi_* X)(q) \in T_q N$ for each $q \in N$, which is obvious, and (b) that $\varphi_* X\colon N \to TN$ is smooth. The latter holds because it is simply the composition

$$N \xrightarrow{\varphi^{-1}} M \xrightarrow{X} TM \xrightarrow{D\varphi} TN$$

of smooth maps, and hence is smooth.

The map φ_* is again linear:

$$\varphi_*(X+Y) = \varphi_* X + \varphi_* Y, \qquad \forall\, X, Y \in \mathfrak{X}(M).$$

Moreover one has

$$\varphi_*(fX) = \varphi_*(f)\varphi_* X, \qquad \forall\, X \in \mathfrak{X}(M),\ \forall\, f \in C^\infty(M).$$

Remark 8.17. It may at first seem confusing that we have defined two different maps (one from functions to functions and one from vector fields to vector fields) and called them *both* φ_*. The reason for this will become clear when we discuss the **tensor algebra** $\mathcal{T}(M)$ of a manifold. Roughly speaking, the tensor algebra is a big direct sum:

$$\mathcal{T}(M) = \bigoplus_{h,k\geq 0} \mathcal{T}^{h,k}(M)$$

where $\mathcal{T}^{h,k}(M)$ denotes the *tensors of type* (h, k). As we will eventually see, a tensor of type $(0, 0)$ is simply a function, i.e.

$$\mathcal{T}^{0,0}(M) = C^\infty(M),$$

whereas a tensor of type $(1, 0)$ is a vector field:

$$\mathcal{T}^{1,0}(M) = \mathfrak{X}(M).$$

Given a diffeomorphism $\varphi\colon M \to N$, in Chapter 21 we will construct a single isomorphism

$$\varphi_*\colon \mathcal{T}(M) \to \mathcal{T}(N) \tag{8.7}$$

that preserves type, i.e.

$$\varphi_* \mathcal{T}^{h,k}(M) \subset \mathcal{T}^{h,k}(N).$$

The map φ_* from Definition 8.15 is the restriction of the isomorphism φ_* from (8.7) to $\mathcal{T}^{0,0}(M) \subset \mathcal{T}(M)$ and the map φ_* from Definition 8.16 is the restriction of the master φ_* from (8.7) to $\mathcal{T}^{1,0}(M) \subset \mathcal{T}(M)$. Thus it makes sense to denote them both by φ_*.

Definition 8.18. Let $\mathfrak{g}$ and $\mathfrak{h}$ be two Lie algebras. A **Lie algebra homomorphism** is a linear map $\ell\colon \mathfrak{g} \to \mathfrak{h}$ which respects the Lie brackets, i.e.

$$[\ell v, \ell w] = \ell[v, w], \qquad \forall\, v, w \in \mathfrak{g},$$

where the left-hand side is the Lie bracket in $\mathfrak{h}$ and the right-hand side is the Lie bracket in $\mathfrak{g}$. A **Lie algebra isomorphism** is a bijective Lie algebra homomorphism whose inverse is also a Lie algebra homomorphism.

Proposition 8.19. *Let $\varphi\colon M \to N$ be a diffeomorphism. Then $\varphi_*\colon \mathfrak{X}(M) \to \mathfrak{X}(N)$ is a Lie algebra isomorphism.*

Proposition 8.19 is a special case of part (ii) of Problem D.6.

Chapter 9
Flows

In this chapter we make contact with the theory of ordinary differential equations. A vector field defines an ordinary differential equation on a manifold, and just as in the Euclidean case, solutions to this ordinary differential equation exist (at least for a short time) and are unique. As with the Implicit Function Theorem, the local version of this statement follows readily from the corresponding Euclidean statement, but the global version is deeper. We begin by recalling two theorems from the theory of ordinary differential equations.

Theorem 9.1 (Existence of Solutions) *Let $O \subset \mathbb{R}^m$ be open and let $f\colon O \to \mathbb{R}^m$ be smooth. For any $p \in O$ there exists a neighbourhood V of p and an open interval (a, b) with $a < 0 < b$, together with a smooth map $h\colon (a, b) \times V \to O$ such that:*

(i) *$h(0, q) = q$, for all $q \in V$,*
(ii) *If we write*

$$\frac{d}{dt}h(t, q) := \lim_{s \to 0} \frac{h(t + s, q) - h(t, q)}{s}$$

then

$$\frac{d}{dt}h(t, q) = f(h(t, q)), \qquad \forall\, (t, q) \in (a, b) \times V.$$

We deliberately write $\frac{d}{dt}$ and not $\frac{\partial}{\partial t}$ to emphasise that this is ordinary differentiation.

Theorem 9.1 can be interpreted as follows. Suppose

$$\gamma = (\gamma^1, \dots, \gamma^m)\colon (a, b) \to O$$

is a smooth curve. One calls γ an **integral curve** of $f = (f^1, \dots, f^m)$ if

$$(\gamma^i)'(t) = f^i \circ \gamma(t), \qquad \forall\, 1 \le i \le n. \tag{9.1}$$

W. J. Merry, *Lectures on Differential Geometry I*,
https://doi.org/10.1007/978-3-032-03733-6_9

Thus Theorem 9.1 tells us that integral curves $\gamma(t) = h(t, q)$ exist for arbitrary initial conditions $\gamma(0) = q$, and depend smoothly on their initial conditions. Moreover, they all locally exist for a common time (i.e. for every q in V, the integral curve with initial condition q lasts for all $t \in (a, b)$. Next, we address uniqueness of solutions.

Theorem 9.2 (Uniqueness of Solutions) *Let $\mathcal{O} \subset \mathbb{R}^m$ be open and let $f : \mathcal{O} \to \mathbb{R}^m$ be smooth. If $\gamma, \delta : (a, b) \to \mathcal{O}$ are two integral curves of f with $\gamma(t) = \delta(t)$ for some $t \in (a, b)$ then $\gamma \equiv \delta$.*

We will not prove either Theorem 9.1 or Theorem 9.2. They are both hopefully familiar to you from previous courses on ordinary differential equations. Instead, we will generalise them to manifolds.

DEFINITION 9.3. Let M be a manifold and let X be a vector field on M. Let $(a, b) \subset \mathbb{R}$ be an interval, and suppose $\gamma : (a, b) \to M$ is a smooth map. We say that γ is an **integral curve** of X if

$$\dot{\gamma}(t) = X(\gamma(t)), \qquad \forall\, t \in (a, b). \tag{9.2}$$

This definition is consistent with the usual one (9.1) in the special case where $M = \mathcal{O}$ is an open subset of $\mathbb{R}^m$. See Problem D.1. Before stating the next result, let us introduce a convention.

DEFINITION 9.4. If M is a manifold and (a, b) is an interval then $(a, b) \times M$ is also a manifold. Given $p \in M$ we denote by $c_p : (a, b) \to (a, b) \times M$ the smooth curve in $(a, b) \times M$ defined by

$$c_p(t) := (t, p).$$

We denote by

$$\frac{\partial}{\partial t}\bigg|_{(t,p)} := Dc_p(t)\left(\frac{\partial}{\partial t}\bigg|_t\right) = \dot{c}_p(t) \tag{9.3}$$

the tangent vector in $T_{(t,p)}((a, b) \times M)$ obtained from the canonical generator $\frac{\partial}{\partial t}\big|_t \in T_t\mathbb{R}$. One can think of $(t, p) \mapsto \frac{\partial}{\partial t}\big|_{(t,p)}$ as defining a vector field on $(a, b) \times M$. (Exercise: Check this.)

Theorem 9.5 (Local Flow) *Let M be a smooth manifold and let $X \in \mathfrak{X}(M)$. For any $p \in M$ there exists a neighbourhood W of p and an interval (a, b) with $a < 0 < b$, together with a smooth map*

$$\Phi^{\mathrm{loc}} : (a, b) \times W \to M.$$

such that,

(i) $\Phi^{\mathrm{loc}}(0, q) = q$, *for all* $q \in W$.
(ii) *For all* $(t, q) \in (a, b) \times W$ *one has*

$$D\Phi^{\mathrm{loc}}(t, q)\left(\frac{\partial}{\partial t}\bigg|_{(t,q)}\right) = X(\Phi^{\mathrm{loc}}(t, q)). \tag{9.4}$$

We call Φ^{loc} a **local flow** of X. We will shortly get rid of the "loc".

Proof. Let $x: U \to O$ be a chart around p with local coordinates (x^i). Let $\tilde{x}: \pi^{-1}(U) \to O \times \mathbb{R}^m$ denote the corresponding chart on TM. Then we can write (cf. (8.5))

$$\tilde{x} \circ X \circ x^{-1} = (\mathrm{id}, f)$$

where $f: O \to \mathbb{R}^m$ is smooth.[1] Theorem 9.1 gives us a neighbourhood V of $x(p)$, an interval (a, b), and a smooth map $h: (a, b) \times V \to O$ such that the two stated conditions hold. To complete the proof, set $W := x^{-1}(V)$ and define

$$\Phi^{\mathrm{loc}}(t, q) := x^{-1} \circ h(t, x(q)), \qquad (t, q) \in (a, b) \times W.$$

That Φ^{loc} satisfies the two required conditions is immediate from the fact that h did. ■

Remark 9.6. The condition (9.4) is simpler than it looks. Given $q \in W$, set $\gamma_q(t) := \Phi^{\mathrm{loc}}(t, q)$, so that $\gamma_q: (a, b) \to U$ is a curve in M. Then by definition

$$\dot{\gamma}_q(t) = D\Phi^{\mathrm{loc}}(t, q)\left(\frac{\partial}{\partial t}\bigg|_{(t,q)} \right),$$

and so (9.4) asserts that γ_q is an integral curve of X.

A similar argument also proves the manifold version of Theorem 9.2:

Theorem 9.7 *Let M be a smooth manifold and let $X \in \mathfrak{X}(M)$. If $\gamma, \delta: (a, b) \to M$ are two integral curves of X with $\gamma(t) = \delta(t)$ for some $t \in (a, b)$ then $\gamma \equiv \delta$.*

Thanks to Theorem 9.7, it makes sense to talk about the maximal integral curve through a given point.

Definition 9.8. Let X be a vector field on M. Given a point $p \in M$, we denote by $\big(t^-(p), t^+(p)\big)$ the maximal interval around 0 on which the (unique by Theorem 9.7) integral curve $\gamma_p: \big(t^-(p), t^+(p)\big) \to M$ of X whose initial condition is $\gamma_p(0) = p$ is defined. We call γ_p the **maximal integral curve through** p.

Lemma 9.9. *Let X be a vector field on M. Fix $p \in M$ and $s \in (t^-(p), t^+(p))$ and set $q := \gamma_p(s)$. Then*

$$t^{\pm}(\gamma_p(s)) = t^{\pm}(p) - s, \qquad \forall\, s \in \big(t^-(p), t^+(p)\big). \tag{9.5}$$

and

$$\gamma_q(t) = \gamma_p(t + s), \qquad \forall\, t \in (t^-(q), t^+(q)). \tag{9.6}$$

Proof. This follows from uniqueness. The curve $t \mapsto \gamma_p(t + s)$ is an integral curve for X with initial condition q, and hence it is γ_q. Thus (9.6) follows, and hence so does (9.5). ■

[1] Compare this to Problem D.1.

We emphasise that $\big(t^-(p), t^+(p)\big)$ typically will be larger than the interval (a, b) given by Theorem 9.5—indeed, by construction $\Phi^{\mathrm{loc}}(t, p)$ never leaves the open set U that the chart x was defined on. Thus whilst $\gamma_p(t) = \Phi^{\mathrm{loc}}(t, p)$ for small enough t, in general the curve γ_p could wander all over the manifold.

Here is the global version of Theorem 9.5, which extends Φ to all of M so that the equality $\gamma_p(t) = \Phi(t, p)$ holds whenever the former is defined.

Theorem 9.10 (Maximal Flow) *Let M be a smooth manifold and let $X \in \mathfrak{X}(M)$. There exists a unique open set $\mathcal{D}(X) \subset \mathbb{R} \times M$ and a unique smooth map $\Phi \colon \mathcal{D}(X) \to M$ such that*

(i) *For all $p \in M$ one has*

$$\mathcal{D}(X) \cap \big(\mathbb{R} \times \{p\}\big) = \big(t^-(p), t^+(p)\big) \times \{p\} .$$

(ii) $\Phi(t, p) = \gamma_p(t)$ *for all* $(t, p) \in \mathcal{D}(X)$.

We call Φ the **flow** of X. The proof is relegated to the bonus section at the end of this chapter, since it is rather fiddly.

We can also reverse the roles of t and p. Given $t \in \mathbb{R}$, set

$$M_t := \{p \in M \mid (t, p) \in \mathcal{D}(X)\} .$$

Then M_t is open in M and $M = \bigcup_{t>0} M_t$. Moreover there is a well-defined smooth map $\Phi_t \colon M_t \to M_{-t}$ given by

$$\Phi_t(p) := \Phi(t, p), \qquad p \in M_t$$

—the fact that Φ_t takes values in M_{-t} follows from Lemma 9.9. This map Φ_t is a diffeomorphism, since $\Phi_{-t} \colon M_{-t} \to M_t$ is an inverse. More generally, if $s, t \in \mathbb{R}$ then the domain of $\Phi_s \circ \Phi_t$ is contained in (though not necessarily equal to) M_{s+t}. If s and t have the same sign then we have equality. In any case, by Lemma 9.9 again one has $\Phi_s \circ \Phi_t = \Phi_{s+t}$ on the domain of $\Phi_s \circ \Phi_t$.

Restricting the maps Φ_t to open subsets of M is annoying. The next condition rules this out.

Definition 9.11. A vector field X is **complete** if the set $\mathcal{D}(X)$ from Theorem 9.10 is all of $\mathbb{R} \times M$. Equivalently, a vector field is complete if either (a) its integral curves exist for all time or (b) the maps $\Phi_t := \Phi(t, \cdot)$ are all diffeomorphisms of the entire manifold M.

Definition 9.12. We write $\mathrm{Diff}(M)$ for the set of diffeomorphisms $\varphi \colon M \to M$. Note that $\mathrm{Diff}(M)$ is actually a group under composition, where the identity element is just the identity map.

Remark 9.13. Assume that M is compact. Then more is true: the group $\mathrm{Diff}(M)$ can itself be given a (Fréchet) manifold structure. We will say more about this in the bonus section to Chapter 13.

Definition 9.14. A **one-parameter group of diffeomorphisms** is a smooth group homomorphism $\mathbb{R} \to \mathrm{Diff}(M)$. Writing this as $t \mapsto \Phi_t$, the smoothness property tells us that $t \mapsto \Phi_t$ is a smooth map from the manifold $\mathbb{R}$ to the (infinite-dimensional) manifold $\mathrm{Diff}(M)$,[2] and the group property tells us that

$$\Phi_0 = \mathrm{id}, \qquad \Phi_{s+t} = \Phi_s \circ \Phi_t, \qquad \forall\, s, t \in \mathbb{R}.$$

If $\{\Phi_t\}$ is a one-parameter group of diffeomorphisms then we define its **infinitesimal generator** as the (necessarily complete) vector field

$$X(p) := D\Phi(0, p)\left(\frac{\partial}{\partial t}\Big|_{(0,p)}\right), \tag{9.7}$$

where we wrote $\Phi(t, p) := \Phi_t(p)$ and used the convention from (9.3). Then the flow of X is simply the one-parameter group Φ_t.

Theorem 9.10 and Lemma 9.9 thus give us:

Proposition 9.15. *Let M be a smooth manifold. Then there is a bijective correspondence between one-parameter subgroups of diffeomorphisms and complete vector fields.*

Example 9.16. Perhaps the easiest example of a *non-complete* vector field is given by taking $M = \mathbb{R}^2 \setminus 0$ and taking $X = \frac{\partial}{\partial u^1}$. If $(u^1, u^2) \in \mathbb{R}^2 \setminus 0$ then the flow line passing through (u^1, u^2) takes the form: $(u^1, u^2) \mapsto (t + u^1, u^2)$. It is then obvious that something must go wrong if you take $(u^1, u^2) = (-1, 0)$ and try and flow forwards—indeed, if the flow existed for all time then at time $t = 1$ you would fall out the manifold through the hole...[3]

> From now on we will switch between the notations $\Phi(t, p)$, $\Phi_t(p)$ and $\gamma_p(t)$ whenever convenient.

Here is an easy way to guarantee completeness.

Lemma 9.17. *Let X be a vector field on M. Assume there exists $\varepsilon > 0$ such that $(-\varepsilon, \varepsilon) \subset \big(t^-(p), t^+(p)\big)$ for all $p \in M$. Then X is complete.*

Proof. If not, there exists some $p \in M$ such that either $t^+(p) < \infty$ or $t^-(p) > -\infty$. Assume the former (the proof in the other case is almost identical). Choose a number t_0 such that $t^+(p) - \varepsilon < t_0 < t^+(p)$. Set $p_0 := \gamma_p(t_0)$. By assumption $\gamma_{p_0}(t)$ is defined for all $t \in (-\varepsilon, \varepsilon)$. Now consider the curve

$$\gamma(t) := \begin{cases} \gamma_p(t), & t^-(p) < t < t^+(p), \\ \gamma_{p_0}(t - t_0), & t_0 - \varepsilon < t < t_0 + \varepsilon. \end{cases}$$

These two definitions agree on the overlap, since

[2] In more down-to-earth language, this just means that $(t, p) \mapsto \Phi_t(p)$ is a smooth function $\mathbb{R} \times M \to M$.

[3] Exercise: Make this rigorous.

$$\begin{aligned}\gamma_{p_0}(t-t_0) &= \Phi_{t-t_0}(p_0)\\ &= \Phi_{t-t_0}\circ\Phi_{t_0}(p)\\ &= \Phi_t(p)\\ &= \gamma_p(t).\end{aligned}$$

This shows that γ is an integral curve for X with initial condition p which is defined on $\big(t^-(p), t_0+\varepsilon\big)$. Since $t_0+\varepsilon > t^+(p)$, this contradicts the maximality of $t^+(p)$. ■

We define the **support** of a vector field X in exactly the same way as we define the support of a function:

$$\operatorname{supp}(X) := \overline{\big\{p \in M \mid X(p) \neq 0 \in T_pM\big\}}.$$

COROLLARY 9.18. *Let X be a vector field with compact support. Then X is complete.*

Proof. By Theorem 9.5 for each $p \in \operatorname{supp}(X)$ there exists a neighbourhood U_p of p and an interval $(-\varepsilon_p, \varepsilon_p)$ such that the flow is defined on $(-\varepsilon_p, \varepsilon_p) \times U_p$. Since $\operatorname{supp}(X)$ is compact, we may select finitely many points $p_1, \dots, p_k$ such that

$$\operatorname{supp}(X) \subset \bigcup_{i=1}^{k} U_{p_i}.$$

Now set

$$\varepsilon := \min_{i=1,\dots,k} \varepsilon_{p_i}.$$

Then for every $p \in \operatorname{supp}(X)$ one has $(-\varepsilon, \varepsilon) \subset \big(t^-(p), t^+(p)\big)$. Since X is identically zero on $M \setminus \operatorname{supp}(X)$, every integral curve of X starting at some point in $M \setminus \operatorname{supp}(X)$ is trivially defined for all $t \in \mathbb{R}$ (it is the constant curve). Thus the hypotheses of Lemma 9.17 are satisfied, and the proof is complete. ■

COROLLARY 9.19. *If M is compact then every vector field on M is complete.*

Proof. If M is compact then certainly every vector field has compact support. ■

We conclude with another variant on Lemma 9.17, which is sometimes more useful.

LEMMA 9.20. *Let X be a vector field on M. If the maximal domain of an integral curve γ_p is not all of $\mathbb{R}$, then the image of that curve cannot be contained in any compact subset of M.*

Proof. Assume for instance that $t^+(p) < \infty$ and that γ_p is contained in a compact set K. Choose a sequence t_n such that $t_i \to t^+(p)$ from below. By compactness, $\gamma_p(t_i)$ converges to some point p_0. By Theorem 9.5, a local flow Φ^{loc} of X is defined on $(-\varepsilon, \varepsilon) \times U$ for some $\varepsilon > 0$ and some neighbourhood U of p_0. Choose i large enough so that $\gamma_p(t_i) \in U$ and $t_i + \varepsilon > t^+(p)$. Then arguing just as in the proof of Lemma 9.17, the curve

$$\gamma(t) := \begin{cases} \gamma_p(t), & t^-(p) < t < t^+(p), \\ \Phi^{\mathrm{loc}}(t - t_i, \gamma_p(t_i)), & t_i - \varepsilon < t < t_i + \varepsilon. \end{cases}$$

is a well-defined integral curve of X starting at p, and thus contradicts the maximality of $t^+(p)$. ■

Bonus Material for Chapter 9

In this bonus section we prove Theorem 9.10.

Proof of Theorem 9.10. Note that (i) determines $\mathcal{D}(X)$ uniquely, and (ii) does the same for Φ. It remains therefore to show that $\mathcal{D}(X)$ is open and Φ is smooth. This however is somewhat trickier than it looks.

Fix $p \in M$ and let I denote the set of all $t \in \big(t^-(p), t^+(p)\big)$ for which there exists some neighbourhood of (t, p) contained in $\mathcal{D}(X)$ on which Φ is smooth. Since smoothness is an open condition, I is an open set. We will prove that I is non-empty and closed, whence it follows that $I = \big(t^-(p), t^+(p)\big)$.

Firstly, I is non-empty, since $0 \in I$ by Theorem 9.5. Now suppose $t_0 \in \overline{I}$. Set $p_0 := \gamma_p(t_0)$. We apply Theorem 9.5 at the point p_0 to obtain a local flow $\Phi^{\mathrm{loc}} \colon (a, b) \times U_0 \to M$ about p_0. Since t_0 belongs to the closure of I, we may choose $t_1 \in I$ close enough to t_0 such that $t_0 - t_1$ belongs to (a, b) and such that $\gamma_p(t_1)$ belongs to U_0 (here we are using the fact that γ_p is continuous at t_0 and that U_0 is a neighbourhood of p_0).

Since (a, b) is an interval, we can do a little better: we can choose an interval I_0 about t_0 such that $t - t_1 \in (a, b)$ for all $t \in I_0$. Finally, by continuity of Φ at (t_1, p), there exists a neighbourhood V of p such that $\Phi(\{t_1\} \times V) \subset U_0$.

We now claim that our original Φ is defined and smooth on all of $I_0 \times V$, so that in particular $t_0 \in I$. Indeed, if $t \in I_0$ and $q \in V$ then $t - t_1 \in (a, b)$ and $\Phi(t_1, q) \in U_0$. Thus $\Phi^{\mathrm{loc}}(t - t_1, \Phi(t_1, q))$ is defined and smooth. But the curve $s \mapsto \Phi^{\mathrm{loc}}(s - t_1, \Phi(t_1, q))$ is an integral curve of X which passes through $\Phi(t_1, q)$ at t_1. By uniqueness, this curve is $\Phi(t, q)$. Therefore

$$\Phi(t, q) = \Phi^{\mathrm{loc}}(t - t_1, \Phi(t_1, q))$$

is defined and smooth at (t, q).

We have thus shown that for all $p \in M$ and for all $t \in \big(t^-(p), t^+(p)\big)$, there exists a neighbourhood of (t, p) in M contained in $\mathcal{D}(X)$ on which Φ is smooth. Thus $\mathcal{D}(X)$ is open and $\Phi \colon \mathcal{D}(X) \to M$ is smooth. This completes the proof. ■

Problem Sheet D

A (♣) indicates the problem is non-examinable. This does not necessarily imply it is difficult.

PROBLEM D.1 Let $\mathcal{O} \subset \mathbb{R}^m$ be an open set.

(i) Prove that the dash-to-dot maps induce a diffeomorphism between $T\mathcal{O}$ and $\mathcal{O}\times\mathbb{R}^m$.
(ii) Prove that there is a bijective correspondence between vector fields on $\mathcal{O}$ and smooth functions $\mathcal{O} \to \mathbb{R}^m$. Namely, given a vector field X associate the function f defined by

$$f(p) := \mathcal{J}_p^{-1}(X(p)), \qquad \forall\, p \in \mathcal{O}.$$

(iii) Let X and f be associated as above, and let γ be a smooth curve in $\mathcal{O}$. Prove that γ is an integral curve of f in the sense of (9.1), i.e. $\gamma' = f(\gamma)$, if and only if γ is an integral curve of X in the sense of (9.2), i.e. $\dot{\gamma} = X(\gamma)$.

PROBLEM D.2 Let M be a smooth manifold, let $p \in M$, and let $\xi \in T_pM$. Let U be any open set containing p. Prove that there exists a vector field $X \in \mathfrak{X}(U)$ such that $X(p) = \xi$.

PROBLEM D.3 Let M be a smooth manifold and let $W \subset M$ be a non-empty open set. Prove that the Lie bracket $[\cdot,\cdot]$ on $\mathfrak{X}(W)$ satisfies the Jacobi identity.

PROBLEM D.4 Let M be a smooth manifold and let (U, x) be a chart on M with local coordinates (x^i). Fix $X, Y \in \mathfrak{X}(U)$, and write $X = X^i \frac{\partial}{\partial x^i}$ and $Y = Y^i \frac{\partial}{\partial x^i}$. Prove that

$$[X, Y] = \left(X^i \frac{\partial Y^j}{\partial x^i} - Y^i \frac{\partial X^j}{\partial x^i} \right) \frac{\partial}{\partial x^j},$$

where $\frac{\partial Y^j}{\partial x^i}$ and $\frac{\partial X^j}{\partial x^i}$ are the functions from Definition 8.4.

PROBLEM D.5 Let M be a smooth manifold and let $W \subset M$ be a non-empty open set. Let $X, Y \in \mathfrak{X}(W)$, and let $f, g \in C^\infty(W)$. Prove that

$$[fX, gY] = fg[X, Y] + fX(g)Y - gY(f)X.$$

PROBLEM D.6 Let $\varphi\colon M \to N$ be a smooth map. Let $X \in \mathfrak{X}(M)$ and $Y \in \mathfrak{X}(N)$. We say that X and Y are φ-**related** if

$$D\varphi(p)X(p) = Y(\varphi(p)), \qquad \forall\, p \in M.$$

Note that if φ is a diffeomorphism then any $X \in \mathfrak{X}(M)$ is φ-related to $\varphi_* X$.

(i) Prove that $X \in \mathfrak{X}(M)$ and $Y \in \mathfrak{X}(N)$ are φ-related if and only if for every open set $V \subset N$ and every smooth function $f \in C^\infty(V)$, one has

$$X(f \circ \varphi) = Y(f) \circ \varphi.$$

(ii) Let $X_i \in \mathfrak{X}(M)$ and $Y_i \in \mathfrak{X}(N)$ for $i = 1, 2$ be vector fields. Assume X_i is φ-related to Y_i for each $i = 1, 2$. Prove that $[X_1, X_2]$ is φ-related to $[Y_1, Y_2]$.

(♣)PROBLEM D.7 Let $M \subset N$ be an (immersed or embedded) submanifold and let $p \in M$. We say that a vector field $Y \in \mathfrak{X}(N)$ is **tangent to** M **at** p if $Y(p) \in T_pM \subset T_pN$. We say Y is **tangent to** M if it is tangent to M at every point $p \in M$.

(i) Assume $M \subset N$ is an embedded submanifold. Prove that $Y \in \mathfrak{X}(N)$ is tangent to M if and only if $Y(f)|_M \equiv 0$ for every function $f \in C^\infty(N)$ such that $f|_M \equiv 0$.
(ii) Now assume $M \subset N$ is merely an immersed submanifold. Let $\iota\colon M \hookrightarrow N$ denote the inclusion. Assume that $Y \in \mathfrak{X}(N)$ is tangent to M. Prove there exists a unique $X \in \mathfrak{X}(M)$ such that X is ι-related to Y.
(iii) Continue to assume that $M \subset N$ is an immersed submanifold. Suppose $Y_1, Y_2 \in \mathfrak{X}(N)$ are tangent to M. Prove that $[Y_1, Y_2]$ is tangent to M.

Bonus Problem(s) for Sheet D

These problem(s) are hard, and are included for enthusiasts only. Solutions will not be provided.

This question asks you to remove the compactness hypothesis from Theorem 7.2. This result is commonly known as the Weak Whitney Embedding Theorem.

(♣)PROBLEM D.8 Let M be a smooth manifold of dimension m. Prove that there exists a proper embedding $\varphi\colon M \to \mathbb{R}^{2m+1}$.

The next problem shows that the Strong Whitney Embedding Theorem 7.1 is sharp.

(♣)PROBLEM D.9 Let $m = 2^k$. Prove that $\mathbb{R}P^m$ does not smoothly embed in $\mathbb{R}^{2m-1}$.

Chapter 10
Lie Groups

We now move on to defining the **Lie derivative** associated to a vector field X. As with maps φ_* from the last chapter, we will actually give two definitions, one for the Lie derivative eating a function, and one for the Lie derivative eating a vector field. In Chapter 22, after we have discussed tensors (cf. Remark 8.17), we will unify these two definitions into a single Lie derivative that eats any tensor and spits out another tensor of the same type.

DEFINITION 10.1. Let $X \in \mathfrak{X}(M)$ with flow Φ_t. We define the **Lie derivative** of X to be the map

$$\mathcal{L}_X : C^\infty(M) \to C^\infty(M)$$

given by

$$(\mathcal{L}_X f)(p) := \lim_{t \to 0} \frac{f \circ \Phi_t(p) - f(p)}{t}.$$

To see that this is well-defined (i.e. why the limit exists and defines a smooth function), we prove:

LEMMA 10.2. $\mathcal{L}_X f = X(f)$.

Proof. From the definitions one has

$$\begin{aligned} X(f)(p) &= X(p)(f) \\ &= \dot{\gamma}_p(0)(f) \\ &= (f \circ \gamma_p)'(0). \end{aligned}$$

But then clearly

W. J. Merry, *Lectures on Differential Geometry I*,
https://doi.org/10.1007/978-3-032-03733-6_10

$$\begin{aligned}(f \circ \gamma_p)'(0) &= \lim_{t \to 0} \frac{f \circ \gamma_p(t) - f(p)}{t} \\ &= \lim_{t \to 0} \frac{f \circ \Phi_t(p) - f(p)}{t}.\end{aligned}$$

■

Now we define the Lie derivative on vector fields.

DEFINITION 10.3. Let $X \in \mathfrak{X}(M)$ have flow Φ_t. We define the **Lie derivative** of X to be the map

$$\mathcal{L}_X : \mathfrak{X}(M) \to \mathfrak{X}(M)$$

where

$$(\mathcal{L}_X Y)(p) := \lim_{t \to 0} \frac{D\Phi_{-t}(\Phi_t(p))Y(\Phi_t(p)) - Y(p)}{t}.$$

To see that this is well-defined (i.e. why the limit exists and defines a vector field) we prove:

PROPOSITION 10.4. *For $X, Y \in \mathfrak{X}(M)$ one has $\mathcal{L}_X Y = [X, Y]$.*

Proposition 10.4 also explains the name "Lie derivative". Before going any further, let us emphasise once more: the main point of the Lie derivative is that we will eventually extend this to an operator $\mathcal{L}_X : \mathcal{T}(M) \to \mathcal{T}(M)$ on the tensor algebra of M.[1] For now though, we can simply think of the Lie derivative as giving more insight into the Lie bracket; an example of this is Proposition 10.6 below.

The proof of Proposition 10.4 requires a preliminary lemma, which can be thought of as a manifold version of Lemma 3.7.

LEMMA 10.5. *Let $U \subset M$ be open and let $a < 0 < b$. Let $f : (a, b) \times U \to \mathbb{R}$ be a smooth function such that $f(0, p) = 0$ for all $p \in U$. Then there exists another smooth function $h : (a, b) \times U \to \mathbb{R}$ such that*

$$f(t, p) = t h(t, p), \qquad \left.\frac{\partial}{\partial t}\right|_{(0,p)} (f) = h(0, p), \qquad \forall\, (t, p) \in (a, b) \times U.$$

Proof. Simply define

$$h(t, p) := \int_0^1 \left.\frac{\partial}{\partial t}\right|_{(st,p)} (f)\, ds.$$

Then h is smooth. To see that $f(t, p) = t h(t, p)$ one considers the curve $\gamma(s) := f(st, p)$. Then

$$\begin{aligned}f(t, p) &= f(t, p) - f(0, p) \\ &= \gamma(1) - \gamma(0) \\ &= \int_0^1 \gamma'(s)\, ds.\end{aligned}$$

---

[1] See Chapter 22.

But by definition

$$\gamma'(s) = t \frac{\partial}{\partial t}\bigg|_{(st,p)}(f).$$

This completes the proof. ∎

We now prove Proposition 10.4.

Proof of Proposition 10.4. Fix $p \in M$. By Theorem 9.5, there exists $a < 0 < b$ and a neighbourhood U of p such that $(a, b) \times U \subset \mathcal{D}(X)$, the domain of Φ. Now fix $g \in C^\infty(M)$. We apply Lemma 10.5 to the function

$$f(t, y) := g(\Phi_t(y)) - g(y)$$

to obtain a function h, which, writing $h_t(q) := h(t, q)$, satisfies:

$$g \circ \Phi_t = g + th_t, \qquad h_0 = X(g),$$

where we used Lemma 10.2. Thus for another vector field Y we have

$$\begin{aligned} D\Phi_{-t}(\Phi_t(p))Y(\Phi_t(p))(g) &= Y(\Phi_t(p))(g \circ \Phi_{-t}) \\ &= Y(\Phi_t(p))(g - th_{-t}) \\ &= Y(g)(\Phi_t(p)) - tY(h_{-t})(\Phi_t(p)). \end{aligned}$$

We therefore have

$$\begin{aligned} (\mathcal{L}_X Y)(g)(p) &= \lim_{t\to 0} \frac{Y(g)(\Phi_t(p)) - (Y(g))(p)}{t} - \lim_{t\to 0} Y(h_{-t})(\Phi_t(p)) \\ &= \mathcal{L}_X(Y(g)(p) - Y(h_0)(p) \\ &= X(Y(g))(p) - Y(X(g))(p) \\ &= [X, Y](g)(p), \end{aligned}$$

where the second equality used the definition of the Lie derivative $\mathcal{L}_X$ applied to the function $Y(g)$ and the third equality used Lemma 10.2. Since p and g were arbitrary, this completes the proof. ∎

An application of Theorem 10.4 is the following result, whose proof is deferred to Problem Sheet E.

PROPOSITION 10.6. *Let X and Y be vector fields on M with flows Φ_t and Ψ_t respectively. Then $[X, Y] \equiv 0$ if and only if the two flows commute, i.e.*

$$\Phi_t \circ \Psi_s(p) = \Psi_s \circ \Phi_t(p)$$

for all s, t small and for all p in the domains of both $\Phi_t \circ \Psi_s$ and $\Psi_s \circ \Phi_t$.[2]

[2] In general the maps $\Phi_t \circ \Psi_s$ and $\Psi_s \circ \Phi_t$ will not have the same domain, but their restrictions to compact sets coincide for small s and t.

We now move onto to our next major topic—*Lie groups*—which will take most of the next four chapters. Lie groups are important in many areas of mathematics (not just geometry!), including representation theory, harmonic analysis, differential equations and more. Lie groups also crop up naturally in physics. On the classical side one has Noether's theorem [10], which states that every smooth symmetry of a physical system has a corresponding conservation law. In high-energy particle physics, Lie groups are an important ingredient of *gauge theory*. We will come back to gauge theory in Volume II when we study connections on principal bundles (in Chapter 12).

DEFINITION 10.7. A **Lie group** G is a smooth manifold that is also a group in the algebraic sense, with the property that the group multiplication

$$\mu \colon G \times G \to G, \qquad \mu(g, h) = gh,$$

and group inversion

$$i \colon G \to G, \qquad i(g) = g^{-1},$$

are both smooth maps.

CONVENTION. We typically use the letters G, H for Lie groups. We also follow the standard group theory notation and use the letters g, h to denote points, and e for the identity element.[3] This has two important consequences for our notational conventions:

- The dimension of a Lie group is *not* given by the corresponding lower-case letter. When necessary, the dimension will always be explicitly stated.
- When discussing Lie groups, we will never use the letters g, h to denote functions.

DEFINITION 10.8. A **Lie group homomorphism** $\varphi \colon G \to H$ is a smooth map $G \to H$ which is also a group homomorphism. A **Lie group isomorphism** is a Lie group homomorphism which is also a diffeomorphism.[4]

EXAMPLES 10.9. Here are some examples of Lie groups.

(i) $\mathbb{R}^m$ is a Lie group under addition.
(ii) $\mathbb{R} \setminus \{0\}$ is a Lie group under multiplication.
(iii) The set $\mathrm{GL}(m)$ of invertible $m \times m$ matrices is a Lie group under matrix multiplication. Indeed, it is a manifold of dimension m^2 (cf. Problem A.2). Multiplication is smooth because the matrix entries of a product AB are given by polynomials in the entries of A and B, and inversion is smooth by Cramer's rule.
(iv) If G is a Lie group and $H \subset G$ is an **open subgroup** (that is, a subgroup which is also an open set in G) then H naturally inherits a Lie group structure (cf. Proposition 1.15). Thus the set $\mathrm{GL}^+(m)$ of invertible matrices with positive determinant is a Lie group.
(v) The m-torus $T^m = \mathbb{R}^m / \mathbb{Z}^m$ is an abelian Lie group, where the group structure is induced by addition on $\mathbb{R}^m$. In fact, one can show that any compact abelian Lie group is (isomorphic to) a torus.

---

[3] **Exception**: If G is a matrix Lie group we use the letters A, B to denote matrices and denote the identity matrix by I.

[4] A bijective group homomorphism is necessarily a group isomorphism; hence a Lie group isomorphism is in particular a group isomorphism.

(vi) The same underlying smooth manifold can carry multiple Lie group structures. For instance, a different Lie group structure on $\mathbb{R}^3$ is given by

$$\mu(u, v) := \left(u^1 + v^1, u^2 + v^2, u^3 + v^3 + u^1 v^2\right).$$

This Lie group is known as the **Heisenberg group**. In order to see that this does indeed define a group structure, we identify $\mathbb{R}^3$ with upper triangular 3×3 matrices:

$$(u^1, u^2, u^3) \qquad \longleftrightarrow \qquad \begin{pmatrix} 1 & u^1 & u^3 \\ 0 & 1 & u^2 \\ 0 & 0 & 1 \end{pmatrix}$$

The group multiplication μ corresponds to normal matrix multiplication.

(vii) Not all smooth manifolds can be made into Lie groups. For instance, S^m admits a Lie group structure only for $m = 0, 1$ and 3. The reason for this is briefly discussed in Remark 13.15.

Definition 10.10. Let G be a Lie group and let $g \in G$. We define diffeomorphisms

$$l_g : G \to G, \qquad r_g : G \to G,$$

called the **left translation** by g and the **right translation** by g respectively, defined by

$$l_g(h) := gh, \qquad r_g(h) := hg.$$

To see that these maps are diffeomorphisms, note that l_g is the composition of the smooth map $h \mapsto (g, h)$ and the group multiplication $\mu : G \times G \to \mathbb{R}$. The inverse of l_g is $l_{g^{-1}}$.

Remark 10.11. Throughout this chapter we will almost exclusively work with left translations. This is purely a convention—everything we do could be reformulated (with appropriate modifications) to work with right translations instead.

Proposition 10.12. *Every Lie group homomorphism has constant rank.*

Proof. Let $\varphi : G \to H$ be a Lie group homomorphism. Fix $g \in G$. We show that φ has the same rank at g as it does at e. Since φ is a group homomorphism, the following commutes:

$$\begin{array}{ccc} G & \xrightarrow{\;l_g\;} & G \\ {\scriptstyle\varphi}\downarrow & & \downarrow{\scriptstyle\varphi} \\ H & \xrightarrow[\;l_{\varphi(g)}\;]{} & H \end{array}$$

Indeed, if we fix $h \in G$ then

$$\varphi(l_g(h)) = \varphi(gh) = \varphi(g)\varphi(h) = l_{\varphi(g)}(\varphi(h)).$$

Since h was arbitrary this shows that the diagram commutes. Differentiating the equation $\varphi \circ l_g = l_{\varphi(g)} \circ \varphi$ at e and applying the chain rule for manifolds (Proposition 4.2) we obtain

$$D\varphi(g) \circ Dl_g(e) = Dl_{\varphi(g)}(e) \circ D\varphi(e). \tag{10.1}$$

Since l_g and $l_{\varphi(g)}$ are diffeomorphisms, both $Dl_g(e)$ and $Dl_{\varphi(g)}$ are linear isomorphisms. The claim follows. ∎

COROLLARY 10.13. *A Lie group homomorphism is a Lie group isomorphism if and only if it is bijective.*

Proof. This follows immediately from Proposition 10.12 and Problem C.7. ∎

DEFINITION 10.14. Let G be a Lie group. A **Lie subgroup** of G is a subgroup H endowed with a topology and a smooth structure that simultaneously makes H into a Lie group and into an *immersed* submanifold of G.

In fact, *embedded* submanifolds are automatically Lie subgroups.

PROPOSITION 10.15. *Let G be a Lie group, and let $H \subset G$ be a subgroup which is an embedded submanifold. Then H is a Lie subgroup.*

Proof. We need to check that H is a Lie group in its own right. Thus for instance we must show that the group multiplication $\mu \colon H \times H \to H$ is smooth. For this we need the following two facts:

- If $M \subset N$ is an immersed submanifold and $\varphi \colon N \to L$ is smooth then $\varphi|_M \colon M \to L$ is also smooth. (*Proof*: the inclusion map $\iota \colon M \to N$ is smooth by definition of an immersed submanifold, and $\varphi|_M = \varphi \circ \iota$.)
- If $M \subset N$ is an embedded submanifold[5] and $\varphi \colon L \to N$ is a smooth map with $\varphi(L) \subset M$ then $\varphi \colon L \to M$ is also smooth. (*Proof*: This is immediate from the definition of the subspace topology.)

Going back to the proof, from the first bullet point, $\mu|_{H\times H} \colon H \times H \to G$ is smooth. Since H is a subgroup, $\mu(H \times H) \subset H$. By the second bullet point, $\mu|_{H\times H} \colon H \times H \to H$ is smooth. A similar argument applies for inversion. ∎

The following result is much deeper. It is not that difficult to prove, but it would take the entire chapter (and then some), so we will skip it.

Theorem 10.16 (The Closed Subgroup Theorem) *Let G be a Lie group and suppose H is any subgroup of G. The following are equivalent*:

(i) *H is a closed subgroup (i.e. H is a closed set in G).*
(ii) *H is an embedded submanifold of G.*
(iii) *H is an embedded Lie subgroup of G.*

Clearly (iii) implies (ii). Proposition 10.15 proved that (ii) implies (iii). On Problem Sheet E you are asked to prove that (ii) implies (i). The trickier bit is to show that (i) implies (ii), and this is what we will skip.

[5] This property will be formalised in Definition 15.1.

Definition 10.17. A **matrix Lie group** is a closed subgroup of $\mathrm{GL}(m)$.

The Closed Subgroup Theorem 10.16 tells us that any matrix Lie group is a Lie subgroup of $\mathrm{GL}(m)$, and hence a Lie group in its own right. Nevertheless, quoting the Closed Subgroup Theorem is overkill here, since one can typically reduce the problem to a simple application of the Implicit Function Theorem. We illustrate this principle using the example of the orthogonal group $\mathrm{O}(m)$.

Proposition 10.18. *The set of orthogonal matrices* $\mathrm{O}(m)$ *is a Lie subgroup of* $\mathrm{GL}(m)$ *of dimension* $\frac{1}{2}m(m-1)$.

Proof. By Proposition 10.15 we need only show that $\mathrm{O}(m)$ is an embedded submanifold. For this first consider the set $\mathrm{Sym}(m)$ of *symmetric* matrices. We can identify $\mathrm{Sym}(m) \cong \mathbb{R}^{\frac{m(m+1)}{2}}$, and thus $\mathrm{Sym}(m)$ is naturally a smooth manifold. Now consider the (obviously smooth) map $\varphi\colon \mathrm{GL}(m) \to \mathrm{Sym}(m)$ given by

$$\varphi(A) := AA^T,$$

where A^T denotes the transpose of A. Then $\mathrm{O}(m) = \varphi^{-1}(I)$, where I is the $m \times m$ identity matrix. Thus by the Implicit Function Theorem 6.10, we need only show that I is a regular value of φ, whence $\mathrm{O}(m)$ is an embedded submanifold of $\mathrm{GL}(m)$ of dimension $m^2 - \frac{1}{2}m(m+1) = \frac{1}{2}m(m-1)$. For $A \in \mathrm{O}(m)$ one has $\varphi \circ r_A = \varphi$, and thus

$$D\varphi(A) \circ Dr_A(I) = D\varphi(I).$$

Since r_A is a diffeomorphism, it follows that the rank of φ at A is the same as the rank of φ at I. Thus we need only show that φ has maximal rank at I, i.e. that $D\varphi(I)$ is surjective.

Since $\mathrm{GL}(m)$ is an open subset of the vector space $\mathrm{Mat}(m)$, its tangent space at I is canonically identified with $\mathrm{Mat}(m)$ via the dash-to-dot map $\mathcal{J}_I$, and similarly $T_I\,\mathrm{Sym}(m) \cong \mathrm{Sym}(m)$. Thus $D\varphi(I)$ induces a canonical linear map $\ell\colon \mathrm{Mat}(m) \to \mathrm{Sym}(m)$ such that the following commutes

$$\begin{array}{ccc} \mathrm{Mat}(m) & \xrightarrow{\ \ell\ } & \mathrm{Sym}(m) \\ \big\downarrow \mathcal{J}_I & & \big\downarrow \mathcal{J}_I \\ T_I\,\mathrm{Mat}(m) & \xrightarrow[D\varphi(I)]{} & T_I\,\mathrm{Sym}(m) \end{array}$$

Since $\mathcal{J}_I$ is an isomorphism, it suffices to show that ℓ is surjective. Take $A \in \mathrm{Mat}(m)$ and compute[6]

[6] Sanity check: $I + tA$ belongs to $\mathrm{GL}(m)$ for t small enough, so this expression makes sense.

$$
\begin{aligned}
D\varphi(I)\mathcal{J}_I(A) &= \frac{d}{dt}\bigg|_{t=0}\varphi(I+tA) \\
&= \frac{d}{dt}\bigg|_{t=0}\left(I + t(A+A^T) + t^2AA^T\right) \\
&= \mathcal{J}_I(A+A^T)
\end{aligned}
$$

Thus ℓ is the map $A \mapsto A + A^T$. This map is surjective, since if $S \in \mathrm{Sym}(m)$ then $\ell(\frac{1}{2}S) = S$. This completes the proof. ■

This technique works for all matrix Lie groups—the only trick is to find the right map φ. On Problem Sheet E you are asked to carry this out for the symplectic linear group.

Chapter 11
The Lie Algebra of a Lie Group

In this chapter we make the connection between Lie groups and Lie algebras. We begin by explaining how to produce a Lie algebra from a Lie group. The converse—producing a Lie group from a Lie algebra—is harder. This is the content of the famous *Lie Correspondence Theorem*, stated as Theorem 11.11 below. The proof of the Lie Correspondence Theorem will be given in Chapter 14.

DEFINITION 11.1. Let G be a Lie group. We define the **Lie algebra** of G, written as $\mathfrak{g}$, as the tangent space to G at the identity element e:

$$\mathfrak{g} := T_e G.$$

Of course, for this definition not to be completely insane, the Lie algebra of a Lie group had better be a Lie algebra in the sense of Definition 8.11. Luckily this is indeed the case, as we will prove in Corollary 11.6 below.

CONVENTION. The standard convention is that the Lie algebra of a given Lie group is written with the corresponding lower case Fraktur letter. This convention will be used throughout, often without comment. Thus for instance if H is a Lie group, the symbol $\mathfrak{h}$ should always be understood to denote $T_e H$, even if this is not explicitly stated. A similar convention applies for matrix Lie groups: we write $\mathfrak{gl}(m)$ for the Lie algebra of $\mathrm{GL}(m)$, $\mathfrak{o}(m)$ for the Lie algebra of $\mathrm{O}(m)$, and so on.

EXAMPLES 11.2. Here are some examples of Lie algebras of Lie groups.

(i) The Lie algebra of $\mathrm{GL}(m)$ is $\mathfrak{gl}(m) \cong \mathrm{Mat}(m)$.
(ii) The Lie algebra of $\mathrm{O}(m)$ is

$$\mathfrak{o}(m) := \left\{ A \in \mathfrak{gl}(m) \mid A + A^T = 0 \right\}.$$

This follows from Proposition 10.18 together with Proposition 6.15.

W. J. Merry, *Lectures on Differential Geometry I*,
https://doi.org/10.1007/978-3-032-03733-6_11

(iii) The Lie algebra of T^m is $\mathbb{R}^m$. Indeed, for $m = 1$ this is clear, and for $m > 1$ this follows from Problem C.1. More generally, the Lie algebra of any abelian Lie group is an abelian Lie algebra (and the converse holds if the Lie group is connected), as you will prove on Problem Sheets E and G.

The key to proving that the Lie algebra of a Lie group is indeed a Lie algebra is the following concept.

DEFINITION 11.3. Let G be a Lie group. A vector field $X \in \mathfrak{X}(G)$ is said to be **left-invariant** if $(l_g)_* X = X$ for all $g \in G$. Equivalently, this means that

$$Dl_g(h)X(h) = X(gh), \qquad \forall\, g, h \in G.$$

We denote by $\mathfrak{X}_l(G) \subset \mathfrak{X}(G)$ the set of left-invariant vector fields.

It is immediate that $\mathfrak{X}_l(G)$ is a linear subspace of $\mathfrak{X}(G)$. In fact, much more is true - the Lie bracket of two left-invariant vector fields is again left-invariant:

PROPOSITION 11.4. *Let G be a Lie group and let $X, Y \in \mathfrak{X}_l(G)$. Then $[X, Y]$ also belongs to $\mathfrak{X}_l(G)$. Consequently $\mathfrak{X}_l(G)$ is a Lie subalgebra of $\mathfrak{X}(G)$.*

Proof. Fix $g \in G$. Then by Proposition 8.19 one has

$$(l_g)_*[X, Y] = [(l_g)_* X, (l_g)_* Y] = [X, Y].$$

Since g was arbitrary, the result follows. ■

The next result is the main step needed to show that $\mathfrak{g}$ is a Lie algebra.

Theorem 11.5 *Let G be a Lie group with Lie algebra $\mathfrak{g}$. The evaluation map*

$$\mathrm{eval}_e \colon \mathfrak{X}_l(G) \to \mathfrak{g}, \qquad \mathrm{eval}_e(X) := X(e)$$

is a vector space isomorphism. Thus $\mathfrak{X}_l(G)$ is a vector space of the same dimension as G.

Proof. The map eval_e is clearly linear. If $\mathrm{eval}_e(X) = 0$ then X is identically zero, since for any $g \in G$ one has by left-invariance:

$$X(g) = Dl_g(e)X(e) = 0.$$

Thus we need only show that eval_e is surjective. For this, fix an arbitrary $\xi \in \mathfrak{g} = T_e G$. We define a map $X_\xi \colon G \to TG$ by

$$X_\xi(g) := Dl_g(e)\xi.$$

Then X_ξ certainly satisfies the section property (8.1), since $Dl_g(e)$ is a map $T_e G \to T_g G$. To show that X_ξ is a vector field, it suffices to show that $X_\xi(f)$ is smooth for any $f \in C^\infty(G)$. Fix such an f, and choose a smooth curve $\gamma \colon (-\varepsilon, \varepsilon) \to G$ such that $\gamma(0) = e$ and $\dot\gamma(0) = \xi$. Define a smooth map

$$\delta\colon (-\varepsilon,\varepsilon)\times G \to \mathbb{R}, \qquad \delta(t,g) := f\big(\mu(g,\gamma(t))\big),$$

where $\mu\colon G\times G \to G$ is the group multiplication. Write $\delta_g := \delta(\cdot, g)$, so $\delta_g = f \circ l_g \circ \gamma$; we think of δ_g as a family of smooth curves in $\mathbb{R}$ which depend smoothly on g. Following through the definitions, for any $g \in G$ one has

$$\begin{aligned} X_\xi(f)(g) &= X_\xi(g)(f) \\ &= \big(Dl_g(e)\xi\big)(f) \\ &= \xi(f \circ l_g) \\ &= \dot\gamma(0)(f \circ l_g) \\ &= \dot\delta_g(0). \end{aligned}$$

Since $g \mapsto \dot\delta_g(0)$ is smooth, this shows that $X_\xi(f)$ is smooth. Since f was arbitrary, it follows from Proposition 8.2 that X_ξ is smooth, and hence a vector field. Next, we claim X_ξ is left-invariant. Indeed, if $g, h \in G$ then

$$\begin{aligned} Dl_g(h)X_\xi(h) &= Dl_g(h) \circ Dl_h(e)\xi \\ &= D(l_g \circ l_h)(e)\xi \\ &= Dl_{gh}(e)\xi \\ &= X_\xi(gh). \end{aligned}$$

Thus $X_\xi \in \mathfrak{X}_l(G)$. Since $\mathrm{eval}_e(X_\xi) = X_\xi(e) = \xi$, this shows eval_e is surjective. The proof is complete. ■

Corollary 11.6. *Let G be a Lie group of dimension m. Then its Lie algebra is a Lie algebra (!) of dimension m.*

Proof. We need only define a Lie bracket on $\mathfrak{g}$. For this, using the notation from Theorem 11.5, we define

$$[\xi,\zeta] := \mathrm{eval}_e\big([X_\xi, X_\zeta]\big), \qquad \xi,\zeta \in \mathfrak{g}.$$

This works by Theorem 11.5 and Proposition 11.4. ■

For the next result, let us recall from Problem D.6 that if $\varphi\colon M \to N$ is a smooth map between manifolds, and $X \in \mathfrak{X}(M)$ and $Y \in \mathfrak{X}(N)$ then we say that X and Y are **φ-related** if

$$D\varphi(p)X(p) = Y(\varphi(p)), \qquad \forall\, p \in M.$$

Proposition 11.7. *Let $\varphi\colon G \to H$ be a Lie group homomorphism between two Lie groups. Then $D\varphi(e)\colon \mathfrak{g} \to \mathfrak{h}$ is a Lie algebra homomorphism.*

Proof. Let $\xi \in \mathfrak{g}$ and let $X_\xi \in \mathfrak{X}_l(G)$ denote the unique left-invariant vector field such that $X_\xi(e) = \xi$. Let $\zeta := D\varphi(e)\xi$ and let $Y_\zeta \in \mathfrak{X}_l(H)$ denote the unique left-invariant vector field such that $Y_\zeta(e) = \zeta$. We claim that X_ξ and Y_ζ are φ-related. Indeed, by (10.1) one has

$$
\begin{aligned}
D\varphi(g)X_{\xi}(g) &= D\varphi(g) \circ Dl_g(e)\xi \\
&= Dl_{\varphi(g)}(e) \circ D\varphi(e)\xi \\
&= Dl_{\varphi(g)}(e)\zeta \\
&= Y_{\zeta}(\varphi(g)).
\end{aligned}
$$

Now by part (ii) of Problem D.6, if $\xi_1, \xi_2 \in \mathfrak{g}$ and $\zeta_i := D\varphi(e)\xi_i$ then $[X_{\xi_1}, X_{\xi_2}]$ is φ-related to $[Y_{\zeta_1}, Y_{\zeta_2}]$. Evaluating both sides at e gives

$$D\varphi(e)[\xi_1, \xi_2] = [\zeta_1, \zeta_2].$$

This completes the proof. ∎

Suppose now H is a Lie subgroup of G. Let $\iota\colon H \hookrightarrow G$ denote the inclusion. Then since $D\iota(e)\colon \mathfrak{h} = T_eH \to \mathfrak{g} = T_eG$ is injective, we can regard $\mathfrak{h}$ as a linear subspace of $\mathfrak{g}$. A priori however, this identification might not respect the Lie brackets. Thanks to Proposition 11.7, however, it does:

Corollary 11.8. *Let $H \subset G$ be a Lie subgroup, and identify $\mathfrak{h}$ with its image in $\mathfrak{g}$. The Lie bracket on $\mathfrak{h}$ is simply the restriction of the Lie bracket on $\mathfrak{g}$ to $\mathfrak{h}$. Thus $\mathfrak{h}$ is a Lie subalgebra of $\mathfrak{g}$.*

Proof. Apply Proposition 11.7 with φ the inclusion (note that the roles of H and G have been reversed!) ∎

Let us go back to $\mathrm{GL}(m)$. We now have potentially two different Lie brackets on $\mathfrak{gl}(m) \cong \mathrm{Mat}(m)$: the one coming from Corollary 11.6, and the commutator bracket (cf. part (ii) of Example 8.12). The next result, whose proof is deferred to Problem Sheet E, tells us that these coincide.

Proposition 11.9. *The Lie bracket on $\mathfrak{gl}(m)$ is given by matrix commutation, i.e.*

$$[A, B] = AB - BA, \qquad \forall\, A, B \in \mathfrak{gl}(m).$$

Combining Corollary 11.8 and Proposition 11.9, we end up with:

Corollary 11.10. *Let G be a matrix Lie group. Then the Lie bracket on $\mathfrak{g}$ is given by matrix commutation.*

The Lie group-Lie algebra correspondence established in Corollary 11.8 goes both ways. This is the content of the next famous result. The proof will be given in Chapter 13 after we have proved the Frobenius Theorem.

Theorem 11.11 (The Lie Correspondence Theorem) *Let G be a Lie group with Lie algebra $\mathfrak{g}$. If $\mathfrak{h}$ is a Lie subalgebra of $\mathfrak{g}$ then there is a unique connected Lie subgroup H of G whose Lie algebra is $\mathfrak{h}$.*

We conclude this chapter by defining a special type of Lie subgroup.

DEFINITION 11.12. Let G be a Lie group. A **one-parameter subgroup** of G is a Lie group homomorphism $\mathbb{R} \to G$, where $\mathbb{R}$ is thought of as a Lie group under addition.

The relation between Definition 9.14 and Definition 11.12 will be explored in Chapter 13.

It turns out that one-parameter subgroups are easy to classify. The starting point for this classification is the following statement.

PROPOSITION 11.13. *Let G be a Lie group and let $X \in \mathfrak{X}_l(G)$. Then X is complete.*

Proof. We will show that there exists $\varepsilon > 0$ such that for all $g \in G$ the integral curve γ_g of X with initial condition is defined on at least $(-\varepsilon, \varepsilon)$. This implies the result, by Lemma 9.17.

Choose $\varepsilon > 0$ such that the integral curve γ_e is defined on $(-\varepsilon, \varepsilon)$. Fix $g \in G$ and let $\delta := l_g \circ \gamma_e$. Then $\delta(0) = g$, and moreover

$$\begin{aligned}\dot{\delta}(t) &= \frac{d}{ds}\Big|_{s=t} l_g(\gamma_e(s)) \\ &= Dl_g(\gamma_e(t))\dot{\gamma}_e(t) \\ &= Dl_g(\gamma_e(t))X(\gamma_e(t)) \\ &= X(g\gamma_e(t)) \\ &= X(\delta(t)),\end{aligned}$$

where the penultimate line used left-invariance. By uniqueness of integral curves (Theorem 9.7), it follows that $\delta \equiv \gamma_g$ where defined. Thus γ_g is defined on (at least) $(-\varepsilon, \varepsilon)$. The result follows. ■

NOTATION. For $\xi \in \mathfrak{g}$ we denote by $\gamma^\xi : \mathbb{R} \to G$ the integral curve of the left-invariant vector field X_ξ with initial condition $\gamma^\xi(0) = e$.

PROPOSITION 11.14. *Let G be a Lie group with Lie algebra $\mathfrak{g}$. Then each curve $\gamma^\xi : \mathbb{R} \to G$ is a one-parameter subgroup. Moreover any one-parameter subgroup is of this form.*

Proof. Fix $\xi \in \mathfrak{g}$. To show that γ^ξ is a one-parameter subgroup we must show that

$$\gamma^\xi(s + t) = \gamma^\xi(s)\gamma^\xi(t)$$

for all $s, t \in \mathbb{R}$, where on the right-hand side we use multiplication in G. For this, fix $s \in \mathbb{R}$ and consider the curve

$$\delta(t) := \gamma^\xi(s)^{-1}\gamma^\xi(s + t).$$

Then $\delta(0) = e$, and by the chain rule

$$\begin{aligned}\dot{\delta}(t) &= Dl_{\gamma^\xi(s)^{-1}}(\gamma^\xi(s+t))\dot{\gamma}^\xi(s+t)\\ &= Dl_{\gamma^\xi(s)^{-1}}(\gamma^\xi(s+t))X_\xi(\gamma^\xi(s+t))\\ &= X_\xi\big(\gamma^\xi(s)^{-1}\gamma^\xi(s+t)\big)\\ &= X_\xi(\delta(t)).\end{aligned}$$

where the penultimate line used left-invariance again. Thus by uniqueness of integral curves, one must have $\delta(t) = \gamma^\xi(t)$.

Conversely, suppose γ is a one-parameter subgroup. Let $\xi := \dot{\gamma}(0) \in \mathfrak{g}$. We claim that $\dot{\gamma}(t) = X_\xi(\gamma(t))$. Since $\gamma(t+s) = \gamma(t)\gamma(s) = l_{\gamma(t)}(\gamma(s))$, we have

$$\begin{aligned}\dot{\gamma}(t) &= \frac{d}{ds}\Big|_{s=0}\gamma(t+s)\\ &= \frac{d}{ds}\Big|_{s=0} l_{\gamma(t)}(\gamma(s))\\ &= Dl_{\gamma(t)}(\gamma(0))\dot{\gamma}(0)\\ &= Dl_{\gamma(t)}(e)\xi\\ &= X_\xi(\gamma(t)).\end{aligned}$$

where the last line used the definition of X_ξ. Thus again by uniqueness of integral curves we have $\gamma \equiv \gamma^\xi$. This completes the proof. ■

We can play a similar game by replacing $\xi \in \mathfrak{g}$ with a scalar multiple $s\xi$.

LEMMA 11.15. *For any $s, t \in \mathbb{R}$ one has*

$$\gamma^\xi(st) = \gamma^{s\xi}(t).$$

Proof. First note as $Dl_g(e)$ is a linear map one has for any $g \in G$ that

$$X_{s\xi}(g) = Dl_g(e)(s\xi) = sDl_g(e)\xi = sX_\xi(g).$$

Thus $X_{s\xi} = sX_\xi$. Now by the chain rule

$$\frac{d}{dt}\Big|_{t=t_0}\gamma^\xi(st) = s\dot{\gamma}^\xi(st_0) = sX_\xi(\gamma^\xi(st_0)) = X_{s\xi}(\gamma^\xi(st_0)).$$

Thus $t \mapsto \gamma^\xi(st)$ is an integral curve of $X_{s\xi}$ with initial condition e, and hence by uniqueness of integral curves once more, one has $\gamma^\xi(st) \equiv \gamma^{s\xi}(t)$. ■

Problem Sheet E

A (♣) indicates the problem is non-examinable. This does not necessarily imply it is difficult.

PROBLEM E.1 Let $J_0 \in \mathrm{Mat}(2n)$ denote the matrix

$$J_0 := \begin{pmatrix} 0 & -I \\ I & 0 \end{pmatrix},$$

where I is the $n \times n$ identity matrix. The **symplectic linear group** $\mathrm{Sp}(2n)$ consists of the matrices A such that $A^T J_0 A = J_0$. Prove that $\mathrm{Sp}(2n)$ is a Lie group. Compute its dimension, and compute its Lie algebra $\mathfrak{sp}(2n)$.

PROBLEM E.2 Let G be a Lie group with Lie algebra $\mathfrak{g}$. Suppose $\gamma \colon \mathbb{R} \to G$ is a smooth curve with $\gamma(0) = e$. Set $\xi := \dot{\gamma}(0)$. Prove that γ is a one-parameter subgroup if and only if

$$\Phi_t^{\xi} = r_{\gamma(t)}$$

where $\Phi_t^{\xi} : G \to G$ denotes the flow of the unique left-invariant vector field X_ξ satisfying $X_\xi(e) = \xi$, and this equality is an equality of diffeomorphisms of G.

PROBLEM E.3 Prove that the Lie bracket on $\mathfrak{gl}(n)$ is given by matrix commutation, i.e.

$$[A, B] = AB - BA, \qquad \forall\, A, B \in \mathfrak{gl}(n) = \mathrm{Mat}(n).$$

PROBLEM E.4 Let $\varphi \colon M \to N$ be a smooth map between two smooth manifolds. Let $X \in \mathfrak{X}(M)$ and $Y \in \mathfrak{X}(N)$, and assume X and Y are φ-related in the sense of Problem D.6. Let Φ_t and Ψ_t denote the respective flows, with domains $M_t \subset M$ and $N_t \subset N$ respectively. Prove that $\varphi(M_t) \subset N_t$ and that

$$\Psi_t \circ \varphi = \varphi \circ \Phi_t, \qquad \text{on } M_t.$$

Deduce that if φ is a diffeomorphism then for any vector field X with flow Φ_t, the flow of $\varphi_* X$ is given by $\varphi \circ \Phi_t \circ \varphi^{-1}$.

(♣)PROBLEM E.5 Let X and Y be vector fields on a smooth manifold M with flows Φ_t and Ψ_t respectively. Prove that $[X, Y] \equiv 0$ if and only if the two flows commute, i.e.

$$\Phi_t \circ \Psi_s(p) = \Psi_s \circ \Phi_t(p)$$

for all s, t small and for all p in the domains of both $\Phi_t \circ \Psi_s$ and $\Psi_s \circ \Phi_t$.

PROBLEM E.6 Prove that if a Lie group is abelian then its Lie algebra is abelian.

(♣)PROBLEM E.7 Let G be a Lie group and suppose H is a subgroup of G which is also an embedded submanifold. Prove that H is closed in G (as a subspace).

Bonus Problem(s) for Sheet E

These problem(s) are hard, and are included for enthusiasts only. Solutions will not be provided.

Both $\mathbb{R}^m$ and the torus T^m have the abelian Lie algebra $\mathbb{R}^m$. This shows that the functor $G \mapsto \mathfrak{g}$ from the category of Lie groups to the category of Lie algebras does not reflect isomorphisms. If however we restrict to the subcategory of simply connected Lie groups, this problem goes away:

(♣)PROBLEM E.8 Let G and H be simply connected Lie groups with Lie algebras $\mathfrak{g}$ and $\mathfrak{h}$. Assume $\mathfrak{g}$ and $\mathfrak{h}$ are isomorphic (as Lie algebras). Prove that G and H are isomorphic as Lie groups.

Chapter 12
Smooth Actions of Lie Groups

We begin this chapter by introducing the **exponential map** of a Lie group G, which will be a smooth map $\exp\colon \mathfrak{g} \to G$. The reason for the name will become apparent in Proposition 12.7—namely, the exponential map of a matrix Lie group is given by matrix exponentiation. As in the previous chapter, given $\xi \in \mathfrak{g}$, we denote by $\gamma^{\xi} : \mathbb{R} \to G$ the integral curve of X_{ξ} with initial condition $\gamma^{\xi}(0) = e$.

DEFINITION 12.1. Let G be a Lie group with Lie algebra $\mathfrak{g}$. The **exponential map** is the map

$$\exp\colon \mathfrak{g} \to G, \qquad \xi \mapsto \gamma^{\xi}(1).$$

The following result is an immediate corollary of Proposition 11.14, Lemma 11.15, and Problem E.2.

PROPOSITION 12.2 (PROPERTIES OF THE EXPONENTIAL MAP). *The exponential map* $\exp\colon \mathfrak{g} \to G$ *satisfies*:

(i) $\exp((s+t)\xi) = \exp(s\xi)\exp(t\xi)$ *for all* $\xi \in \mathfrak{g}$ *and* $s, t \in \mathbb{R}$,
(ii) $\exp(-t\xi) = (\exp(t\xi))^{-1}$ *for all* $\xi \in \mathfrak{g}$ *and* $t \in \mathbb{R}$,
(iii) *The map* $t \mapsto \exp(t\xi)$ *is precisely the one-parameter subgroup* $\gamma^{\xi}(t)$.
(iv) *The flow* Φ_t^{ξ} *of* X_{ξ} *is given by* $\Phi_t^{\xi} = r_{\exp(t\xi)}$.

We now prove that the exponential map is smooth, and compute its derivative at the origin.

Theorem 12.3 *The exponential map* $\exp\colon \mathfrak{g} \to G$ *is smooth. Moreover the following commutes*:

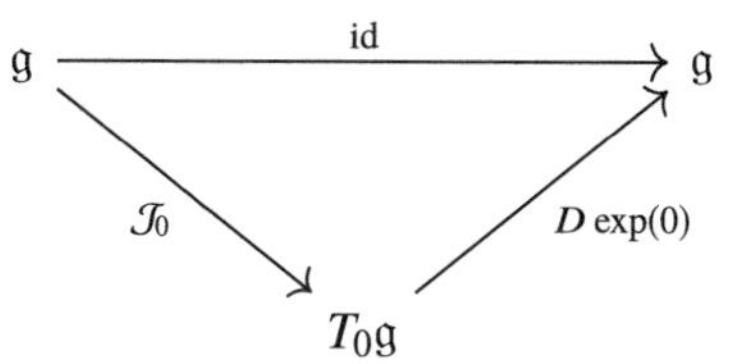

W. J. Merry, *Lectures on Differential Geometry I*,
https://doi.org/10.1007/978-3-032-03733-6_12

This is often imprecisely stated as $D\exp(0) = \mathrm{id}$.

Proof. We prove the result in three steps.

1. The space $G\times\mathfrak{g}$ is a smooth manifold by Problem A.3. We define a map $\widehat{X}\colon G\times\mathfrak{g} \to T(G\times\mathfrak{g})$ by

$$\widehat{X}(g,\xi) := \big(X_\xi(g), 0\big) \in T_g G \times T_\xi \mathfrak{g} \cong T_{(g,\xi)}(G\times\mathfrak{g}),$$

where the last isomorphism used Problem C.1. We claim that $\widehat{X}$ is a vector field on $G\times\mathfrak{g}$. It clearly satisfies the section property (8.1), and thus we need only check that $\widehat{X}$ is smooth. For this, suppose $f \in C^\infty(G\times\mathfrak{g})$ is smooth. Given $\xi \in \mathfrak{g}$, let $f_\xi := f(\cdot,\xi)\colon G \to \mathbb{R}$, so that f_ξ is a smooth function on G. Then by definition

$$\widehat{X}(f)(g,\xi) = X_\xi(f_\xi)(g).$$

The vector field X_ξ depends linearly on ξ by Theorem 11.5. Since linear functions are always smooth, $\xi \mapsto X_\xi$ is also smooth in ξ. The function f_ξ depends smoothly on ξ as f is smooth as a function of both g and ξ. Thus the expression $(g,\xi) \mapsto X_\xi(f_\xi)(g)$ depends smoothly on both g and ξ. Thus by Proposition 8.2, $\widehat{X}$ is indeed a vector field. Thus the flow $\widehat{\Phi}$ of $\widehat{X}$ is also smooth.

2. In this step we compute the flow $\widehat{\Phi}_t$ of $\widehat{X}$. Fix $(g,\xi) \in G\times\mathfrak{g}$. Define $\gamma\colon \mathbb{R} \to G\times\mathfrak{g}$ by

$$\gamma(t) := \big(g\exp(t\xi), \xi\big) = (l_g(\gamma^\xi(t)), \xi)$$

Then $\gamma(0) = (g,\xi)$ and

$$\begin{aligned}
\dot{\gamma}(t) &= \big(Dl_g(\gamma^\xi(t))\dot{\gamma}^\xi(t), 0\big) \\
&= \big(Dl_g(\gamma^\xi(t))X_\xi(\gamma^\xi(t)), 0\big) \\
&= \big(X_\xi(g\gamma^\xi(t)), 0\big) \\
&= \widehat{X}(\gamma(t)).
\end{aligned}$$

Thus γ is an integral curve of $\widehat{X}$. By uniqueness of integral curves, it follows that $\widehat{X}$ is complete and its flow is given by

$$\widehat{\Phi}_t(g,\xi) := \big(g\exp(t\xi), \xi\big).$$

In particular, $\widehat{\Phi}_1(e,\cdot)\colon \mathfrak{g} \to G\times\mathfrak{g}$ is smooth. This is the map $\xi \mapsto (\exp(\xi),\xi)$. Thus exp is smooth.

3. In this last step we compute $D\exp(0)$. Take $\xi \in \mathfrak{g}$. Then

$$\begin{aligned}
D\exp(0)\mathcal{J}_0(\xi) &= \frac{d}{dt}\Big|_{t=0} \exp(t\xi) \\
&= \frac{d}{dt}\Big|_{t=0} \gamma^\xi(t) \\
&= \xi.
\end{aligned}$$

This completes the proof. ∎

COROLLARY 12.4. *The exponential map is a diffeomorphism of some neighbourhood of the origin in* $\mathfrak{g}$ *onto its image in* G.

Proof. Since exp has maximal rank at 0 by Theorem 12.3, this follows immediately from the Inverse Function Theorem 5.10. ■

Now let us investigate how the exponential map behaves with respect to Lie group homomorphisms.

PROPOSITION 12.5. *Let* G *and* H *be Lie groups with Lie algebras* $\mathfrak{g}$ *and* $\mathfrak{h}$. *Let* $\varphi\colon G \to H$ *be a Lie group homomorphism. Then the following diagram commutes*:

$$\begin{array}{ccc} \mathfrak{g} & \xrightarrow{D\varphi(e)} & \mathfrak{h} \\ {\scriptstyle \exp}\downarrow & & \downarrow{\scriptstyle \exp} \\ G & \xrightarrow[\varphi]{} & H \end{array}$$

Proof. If $\gamma\colon \mathbb{R} \to G$ is a homomorphism then since φ is a homomorphism so is $\varphi \circ \gamma\colon \mathbb{R} \to H$. Applying this with $\gamma(t) = \exp(t\xi)$ shows that $t \mapsto \varphi(\exp(t\xi))$ is a one-parameter subgroup of H. Set $\delta(t) := \varphi(\gamma(t))$. Then

$$\begin{aligned} \dot{\delta}(0) &= \frac{d}{dt}\Big|_{t=0} \varphi(\exp(t\xi)) \\ &= D\varphi(e) \circ D\exp(0)\mathcal{J}_0(\xi) \\ &= D\varphi(e)\xi, \end{aligned}$$

where we used the chain rule and Theorem 12.3. Thus by uniqueness of integral curves it follows that

$$\varphi(\exp(t\xi)) = \exp(tD\varphi(e)\xi),$$

which is what we wanted to prove. ■

Applying Proposition 12.5 to an inclusion of a subgroup, as in Corollary 11.8 tells us:

COROLLARY 12.6. *Let* G *be a Lie group with Lie algebra* $\mathfrak{g}$ *and* $H \subset G$ *a Lie subgroup with Lie subalgebra* $\mathfrak{h} \subset \mathfrak{g}$. *Then the exponential map* $\exp\colon \mathfrak{h} \to H$ *is the restriction of* $\exp\colon \mathfrak{g} \to G$ *to* $\mathfrak{h}$.

The next result identifies the exponential map for $G = \mathrm{GL}(m)$.

PROPOSITION 12.7. *Let* $A \in \mathfrak{gl}(m)$. *Then the matrix exponential*

$$e^A := \sum_{k=0}^{\infty} \frac{1}{k!} A^k$$

converges and defines an element of $\mathrm{GL}(m)$. *Moreover* $A \mapsto e^A$ *is the exponential map of* $\mathrm{GL}(m)$.

The proof is deferred to Problem Sheet F. As with Corollary 11.10, this also allows us to characterise the matrix exponential for matrix Lie groups.

COROLLARY 12.8. *Let G be a matrix Lie group with Lie algebra* $\mathfrak{g}$. *Then the exponential map* $\exp\colon \mathfrak{g} \to G$ *is given by matrix exponentiation*: $\exp(A) = e^A$.

Proof. Apply Corollary 12.6 and Proposition 12.7. ■

This concludes our introductory treatment of Lie groups.

We now explore the manifold version of another key algebraic idea. If you think back to your introductory course on group theory, you will no doubt remember that one of the most important reasons to study groups is because we can let them act on sets. In fact, the notion of a group acting on a set is arguably more fundamental than that of a group itself, and certainly historically, the idea of a "transformation group" (i.e. an action of a group on a given set) predates that of an abstract group. One might even go so far as to say that groups are interesting precisely *because* of their actions.

Lie groups are no exception. Since we are working in the smooth category, we restrict our attention to smooth actions of Lie groups on manifolds.

DEFINITION 12.9. Let G be a Lie group and let M be a manifold. For a smooth map $\sigma\colon G \times M \to M$ let σ_g denote $\sigma(g, \cdot)$. Then if σ satisfies

$$\sigma_{gh} = \sigma_g \circ \sigma_h \qquad \sigma_e = \mathrm{id}, \tag{12.1}$$

for all $g, h \in G$ σ is called **a smooth left action of** G **on** M. For fixed $g \in G$, this implies that $\sigma_g : p \mapsto \sigma(g, p)$ is a diffeomorphism of M.

If one replaces (12.1) with

$$\sigma_{gh} = \sigma_h \circ \sigma_g \qquad \sigma_e = \mathrm{id},$$

one arrives at a **smooth right action of** G **on** M.

REMARK 12.10. The difference between right actions and left actions is purely notational, since we can always convert one into the other. Indeed, if σ is a left action then we can define a right action $\tilde{\sigma}$ by setting $\tilde{\sigma}_g := \sigma_{g^{-1}}$, and conversely.

> In the definitions that follow, by a **smooth action** we mean either a smooth left action or a smooth right action.

As you no doubt remember from algebra, there are various different types of group actions one can study. We define the **fixed point set of** σ_g as

$$\mathsf{fix}(\sigma_g) := \left\{ p \in M \mid \sigma_g(p) = p \right\}.$$

and the **orbit** of p as

$$\mathsf{orb}_\sigma(p) := \{\sigma_g(p) \mid g \in G\}.$$

Finally the **stabiliser** of p is the set

$$\mathsf{stab}_\sigma(p) := \{g \in G \mid p \in \mathsf{fix}(\sigma_g)\}.$$

DEFINITIONS 12.11. Let $\sigma : G \times M \to M$ be a smooth action of G on M.

(i) We say that σ is an **effective** action if

$$\sigma_g = \mathrm{id} \qquad \Rightarrow \qquad g = e.$$

(ii) We say that σ is a **free** action if

$$\mathsf{fix}(\sigma_g) \neq \emptyset \qquad \Rightarrow \qquad g = e.$$

Thus free $\Rightarrow$ effective.

(iii) The action σ is said to be **transitive** if

$$\mathsf{orb}_\sigma(p) = M, \qquad \forall\, p \in M.$$

(iv) The action σ is said to be **proper** if the map

$$(\sigma, \mathrm{id}) : G \times M \to M \times M \qquad (g, p) \mapsto (\sigma_g(p), p)$$

is a proper map in the topological sense.[1]

An alternative characterisation of proper actions is the following statement, whose proof is left for you on Problem Sheet F.

LEMMA 12.12. *Let σ be a smooth action of G on M. Then σ is proper if and only if for every pair of points $p, q \in M$ and every sequence $p_k \to p$, if (g_k) is a sequence in G such that $\sigma_{g_k}(p_k) \to q$ then (g_k) admits a convergent subsequence. Moreover if $g_k \to g$ then $\sigma_g(p) = q$.*

EXAMPLES 12.13. Here are some examples of smooth actions.

(i) Let G be a Lie group. We can think of G acting on itself via both left and right translations (which are left and right actions respectively). These actions—denoted by l and r—are obviously both free and transitive. In fact, they are also proper, as you will prove on Problem Sheet F.

(ii) A Lie group can also act on itself via **conjugation**. Define

$$c_g(h) := ghg^{-1},$$

so that $c_g = l_g \circ r_{g^{-1}} = r_{g^{-1}} \circ l_g$. This left action is not proper unless G is compact. It is also never free or transitive (unless $G = \{e\}$). If G is commutative it also fails to be effective.

[1] i.e. the preimage of a compact set is compact.

(iii) A smooth[2] left action σ of a Lie group G on a vector space V is called a **representation** of G on V if σ_g is a linear map for each $g \in G$. Thus one can think of $g \mapsto \sigma_g$ as Lie group homomorphism $G \to \mathrm{GL}(V)$. The study of such actions is called *Representation Theory*, and is a major field of mathematics in itself. In the context of representation theory, an effective representation is usually called a **faithful representation**.

(iv) As a special case of the preceding example, if we take $G = \mathrm{GL}(V)$ and $\sigma = \mathrm{id}$, we obtain the **canonical representation** of $\mathrm{GL}(V)$ on V.

(v) The canonical representation of $\mathrm{GL}(m)$ on $\mathbb{R}^m$ restricts to define a transitive left action of $\mathrm{O}(m)$ on S^{m-1}. (Exercise: Prove this.)

It will be useful in Chapter 16 when discussing fibre bundles to know that we can always convert an action into an effective action.

PROPOSITION 12.14. *Let σ be an action of a Lie group G on M. Let*

$$H := \left\{ g \in G \mid \sigma_g = \mathrm{id} \right\}.$$

Then H is a closed normal subgroup of G. If $K := G/H$ then σ induces an action $\bar{\sigma}$ of K on M such that

$$\left\{ \sigma_g \mid g \in G \right\} = \left\{ \bar{\sigma}_k \mid k \in K \right\}, \tag{12.2}$$

where both sides are thought of as subgroups of $\mathrm{Diff}(M)$.

Proof. Suppose $h \in H$ and $g \in G$. Then

$$\begin{aligned} \sigma_{ghg^{-1}} &= \sigma_g \circ \sigma_h \circ \sigma_{g^{-1}} \\ &= \sigma_g \circ \mathrm{id} \circ \sigma_{g^{-1}} \\ &= \mathrm{id}. \end{aligned}$$

Thus H is normal and the quotient group $K := G/H$ is well defined. We identify an element $k \in K$ with the coset $kH \subset G$. We define

$$\bar{\sigma}_k := \sigma_g,$$

where g is any element of kH. This is well-defined, since if g_1 is another element of kH then $g_1 g^{-1} \in H$ and hence

$$\sigma_{g_1} = \sigma_{g_1 g^{-1} g} = \sigma_{g_1 g^{-1}} \circ \sigma_g = \sigma_g.$$

Since the map $G \to G/K$ is a submersion the action $\bar{\sigma}$ is smooth; it is obviously effective, and (12.2) holds by definition.[3] ■

It is also convenient to regard $p \in M$ as fixed and consider the map $\sigma(\cdot, p)$. We denote this map by σ^p and call it an **orbit map**:

---

[2] If σ is continuous and linear then σ is automatically smooth—this can be proved using Problem F.9.

[3] In Proposition 13.8 we shall show that the quotient G/H is a Lie Group.

$$\sigma^p : G \to M, \qquad \sigma^p(g) := \sigma(g, p).$$

Thus by definition

$$\operatorname{im} \sigma^p = \mathsf{orb}_\sigma(p).$$

Immediately from the definition we obtain:

Lemma 12.15. *Let σ be a smooth action of G on M. Then*

(i) *The action is free if and only if σ^p is injective for all $p \in M$.*
(ii) *The action is transitive if and only if σ^p is surjective for all $p \in M$.*

Definition 12.16. Suppose $\sigma : G \times M \to M$ and $\tau : G \times N \to N$ are two smooth actions of the same Lie group G. A smooth map $\varphi : M \to N$ is said to be **(σ, τ)-equivariant** if the following commutes for all $g \in G$:

$$\begin{array}{ccc} M & \xrightarrow{\varphi} & N \\ \sigma_g \downarrow & & \downarrow \tau_g \\ M & \xrightarrow[\varphi]{} & N \end{array}$$

If the actions σ and τ are clear from the context, we simply say that φ is **equivariant**.

Examples 12.17. Here are two examples of equivariant maps.

(i) Let G and H be Lie groups and $\varphi : G \to H$ a Lie group homomorphism. Define a left action σ of G on H by

$$\sigma_g(h) := \varphi(g)h.$$

Then φ is (l, σ) equivariant, where l is the action of G on itself by left translations.
(ii) Let σ be a smooth left action of G on M. Then for every point $p \in M$, the orbit map $\sigma^p : G \to M$ is (l, σ)-equivariant, where l is the action of G on itself by left translations. If instead σ is a smooth right action on M, then σ^p is (r, σ)-equivariant.

The next result is analogous to Proposition 10.12.

Proposition 12.18. *Suppose $\sigma : G \times M \to M$ and $\tau : G \times N \to N$ are two smooth actions such that σ is transitive. Then any (σ, τ)-equivariant smooth map $\varphi : M \to N$ has constant rank.*

Proof. Fix $p, q \in M$. We show that the rank of φ at p is the same as the rank of φ at q. Since σ is transitive, there exists $g \in G$ such that $\sigma_g(p) = q$. We differentiate the equality $\varphi \circ \sigma_g = \tau_g \circ \varphi$ at p to obtain

$$D\varphi(q) \circ D\sigma_g(p) = D\tau_g(\varphi(p)) \circ D\varphi(p).$$

Since σ_g and τ_g are both diffeomorphisms, $D\sigma_g(p)$ and $D\tau_g(\varphi(p))$ are linear isomorphisms. The result follows. ■

Corollary 12.19. *Let σ by a smooth action of G on M. Fix $p \in M$.*

(i) *If σ^p is injective then σ^p is an immersion, and thus* $\mathsf{orb}_\sigma(p)$ *is an immersed submanifold.*
(ii) *If σ^p is surjective then σ^p is a submersion, and thus σ^p is a quotient map which admits smooth local sections.*

Proof. By part (ii) of Examples 12.17 the orbit maps σ^p are equivariant, and hence by Proposition 12.18 they have constant rank. The result follows from Problem C.7 and Proposition 6.13. ■

Chapter 13
Homogeneous Spaces

In this chapter we state the *Quotient Manifold Theorem*, which tells us that the quotient space of a smooth proper free action is naturally a smooth manifold. We use this to define **homogeneous spaces**. Finally we investigate the **adjoint** representation of a Lie group.

PROPOSITION 13.1. *Let σ be a smooth proper action of G on M. Then for every $p \in M$, the orbit $\mathsf{orb}_\sigma(p)$ is an embedded submanifold.*

Proof. The orbit map σ^p is (l, σ)-equivariant (cf. part (ii) of Example 12.17), and hence by Proposition 12.18 has constant rank, say n. By Corollary 6.19[1] there exists a neighbourhood U of $e \in G$ and an open set $O \subset \mathbb{R}^n$, together with an immersion $\psi : O \to G$ such that $\sigma^p \circ \psi : O \to M$ is an embedding with

$$\sigma^p(\psi(O)) = \sigma^p(U). \tag{13.1}$$

Since $\sigma^p \circ \psi$ is an embedding there exists by Proposition 6.3 a slice chart (W, y) about p such that

$$\sigma^p(\psi(O)) \cap W = \left\{q \in W \mid y^{n+1}(q) = \cdots = y^m(q) = 0\right\},$$

and hence by (13.1) we have

$$\sigma^p(U) \cap W = \left\{q \in W \mid y^{n+1}(q) = \cdots = y^m(q) = 0\right\}. \tag{13.2}$$

[1] When the action is free—which is the main case of interest—one does not need to quote the (unproven) Corollary 6.19, and can instead use Corollary 12.19.

W. J. Merry, *Lectures on Differential Geometry I*,
https://doi.org/10.1007/978-3-032-03733-6_13

We now claim there exists an open set $V \subset W$ containing p such that

$$\mathsf{orb}_\sigma(p) \cap V = \sigma^p(U) \cap V. \tag{13.3}$$

One always has $\supseteq$ in (13.3), so if no such neighbourhood V exists this means we can find a sequence $(p_k) \in \mathsf{orb}_\sigma(p)$ such that $p_k \to p$ and $p_k \notin \sigma^p(U)$. Let $g_k \in G$ be such that $\sigma_{g_k}(p) = p_k$. Then by Lemma 12.12, up to passing to a subsequence one has $g_k \to g$ for some $g \in G$, and moreover $\sigma_g(p) = p$. Since $r_g(U)$ is a neighbourhood of g in G, we must have $g_k \in r_g(U)$ for all k large. Thus

$$p_k = \sigma_{g_k}(p) \in \sigma^p(r_g(U))$$

for all large k. But by equivariance and the definition of an orbit

$$\sigma^p(r_g(U)) = \sigma^{\sigma_g(p)}(U) = \sigma^p(U).$$

This contradicts the claim that $p_k \notin \sigma^p(U)$ for all k, and hence (13.3) is proved. Combining (13.2) and (13.3) tells us that

$$\mathsf{orb}_\sigma(p) \cap V = \left\{ q \in V \mid y^{n+1}(q) = \cdots = y^m(q) = 0 \right\}.$$

This shows that $\mathsf{orb}_\sigma(p)$ is embedded in a neighbourhood of p. However the same argument also works for an arbitrary point $p_0 \in \mathsf{orb}_\sigma(p)$, since $\mathsf{orb}_\sigma(p_0) = \mathsf{orb}_\sigma(p)$ by definition. Thus we have produced slice charts at every point of $\mathsf{orb}_\sigma(p)$, and thus the result follows from Proposition 6.7. ∎

The hypotheses of Proposition 13.1 do *not* imply that the orbit $\mathsf{orb}_\sigma(p)$ is diffeomorphic to G—for instance, consider the trivial action $\sigma_g \equiv \mathrm{id}$. If however we additionally assume the action is free, then the orbits are all diffeomorphic to G.

Corollary 13.2. *Let σ be a free and proper action of G on M. Then for every point $p \in M$ the orbit map $\sigma^p : G \to M$ is an embedding, and thus* $\dim \mathsf{orb}_\sigma(p) = \dim G$.

Proof. Since $\mathsf{orb}_\sigma(p)$ is embedded, the orbit map is smooth as a map $G \to \mathsf{orb}_\sigma(p)$ (cf. the second bullet point in the proof of Proposition 10.15). The map $\sigma^p : G \to \mathsf{orb}_\sigma(p)$ is always surjective, and if the action is free, it is also injective by part (i) of Lemma 12.15. Thus σ^p is a bijective smooth map of constant rank, and hence by Problem C.7 it is a diffeomorphism. ∎

The stabilisers are also always embedded Lie subgroups of G.

Proposition 13.3. *Let σ be a smooth action of G on M. Then for every $p \in M$ the set* $\mathsf{stab}_\sigma(p)$ *is an embedded Lie subgroup of G.*

Proof. It is clear $\mathsf{stab}_\sigma(p)$ is a closed subgroup. The claim now follows from the Closed Subgroup Theorem 10.16. ∎

We now move onto the Quotient Manifold Theorem.

Definition 13.4. Let σ be a smooth action of G on M. Define an equivalence relation on M by saying that $p \sim q$ if and only if $\mathsf{orb}_\sigma(p) = \mathsf{orb}_\sigma(q)$. Denote the quotient space by M/G and let $\rho \colon M \to M/G$ denote the projection. We call M/G the **quotient space** of M by the G action.[2]

We endow M/G with the quotient topology. In general this topology can be very badly behaved, but if the action is proper then it is at least Hausdorff.

Lemma 13.5. *Suppose σ is a proper smooth action of G on M. Then the quotient space M/G is Hausdorff.*

Proof. The quotient map ρ is an open map for the quotient topology, as if $U \subset M$ is open then

$$\rho^{-1}(\rho(U)) = \bigcup_{g \in G} \sigma_g(U)$$

is open in G. Since $(\sigma, \mathrm{id}) \colon G \times M \to M \times M$ is proper, its image—call it C—is a closed subset of $M \times M$. Now suppose $p, q \in M$ are such that $\rho(p) \neq \rho(q)$. This means that $(p, q) \notin C$. Since C is closed, there exist neighbourhoods U and V of p and q respectively such that $(U \times V) \cap C = \emptyset$. Then $\rho(U)$ and $\rho(V)$ are open neighbourhoods of $\rho(p)$ and $\rho(q)$ in M/G such that $\rho(U) \cap \rho(V) = \emptyset$. Thus M/G is Hausdorff. ■

In fact, more is true.

Theorem 13.6 (The Quotient Manifold Theorem) *Let σ be a smooth action of G on M which is both proper and free. Then the quotient space M/G is a topological manifold of dimension* $\dim M - \dim G$. *Moreover there exists a unique smooth structure on M/G such that the quotient map $\rho \colon M \to M/G$ is a smooth submersion.*

Just as with Theorem 11.11, the proof of Theorem 13.6 requires the Frobenius Theorem, and hence the proof is deferred until Chapter 14.

Corollary 13.7. *Let H be a closed subgroup*[3] *of G. Then the space G/H of left cosets of H in G is a topological manifold of dimension* $\dim G - \dim H$. *Moreover there exists a unique smooth structure on G/H such that the quotient map $\rho \colon G \to G/H$ is a smooth submersion.*

Proof. By Problem F.5 the right action $r \colon H \times G \to G$ is both free and proper. The statement now follows from the Quotient Manifold Theorem. ■

Corollary 13.7 does not assert that G/H is a Lie group. Indeed, G/H is not necessarily even a group! If H is a *normal* subgroup however then G/H is a group, and in fact in this case:

---

[2] This notation is somewhat imprecise, since the space depends on the choice of action σ. Where necessary we write $M/_\sigma G$ to indicate this dependence.

[3] Recall by the Closed Subgroup Theorem 10.16 this is equivalent to asking that H be an embedded Lie subgroup of G.

Proposition 13.8. *Let G be a Lie group and let H be a closed normal subgroup. Then the smooth manifold G/H with its natural group structure is a Lie group.*

Proof. The map $\rho\colon G \to G/H$ is given by $\rho(g) = gH$ (the coset). The multiplication on G/H is given by

$$\bar{\mu}(g_1 H, g_2 H) := g_1 g_2 H.$$

This is only well-defined when H is normal.

To check this is smooth, fix $g_1, g_2 \in G$ and let ψ_1 and ψ_2 be smooth local sections of ρ near $g_1 H$ and $g_2 H$ respectively.[4] Then near the point $(g_1 H, g_2 H) \in G/H \times G/H$, one has

$$\bar{\mu} = \rho \circ \mu \circ (\psi_1, \psi_2),$$

where $\mu\colon G \times G \to G$ is the multiplication on G. Thus near $(g_1 H, g_2 H)$, $\bar{\mu}$ is the composition of smooth maps, and hence is smooth. Since $g_1 H$ and $g_2 H$ were arbitrary, $\bar{\mu}$ is smooth everywhere. A similar argument works for the inversion map. ∎

Definition 13.9. A **homogeneous space** is a smooth manifold M which is diffeomorphic to a smooth manifold of the form G/H, where G is a Lie group, H is a closed subgroup, and G/H is given the smooth structure from Corollary 13.7.

Many manifolds are homogeneous spaces (we will shortly see some examples). The key tool used to prove a given manifold is a homogeneous space is Theorem 13.12 below, which needs a few preliminary definitions.

Definition 13.10. Let σ be a smooth action of G on M. A point $p \in M$ is called a **stationary point** of σ if

$$p \in \bigcap_{g \in G} \mathsf{fix}(\sigma_g), \qquad \text{that is,} \qquad \mathsf{stab}_\sigma(p) = G.$$

Proposition 13.11. *Let σ be a smooth left action of G on M. Assume that p is a stationary point of σ. Then the map*

$$\tau\colon G \to \mathrm{GL}(T_p M), \qquad \tau(g) := D\sigma_g(p)$$

is a Lie group homomorphism (i.e. a representation).

Proof. Let us first check τ is a homomorphism. For this, observe

$$\tau(gh) = D\sigma_{gh}(p) = D(\sigma_g \circ \sigma_h)(p) = \tau(g)\tau(h).$$

The smoothness issue is a little more delicate. At first glance, it would appear obvious—σ is smooth, and hence so is $D\sigma$. But herein lies the problem: $\mathrm{GL}(T_p M) \cong \mathrm{GL}(m)$ has its own topology, and a priori it is not clear how smoothness of σ helps.

By definition of the topology on $\mathrm{GL}(T_p M) \cong \mathrm{GL}(m)$, it suffices to show that for a fixed $\xi \in T_p M$ the map

[4] These exist by Proposition 6.13.

$$g \mapsto D\sigma_g(p)\xi \tag{13.4}$$

is smooth as a map $G \to T_pM$. Since T_pM is an embedded submanifold of TM,[5] it suffices to show that (13.4) is smooth as a map $G \to TM$. This however is the composition

$$G \xrightarrow{Z} TG \times TM \cong T(G \times M) \xrightarrow{D\sigma} TM,$$

where Z is the smooth map $g \mapsto \big((g,0),(p,\xi)\big)$ and the second map is the canonical identification coming from Problem C.1. This is the composition of smooth maps, and hence is smooth. ■

In general an action may have no stationary points. But if we restrict attention to the stabiliser group of a point (which is another smooth Lie group by Proposition 13.3) then that point is automatically stationary.

Theorem 13.12 *Let σ be a transitive smooth left action of G on M. Fix $p \in M$ and let $H := \mathsf{stab}_\sigma(p)$. Let $\rho\colon G \to G/H$ denote the quotient map, and endow G/H with the smooth structure from Corollary* 13.7. *Define*

$$\varphi\colon G/H \to M, \qquad \varphi(gH) := \sigma_g(p).$$

Then φ is a diffeomorphism, and hence M is a homogeneous space.

Proof. First observe that φ is well defined, since if $h \in H$ then $\sigma_{gh}(p) = \sigma_g \circ \sigma_h(p) = \sigma_g(p)$. We claim that φ is a bijection. Surjectivity follows from transitivity of σ. If $\varphi(g_1H) = \varphi(g_2H)$ then $\sigma_{g_1^{-1}g_2}(p) = p$, whence $g_1^{-1}g_2 \in H$ and thus $g_1H = g_2H$. This shows injectivity.

To show that φ is smooth in a neighbourhood of a point gH, it suffices to show that $\varphi \circ \rho$ is smooth near g. Indeed, if $\varphi \circ \rho$ is smooth at g and $\psi\colon G/H \to G$ is a smooth local section of ρ at gH then $\varphi = (\varphi \circ \rho) \circ \psi$ is the composition of smooth maps. Now observe that $\varphi \circ \rho = \sigma \circ i_p$, where $\iota_p\colon G \to G \times M$ is the smooth map $g \mapsto (g,p)$. Thus $\varphi \circ \rho$ is the composition of smooth maps, and hence is smooth.

We now show that φ is a diffeomorphism. We will do this by showing that φ has constant rank. Left translation induces a transitive smooth left action $\bar{l}$ of G on G/H:

$$\bar{l}_g(g_1H) := gg_1H.$$

The map φ is $(\bar{l}, \sigma)$-equivariant, since

$$\begin{aligned} \varphi \circ \bar{l}_g(g_1H) &= \varphi(gg_1H) \\ &= \sigma_{gg_1}(p) \\ &= \sigma_g \circ \sigma_{g_1}(p) \\ &= \sigma_g \circ \varphi(g_1H). \end{aligned}$$

[5] The projection map $\pi\colon TM \to M$ is a smooth submersion (Theorem 5.6). Thus $T_pM = \pi^{-1}(p)$ is an embedded submanifold by the Implicit Function Theorem 6.10.

Thus by Proposition 12.18 the map φ has constant rank, and then by Problem C.7 φ is a diffeomorphism. ■

Theorem 13.12 tells us that we can define a homogeneous space as a smooth manifold that admits a transitive Lie group action. We emphasise however that a given smooth manifold can sometimes be made into a homogeneous space in multiple ways.

Let us return to part (v) of Examples 12.13 and see how this fits into the homogeneous space picture.

EXAMPLE 13.13. The Lie group $\mathrm{GL}(m)$ acts on $\mathbb{R}^m$. This in itself is not very interesting, but observe the action of $\mathrm{O}(m) \subset \mathrm{GL}(m)$ restricts to a *transitive* action on $S^{m-1} \subset \mathbb{R}^m$ by elementary linear algebra. Moreover the isotropy subgroup of $e_m = (0, 0, \ldots, 0, 1) \in S^{m-1}$ is given by those matrices $A \in \mathrm{O}(m)$ of the form

$$A = \begin{pmatrix} \begin{pmatrix} & & \\ & B & \\ & & \end{pmatrix} & \begin{matrix} 0 \\ \vdots \\ 0 \end{matrix} \\ 0 \cdots 0 & 1 \end{pmatrix}$$

where $B \in \mathrm{O}(m-1)$. We conclude that S^{m-1} is the homogeneous space

$$S^{m-1} \cong \mathrm{O}(m)/\,\mathrm{O}(m-1).$$

The same argument works to show that

$$S^{m-1} \cong \mathrm{SO}(m)/\,\mathrm{SO}(m-1).$$

EXAMPLE 13.14. Let $\mathrm{U}(m) \subset \mathrm{GL}(m; \mathbb{C})$ denote the unitary group and $\mathrm{SU}(m) \subset \mathrm{U}(m)$ the special unitary group. If we regard S^{2m-1} as the unit sphere in $\mathbb{C}^m$ then a similar argument shows that

$$S^{2m-1} \cong \mathrm{U}(m)/\,\mathrm{U}(m-1) \qquad \text{and} \qquad S^{2m-1} \cong \mathrm{SU}(m)/\,\mathrm{SU}(m-1).$$

Since $\mathrm{SU}(1)$ is just the 1×1 identity matrix, taking $m = 2$ shows that S^3 is diffeomorphic to $\mathrm{SU}(2)$, and hence S^3 can be given a Lie group structure.

REMARK 13.15. Not all smooth manifolds admit the structure of a Lie group. For instance, S^m admits a Lie group structure only for $m = 0, 1$ or $m = 3$. For $m = 0$ this is trivial. For $m = 1$, this was part (vi) from Examples 10.9 above, and we just did the case of S^3 in Example 13.14. The proof that no other sphere admits a Lie group structure is quite tricky, but roughly speaking proceeds as follows: suppose S^m admits a Lie group structure for $n > 1$. Since S^m is simply connected for $m > 1$, the Lie group structure is necessarily non-abelian. Next, one can show that any compact non-abelian Lie group G carries a natural closed but not exact bi-invariant differential 3-form. Thus $H^3(G; \mathbb{R}) \neq 0$. For S^m this forces $n = m = 3$.

Next we return to part (ii) of Examples 12.13.

Definition 13.16. Let G be a Lie group and let c denote the conjugation action of G on itself. The identity e is a stationary point of this action, and hence by Proposition 13.11 we obtain a Lie group homomorphism $G \to \mathrm{GL}(\mathfrak{g})$. This is called the **adjoint representation** and is denoted by

$$\mathrm{Ad}\colon G \to \mathrm{GL}(\mathfrak{g}).$$

We usually write $\mathrm{Ad}(g) = \mathrm{Ad}_g : \mathfrak{g} \to \mathfrak{g}$.

We can then go one step further and differentiate Ad. This requires us to look at the Lie algebra of $\mathrm{GL}(\mathfrak{g})$, which we write as

$$\mathfrak{gl}(\mathfrak{g}) = \{\text{all linear maps } \mathfrak{g} \to \mathfrak{g}\}\,.$$

Definition 13.17. The derivative of the adjoint representation is denoted by

$$\mathrm{ad} := D(\mathrm{Ad})(e)\colon \mathfrak{g} \to \mathfrak{gl}(\mathfrak{g}).$$

We usually write $\mathrm{ad}(\xi) = \mathrm{ad}_\xi : \mathfrak{g} \to \mathfrak{g}$.

By Proposition 11.7 the map ad is a Lie algebra homomorphism. Moreover Proposition 12.5 gives us a commutative diagram:

$$\begin{array}{ccc} \mathfrak{g} & \xrightarrow{\ \mathrm{ad}\ } & \mathfrak{gl}(\mathfrak{g}) \\ {\scriptstyle \exp}\big\downarrow & & \big\downarrow{\scriptstyle \exp} \\ G & \xrightarrow[\ \mathrm{Ad}\ ]{} & \mathrm{GL}(\mathfrak{g}) \end{array}$$

The map ad has a pleasing description. The proof of the next result is deferred to Problem Sheet F.

Proposition 13.18. *Let G be a Lie group with Lie algebra $\mathfrak{g}$. Then for $\xi, \zeta \in \mathfrak{g}$ one has $\mathrm{ad}_\xi(\zeta) = [\xi, \zeta]$.*

Bonus Material for Chapter 13

In this bonus section we survey one the most important infinite-dimensional Lie groups: namely, the diffeomorphism group of a compact manifold.

Let M be a compact manifold. The group $\mathrm{Diff}(M)$ can itself be given a Fréchet manifold structure.

Under this Fréchet manifold structure, one can show that composition

$$\mu\colon \mathrm{Diff}(M)\times \mathrm{Diff}(M) \to \mathrm{Diff}(M), \qquad \mu(\varphi,\psi) := \varphi\circ\psi \tag{13.5}$$

is smooth. Similarly the map $\varphi \mapsto \varphi^{-1}$ is smooth. This means that $\mathrm{Diff}(M)$ is an **infinite-dimensional Fréchet Lie group**.

A Fréchet manifold is a weaker and less useful concept than that of a Banach manifold. The difference is that a Fréchet manifold is locally modelled on a Fréchet space rather than a Banach space. The reason they are less useful is that the Inverse and Implicit Function Theorems are valid for Banach manifolds, but not for Fréchet manifolds.

Sadly we have no choice in the matter. Even if we wanted to work with lower regularity, whilst the space of C^k-diffeomorphisms $C^k(M,M)$ does have a nice Banach manifold structure, it is *not* a Lie group. Indeed, with μ as in (13.5), whilst the map

$$\mu(\cdot,\psi)\colon C^k(M,M)\to C^k(M,M), \qquad \varphi\mapsto\varphi\circ\psi.$$

is smooth, the map

$$\mu(\varphi,\cdot)\colon C^k(M,M)\to C^k(M,M), \qquad \psi\mapsto\varphi\circ\psi.$$

is not even C^1! (Exercise: Why?)

In any case, if we give $\mathrm{Diff}(M)$ its Fréchet smooth structure, then one can show that

$$T_{\mathrm{id}}\mathrm{Diff}(M) = \mathfrak{X}(M),$$

(as one would expect, the tangent space to an infinite-dimensional manifold is itself infinite-dimensional).

We now go through a few of the Lie-theoretic concepts we have studied, and see how they fit into the infinite dimensional picture:

(i) One-parameter subgroups of $\mathrm{Diff}(M)$ are precisely one-parameter groups of diffeomorphisms in the sense of Definition 11.12, i.e. paths $t\mapsto\Phi_t$ such that $\Phi_0=\mathrm{id}$ and $\Phi_{s+t}=\Phi_s\circ\Phi_t$. Denote by X the infinitesimal generator of $\{\Phi_t\}$, defined in (9.7) If we adopt the notation from introduced before Proposition 11.14 then the curve $t\mapsto\Phi_t$ is the maximal integral curve γ^X.

(ii) The exponential map $\exp\colon\mathfrak{X}(M)\to\mathrm{Diff}(M)$ assigns to a vector field X its time one flow Φ_1—this is well-defined by Corollary 9.19.

(iii) The conjugation action

$$c_\varphi(\psi) := \varphi\circ\psi\circ\varphi^{-1}$$

gives rise to the adjoint map

$$\mathrm{Ad}_\varphi\colon \mathfrak{X}(M)\to\mathfrak{X}(M)$$

which one easily sees is given by

$$\mathrm{Ad}_\varphi(X) = \varphi_\star X.$$

(iv) If we differentiate this to get $\mathrm{ad}\colon \mathfrak{X}(M) \to \mathfrak{gl}(\mathfrak{X}(M))$, we find that

$$\mathrm{ad}_X(Y) = \mathcal{L}_Y X = [Y, X], \qquad \forall\, X, Y \in \mathfrak{X}(M),$$

Equation (iv) is somewhat problematic, since this sign error would appear to *contradict* Proposition 13.18!

Of course there is no actual contradiction, since this is all a matter of conventions. What we have learnt is:

If we want to think of $\mathfrak{X}(M)$ as the Lie algebra of the infinite-dimensional Lie group $\mathrm{Diff}(M)$ then the Lie bracket should have been defined with the *opposite sign convention*:

$$[X, Y] := \mathcal{L}_Y X.$$

Some brave authors do indeed define the Lie bracket of vector fields in this way. Nevertheless we have chosen the "incorrect" sign convention so as to be consistent with the vast majority of the literature.

Chapter 14
Distributions and Integrability

In this chapter we introduce *distributions* on manifolds and prove the local version of the famous *Frobenius Theorem*. The global version of this theorem—which will be proved next chapter—is the cornerstone of an area of differential geometry called *foliation theory*. In this book we will use the (global) Frobenius Theorem to prove the Lie Correspondence Theorem 11.11 and the Quotient Manifold Theorem 13.6. In Volume II we will use the Frobenius Theorem to show that flat connections on vector bundles have trivial restricted holonomy groups (Volume II Theorem 6.8).

We begin with the following preliminary result.

PROPOSITION 14.1. *Let M be a smooth manifold and $W \subset M$ a non-empty open set. Suppose $X_1, \ldots, X_l \in \mathfrak{X}(W)$ are vector fields such that*

(i) *There exists $p \in W$ such that the vectors $X_i(p)$ are all linearly independent in T_pM (and thus necessarily $l \leq m$)*
(ii) *For all i, j one has $[X_i, X_j] \equiv 0$.*

Then there exists a chart (U, x) about p with $U \subset W$ such that

$$\frac{\partial}{\partial x^i} = X_i|_U, \qquad \forall\, 1 \leq i \leq l.$$

An immediate corollary is the following extension of Problem D.2.

COROLLARY 14.2. *Let M be a smooth manifold and $W \subset M$ a non-empty open set. Let $X \in \mathfrak{X}(W)$ and suppose $X(p) \neq 0$ for some $p \in W$. Then there exists a chart (U, x) about p with $U \subset W$ such that $X|_U = \frac{\partial}{\partial x^1}$.*

Proof of Proposition 14.1. We prove the result in two steps. The first step reduces the problem to $\mathbb{R}^m$. That this is possible should be clear from the statement, since the assertion is visibly local.

W. J. Merry, *Lectures on Differential Geometry I*,
https://doi.org/10.1007/978-3-032-03733-6_14

1. If $x\colon U \to O$ is any chart about p then the map $x_*\colon \mathfrak{X}(U) \to \mathfrak{X}(O)$ satisfies

$$x_*\left(\frac{\partial}{\partial x^i}\right) = \frac{\partial}{\partial u^i}, \qquad \forall\, 1 \le i \le m.$$

where (u^i) are the canonical local coordinates on $O \subset \mathbb{R}^m$. Since x_* is an isomorphism, it is sufficient to find such a chart x so that

$$x_*(X_i|_U) = \frac{\partial}{\partial u^i}, \qquad \forall\, 1 \le i \le l. \tag{14.1}$$

Now let $y\colon U \to O$ be an arbitrary chart about p such that $y(p) = 0$. We will modify y to produce a chart x satisfying (14.1). Let $Y_i \in \mathfrak{X}(O)$ denote the unique vector field such that

$$y_*(X_i|_U) = Y_i.$$

Since the X_i are linearly independent at p, the Y_i are linearly independent at 0. Thus there exists a linear isomorphism $\lambda\colon \mathbb{R}^m \to \mathbb{R}^m$ that maps $\mathcal{J}_0^{-1}(Y_i(0))$ to the standard basis vector e_i for each $1 \le i \le l$. Set $\tilde{y} := \lambda \circ y$ and set $\tilde{Y}_i = \tilde{y}_*(X_i|_U)$. Then

$$\tilde{Y}_i(0) = \frac{\partial}{\partial u^i}\bigg|_0 \qquad \forall\, 1 \le i \le l. \tag{14.2}$$

We emphasise this identity only holds at the point 0. The aim now is to construct a local diffeomorphism h defined on a neighbourhood $V \subset \mathbb{R}^m$ about 0 such that $h(0) = 0$ and such that on V

$$h_*(\tilde{Y}_i) = \frac{\partial}{\partial u^i} \qquad \forall\, 1 \le i \le l. \tag{14.3}$$

Then setting $x := h \circ \tilde{y}$ one has where defined that

$$x_*(X_i) = h_* \circ \tilde{y}_*(X_i) = h_*(\tilde{Y}_i) = \frac{\partial}{\partial u^i}.$$

2. In this second step we construct such an h. Note by Proposition 8.19 that the vector fields $\tilde{Y}_i$ satisfy $[\tilde{Y}_i, \tilde{Y}_j] \equiv 0$. Let Φ_t^i denote the flow of $\tilde{Y}_i$. For a sufficiently small neighbourhood Ω of 0 in $\mathbb{R}^m$ there is a well defined smooth function $f\colon \Omega \to \mathbb{R}^m$ given by the following somewhat improbable looking formula:

$$f(u^1, \dots, u^m) := \left(\Phi^1_{u^1} \circ \cdots \circ \Phi^l_{u^l}\right)(0, \dots, 0, u^{l+1}, \dots, u^m).$$

Let $g \in C^\infty(\mathbb{R}^m)$ and fix $q = (a^1, \dots, a^m) \in \Omega$. We first consider what $Df(q)$ does to $\frac{\partial}{\partial u^1}\big|_q$. Namely:

$$\begin{aligned}
Df(q)\left(\frac{\partial}{\partial u^1}\bigg|_q\right)(g) &= \frac{\partial}{\partial u^1}\bigg|_q (g \circ f) \\
&= \lim_{t \to 0} \frac{(g \circ \Phi^1_{a^1+t} \circ \cdots \circ \Phi^k_{u^k})(0, \dots, 0, a^{l+1}, \dots, a^m) - (g \circ f)(q)}{t} \\
&= \lim_{t \to 0} \frac{g \circ \Phi^1_t(f(q)) - g(f(q))}{t} \\
&= \tilde{Y}_1(f(q))(g).
\end{aligned}$$

Since g and q were arbitrary, this shows that $f_*\left(\frac{\partial}{\partial u^1}\right) = \tilde{Y}_1$. Since the Lie brackets vanish, using induction and Proposition 10.6 we have for any $1 \leq i \leq l$ that

$$\Phi^1_{u^1} \circ \cdots \circ \Phi^i_{u^i} \circ \cdots \circ \Phi^l_{u^l} = \Phi^i_{u^i} \circ \cdots \circ \Phi^1_{u^1} \circ \cdots \circ \Phi^l_{u^l},$$

and thus exactly the same argument shows that

$$f_*\left(\frac{\partial}{\partial u^i}\right) = \tilde{Y}_i, \qquad \forall\, 1 \leq i \leq l. \tag{14.4}$$

In particular, since $\tilde{Y}_i(0) = \frac{\partial}{\partial u^i}\big|_0$ by (14.2),

$$Df(0)\left(\frac{\partial}{\partial u^i}\bigg|_0\right) = \frac{\partial}{\partial u^i}\bigg|_0, \qquad \forall\, 1 \leq i \leq l, \tag{14.5}$$

In fact, we claim that (14.5) holds for all $1 \leq i \leq m$, and not just $1 \leq i \leq l$. To see this take $l < i \leq m$ and observe with g as above that

$$\begin{aligned} Df(0)\left(\frac{\partial}{\partial u^i}\bigg|_0\right)(g) &= \frac{\partial}{\partial u^i}\bigg|_0 (g \circ f) \\ &= \lim_{t \to 0} \frac{g(0, \dots, 0, t, 0 \dots, 0) - g(0)}{t} \\ &= \frac{\partial}{\partial u^i}\bigg|_0 (g). \end{aligned}$$

This shows that (14.5) holds for $l < i \leq m$ as well, and hence $Df(0)$ is the identity. Thus by the Inverse Function Theorem 5.9 there exists a neighbourhood $V \subset \Omega$ containing 0 such that $f|_V$ is a diffeomorphism. Set $h := f|_V^{-1}$. Then (14.4) implies that the diffeomorphism h satisfies (14.3) and the proof is complete. ■

We now introduce the notion of a distribution.

Definition 14.3. Let M be a smooth manifold of dimension m, and let $l \leq m$. A **distribution Δ on M of dimension** l is a choice of l-dimensional linear subspace $\Delta_p \subset T_pM$ for each $p \in M$ that varies smoothly with p in the following sense: For each point $p \in M$ there exists a neighbourhood U of p and l vector fields $X_1, \dots, X_l \in \mathfrak{X}(U)$ such that

$$\Delta_q = \operatorname{span}_{\mathbb{R}}\{X_1(q), \dots, X_l(q)\}, \qquad \forall\, q \in U.$$

The simplest example is $l = 1$.

Example 14.4. A vector field X is **non-vanishing** if $X(p) \neq 0$ for all $p \in M$. A non-vanishing vector field X defines a one-dimensional distribution by setting $\Delta_p := \operatorname{span}_{\mathbb{R}}\{X(p)\}$ for each $p \in M$.

Remark 14.5. Not every manifold admits such a vector field. Indeed, if m is even then every vector field on S^m vanishes in at least one point. This is the so-called "Hairy Ball Theorem", which you will be asked to prove later on in the book.[1] In fact, the Hairy

[1] In part (ii) of Problem K.8..

Ball Theorem is a purely topological result, and thus the smoothness assumption is not necessary: if m is even then any continuous map $S^m \to TS^m$ satisfying the section property 8.1 must vanish somewhere. This can be proved by applying the Whitney Approximation Theorem 7.12 to the smooth case, but it is also easy to show using some basic algebraic topology.

DEFINITION 14.6. Let Δ be an l-dimensional distribution on M, and suppose $L \subset M$ is an l-dimensional immersed submanifold. We say that L is an **integral manifold** of Δ if

$$D\iota(p)T_pL = \Delta_p, \qquad \forall\, p \in L,$$

where $\iota\colon L \hookrightarrow M$ is the inclusion.

In the one-dimensional case, integral manifolds always exist about every point. Indeed, suppose Δ is a one-dimensional distribution. Given any $p \in M$ there exists a neighbourhood W of p and a vector field $X \in \mathfrak{X}(W)$ such that $\Delta_q = \operatorname{span}_{\mathbb{R}}\{X(q)\}$ for all $q \in W$. Since in particular $X(p) \neq 0$, by Corollary 14.2 there exists a chart (U, x) about p with $U \subset W$ such that $X|_U = \frac{\partial}{\partial x^1}$. Then the set

$$L := \left\{q \in U \mid x^2(q) = \cdots = x^m(q) = 0\right\}$$

is an embedded one-dimensional submanifold of M by Proposition 6.7. Moreover the proof of Proposition 6.7 shows that $D\iota(q)(T_qL) = \frac{\partial}{\partial x^1}\big|_q$ for all $q \in L$, where $\iota\colon L \hookrightarrow M$ is the inclusion. Thus L is an integral manifold of Δ containing p.

The case $l = 1$ of Theorem 14.13 shows that every connected integral manifold of Δ contained in U is of this form.

For higher dimensional distributions, integral manifolds need not exist. Here is an example.

EXAMPLE 14.7. Consider the distribution on $\mathbb{R}^3$ spanned by the vector field

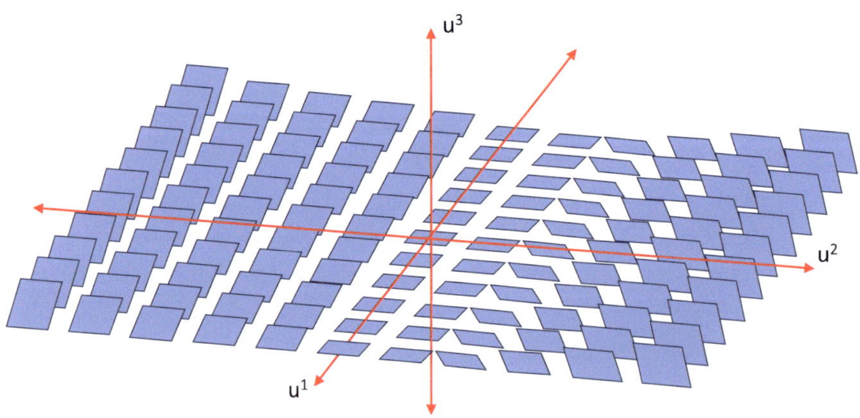

Figure 14.1: The standard contact distribution on $\mathbb{R}^3$, based on the image in [6]

$$X := \frac{\partial}{\partial u^1} + u^2 \frac{\partial}{\partial u^3}, \qquad Y := \frac{\partial}{\partial u^2}.$$

We claim this distribution has no integral manifolds through the origin. Indeed, suppose such an L existed. Then (as Figure 14.1 indicates), one would have $T_{(0,0,0)}L$ equal to the (u^1, u^2)-plane. But now suppose $\gamma\colon S^1 \to L$ is a closed curve in L that circles round the u^3-axis. Since γ is tangent to Δ, one readily sees that the u^3-component of γ is an increasing function. But then γ endlessly spirals upwards, and hence cannot close up—contradiction.

Example 14.7 is the starting point for the field of geometry called **contact geometry**. In general a **contact distribution** on a manifold is a distribution which is "maximally" non-integrable. Such a manifold is necessarily odd-dimensional. Contact manifolds are the odd-dimensional cousins of **symplectic manifolds**. Sadly we won't have time to study either contact or symplectic manifolds in this course.

We now formulate a condition that rules out such "pathologies".

DEFINITION 14.8. Let Δ be a distribution on M and let X be a vector field on M. We say that X **belongs to** Δ if $X(p) \in \Delta_p$ for each $p \in M$. Let $\mathfrak{X}(\Delta, M) \subset \mathfrak{X}(M)$ denote the set of vector fields belonging to Δ.

Since Δ_p is a linear subspace of T_pM for each $p \in M$, $\mathfrak{X}(\Delta, M)$ is a linear subspace of the infinite-dimensional vector space $\mathfrak{X}(M)$.

DEFINITION 14.9. A distribution Δ is said to be **integrable**[2] if $\mathfrak{X}(\Delta, M)$ is a Lie subalgebra of $\mathfrak{X}(M)$, that is

$$X, Y \in \mathfrak{X}(\Delta, M) \qquad \Rightarrow \qquad [X, Y] \in \mathfrak{X}(\Delta, M)$$

Here are two conditions that guarantee integrability.

LEMMA 14.10. *Let Δ be an l-dimensional distribution on M. Suppose for every $p \in M$ there exists a neighbourhood U of p and l vector fields $X_1, \dots, X_l \in \mathfrak{X}(\Delta, U)$ such that Δ is spanned by the X_i over U and such that $[X_i, X_j] \in \mathfrak{X}(\Delta|_U, U)$ for all $1 \le i, j \le l$. Then Δ is integrable.*

Proof. Let $p \in M$ and let X and Y be vector fields that belong to Δ. Choose a neighbourhood U of p for which there exist vector fields spanning Δ as in the hypotheses of the Lemma. Then on U we can write

$$X|_U = f^i X_i, \qquad Y|_U = g^i X_i$$

for some smooth[3] functions $f^i, g^i : U \to \mathbb{R}$. By Problem D.5 one has on U that

$$\begin{aligned} [X, Y]|_U &= [f^i X_i, g^j X_j] \\ &= f^i g^j [X_i, X_j] + f^i X_i(g^j) X_j - g^j X_j(f^i) X_i. \end{aligned}$$

[2] Some authors use the word *involutive* instead of integrable to describe a distribution satisfying the conditions of Definition 14.9.

[3] If you are worried why these functions are smooth, see Remark 20.10.

Since $[X_i, X_j](q) \in \Delta_q$ for all $q \in U$, this shows that $[X, Y]$ belongs to Δ for every point in U. Since p was arbitrary, it follows that $[X, Y]$ belongs to Δ. ■

LEMMA 14.11. *Let Δ be a distribution on M. Assume that for every $p \in M$ there exists an integral manifold L_p of Δ with $p \in L_p$. Then Δ is integrable.*

Proof. Let X and Y belong to Δ. Fix an arbitrary point $p \in M$, and let $\iota_p \colon L_p \hookrightarrow M$ denote the inclusion. In the language of Problem D.7, X and Y are tangent to L_p. By part (iii) of Problem D.7, $[X, Y]$ is also tangent to L_p, or equivalently, $[X, Y](p) \in D\iota_p(p)(T_p L_p) = \Delta_p$. Since p was arbitrary, we conclude $[X, Y]$ belongs to Δ. ■

A more difficult result states that the converse to Lemma 14.11 holds.

NOTATION. Let $\mathbb{I}^l := (-1, 1)^l$ denote the l-dimensional open unit cube, and write an element of $\mathbb{I}^l$ as a tuple $a = (a^1, \dots, a^l)$.

DEFINITION 14.12. A **shifted slice** in M of dimension l is an embedded submanifold of the form

$$L(a) := \{p \in U \mid x^{l+1}(p) = a^1, \dots, x^m(p) = a^{m-l}\},$$

for some element $a = (a^1, \dots, a^{m-l}) \in \mathbb{I}^{m-l}$.

Thus the difference between a shifted slice and a normal slice is that instead of requiring the last $m - l$ coordinates to all be zero, we merely require them to be some fixed element in $\mathbb{I}^{m-l}$. Shifted slices are no more general than normal slices; nevertheless, they are a useful bookkeeping tool.

Theorem 14.13 (The Local Frobenius Theorem) *Let M be a smooth manifold and let Δ be an integrable l-dimensional distribution on M. Then for every $p \in M$ there exists a chart $x \colon U \to \mathbb{I}^m$ with $x(p) = 0$ and such that for any $a \in \mathbb{I}^{m-l}$, the shifted slice*

$$L(a) := \{q \in U \mid x^{l+1}(q) = a^1, \dots, x^m(q) = a^{m-l}\}$$

is an integral manifold of Δ. Moreover any connected integral manifold of Δ contained in U is contained in such a shifted slice.

Proof. Once again, the statement is purely local, so by arguing as in Step 1 of Proposition 14.1, we may assume that $M = \mathbb{R}^m$, $p = 0$, and Δ_0 is spanned by the vectors $\frac{\partial}{\partial u^i}\big|_0$ for $i = 1, \dots, l$.[4]

We argue in three steps.

1. Write $\mathbb{R}^m = \mathbb{R}^l \times \mathbb{R}^{m-l}$. Let ρ_1 and ρ_2 denote the two projections $\mathbb{R}^m \to \mathbb{R}^l$ and $\mathbb{R}^{m-l}$ respectively:

$$\rho_1(u^1, \dots, u^m) := (u^1, \dots, u^l), \qquad \rho_2(u^1, \dots, u^m) := (u^{l+1}, \dots, u^m),$$

[4] Exercise: Fill in these details!

Let

$$\delta_q := D\rho_1(q)|_{\Delta_q} : \Delta_q \to T_q\mathbb{R}^l$$

Then $q \mapsto \delta_q$ is a smooth family of linear maps, whose domain ranges smoothly with q. By assumption δ_0 is an isomorphism. Since being invertible is an open condition,[5] it follows that there is a neighbourhood W of 0 in $\mathbb{R}^m$ such that δ_q is an isomorphism for all $q \in W$.

Thus up to possibly shrinking W, there exist unique vector fields $X_i \in \mathfrak{X}(\Delta, W)$ that are ρ_1-related to $\frac{\partial}{\partial u^i}$ for $i = 1, \dots, l$. By part (ii) of Problem D.6 one has that $[X_i, X_j]$ is ρ_1-related to $[\frac{\partial}{\partial u^i}, \frac{\partial}{\partial u^j}]$. By Proposition 8.10, $[\frac{\partial}{\partial u^i}, \frac{\partial}{\partial u^j}] = 0$, and thus $[X_i, X_j]$ is ρ_1-related to the zero vector field. Now since Δ is integrable,[6] $[X_i, X_j]$ belongs to Δ, and since $D\rho_1(q)|_{\Delta_q} = \delta_q$ is injective for $q \in W$, it follows that $[X_i, X_j] = 0$.

2. Thus we can apply Proposition 14.1 to obtain a chart $x \colon U \to \mathbb{I}^m$ defined on $U \subset W$ such that $X_i|_U = \frac{\partial}{\partial x^i}$. Now let

$$\varphi := \rho_2 \circ x \colon U \to \mathbb{I}^{m-l}$$

Then φ is a smooth surjective submersion, and thus by the Implicit Function Theorem 6.10, for any $a \in \mathbb{I}^{m-l}$, the set $L(a) := \varphi^{-1}(a)$ is an embedded submanifold of M, and any $q \in U$ belongs to a unique $L(a)$—namely, $a = \varphi(q)$. Moreover by Proposition 6.15, if we denote by $\iota \colon L(a) \hookrightarrow U$ the inclusion then for any $q \in L(a)$ one has

$$\begin{aligned} D\iota(q)\, T_q L(a) &= \ker D\varphi(q) \\ &= \{\xi \in T_q U \mid \xi(x^i) = 0 \text{ for } i = l+1, \dots, m\} \\ &= \operatorname{span}_{\mathbb{R}} \left\{ \left.\frac{\partial}{\partial x^i}\right|_q \mid 1 \le i \le l \right\} \\ &= \Delta_q . \end{aligned}$$

3. It remains to prove the last sentence of the theorem. Suppose L is an arbitrary integral manifold of Δ contained in U. Then for any $q \in L$ and $\xi \in T_q L$, one has $(D\iota(q)\xi)(x^i) = 0$ for $i = l+1, \dots, m$. Thus $D(x^i \circ \iota)(q)$ is the zero map for each $i = l+1, \dots, m$, and hence $q \mapsto x^i(\iota(q))$ is a locally constant function. If L is connected, then it is constant, and thus L is contained in a single shifted slice. This completes the proof. ■

[5] This is because $q \mapsto \det \delta_q$ is a continuous function.

[6] This is the only place in the proof where we use integrability of Δ!

Problem Sheet F

A (♣) indicates the problem is non-examinable. This does not necessarily imply it is difficult.

PROBLEM F.1

(i) Let M be a smooth manifold. Assume there exist vector fields $X_1, \ldots, X_m \in \mathfrak{X}(M)$ such that $\{X_i(p)\}$ is a basis of T_pM for every $p \in M$. Prove that the tangent bundle TM is diffeomorphic to $M \times \mathbb{R}^m$.

(ii) Let G be a Lie group with Lie algebra $\mathfrak{g}$. Prove that TG is diffeomorphic to $G \times \mathfrak{g}$.

PROBLEM F.2 Let $A \in \mathfrak{gl}(m)$. Prove that the matrix exponential

$$e^A := \sum_{k=0}^{\infty} \frac{1}{k!} A^k$$

converges and defines an element of $\mathrm{GL}(m)$. Prove that $A \mapsto e^A$ is the exponential map of $\mathrm{GL}(m)$.

PROBLEM F.3 Let G be a Lie group with Lie algebra $\mathfrak{g}$. Prove that for $\xi, \zeta \in \mathfrak{g}$ one has $\mathrm{ad}_\xi(\zeta) = [\xi, \zeta]$.

PROBLEM F.4 Let σ be a smooth action of a Lie group G on a smooth manifold M.

(i) Prove that σ is proper if and only if the following condition holds: if (p_k) is a sequence in M and (g_k) is a sequence in G such that both (p_k) and $(\sigma_{g_k}(p_k))$ converge, then a subsequence of (g_k) converges.

(ii) Deduce that if G is compact then every smooth action is proper.

PROBLEM F.5 Let G be a Lie group and H be a closed subgroup (possibly equal to G). Let H act on G via left (or right) translations. Prove that this action is proper.

PROBLEM F.6 Let σ be a proper smooth action of a Lie group G on a smooth manifold M. Prove that the orbits $\mathsf{orb}_\sigma(p)$ are closed subsets of M.

PROBLEM F.7 Let Δ be an integrable distribution on a smooth manifold M, and let L be a connected integral manifold of Δ. Assume that L is closed in M. Prove that L is a leaf of foliation induced by Δ.

PROBLEM F.8 Let $\varphi \colon M \to N$ be a surjective submersion. Prove that the connected components of the preimages $\varphi^{-1}(p)$ as p ranges over N defines an $(m-n)$-dimensional foliation of M.

Bonus Problem(s) for Sheet F

These problem(s) are hard, and are included for enthusiasts only. Solutions will not be provided.

A **topological group** G is a topological space that is also a group in the algebraic sense, with the property that the group multiplication

$$\mu\colon G \times G \to G, \qquad \mu(g, h) = gh,$$

and group inversion

$$i\colon G \to G, \qquad i(g) = g^{-1},$$

are both **continuous** maps. The goal of the next problem is to show that if G is a topological space that simultaneously carries the structure of a topological manifold and a topological group, then G admits at most one diffeomorphism class of smooth structures that turns G into a Lie group.

(♣)PROBLEM F.9

(i) Let G be a Lie group. Suppose $\gamma\colon \mathbb{R} \to G$ is a continuous group homomorphism. Prove that γ is necessarily smooth, and hence is a one-parameter subgroup.
(ii) Let G and H be Lie groups, and suppose $\varphi\colon G \to H$ is a continuous group homomorphism. Prove that φ is automatically smooth, and hence is a Lie group homomorphism. *Hint*: Use the previous part.
(iii) Let G be a topological space which is simultaneously a topological group and a topological manifold. Prove that G admits at most one diffeomorphism class of smooth structures that turns G into a Lie group.

Remark: The converse to (iii) was Hilbert's Fifth Problem [8], famously posed by David Hilbert in 1900. It was eventually proved in 1952 by Montgomery and Zippen.

(♣)PROBLEM F.10 Let G be a connected space that satisfies all the conditions for a Lie group apart from not necessarily being second countable. Prove that G is automatically second countable (and hence a Lie group).

(♣)PROBLEM F.11 Let G be a connected Lie group. Let $\mathrm{Aut}(G)$ denote the set of Lie group isomorphisms $\varphi\colon G \to G$. Prove that $\mathrm{Aut}(G)$ admits the structure of a Lie group. *Hint*: First prove this in the case where G is simply connected.

Chapter 15
Foliations and the Frobenius Theorem

In this chapter we will globalise the Local Frobenius Theorem 14.13, and then use the Global Frobenius Theorem to prove the two outstanding results from Chapters 11 and 13: the Lie Correspondence Theorem 11.11 and the Quotient Manifold Theorem 13.6.

We begin with an observation which will be useful later. The integral manifolds produced by the Local Frobenius Theorem are always embedded, despite the fact that Definition 14.6 only required them to be immersed. This is because we have only looked at integral submanifolds contained in some (small) set U. In general, integral manifolds do not always have to be embedded. Nevertheless, the Local Frobenius Theorem 14.13 shows that an arbitrary integral manifold retains one of the important properties of embedded submanifolds. We now formalise the condition used in the second bullet point in the proof of Proposition 10.15 as a definition.

Definition 15.1. Let $L \subset M$ be an immersed submanifold. We say that L is **weakly embedded** if for every smooth manifold N and every smooth map $\varphi\colon N \to M$ such that $\varphi(N) \subset L$, the map φ is also smooth as a map $N \to L$.

Thus embedded submanifolds are automatically weakly embedded. Integral manifolds are too, as the following result shows.

Proposition 15.2. *Let Δ be an integrable distribution on a smooth manifold M. Every integral manifold L of Δ is a weakly embedded submanifold of M.*

Proof. Assume that $\varphi\colon N \to M$ is a smooth map such that $\varphi(N) \subset L$. Fix a point $p \in N$. By the Local Frobenius Theorem there exists a chart $x\colon U \to \mathbb{I}^m$ with $x(\varphi(p)) = 0$ such that all connected integral submanifolds of Δ contained in U are contained in shifted slices

$$L(a) = \left\{q \in U \mid x^{l+1}(q) = a^1, \dots, x^m(q) = a^{m-l}\right\},$$

for $a \in \mathbb{I}^{m-l}$. Now consider $L \cap U$. Since U is open, this is another immersed submanifold, and hence—by definition of a manifold—has at most countably many connected

W. J. Merry, *Lectures on Differential Geometry I*,
https://doi.org/10.1007/978-3-032-03733-6_15

components.[1] Each such component is then a connected integral submanifold of Δ contained in U, and so by the Local Frobenius Theorem is contained in some shifted slice. Thus there are countably many $a_k \in \mathbb{I}^{m-l}$ such that

$$L \cap U \subset \bigcup_k L(a_k). \tag{15.1}$$

Now choose a chart (V, y) on N about p such that V is connected and $\varphi(V) \subset L \cap U$. Then the function

$$f := x \circ \varphi \circ y^{-1} \colon y(V) \to \mathbb{I}^m$$

is smooth. Write $f^i := u^i \circ f$ as usual so that $f = (f^1, \dots, f^m)$. Then by (15.1) the last $(m-l)$ functions $(f^{l+1}, \dots, f^m)$ can only take values in the countable set $\{a_k\}_{k=1}^{\infty}$, and therefore they are locally constant. Since V is connected, they are constant. Thus $\varphi(V)$ is contained in a single shifted slice $L(a_k)$. Since $L \cap L(a_k)$ is an open subset of L that is embedded in M, it follows that $\varphi|_V \colon V \to L \cap L(a_k)$ is smooth. Thus also the composition $\varphi|_V \colon V \to L \cap L(a_k) \hookrightarrow L$ is smooth. Since p was an arbitrary point of N the claim follows. ■

We now globalise Theorem 14.13. We begin with a definition.

Definition 15.3. Let M be a smooth manifold. An **l-dimensional foliation** $\mathcal{F}$ of M is a partition of M into l-dimensional connected immersed submanifolds, called the **leaves** of the foliation, such that:

(i) The collection of tangent spaces to the leaves defines a distribution Δ on M.
(ii) Any connected integral manifold of Δ is contained in a leaf of $\mathcal{F}$.

Each leaf L of $\mathcal{F}$ is called a **maximal integral manifold** of Δ. One says that the distribution Δ is **induced by** $\mathcal{F}$.

Here is the global version of Theorem 14.13.

Theorem 15.4 (The Global Frobenius Theorem) *Let Δ be an integrable distribution on M. Then Δ is induced by a foliation.*

Proof. Let Δ be an integrable distribution. By the Local Frobenius Theorem 14.13 for any point $p \in M$ there is a chart $x \colon U \to \mathbb{I}^m$ such that the slices

$$L(a) := \left\{ q \in U \mid x^{l+1}(q) = a^1, \dots, x^m(q) = a^{m-l} \right\} \tag{15.2}$$

for $a \in \mathbb{I}^{m-l}$ are integral manifolds of Δ. Since M is a separable metric space, we may choose a countable collection $(U_k, x_k)_{k \in \mathbb{N}}$ of charts whose domains U_k form an open cover of M. Now let $\mathcal{L}$ denote the collection of all slices $L(a)$ of the form (15.2) for all of the charts x_k. Define an equivalence relation on $\mathcal{L}$ by declaring that $L \sim L'$ if there exists a finite sequence $L = L_0, L_1, \dots, L_h = L'$ such that $L_i \cap L_{i+1} \neq \emptyset$ for $i = 0, \dots, h-1$.

[1] A topological space with uncountably many components can never be separable.

Suppose $L \subset U_i$ is a shifted slice such that $L \cap U_j \neq \emptyset$. Since U_j is open, $L \cap U_j$ is a manifold, and hence $L \cap U_j$ has at most countably many components (we already used this argument in the proof of Proposition 15.2). It follows that each equivalence class can only contain countably many shifted slices $L \in \mathcal{L}$.

Now fix one such equivalence class $[L]$ and enumerate the countably many elements belonging to L as $\{K_i\}_{i \in \mathbb{N}}$, with $K_i \cap K_{i+1} \neq \emptyset$ for each i. For fixed $j \in \mathbb{N}$, if

$$\tilde{L}_j := \bigcup_{i=1}^{j} K_i$$

then $\tilde{L}_j$ is a finite union of connected embedded integral manifolds of Δ, and thus is itself an connected embedded integral manifold of Δ. This means that if we set

$$\tilde{L} := \bigcup_{j=1}^{\infty} \tilde{L}_j$$

then $\tilde{L}$ is a union of an increasing sequence of connected embedded integral manifolds of Δ, and hence is itself a connected *immersed* integral manifold[2] of Δ. If $[L] \neq [L']$ are two distinct equivalence classes then the corresponding unions $\tilde{L}$ and $\tilde{L}'$ are disjoint. Since by definition any connected integral manifold of Δ is contained in such a union, this shows that the set of these unions form a foliation of M which is induced by Δ. This completes the proof. ■

We now provide the promised proofs of the Lie Algebra Correspondence Theorem 11.11 and the Quotient Manifold Theorem 13.6. For convenience, we restate both results here.

Theorem 15.5 (The Lie Correspondence Theorem) *Let G be a Lie group with Lie algebra $\mathfrak{g}$. If $\mathfrak{h}$ is a Lie subalgebra of $\mathfrak{g}$ then there is a unique connected Lie subgroup H of G whose Lie algebra is $\mathfrak{h}$.*

Proof. Given $g \in G$, let Δ_g denote the subspace of T_gG given by the set of all vectors of the form $X_\xi(g)$, where $X_\xi \in \mathfrak{X}_l(G)$ is a left-invariant vector field such that $\xi = X_\xi(e) \in \mathfrak{h} \subset \mathfrak{g}$. Thus

$$\Delta_g := \{Dl_g(e)\xi \mid \xi \in \mathfrak{h}\}.$$

To see that Δ really is a distribution, note that if $\{\xi_i\}$ is a basis of $\mathfrak{h}$ then the left-invariant vector fields $\{X_{\xi_i}(g)\}$ span Δ_g at every point $g \in G$. Moreover since $\mathfrak{h}$ is a Lie subalgebra, $[\xi_i, \xi_j] \in \mathfrak{h}$ for each i, j and thus $[X_{\xi_i}, X_{\xi_j}] = X_{[\xi_i, \xi_j]}$ belongs to Δ for every i, j. Thus by Lemma 14.10 it follows that Δ is integrable. By the Global Frobenius Theorem 15.4, Δ induces a foliation of G. Let H denote the leaf containing e. For any $g_1 \in G$ we have $Dl_{g_1}(g)(\Delta_g) = \Delta_{g_1 g}$ by construction, and hence Dl_{g_1} leaves the distribution invariant. Thus l_{g_1} permutes the leaves of the foliation, i.e. it maps the leaf passing through g diffeomorphically onto the leaf passing through $g_1 g$. In particular, if $h \in H$ then $l_{h^{-1}}$

[2] Exercise: Why?

maps H to the leaf containing e, which is just H again. Thus $l_{h^{-1}}(H) = H$, which proves that H is a subgroup. It remains to prove that the multiplication map $m\colon H \times H \to H$ is smooth. We know that the multiplication $m\colon H \times H \to G$ is smooth and $m(H \times H) \subset H$. Thus by Proposition 15.2, m is also smooth as a map $H \times H \to H$. This completes the proof. ■

REMARK 15.6. This proof also shows that every Lie subgroup H of a Lie group G is weakly embedded.

Theorem 15.7 (The Quotient Manifold Theorem) *Let σ be a smooth action of G on M which is both proper and free. Then the quotient space M/G is a topological manifold of dimension* $\dim M - \dim G$*. Moreover there exists a unique smooth structure on M/G such that the quotient map $\rho\colon M \to M/G$ is a smooth submersion.*

The proof is non-examinable, and hence is delayed by one more line...

Bonus Material for Chapter 15

...to here.

Proof. We prove the result in five steps. Let

$$m := \dim M, \qquad l := \dim G.$$

We already know from Lemma 13.5 that M/G is Hausdorff. Moreover in the proof of Lemma 13.5 we showed that ρ is an open map. Thus if $\{B_i\}$ is a countable basis for the topology on M then $\{\rho(B_i)\}$ is a countable basis for the quotient topology on M/G. By Proposition 1.32 if we can show that M/G is locally Euclidean, it will follow that M/G is a topological manifold. In fact, we will directly construct a smooth atlas on M/G.

1. We now start the construction of a smooth atlas on M/G. For $p \in M$, let

$$\Delta_p := T_p \mathsf{orb}_\sigma(p)$$

denote the tangent space of the orbit. By Corollary 13.2 the subspace Δ_p has dimension l. We will show that Δ is an l-dimensional distribution on M. The idea is similar to the previous proof: let $\{\xi_i\}$ denote a basis for $\mathfrak{g}$. Define a vector field Y_i on M by

$$Y_i(p) := D\sigma^p(e)\xi_i \in T_p M.$$

The flow of Y_i is given by $\sigma_{\exp t\xi}$. By construction, Y_i belongs to Δ for every $p \in M$. Since $\dim \mathfrak{g} = \dim \Delta$, it follows that $\{Y_i\}$ span Δ everywhere. Thus Δ is a distribution.

Since every point in M is contained in an integral manifold, Lemma 14.11 implies that this distribution is integrable. By the Global Frobenius Theorem 15.4, Δ induces a foliation of M. By Problems F.6 and F.7 the leaf of the foliation containing p is the connected component of $\mathsf{orb}_\sigma(p)$ containing p. (If G is connected, then the leaf is simply the orbit $\mathsf{orb}_\sigma(p)$.) Thus as in the previous proof, we see that σ_g permutes the leaves of this foliation, i.e. it sends the leaf through p to the leaf through $\sigma_g(p)$.

2. Fix $p \in M$. In this step we apply the Local Frobenius Theorem 14.13. This provides us with a chart $x\colon U \to \mathbb{I}^m$ about p, where $\sigma(p) = 0$, such that each shifted slice

$$\{q \in U \mid x^{l+1}(q) = a^1, \dots x^m(q) = a^{m-l}\}$$

for $a = (a^1, \dots, a^{m-l}) \in \mathbb{I}^{m-l}$ is contained in an orbit. Set $V := \rho(U) \subset M/G$. Let

$$K := \{q \in U \mid x^1(q) = \cdots = x^l(q) = 0\},$$

so that K is a connected embedded submanifold of M. Consider now the restriction of the action σ to $G \times K$:

$$\sigma\colon G \times K \to M.$$

Since $\dim K = m - l$, $G \times K$ has dimension m. Note that $\sigma|_{\{e\}\times K}$ is just the inclusion $\iota\colon K \hookrightarrow M$. Under the identification $T_{(e,p)}(G \times K) \cong \mathfrak{g} \oplus T_pK$ given by Problem C.1 the differential of σ at a point (e, p) is given by

$$D\sigma(e, p)(\xi, \zeta) = D\sigma^p(e)\xi + D\iota(p)\zeta, \qquad \xi \in \mathfrak{g},\ \zeta \in T_pM.$$

Since K is an immersed submanifold, $D\iota(p)$ is injective. By Corollary 13.2 the map $D\sigma^p(g)$ is an isomorphism. Thus $D\sigma(g, p)$ is injective linear map between two vector spaces of the same dimension, and hence is a linear isomorphism. Thus the Inverse Function Theorem 5.10 implies that there exists a neighbourhood W of (e, p) in $G \times K$ such that $\sigma|_W$ is a diffeomorphism.

3. In this step we prove that, up to replacing U with a smaller neighbourhood of p if necessary, the map $\rho|_K\colon K \to V$ is in fact a homeomorphism. We begin by showing $\rho|_K$ is a bijective.

- **Surjective:** This is clear, since K intersects every shifted slice in U, so that $\rho(K) = \rho(U) = V$.
- **Injective:** We argue by contradiction: if the claim is false[3] then we can find two sequences (p_k), (q_k) in U such that $p_k \to p$ and $q_k \to p$ and

$$p_k \neq q_k, \quad \text{but} \quad \rho(p_k) = \rho(q_k), \qquad \forall\, k \in \mathbb{N}.$$

Since $\rho(p_k) = \rho(q_k)$ the points p_k, q_k lie in the same orbit. Thus there exists $g_k \in G$ such that $q_k = \sigma_{g_k}(p_k)$. By Lemma 12.12, up to passing to a subsequence the sequence g_k converges to some $g \in G$. Then $\sigma_g(p) = p$, and hence as the action is free we must have $g = e$. For sufficiently large k, we therefore have

[3] Explicitly: if there exists no neighbourhood $U_0 \subset U$ of p such that $\rho|_{K\cap U_0}$ is injective, then ...

$$(e, p_k) \text{ and } (g_k, q_k) \text{ in } W, \qquad \sigma(e, p_k) = \sigma(g_k, q_k).$$

This contradicts our assumption that σ is a diffeomorphism (and thus in particular injective) on W.

Thus $\rho|_K$ is bijective, as claimed. Since ρ is an open map, it follows that $\rho|_K$ is a homeomorphism.

4. We are now ready to construct our smooth atlas, and thus prove that M/G is a smooth manifold. With K as in the previous step, we define the homeomorphism

$$y := \rho_2 \circ x \circ \rho|_K^{-1} : V \to \mathbb{I}^{m-l}.$$

where $\rho_2 := \mathbb{I}^m \to \mathbb{I}^{m-l}$ is the projection to the last $m - l$ factors.

The existence of y shows that M/G is locally Euclidean, and hence M/G is a topological manifold of dimension $m - l$. We will take y as our chart on M/G around $\rho(p)$. To show that the collection of charts (V, y) form a smooth atlas on M/G, we must check that the transition functions are smooth. So suppose (V_1, y_1) and (V_2, y_2) are constructed above with $V_1 \cap V_2 \neq \emptyset$ (with corresponding sets K_i, U_i and charts x_i). We must show

$$y_2 \circ y_1^{-1} : y_1(V_1 \cap V_2) \to y_2(V_1 \cap V_2)$$

is smooth. Choose $p \in V_1 \cap V_2$; we need to show smoothness near p. If there is $q \in U_1 \cap U_2$ such that $\rho(q) = p$ then the claim follows, since the composition $x_2^{-1} \circ x_1$ is smooth where defined.

Otherwise pick $q_1, q_2 \in U_1, U_2$ with $\rho(q_1) = \rho(q_2) = p$. Then there exists $g \in G$ such that $\sigma_g(q_1) = q_2$. Let $\tilde{x}_1 = x_1 \circ \sigma_g^{-1}$ and $\tilde{y}_1 = y_1 \circ \sigma_g^{-1}$. Then since $x_2 \circ \tilde{x}_1^{-1}$ is smooth we see that $y_2 \circ \tilde{y}_1^{-1}$ is also smooth. However by expanding the definitions one sees that actually $\tilde{y}_1 = y_1$ near p, and hence $y_2 \circ y_1^{-1}$ is smooth near p, as required.

5. We have now shown that M/G is a smooth manifold. The map $\rho: M \to M/G$ is smooth, since with the notation as above,

$$y \circ \rho \circ x^{-1} = \rho_2,$$

which is smooth. It remains to show that this is the unique smooth structure on M/G for which $\rho: M \to M/G$ is a smooth submersion. Suppose $(M/G)'$ is the same topological manifold, but endowed with a different smooth atlas for which ρ is a smooth submersion. We claim that $\mathrm{id}: M/G \to (M/G)'$ is a diffeomorphism:

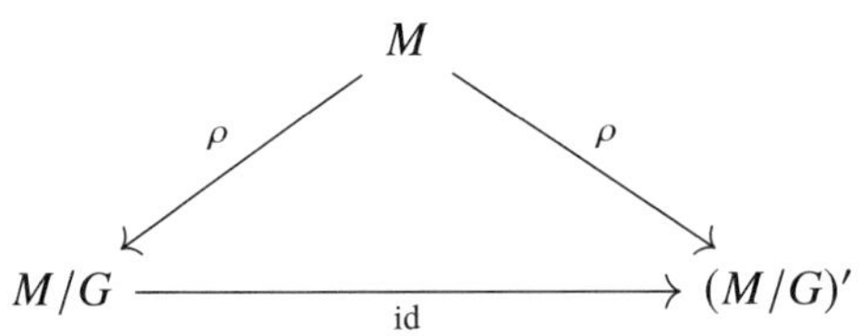

Fix $p \in M$. By Proposition 6.13 there exists a neighbourhood U of p and a neighbourhood V of $\rho(p)$ together with a smooth (with respect to the smooth structure on M/G)

map $\psi: V \to U$ such that $\rho \circ \psi = \mathrm{id}_V$. Thus the identity map $\mathrm{id}|_U : U \subset M/G \to U \subset (M/G)'$ is smooth. Since p was arbitrary, $\mathrm{id}: M/G \to (M/G)'$ is smooth. Reversing the roles of M/G and $(M/G)'$ shows that $\mathrm{id}: (M/G)' \to M/G$ is also smooth, and hence a diffeomorphism. Thus the smooth atlases on M/G and $(M/G)'$ both define the same smooth structure. This completes the proof. ∎

Chapter 16
Bundles

In this chapter we define the general notion of a **fibre bundle**. This is, roughly speaking, a space that locally looks like a product. Whilst fibre bundles are important in many areas of topology, they are slightly too vague to be useful for us. We therefore quickly specialise to the two special types of fibre bundles used in differential geometry: **vector bundles** and **principal bundles**. The study of such bundles will make up the majority of the rest of Volume I.

DEFINITIONS 16.1. Let E, M and L be smooth manifolds, and suppose $\pi\colon E \to M$ is a smooth surjective map. We say that $\pi\colon E \to M$ is a **fibre bundle over** M **with fibre** L if for every point $p \in M$ there exists a neighbourhood U of p and a smooth map

$$\varepsilon\colon \pi^{-1}(U) \to L$$

such that

$$(\pi, \varepsilon)\colon \pi^{-1}(U) \to U \times L$$

is a diffeomorphism. We call ε a **bundle chart** for E. A **bundle atlas** on E is any[1] collection $\{(U_a, \varepsilon_a) \mid a \in A\}$ of bundle charts such that the sets U_a form an open cover of M.

We call E the **total space** of the bundle, M the **base space**, and L the **fibre**. We use the notation $L \to E \xrightarrow{\pi} M$ to denote a fibre bundle E over M with fibre L. This is just notation: the arrow $L \to E$ does not represent any one particular map. When no confusion is possible we shorten the notation $L \to E \xrightarrow{\pi} M$ to simply E.

We should really say "smooth fibre bundle", but since we won't ever have cause in this book to look at non-smooth fibre bundles, we omit the adjective smooth.

[1] There are no compatibility conditions in the definition of a bundle atlas. This is because all the spaces involved are already assumed to be manifolds.

W. J. Merry, *Lectures on Differential Geometry I*,
https://doi.org/10.1007/978-3-032-03733-6_16

If $\varepsilon\colon \pi^{-1}(U) \to L$ is a bundle chart, it is convenient to denote by $\hat{\varepsilon}$ the map (π, ε). Thus $\hat{\varepsilon}$ is a diffeomorphism such that the following commutes:

$$\begin{array}{ccc} \pi^{-1}(U) & \xrightarrow{\quad\hat{\varepsilon}\quad} & U \times L \\ & {\scriptstyle \pi}\searrow \quad \swarrow{\scriptstyle \mathrm{pr}_1} & \\ & U & \end{array}$$

We call $\hat{\varepsilon}$ a **local trivialisation** of E. Thus there is a one-to-one correspondence between local trivialisations and bundle charts. We will use this notation without further comment for the rest of the book.

Convention. The total space of a fibre bundle will usually[2] be denoted by E or F. As with Lie groups, this means that the dimension of such a total space is *not* written with the corresponding lower case letter. However in this case there is no confusion, since if $L \to E \xrightarrow{\pi} M$ is a fibre bundle then the existence of local trivialisations force

$$\dim E = m + l.$$

We normally write points in fibre bundles with the letters u and v.

Examples 16.2. Here are two examples:

(i) The simplest example of a fibre bundle is the product manifold $E = M \times L$ with $\pi\colon M \times L \to M$ the first projection. In this case we can take U to be all of M and define $\hat{\varepsilon}\colon M \times L \to M \times L$ to be the identity map. More generally, any fibre bundle E which is globally diffeomorphic to $M \times L$ is called a **trivial bundle**.

(ii) A **sphere bundle** is a fibre bundle $S^l \to E \xrightarrow{\pi} M$. Sphere bundles are particularly important in algebraic topology. On Problem Sheet G you will show that the Klein bottle is an S^1-bundle over S^1.

Definition 16.3. Given a fibre bundle $L \to E \xrightarrow{\pi} M$, we set $E_p := \pi^{-1}(p)$ for $p \in M$ and call E_p the **fibre** over p.

If (U, ε) is a bundle chart on E, then for $p \in U$ we denote by $\varepsilon_p\colon E_p \to L$ the restriction of ε to the fibre E_p. These maps are diffeomorphisms, as the next lemma shows.

Lemma 16.4. *Let $L \to E \xrightarrow{\pi} M$ be a fibre bundle. Then π is a surjective submersion, and each fibre E_p is an embedded submanifold of E diffeomorphic to L.*

Proof. Fix $p \in M$, and let $\varepsilon\colon \pi^{-1}(U) \to L$ be a bundle chart such that $p \in U$, and let $\mathrm{pr}_1\colon U \times L \to U$ and $\mathrm{pr}_2\colon U \times L \to L$ denote the two projections. These are both

[2] Principal bundles are an exception to this; see Definition 16.13 below.

submersions. Fix $u \in E_p$. Since $\hat{\varepsilon}$ is a diffeomorphism, its derivative at u is a bijection $T_u E \to T_{\hat{\varepsilon}(u)}(M \times L)$. Differentiating the equation $\pi = \mathrm{pr}_1 \circ \hat{\varepsilon}$, we see that $D\pi(u)$ is the composition

$$D\pi(u) = D\,\mathrm{pr}_1(\hat{\varepsilon}(u)) \circ D\hat{\varepsilon}(u),$$

and thus is surjective. Since u was an arbitrary point of E_p, this shows that p is a regular value of π, and since p was arbitrary, we see that π is a submersion. The Implicit Function Theorem 6.10 then tells us that each fibre is naturally an embedded submanifold of E. Finally, $\hat{\varepsilon}$ maps E_p diffeomorphically onto the embedded submanifold $\{p\} \times L$ of $U \times L$, which is itself diffeomorphic to L via pr_2. ∎

REMARK 16.5. Suppose (W, ε) is a bundle chart on E. Let (U, x) and (V, y) be (manifold) charts on M and L respectively with $W \subset U$. Then $(x \circ \pi, y \circ \varepsilon)$ is a manifold chart on an open set in E which is compatible with the given smooth structure on E.

It is often useful to work backwards. Suppose we begin with a *set* E and a surjective map $\pi \colon E \to M$, where M is a smooth manifold. Suppose in addition we are given another smooth manifold L and an open cover $\{U_a \mid a \in A\}$ of M, together with a collection of bijections

$$\hat{\varepsilon}_a \colon \pi^{-1}(U_a) \to U_a \times L$$

such that $\mathrm{pr}_1 \circ \hat{\varepsilon}_a = \pi$. We can then attempt to *define* a smooth structure by declaring that charts on E are of the form $(x \circ \pi, y \circ \varepsilon_a)$, where x is a chart on M defined on an open subset of U_a, and y is some chart on L. Of course, now there is something to check. By Proposition 1.17, if one can verify that the transition functions are diffeomorphisms, this will endow E with a smooth manifold structure in such a way that the (U_a, ε_a) become a fibre bundle atlas.

DEFINITION 16.6. Suppose we have two fibre bundles

$$L_1 \to E \xrightarrow{\pi_1} M, \qquad L_2 \to F \xrightarrow{\pi_2} N$$

such that $L_1 \subset L_2$, $E \subset F$ and $M \subset N$ are all embedded submanifolds. We say that E is a **subbundle** of F if $\pi_2|_E = \pi_1$, that is

$$E_p \subset F_p, \qquad \forall p \in M.$$

EXAMPLE 16.7. If $L \to E \xrightarrow{\pi} M$ is any fibre bundle and $\hat{\varepsilon} \colon \pi^{-1}(U) \to U \times L$ is a local trivialisation, then we can consider $L \to \pi^{-1}(U) \xrightarrow{\pi} U$ as a fibre bundle in its own right. This fibre bundle is trivial and is a subbundle of E. As a result we often say that E is **trivial over** U if there exists a local trivialisation with domain $\pi^{-1}(U)$.

Suppose $L \to E \xrightarrow{\pi} M$ is a fibre bundle and $\{(U_a, \varepsilon_a) \mid a \in A\}$ is a bundle atlas. If $U_a \cap U_b \neq \emptyset$ then for each $p \in U_a \cap U_b$, the fibre parts ε_a and ε_b restrict to define diffeomorphisms $\varepsilon_{a|p}, \varepsilon_{b|p} \colon E_p \to L$. Thus there is a well-defined map

$$\varepsilon_{ab} \colon U_a \cap U_b \to \mathrm{Diff}(L), \qquad p \mapsto \varepsilon_{a|p} \circ \varepsilon_{b|p}^{-1} \tag{16.1}$$

We usually call ε_{ab} the **transition function**[3] from the bundle chart ε_a to the bundle chart ε_b, and refer to the collection $\{\varepsilon_{ab}\}$ of all transition functions arising from the bundle atlas as the transition functions of the bundle atlas. By definition one has

$$\varepsilon_{ab}(p) \circ \varepsilon_{b|p} = \varepsilon_{a|p} \tag{16.2}$$

as maps $E_p \to L$. If $U_a \cap U_b \cap U_c \neq \emptyset$ then the following **cocycle condition**[4] is automatically satisfied:

$$\varepsilon_{ac}(p) = \varepsilon_{ab}(p) \circ \varepsilon_{bc}(p), \qquad \forall\, p \in U_a \cap U_b \cap U_c.$$

The composition on the right-hand side occurs in $\mathrm{Diff}(L)$. In particular,

$$\varepsilon_{ab}(p)^{-1} = \varepsilon_{ba}(p).$$

As remarked at the beginning of the chapter, in this level of generality fibre bundles are not particularly useful in differential geometry. One way to understand this is the following: the transition functions (16.1) take values in the infinite-dimensional manifold $\mathrm{Diff}(L)$. This space is simply "too large" to work with. We therefore seek a way to cut down the possible options for the transition functions, and for this, we introduce a Lie group into the mix.

Suppose σ is an *effective* action of a Lie group G on L. Then the homomorphism $g \mapsto \sigma_g$ is an injective map $G \to \mathrm{Diff}(L)$, and hence we can regard G as a subgroup of $\mathrm{Diff}(L)$.

DEFINITION 16.8. Suppose $L \to E \xrightarrow{\pi} M$ is a fibre bundle and σ is an effective action of G on L. A bundle atlas $\{(U_a, \varepsilon_a) \mid a \in A\}$ is said to be a (G, σ)**-bundle atlas** if the transition functions (16.1) take values in G, i.e., if $U_a \cap U_b$ is non-empty then there exists a smooth map $g_{ab} \colon U_a \cap U_b \to G$ such that

$$\varepsilon_{ab}(p) = \sigma_{g_{ab}(p)}. \tag{16.3}$$

If such an atlas exists, we say that E is a (G, σ)**-fibre bundle**, and we call G the **structure group** of the bundle.

We sometimes refer to a (G, σ)-fibre bundle simply as a G**-fibre bundle**, particularly when the action σ is either unimportant or clear from the context.

REMARK 16.9. Definition 16.8 still makes perfect sense if we drop the assumption that σ is effective. However if σ is not effective then the maps g_{ab} are not uniquely determined—for example, if σ is the trivial action then *any* maps g_{ab} will work. This is not the end of the world, but it is occasionally annoying,[5] and for this reason we will only work with effective actions when discussing fibre bundles.

[3] **Warning:** This is a slightly different meaning of the word "transition function" than was used in Definition 1.10.

[4] The name "cocycle" comes from Čech cohomology. This is not important here.

[5] For example, it complicates the uniqueness part of the Fibre Bundle Construction Theorem 17.5.

Moreover Propositions 12.14 and 13.8 show that if we start with any (not necessarily) effective action of G on L, we can convert it into an effective action without changing its image in $\mathrm{Diff}(L)$. Since the definition of (G,σ)-bundle atlas only uses σ through its image in $\mathrm{Diff}(L)$, this shows that working only with effective actions does not actually involve any loss of generality.

Remark 16.10. Just as with smooth atlases on manifolds, since (G,σ)-bundle atlases come with compatibility conditions, the union of two (G,σ)-bundle atlases may not be still be a (G,σ)-bundle atlas. However we can define an equivalence relation on the set of (G,σ)-bundle atlases by declaring two atlases to be equivalent if their union is another (G,σ)-bundle atlas. We then define a **(G,σ)-bundle structure** to be an equivalence class. Alternatively, a (G,σ)-bundle structure can be thought of as a maximal (G,σ)-bundle atlas. (Compare Remark 1.12). In practice however, just as with smooth atlases versus smooth structures on manifolds, the distinction is usually unimportant.

Remark 16.11. A given fibre bundle $L \to E \xrightarrow{\pi} M$ may have structure group G for many different Lie groups G (and thus we should really say "a structure group" rather than "the structure group"). It is often advantageous to make the structure group as small as possible: if E has structure group G and $H \subset G$ is a Lie subgroup, then sometimes it is possible to find a (G,σ)-bundle atlas such that each transition function ε_{ab} takes image in $\{\sigma_h \mid h \in H\} \subset \mathrm{Diff}(L)$. Then this (G,σ)-bundle atlas is actually an $(H,\sigma|_H)$-bundle atlas, and we say that we have **reduced the structure group to H**. A concrete example of this awaits you on Problem Sheet G.

Passing from general fibre bundles to G-fibre bundles thus replaces the infinite-dimensional group $\mathrm{Diff}(L)$ with the finite-dimensional group G. This is already a major improvement over a general fibre bundle, but it is still not enough. There are two special types of G-fibre bundles that are of particular importance in differential geometry, and we introduce them now.

These two special types of fibre bundles come from the two "canonical" choices of Lie group actions we have met so far:

(i) If V is a vector space, then there is a canonical representation of $\mathrm{GL}(V)$ on V, cf. Part (iv) of Examples 12.13.
(ii) If G is a Lie group, then G acts naturally on itself via left translation.

Option (i) gives rise to *vector bundles*, and option (ii) gives rise to *principal bundles*.

Definition 16.12. Let M be a smooth manifold. A **vector bundle** over M is a $\mathrm{GL}(V)$-fibre bundle $V \to E \xrightarrow{\pi} M$, where V is a vector space and $\mathrm{GL}(V)$ acts on V via the canonical representation. We say that E has **rank** l if $\dim V = l$.

Definition 16.13. Let M be a smooth manifold and G a Lie group. A **principal G-bundle** over M is a G-fibre bundle $G \to P \xrightarrow{\pi} M$, where G acts on itself via left translation.

In contrast to other fibre bundles, principal bundles are usually written with the letters P and Q.

Although it is not obvious from the definitions, the theories of vector bundles and principal bundles are essentially analogous, and it is largely a matter of taste whether one primarily works with vector bundles or principal bundles. Roughly speaking: principal bundles are slightly more general, whereas vector bundles are slightly easier to understand. We will return to this at the end of Chapter 18.

Our canonical example of a vector bundle is the tangent bundle.

EXAMPLE 16.14. Let M be a smooth manifold. Then the tangent bundle $\pi : TM \to M$ is a vector bundle of rank m over M.

It is clear that the fibres T_pM are vector spaces. Let $\{(U_a, x_a) \mid a \in A\}$ denote a smooth atlas on M. Define

$$\hat{\varepsilon}_a : \pi^{-1}(U_a) \to U_a \times \mathbb{R}^m, \qquad \hat{\varepsilon}_a(p, \xi) = \big(p, (dx_a^i)_p(\xi)e_i\big)$$

We need only check that these transition functions are linear.

With our new notation, the corresponding chart $\tilde{x}_a$ on TM constructed in the proof of Theorem 5.6 is given by

$$\tilde{x}_a = (x_a \circ \pi, \varepsilon_a)$$

which is compatible with the first paragraph of Remark 16.5. Moreover if $U_a \cap U_b \neq \emptyset$ then by (5.1) we have

$$\varepsilon_{ab}(p) = D(x_a \circ x_b^{-1})(x_b(p)),$$

which lies in $\mathrm{GL}(m) \subset \mathrm{Diff}(\mathbb{R}^m)$. A similar argument shows that the cotangent bundle T^*M is another vector bundle of rank m over M.

We have also already met many principal bundles in this book, via the Quotient Manifold Theorem 13.6, although at the moment this is not easy to deduce directly from the definition. We will come back to this at the end of the chapter.

Suppose $V \to E \xrightarrow{\pi} M$ is a fibre bundle with fibre a vector space. What does it mean to say that E is a vector bundle? The next result clarifies this.

PROPOSITION 16.15. *Let $\pi : E \to M$ be a fibre bundle with fibre a vector space V. Then E is a vector bundle if and only if it is possible to endow each fibre E_p with a vector space structure and find a bundle atlas $\{(U_a, \varepsilon_a)\}$ with the property that for any $p \in U_a$ the map $\varepsilon_{a|p} : E_p \to V$ is a vector space isomorphism.*

Proof. Sufficiency is clear, for if (U_a, ε_a) and (U_b, ε_b) are two overlapping bundle charts as in the statement then

$$\varepsilon_{ab}(p) = \varepsilon_{a|p} \circ \varepsilon_{b|p}^{-1}$$

is the composition of linear maps, and hence is linear. Conversely if $V \to E \xrightarrow{\pi} M$ is a vector bundle of rank l then Problem B.1 implies that each fibre E_p admits the structure of a vector space, and moreover that this vector space structure has the property that each $\varepsilon_{a|p} : E_p \to V$ is a vector space isomorphism. ∎

Proposition 16.15 allows us to make the following alternative definition of a vector bundle.

Definition 16.16. Let $\pi\colon E \to M$ be a surjective smooth map between two smooth manifolds, and set $E_p := \pi^{-1}(p)$. We say that E is a **vector bundle of rank** n if each E_p admits the structure of an n-dimensional vector space, and any $p \in M$ has a neighbourhood U together with a smooth map $\varepsilon\colon \pi^{-1}(U) \to \mathbb{R}^n$ such that:

(i) $(\pi, \varepsilon)\colon \pi^{-1}(U) \to U \times \mathbb{R}^n$ is a diffeomorphism,
(ii) if $\varepsilon_q := \varepsilon|_{E_q}$ then $\varepsilon_q\colon E_q \to \mathbb{R}^n$ is a vector space isomorphism for all $q \in U$.

We will call such a map ε a **vector bundle chart**, and the collection of such charts is called a **vector bundle atlas**.

Here is the analogous statement for principal bundles.

Proposition 16.17. *Let $G \to P \xrightarrow{\pi} M$ be a fibre bundle with fibre G. Then P is a principal bundle if and only if there exists a smooth fibre preserving*[6] *free right action τ on P, together with a bundle atlas $\{(U_a, \varepsilon_a)\}$ such that each map ε_a is (τ, r)-equivariant.*

The fact that τ is a *right* action is because in Definition 16.12 we chose the convention that G acts on itself by *left* translations.

Proof. Suppose such an action τ exists. Since the action τ is fibre preserving, equivariance also holds in the fibres, that is,

$$\varepsilon_{a|p} \circ \tau_g = r_g \circ \varepsilon_{a|p} \tag{16.4}$$

for all $g \in G$ and $p \in U_a$. Our goal is to show that the map

$$\varepsilon_{ab}\colon U_a \cap U_b \to \mathrm{Diff}(G)$$

is actually of the form

$$\varepsilon_{ab}(p) = l_{g_{ab}(p)},$$

for $g_{ab}\colon U_a \cap U_b \to G$ a smooth function. Define

$$g_{ab}(p) := \varepsilon_{a|p} \circ \varepsilon_{b|p}^{-1}(e).$$

This is the composition of smooth maps and hence is smooth. Moreover it depends smoothly on p since ε_a and ε_b are smooth. Now take an arbitrary element $h \in G$. We compute

$$\begin{aligned}
\varepsilon_{ab}(p)(h) &= \varepsilon_{a|p} \circ \varepsilon_{b|p}^{-1}(h) \\
&= \varepsilon_{a|p} \circ \varepsilon_{b|p}^{-1} \circ r_h(e) \\
&= \varepsilon_{a|p} \circ \tau_h \circ \varepsilon_{b|p}^{-1}(e) \\
&= r_h \circ \varepsilon_{a|p} \circ \varepsilon_{b|p}^{-1}(e) \\
&= r_h(g_{ab}(p)) \\
&= l_{g_{ab}(p)}(h).
\end{aligned}$$

[6] Here by fibre preserving we mean that $\tau_g(P_p) \subset P_p$, i.e. $\pi \circ \tau_g = \pi$.

For the converse, suppose $\{(U_a, \varepsilon_a) \mid a \in A\}$ is a (G, l)-bundle atlas. We define a map

$$\tau\colon G \times P \to P$$

by declaring that

$$\tau_g(u) := \varepsilon_{a|p}^{-1} \circ r_g \circ \varepsilon_{a|p}(u)$$

where $a \in A$ is any element of A such that $\pi(u) \in U_a$. This is well defined, since if $\pi(u) \in U_a \cap U_b$ then

$$\begin{aligned}
\varepsilon_{b|p}^{-1} \circ r_g \circ \varepsilon_{b|p}(u) &= \varepsilon_{a|p}^{-1} \circ \varepsilon_{ab}(p) \circ r_g \varepsilon_{b|p}(u) \\
&= \varepsilon_{a|p}^{-1} \circ l_{g_{ab}(p)} \circ r_g \circ \varepsilon_{b|p}(u) \\
&= \varepsilon_{a|p}^{-1} \circ r_g \circ l_{g_{ab}(p)} \circ \varepsilon_{b|p}(u) \\
&= \varepsilon_{a|p}^{-1} \circ r_g \circ \varepsilon_{a|p}(u),
\end{aligned}$$

where the last line used (16.2). Now that we know τ is well defined, it is immediate that τ is a smooth fibre preserving free right action of G on P. Finally (16.4) holds by definition. ∎

In fact, we can further improve Proposition 16.17.

Proposition 16.18. *Let $\pi\colon P \to M$ be a surjective submersion and let G be a Lie group. Then P is a principal G bundle if and only if there exists a smooth free right action τ of G on P which is fibre preserving and transitive on the fibres.*[7]

Proof. Suppose P is a principal G bundle. By Proposition 16.17 this means there exists a bundle atlas $\{(U_a, \varepsilon_a)\}$ such that each ε_a is (τ, r)-equivariant. Fix $p \in M$ and let $u_1, u_2 \in P_p$. Suppose $p \in U_a$. Let $g_i := (\varepsilon_a)_p(u_i)$ for $i = 1, 2$. Then by definition

$$\hat{\varepsilon}_a\big(\tau_{g_1^{-1}g_2}(u_1)\big) = (p, g_2) = \hat{\varepsilon}_a(u_2).$$

Since $\hat{\varepsilon}_a$ is a diffeomorphism, we have $u_2 = \tau_{g_1^{-1}g_2}(u_1)$. This shows that τ is transitive on the fibres.

Now suppose that τ is a smooth free fibre preserving action of G on P which is transitive on the fibres. Since π is a surjective submersion, by Proposition 6.13, for each $p \in M$ there is a neighbourhood U of p and a smooth local section $\psi\colon U \to P$ of π. Now consider the map

$$\varphi\colon U \times G \to \pi^{-1}(U), \qquad \varphi(q, g) := \tau_g(\psi(q)).$$

By hypothesis the map φ is a smooth injection. Under the splitting $T_{(q,g)}(U \times G) \cong T_qU \oplus T_gG$ from Problem C.1 the derivative of φ is given by

$$D\varphi(q, g) = D\tau_g(\psi(q)) \circ D\psi(q) + D\tau^{\psi(q)}(g).$$

[7] i.e. $P_{\pi(u)} = \mathsf{orb}_\tau(u)$ for all $u \in P$.

By Corollary 13.2 this map has maximal rank $\dim M + \dim G$ at (q, g), and hence by the Inverse Function Theorem 5.10, the map φ is a diffeomorphism. Thus we can write $\varphi^{-1} = (\pi, \varepsilon)$ for a uniquely determined smooth function $\varepsilon\colon \pi^{-1}(U) \to G$. This will form our desired principal bundle chart once we check (τ, r)-equivariance. Let $u \in \pi^{-1}(U)$ and assume that $\pi(u) = q \in U$. Then

$$(\pi, \varepsilon)(\tau_g(u)) = \big(q, \varepsilon(\tau_g(u))\big),$$

and hence

$$\tau_g(u) = \varphi\big(q, \varepsilon(\tau_g(u)\big) \tag{16.5}$$

In particular for $g = e$ we get

$$u = \tau_e(u) = \varphi\big(q, \varepsilon(u)\big). \tag{16.6}$$

Then for $g \in G$ we compute:

$$\begin{aligned} \varphi(q, r_g(\varepsilon(u)) &= \tau_{\varepsilon(u)g}(\psi(q)) \\ &= \tau_g \circ \tau_{\varepsilon(u)}(\psi(q)) \\ &= \tau_g\big(\varphi(q, \varepsilon(u)\big) \\ &= \tau_g(u) \\ &= \varphi\big(q, \varepsilon(\tau_g(u)\big), \end{aligned}$$

where the last two lines used (16.6) and (16.5) respectively. Since φ is a diffeomorphism this shows that $r_g(\varepsilon(u)) = \varepsilon(\tau_g(u))$. ■

Remark 16.19. An action τ satisfying (either) of the hypotheses of Proposition 16.18 is automatically proper. Indeed, the assertion that $\mathsf{orb}_\tau(u) = P_{\pi(u)}$ for all $u \in P$ implies that M is topologically the orbit space P/G. Problem G.2 then implies that τ is automatically proper.

Propositions 16.17 and 16.18 allow us to make the following alternative definition of a principal bundle, which is analogous to Definition 16.16.

Definition 16.20. Let $\pi\colon P \to M$ be a surjective smooth submersion, and set $P_p := \pi^{-1}(p)$. We say that P is a **principal G-bundle** if there exists a smooth free right action τ on P which is both fibre preserving and transitive on the fibres. A bundle chart (U, ε) which is (τ, r)-equivariant is called a **principal bundle chart**—Proposition 16.17 shows that we can find a bundle atlas of such charts, which we call a **principal bundle atlas**.

Definition 16.20 and Remark 16.19 allow us to use Quotient Manifold Theorem 13.6 to produce principal bundles.

Corollary 16.21. *Let τ be a proper free smooth action of a Lie group G on a smooth manifold P. Then $\rho\colon P \to P/G$ is a principal G-bundle.*

As a special case of Corollary 16.21 we have:

Corollary 16.22. *Let $M \cong G/H$ be a homogeneous space. Then M can be seen as the base space of a principal H-bundle $H \to G \xrightarrow{\pi} M$.*

Chapter 17
The Fibre Bundle Construction Theorem

In this chapter we discuss morphisms between G-fibre bundles, give a recipe for constructing such bundles, and show they are determined up to isomorphism by their transition functions. We conclude by showing how a vector bundle canonically determines a principal bundle. Next chapter we will investigate the converse direction: producing vector bundles from principal bundles.

DEFINITIONS 17.1. Let

$$L_1 \to E \xrightarrow{\pi_1} M, \qquad \text{and} \qquad L_2 \to F \xrightarrow{\pi_2} N$$

be two fibre bundles. A **fibre bundle morphism** is a pair (φ, Φ) of smooth maps

$$\varphi \colon M \to N, \qquad \Phi \colon E \to F.$$

such that the following commutes:

$$\begin{array}{ccc} E & \xrightarrow{\Phi} & F \\ {\scriptstyle \pi_1}\downarrow & & \downarrow{\scriptstyle \pi_2} \\ M & \xrightarrow[\varphi]{} & N \end{array}$$

We also say that Φ is a **fibre bundle morphism** along φ.

If (φ, Φ) is a fibre bundle morphism then

$$\Phi_p := \Phi|_{E_p} \colon E_p \to F_{\varphi(p)}$$

W. J. Merry, *Lectures on Differential Geometry I*,
https://doi.org/10.1007/978-3-032-03733-6_17

is a smooth map for each $p \in M$. If $\Phi_p : E_p \to F_{\varphi(p)}$ is a diffeomorphism for each $p \in M$ then we call Φ a **fibre bundle isomorphism along** φ. Since $E_p \cong L_1$ and $F_{\varphi(p)} \cong L_2$, we see that a fibre bundle isomorphism along φ can only exist when $L_1 \cong L_2$.

Of particular interest is the case where $M = N$ and $\varphi = \mathrm{id}$.

DEFINITION 17.2. Let $L_1 \to E \xrightarrow{\pi_1} M$ and $L_2 \to F \xrightarrow{\pi_2} M$ be two fibre bundles over the same base space M. A smooth map $\Phi\colon E \to F$ is said to be **fibre bundle morphism over** M if (id, Φ) is a fibre bundle morphism. If in addition Φ is a fibre bundle isomorphism along id, then we call Φ a **fibre bundle isomorphism**, and we say that E and F are **isomorphic fibre bundles**.

So much for general fibre bundles. The notion of morphisms between G-fibre bundles is rather messier to define, and for simplicity we focus only on isomorphisms. Let

$$L \to E \xrightarrow{\pi_1} M, \qquad \text{and} \qquad L \to F \xrightarrow{\pi_2} N$$

be two (G, σ)-fibre bundles. Let $\{(U_a, \varepsilon_a) \mid a \in A\}$ be a (G, σ)-bundle atlas for E and let $\{(V_b, \gamma_b) \mid b \in B\}$ be a (G, σ)-bundle atlas for F. Let $\varphi\colon M \to N$ be a smooth map and let $\Phi\colon E \to F$ be a fibre bundle isomorphism along φ. If $a \in A$ and $b \in B$ are such that $U_a \cap \varphi^{-1}(V_b) \neq \emptyset$, then for $p \in U_a \cap \varphi^{-1}(V_b)$ we can consider the composition

$$L \xrightarrow{\varepsilon_{a|p}^{-1}} E_p \xrightarrow{\Phi_p} F_{\varphi(p)} \xrightarrow{\gamma_{b|\varphi(p)}} L$$

Denote this composition by $f_a^b(p)$. Thus $f_a^b(p) \in \mathrm{Diff}(L)$ and we can regard f_a^b as a map

$$f_a^b : U_a \cap \varphi^{-1}(V_b) \to \mathrm{Diff}(L), \qquad p \mapsto f_a^b(p). \tag{17.1}$$

So far we have not used the (G, σ)-structure; we can define f_a^b for any fibre bundle. But now recall that the reason we introduced (G, σ)-fibre bundles was to "cut down" the possible values the transition functions from the infinite-dimensional space $\mathrm{Diff}(L)$ to a finite-dimensional subgroup $\{\sigma_g \mid g \in G\}$. It therefore stands to reason that an isomorphism between such bundles along φ should also respect this restriction—in other words, the functions f_a^b from (17.1) should also take values in $\{\sigma_g \mid g \in G\}$.

Here is the formal definition.

DEFINITION 17.3. Let σ be a smooth effective action of a Lie group G on a smooth manifold L. Assume we are given two fibre bundles

$$L \to E \xrightarrow{\pi_1} M, \qquad \text{and} \qquad L \to F \xrightarrow{\pi_2} N$$

together with a (G, σ)-bundle atlas $\{(U_a, \varepsilon_a) \mid a \in A\}$ on E and a (G, σ)-bundle atlas $\{(V_b, \gamma_b) \mid b \in B\}$ on F. Let $\varphi\colon M \to N$ be a smooth map and let $\Phi\colon E \to F$ be a fibre bundle isomorphism along φ. We say that Φ is a (G, σ)**-fibre bundle isomorphism along** φ if for each $a \in A$ and $b \in B$ such that $U_a \cap \varphi^{-1}(V_b) \neq \emptyset$, there exists a smooth map

$$h_a^b : U_a \cap \varphi^{-1}(V_b) \to G$$

such that if f_a^b is defined as in (17.1) then

$$f_b^a(p) = \sigma_{h_a^b(p)}.$$

As before, when $M = N$ and $\varphi = \mathrm{id}$ then we call Φ a (G, σ)-**fibre bundle isomorphism**, and we say that E and F are **isomorphic** (G, σ)-**fibre bundles**.

REMARK 17.4. As a fun exercise, try and correctly write down the definition of a morphism in the most general setting where one has two fibre bundles with different fibres, different Lie groups, and different actions.

We now give a recipe for constructing fibre bundles starting from the transition functions.

Theorem 17.5 (The Fibre Bundle Construction Theorem) *Let $\{U_a \mid a \in A\}$ be an open covering of a manifold M. Let G be a Lie group. Suppose for each $a, b \in A$ such that $U_a \cap U_b \neq \emptyset$, we are given a smooth map $g_{ab} \colon U_a \cap U_b \to G$ such that the following cocycle conditions are satisfied:*

$$\begin{cases} g_{ac}(p) = g_{ab}(p) g_{bc}(p), & \forall\, p \in U_a \cap U_b \cap U_c, \\ g_{aa}(p) = e, & \forall\, p \in U_a,\ \forall\, a \in A. \end{cases} \tag{17.2}$$

Suppose in addition we are given a smooth effective action σ of G on a smooth manifold L. Then there exists a fibre bundle $L \to E \xrightarrow{\pi} M$, which admits a (G, σ)-bundle atlas $\{(U_a, \varepsilon_a) \mid a \in A\}$ such that the transition functions ε_{ab} are given by

$$\varepsilon_{ab}(p) = \sigma_{g_{ab}(p)}.$$

As you might expect from a theorem with such complicated hypotheses (compare Proposition 1.17), the proof is basically trivial—most of the work is in formulating the hypotheses correctly!

Proof. Let

$$E := \left(\bigsqcup_{a \in A} (U_a \times L) \right) \Big/ \sim,$$

where we identify $(p, u) \in U_a \times L$ with $(q, v) \in U_b \times L$ if and only if $p = q$ and

$$u = \sigma_{g_{ab}(p)}(v).$$

Let

$$\rho \colon \bigsqcup_{a \in A} (U_a \times L) \to E$$

denote the quotient map, and endow E with the quotient topology. Let $\pi \colon E \to M$ denote the unique map so that

$$\begin{array}{ccc} \bigsqcup_{a\in A}(U_a \times L) & \xrightarrow{\ \mathrm{pr}_1\ } & \bigsqcup_{a\in A} U_a \\ \Big\downarrow \rho & & \Big\downarrow \text{“id”} \\ E & \xrightarrow[\ \pi\ ]{} & M \end{array}$$

Then $\pi\colon E \to M$ is continuous by definition of the quotient topology. For each $a \in A$, the restriction of ρ to $U_a \times L$ onto its image in E is a homeomorphism. Its inverse is of the form (π, ε_a), where $\varepsilon_a\colon \pi^{-1}(U_a) \to L$. This is our desired bundle atlas on E: we first make E into a smooth manifold using the procedure outlined in the second half of Remark 16.5—the fact that this gives a well-defined smooth structure follows from (17.2). It is then immediate from the definition that the transition functions of this bundle atlas are given by the maps ε_{ab}. This completes the proof. ■

Example 17.6. Take $L = \mathbb{R}$, and identify $\mathrm{GL}(\mathbb{R}) = \mathbb{R} \setminus \{0\}$. We take $M = S^1 \subset \mathbb{C}$. Let $U_1 = S^1 \setminus \{i\}$ and $U_2 := S^1 \setminus \{-i\}$. By the Fibre Bundle Construction Theorem 17.5 a smooth map $g_{12}\colon U_1 \cap U_2 \to \mathbb{R} \setminus \{0\}$ determines a vector bundle of rank 1 over M. If we set

$$g_{21}(z) := \begin{cases} 1, & \Re(z) > 0, \\ -1, & \Re(z) < 0, \end{cases}$$

then the vector bundle so obtained is called the **Möbius bundle**.[1] On Problem Sheet G you will show that there are exactly two rank 1 vector bundles over S^1 (up to isomorphism): the trivial bundle $S^1 \times \mathbb{R}$ and the Möbius bundle.

The next result clarifies the relation between the isomorphism class of a vector or principal bundle and its transition functions.

Proposition 17.7. *Let σ be an effective action of a Lie group G on a smooth manifold L. Assume we are given two (G, σ)-fibre bundles*

$$L \to E \xrightarrow{\pi_1} M, \qquad \textit{and} \qquad L \to F \xrightarrow{\pi_2} M$$

Let $\{U_a \mid a \in A\}$ be an open cover of M such that both E and F admit (G, σ)-bundle atlases over the U_a (this can always be achieved by taking intersections). Let

$$g^1_{ab}\colon U_a \cap U_b \to G, \qquad \textit{and} \qquad g^2_{ab}\colon U_a \cap U_b \to G$$

denote the transition functions of E and F with respect to these bundle atlases. Then E and F are isomorphic as (G, σ)-fibre bundles if and only if there exists a family $f_a\colon U_a \to G$ of smooth functions such that

$$f_a(p) \circ g^1_{ab}(p) = g^2_{ab}(p) \circ f_b(p), \qquad \forall\, p \in U_a \cap U_b,\ \forall\, a, b \in A.$$

[1] The Möbius bundle is a Möbius band of infinite width.

Similarly to the Fibre Bundle Construction Theorem, the most difficult part of Proposition 17.7 is formulating the correct hypotheses. The proof is left to you on Problem Sheet G.

COROLLARY 17.8. *The (G, σ)-fibre bundle constructed in the Fibre Bundle Construction Theorem* 17.5 *is unique up to (G, σ)-fibre bundle isomorphism.*

We now specialise the preceding definitions to vector and principal bundles. The definitions are not quite special cases of what we have already done (since in Definition 17.3 we only looked at (G, σ)-fibre bundle isomorphisms along maps).

DEFINITION 17.9. Let (φ, Φ) be a fibre bundle morphism between two vector bundles $\pi_1 \colon E \to M$ and $\pi_2 \colon F \to N$. We say that (φ, Φ) is a **vector bundle morphism** if $\Phi_p \colon E_p \to F_{\varphi(p)}$ is a *linear* map for each $p \in M$. When φ is fixed, we also say that Φ is a **vector bundle morphism along** φ if (φ, Φ) is a vector bundle morphism. If Φ_p maps each fibre E_p isomorphically onto $F_{\varphi(p)}$ then Φ is called a **vector bundle isomorphism along** φ.

EXAMPLE 17.10. Let $\varphi \colon M \to N$ be a smooth map. Then $D\varphi \colon TM \to TN$ is a vector bundle morphism along φ.

DEFINITION 17.11. Let $\pi_1 \colon E \to M$ and $\pi_2 \colon F \to M$ be two vector bundles over the same base space M. A smooth map $\Phi \colon E \to F$ is said to be **vector bundle morphism over** M if (id, Φ) is a vector bundle morphism. If Φ is in addition a diffeomorphism we say that Φ is a **vector bundle isomorphism**, and that E and F are **isomorphic vector bundles**.

REMARK 17.12. A vector bundle morphism over M is a vector bundle isomorphism if and only if it is a diffeomorphism. This is not true for vector bundle isomorphisms along a map. For instance, if M is a manifold and $p \in M$ then (thinking of $\{p\}$ as a zero-dimensional manifold) we have a smooth map $\iota_p \colon \{p\} \hookrightarrow M$ given by inclusion. If E is any vector bundle over M then the inclusion map $E_p \hookrightarrow E$ is a vector bundle isomorphism along ι_p, but of course it is not a diffeomorphism if $M \neq \{p\}$.

REMARK 17.13. It is also not the case that a vector bundle morphism along a diffeomorphism is necessarily a vector bundle isomorphism. For instance if we take the diffeomorphism to be the identity and the bundle morphism to be the zero map from a ('non-zero') bundle to itself then this is not a vector bundle isomorphism, but it is a vector bundle morphism along a diffeomorphism.

The definition of morphisms for principal bundles is analogous.

DEFINITION 17.14. Let (φ, Φ) be a fibre bundle morphism between two principal G-bundles $\pi_1 \colon P \to M$ and $\pi_2 \colon Q \to N$. Let τ_1 and τ_2 denote the associated right actions on P and Q respectively. We say that (φ, Φ) is a **principal bundle morphism** if Φ is (τ_1, τ_2)-equivariant. When φ is fixed, we also say that Φ is a **principal bundle morphism along** φ if (φ, Φ) is a principal bundle morphism. If Φ is in addition a diffeomorphism then we say Φ is a **principal bundle isomorphism along** φ.

Whilst the definition of principal bundle morphisms looks superficially similar to that of vector bundle morphisms, there is already a major difference between vector and principal bundles.

LEMMA 17.15. *Let* $\pi_1: P \to M$ *and* $\pi_2: Q \to N$ *be two principal G-bundles. Suppose* $\Phi: P \to Q$ *is a principal bundle morphism along a diffeomorphism* $\varphi: M \to N$. *Then* Φ *is automatically a diffeomorphism, and hence a principal bundle isomorphism along* φ.

As Remark 17.13 shows, Lemma 17.15 is *not* true for vector bundle morphisms! The proof of Lemma 17.15 is on Problem Sheet G.

DEFINITION 17.16. Let $\pi_1: P \to M$ and $\pi_2: Q \to M$ be two principal G-bundles over the same base space M. A diffeomorphism $\Phi: P \to Q$ is said to be a **principal bundle isomorphism** if (id, Φ) is a principal bundle morphism. If such a Φ exists we say that P and Q are **isomorphic principal** G**-bundles**.

Note there is no point mimicking Definition 17.11 by first defining a "principal bundle morphism over M", and then declaring that a principal bundle isomorphism is a principal bundle morphism which is in addition a diffeomorphism. Indeed, any principal bundle morphism is automatically an isomorphism by Lemma 17.15, since the identity is a diffeomorphism.

REMARK 17.17. If $\pi: P \to M$ is a principal bundle and $\Phi: P \to P$ is a principal bundle isomorphism from P to itself then we call Φ a **gauge transformation**. The name comes from physics.

Having defined morphisms, we can now define vector and principal subbundles.

DEFINITION 17.18. Let $\pi_1: E \to M$ and $\pi_2: F \to M$ are two vector bundles over the same manifold such that $E \subset F$ is an embedded submanifold. We say that E is a **vector subbundle** of F if the inclusion $E \hookrightarrow F$ is a vector bundle morphism over M.

EXAMPLE 17.19. If Δ is a distribution on M then one can think of Δ as a vector subbundle of TM.

The notion of principal subbundles is slightly subtler, because we also want to allow for a Lie subgroup. Suppose $\psi: G \to H$ is a Lie group homomorphism and τ is a smooth action of H on a space M. Define a new action τ^{ψ} of G on M by

$$\tau_g^{\psi} := \tau_{\psi(g)}, \qquad g \in G.$$

Part (i) of Examples 12.17 is a special case of this.

We now extend the notion of a principal bundle morphism for different Lie groups.

DEFINITION 17.20. Suppose G and H are two Lie groups. Let $\pi_1: P \to M$ be a principal G-bundle and let $\pi_2: Q \to N$ be a principal H-bundle. Suppose $\psi: G \to H$ is a Lie group homomorphism. A **principal bundle morphism from** P **to** Q **with respect to** ψ consists of a pair of smooth maps $\varphi: M \to N$ and $\Phi: P \to Q$ such that $\pi_2 \circ \Phi = \varphi \circ \pi_1$ and such that Φ is (τ_1, τ_2^{ψ})-equivariant. If $M = N$ and $\varphi = \text{id}$ then we call Φ a **principal bundle homomorphism with respect to** ψ.

Warning: In this case the analogue of Lemma 17.15 is *not* true, and so Φ does not need to be a principal bundle isomorphism with respect to ψ.

REMARK 17.21. Definition 17.20 is a special case of the more general setup you were invited to guess in Remark 17.4.

Here is the definition of a principal subbundle.[2]

DEFINITION 17.22. Let G be a Lie group and suppose $H \subset G$ is a Lie subgroup. Suppose $\pi_1 : P \to M$ is a principal H-bundle and $\pi_2 : Q \to M$ is a principal G-bundle such that $P \subset Q$ is a weakly embedded submanifold. We say that P is a **principal H-subbundle** of Q if the inclusion $P \hookrightarrow Q$ is a principal bundle homomorphism with respect to the inclusion $H \hookrightarrow G$.

The Fibre Bundle Construction Theorem 17.5 allows us to produce a principal G-bundle from any G-fibre bundle. Indeed, suppose $L \to E \xrightarrow{\pi} M$ is a (G, σ)-fibre bundle. Let $\{(U_a, \varepsilon_a)\}$ be a (G, σ)-bundle atlas. This means we can write

$$\varepsilon_{ab}(p) = \sigma_{g_{ab}(p)}$$

for functions $g_{ab} : U_a \cap U_b \to G$. The functions $\{g_{ab}\}$ satisfy the cocycle condition, and hence by the Fibre Bundle Construction Theorem 17.5, there exists a principal G-bundle P whose transition functions are given by left translation by the g_{ab}. This principal bundle is unique up to principal bundle isomorphism by Corollary 17.8, and we give it a special name:

DEFINITION 17.23. We call P the **induced principal bundle** of E.

This is a slight abuse of terminology, as P is only unique up to isomorphism.

We conclude this chapter by giving an explicit construction of P in the case where E is a vector bundle.

DEFINITION 17.24. Let $V \to E \xrightarrow{\pi} M$ be a vector bundle. Fix $p \in M$. We call a linear isomorphism $\ell : V \to E_p$ a **frame** at p and let $\mathrm{Fr}(E_p)$ denote the set of such isomorphisms.

Since any two isomorphisms $\ell_1, \ell_2 : V \to E_p$ differ by some element of $\mathrm{GL}(V)$, i.e. there exists $A \in \mathrm{GL}(V)$ such that $\ell_2 = \ell_1 \circ A$. In fact, if we fix our favourite isomorphism ℓ then the map $\mathrm{GL}(V) \to \mathrm{Fr}(E_p)$ given by $A \mapsto \ell \circ A$ is a bijection.

One can equivalently regard $\mathrm{Fr}(E_p)$ as the set of bases of the vector space E_p, since for any $\ell \in \mathrm{Fr}(E_p)$ the vectors (ℓe_i) form a basis of E_p, where e_i are the standard basis vectors in V, and conversely given a basis (v_i) there is a uniquely determined linear isomorphism $\ell : V \to E_p$ such that $\ell e_i = v_i$ for each i.

DEFINITION 17.25. We now form the total space

$$\mathrm{Fr}(E) := \bigsqcup_{p \in M} \mathrm{Fr}(E_p),$$

[2] Principal subbundles will play no role in this book (apart from in Problem G.10). However they will be very important in Volume II, for example in the Ambrose-Singer Holonomy Theorem (Volume II Theorem 15.7).

and let $\Pi\colon \mathrm{Fr}(E) \to M$ denote the map that sends $\mathrm{Fr}(E_p)$ to p. We call $\mathrm{Fr}(E)$ the **frame bundle** of E.

PROPOSITION 17.26. *Let $V \to E \xrightarrow{\pi} M$ be a vector bundle. Then $\Pi\colon \mathrm{Fr}(E) \to M$ is a principal $\mathrm{GL}(V)$-bundle over M. Moreover $\mathrm{Fr}(E)$ is the induced principal bundle of E.*

Proof. Let $\{(U_a, \varepsilon_a) \mid a \in A\}$ denote a vector bundle atlas. For each $p \in U_a$, $\varepsilon^{-1}_{a|p}\colon V \to E_p$ is a linear isomorphism, and thus $\varepsilon^{-1}_{a|p} \in \mathrm{Fr}(E_p)$. Define a map

$$\gamma_a\colon \Pi^{-1}(U_a) \to \mathrm{GL}(V),$$

by declaring that

$$\gamma_a(\varepsilon^{-1}_{a|p} \circ A) = A.$$

We will show that $\{(U_a, \gamma_a) \mid a \in A\}$ is a $\mathrm{GL}(V)$-principal bundle atlas on $\mathrm{Fr}(E)$. As in the proof of the Fibre Bundle Construction Theorem 17.5, we will simultaneously show that $\mathrm{Fr}(E)$ is a smooth manifold and a principal $\mathrm{GL}(V)$-bundle by computing the transition functions γ_{ab}. Fix $p \in U_a \cap U_b$. Then by definition

$$\begin{aligned}\gamma_{ab}(p)(A) &= \gamma_{a|p} \circ \gamma^{-1}_{b|p}(A)\\ &= \gamma_{a|p}\left(\varepsilon^{-1}_{b|p} \circ A\right)\\ &= \varepsilon_{ab}(p) \circ A.\end{aligned}$$

Thus the transition functions of $\mathrm{Fr}(E)$ are just left composition by the transition functions of E. That is,

$$\gamma_{ab}(p) = l_{\varepsilon_{ab}(p)},$$

where l is left translation in the Lie group $\mathrm{GL}(V)$. This means that the transition functions of E play the role of the functions g_{ab} in (16.3). Thus $\mathrm{Fr}(E)$ is indeed a principal $\mathrm{GL}(V)$ bundle, and by definition $\mathrm{Fr}(E)$ is the induced principal bundle of E. ■

We have shown how to produce a principal bundle from a vector bundle. Next chapter we will show how to produce (many) vector bundles from a principal bundle.

Problem Sheet G

A (♣) indicates the problem is non-examinable. This does not necessarily imply it is difficult.

Problem G.1 Show that the real projective space $\mathbb{R}P^{m-1}$ can be seen as the homogeneous space $\mathrm{SO}(m)/\,\mathrm{O}(m-1)$.

Problem G.2 Let σ be a smooth free action of G on M. Assume that the quotient space M/G admits the structure of a smooth manifold such that the quotient map $\rho\colon M \to M/G$ is a smooth submersion. Prove that σ is a proper action.[3]

Problem G.3 Let $\pi_1 : P \to M$ and $\pi_2 : Q \to N$ be two principal G-bundles. Suppose (φ, Φ) is a principal bundle morphism from P to Q such that φ is a diffeomorphism. Prove that Φ is also a diffeomorphism.

Problem G.4 Let σ be an effective action of a Lie group G on a smooth manifold L. Assume we are given two fibre bundles

$$L \to E \xrightarrow{\pi_1} M, \qquad \text{and} \qquad L \to F \xrightarrow{\pi_2} M$$

Let $\{U_a \mid a \in A\}$ be an open cover of M such that both E and F admit G-bundle atlases over the U_a.[4] Let

$$g^1_{ab} : U_a \cap U_b \to G, \qquad \text{and} \qquad g^2_{ab} : U_a \cap U_b \to G$$

denote the transition functions of E and F with respect to these bundle atlases. Prove that E and F are isomorphic as (G, σ)-fibre bundles if and only if there exists a family $f_a : U_a \to G$ of smooth functions such that

$$f_a(p) \circ g^1_{ab}(p) = g^2_{ab}(p) \circ f_b(p), \qquad \forall\, p \in U_a \cap U_b,\ \forall\, a, b \in A.$$

Problem G.5 Let $V \to E \xrightarrow{\pi} M$ be a vector bundle. Prove that it is possible to reduce the structure group from $\mathrm{GL}(V)$ to $\mathrm{O}(V)$. Find an example where it is not possible to reduce the structure group from $\mathrm{GL}(V)$ to $\mathrm{GL}^+(V)$.

(♣)Problem G.6 Prove that there are exactly two vector bundles of rank 1 over S^1 (up to vector bundle isomorphism).

(♣)Problem G.7 Prove that the Klein Bottle [4] is a fibre bundle over S^1 with fibre S^1. Prove however that the Klein bottle is not a principal S^1-bundle over S^1.

---

[3] This question is a partial converse to the Quotient Manifold Theorem 13.6.

[4] This can always be achieved by taking intersections.

(♣)Problem G.8 Find a non-trivial principal S^1-bundle over $\mathbb{R}P^2$.

Problem G.9 Suppose σ is a smooth transitive action of a Lie group G on M, so that M is the homogeneous space G/H for an appropriate subgroup H of G. Prove that the subgroup of G acting trivially on M is the largest normal subgroup $N(H)$ of G contained in H. Let $\overline{G}$ and $\overline{H}$ denote the quotient groups $G/N(H)$ and $H/N(H)$ respectively. Prove that $\overline{G}$ acts effectively and transitively on M, and M is the homogeneous space $\overline{G}/\overline{H}$.

(♣)Problem G.10 Let $\pi\colon P \to M$ be a principal G-bundle with corresponding right action τ. Let $H \subset G$ be a Lie subgroup, and let $Q \subset P$ be a subset such that:

(i) The restriction $\pi|_Q\colon Q \to M$ is surjective.
(ii) If $q \in Q$ and $h \in H$ then $\tau_h(q) \in Q$.
(iii) For all $p \in M$, the action of H on $Q \cap P_p$ is transitive.
(iv) For all $p \in M$, there exists a neighbourhood U of p and a smooth local section $\psi\colon U \to P$ of π such that $\psi(x) \in Q$ for all $x \in U$.

Prove that $\pi|_Q\colon Q \to M$ is a principal H-bundle, and that moreover Q is a principal H-subbundle of P.

Chapter 18
Associated Bundles

We begin this chapter by explaining how to build fibre bundles from principal bundles.

DEFINITION 18.1. Let $\pi\colon P \to M$ be a principal G-bundle, and let σ be a smooth effective left action of G on another smooth manifold L. Define an equivalence relation $\sim$ on $P \times L$ by setting:

$$(\tau_g(u), q) \sim (u, \sigma_g(q)), \qquad u \in P,\ g \in G,\ q \in L, \tag{18.1}$$

Define $P \times_G L$ to be the quotient space $(P \times L)/{\sim}$. Writing $[u, q]$ for the equivalence class of (u, q), we define a map

$$\pi_L\colon P \times_G L \to M, \qquad [u, q] \mapsto \pi(u).$$

We call $P \times_G L$ an **associated bundle** of P.

REMARK 18.2. The notation $\pi_L\colon P \times_G L \to M$ is somewhat ambiguous, since we really should specify the action we are using. When confusion is possible, we will occasionally write $\pi_{L,\sigma}\colon P \times_{G,\sigma} L \to M$ or $\pi_\sigma\colon P \times_\sigma L \to M$ instead. Moreover the assumption that σ is effective is not used anywhere in the proof—it is simply there so as to fit in with the framework of Definition 16.8. As Remark 16.9 shows, restricting to effective actions does not actually involve any loss of generality, and thus assuming it here is harmless.

Theorem 18.3 (The Associated Bundle Theorem) *Let $\pi\colon P \to M$ be a principal G-bundle, and let σ be a smooth effective action of G on L.*

(i) *The associated bundle $\pi_L\colon P \times_G L \to M$ is a (G, σ)-fibre bundle with fibre L, and moreover P is the induced principal bundle of $P \times_G L$.*
(ii) *The quotient map $\wp\colon P \times L \to P \times_G L$ given by $\wp(u, q) := [u, q]$ is another principal G-bundle, and the first projection $\mathrm{pr}_1\colon P \times L \to P$ is a principal bundle morphism along π_L:*

W. J. Merry, *Lectures on Differential Geometry I*,
https://doi.org/10.1007/978-3-032-03733-6_18

$$\begin{array}{ccc} P \times L & \xrightarrow{\mathrm{pr}_1} & P \\ \Big\downarrow \wp & & \Big\downarrow \pi \\ P \times_G L & \xrightarrow[\pi_L]{} & M \end{array}$$

(iii) *For each* $u \in P$, *the map* $\psi_u : L \to P \times_G L$ *given by* $q \mapsto [u, q]$ *is a diffeomorphism from* L *to* $\pi_L^{-1}(\pi(u))$.

Proof. We will prove the result in four steps.

1. In this step we define a tentative bundle chart for $P \times_G L$ and prove that the associated local trivialisation is bijective. Suppose $\varepsilon \colon \pi^{-1}(U) \to G$ is a principal bundle chart over an open set $U \subset M$. We define a map $\gamma \colon \pi_L^{-1}(U) \to L$ by

$$\gamma[u, q] := \sigma_{\varepsilon(u)}(q).$$

This is well defined because ε is (τ, r)-equivariant: if $[u, q] = [u_1, q_1]$ then there exists $g \in G$ such that $\tau_g(u_1) = u$ and $\sigma_g(q) = q_1$. Then

$$\begin{aligned} \gamma[u, q] &= \sigma_{\varepsilon(u)}(q) \\ &= \sigma_{\varepsilon(\tau_g(u_1))}(q) \\ &= \sigma_{\varepsilon(u_1)g}(q) \\ &= \sigma_{\varepsilon(u_1)} \circ \sigma_g(q) \\ &= \sigma_{\varepsilon(u_1)}(q_1) \\ &= \gamma[u_1, q_1]. \end{aligned}$$

We claim that $\hat{\gamma} := (\pi_L, \gamma) \colon \pi_L^{-1}(U) \to U \times L$ is bijective. To see this, for each $p \in U$ let $u_p \in P_p$ denote the unique element such that $\varepsilon(u_p) = e$ (this is well defined as ε_p is a diffeomorphism). Now define $\varphi \colon U \times L \to \pi_L^{-1}(U)$ by $\varphi(p, q) := [u_p, q]$. We claim that φ is an inverse to $\hat{\gamma}$. Indeed,

$$\begin{aligned} \hat{\gamma} \circ \varphi(p, q) &= (\pi_L, \gamma)[u_p, q] \\ &= \big(p, \sigma_{\varepsilon(u_p)}(q)\big) \\ &= (p, \sigma_e(q)) \\ &= (p, q). \end{aligned}$$

Going the other way round, if $p \in U$ and $u \in P_p$ then by equivariance

$$\begin{aligned} \hat{\varepsilon}\big(\tau_{\varepsilon(u)}(u_p)\big) &= \big(p, r_{\varepsilon(u)}\varepsilon(u_p)\big) \\ &= (p, \varepsilon(u)e) \\ &= (p, \varepsilon(u)) \\ &= \hat{\varepsilon}(u) \end{aligned}$$

and thus as $\hat{\varepsilon}$ is a diffeomorphism we must have

$$\tau_{\varepsilon(u)}(u_p) = u. \tag{18.2}$$

We therefore have for $u \in P_p$ that

$$\begin{aligned}\varphi \circ \hat{\gamma}[u, q] &= \varphi(p, \sigma_{\varepsilon(u)}(q)) \\ &= \big[u_p, \sigma_{\varepsilon(u)}(q)\big] \\ &= \big[\tau_{\varepsilon(u)}(u_p), q\big] \\ &= [u, q],\end{aligned}$$

where the last two equalities used (18.1) and (18.2) respectively. Thus $\hat{\gamma}$ is bijective.

2. In this step we prove (i). Let $\{(U_a, \varepsilon_a) \mid a \in A\}$ be a principal bundle atlas for P, and for each $a \in A$, let $\gamma_a \colon \pi_L^{-1}(U_a) \to L$ be defined as in Step 1. We claim that $\{(U_a, \gamma_a)\}$ can serve as a bundle atlas on $P \times_G L$. For this we must investigate the transition functions γ_{ab}. We want to show that

$$\gamma_{ab}(p)(q) = \sigma_{g_{ab}(p)}(q)$$

for some smooth functions $g_{ab} \colon U_a \cap U_b \to G$. This however is immediate from the previous step:

$$\gamma_{ab}(p)(q) = \sigma_{\varepsilon_{ab}(p)}(q), \tag{18.3}$$

that is,

$$g_{ab}(p) = \varepsilon_{ab}(p)$$

This is smooth, and hence as in Remark 16.5, we can endow $P \times_G L$ with a smooth structure by declaring all the maps $\hat{\gamma}_a$ to be diffeomorphisms. Then the collection $\{(U_a, \gamma_a)\}$ form a (G, σ)-bundle atlas, and $P \times_G L$ is a (G, σ)-fibre bundle. Moreover by definition P is the principal bundle induced by $P \times_G L$.

3. We now prove (ii). On the open set $\pi^{-1}(U_a) \times L$ of $P \times L$, the map $\wp$ is given by

$$\wp(u, q) = \hat{\gamma}_a^{-1}\big(\pi(u), \sigma_{\varepsilon_a(u)}(q)\big),$$

This shows that $\wp$ is smooth. If we differentiate this equation we obtain

$$\begin{aligned}&D\wp(u, q) \\ &\quad = D\hat{\gamma}_a^{-1}\big(\pi(u), \sigma_{\varepsilon_a(u)}(q)\big) \circ \Big(D\pi(u), D\sigma_{\varepsilon_a(u)}(q) + D\sigma^q(\varepsilon_a(u)) \circ D\varepsilon_a(u)\Big).\end{aligned}$$

Since $\hat{\gamma}_a$ and $\sigma_{\varepsilon_a(u)}$ are diffeomorphisms, it follows from Proposition 16.4 that $\wp$ is a submersion. We now define a right action $\tilde{\tau}$ of G on $P \times L$ by

$$\tilde{\tau}_g(u, q) = \big(\tau_g(u), \sigma_{g^{-1}}(q)\big).$$

This action is free since τ is. Moreover $\tilde{\tau}$ preserves the fibres of $\wp$:

$$\begin{aligned}\wp(\tilde{\tau}_g(u,q)) &= \wp(\tau_g(u), \sigma_{g^{-1}}(q)) \\ &= [\tau_g(u), \sigma_{g^{-1}}(q)] \\ &= [u,q] \\ &= \wp(u,q)\end{aligned}$$

by the defining relationship (18.1). Finally $\tilde{\tau}$ is transitive on the fibres, since if

$$\wp(u_1, q_1) = \wp(u_2, q_2)$$

then by (18.1) again there exists $g \in G$ such that

$$\tau_g(u_1) = u_2, \qquad \sigma_g(q_2) = q_1,$$

and hence

$$\tilde{\tau}_g(u_1, q_1) = (u_2, q_2).$$

Since $P \times_G L$ is a manifold, Problem G.2 implies that $\tilde{\tau}$ is proper. It now follows from Proposition 16.18 that $\wp\colon P \times L \to P \times_G L$ is another principal G-bundle. This proves (ii). The identity

$$\mathrm{pr}_1(\tau_g(u), \sigma_{g^{-1}}(q)) = \tau_g(u) = \tau_g(\mathrm{pr}_1(u,q))$$

shows that pr_1 is a principal G-bundle morphism along π_L.

4. It remains to prove (iii). Fix $p \in M$ and $u \in P_p$. The map $\psi_u\colon L \to \pi_L^{-1}(p)$ given by $q \mapsto [u,q]$ is smooth because $\wp$ is. Moreover if $p \in U_a$ then near $[u,q]$ the map

$$\sigma_{\varepsilon_a(u)^{-1}} \circ \gamma_a\colon \pi_L^{-1}(p) \to L$$

is a smooth inverse to ψ_u. Thus ψ_u is a diffeomorphism. This completes the proof of part (iii), and hence also the theorem. ∎

Corollary 18.4. *Let $L \to E \xrightarrow{\pi} M$ be a (G,σ)-fibre bundle. Then there exists a principal G-bundle P such that E is isomorphic as a (G,σ)-fibre bundle to the associated bundle $P \times_G L$. Moreover P is unique up to principal bundle isomorphism.*

Proof. Let P denote the principal G-bundle induced by E. Thus the transition functions of E can be identified with the transition functions of P. Now consider the new fibre bundle $P \times_G L$. As the proof of Theorem 18.3 shows, the transition functions of $P \times_G L$ can also be identified with the transition functions of P. Thus E and $P \times_G L$ have the same transition functions, and by Proposition 17.7 it follows that E and $P \times_G L$ are isomorphic as (G,σ)-fibre bundles. ∎

Suppose $L = V$ is a vector space and σ is a linear action. Then (18.3) shows that the transition functions of $P \times_G V$ are linear, and hence $P \times_G V$ is a vector bundle over M. It will be useful later to have an explicit description of the vector space structure on the fibres. Fix $p \in M$, $v, w \in V$ and $c \in \mathbb{R}$. We define addition and scalar multiplication on $\pi_V^{-1}(p)$ as

$$[u, v] + c[u, w] := [u, v + cw],$$

where $u \in P_p$ is any element in the fibre over p and $v + cw$ is addition and scalar multiplication in the vector space V. This is well defined, i.e. independent of the choice of u, since if $g \in G$ then $[u, v] = [\tau_g(u), \sigma_{g^{-1}}(v)]$ and $[u, w] = [\tau_g(u), \sigma_{g^{-1}}(w)]$ and then since $\sigma_{g^{-1}}$ is linear

$$\begin{aligned}\left[\tau_g(u), \sigma_{g^{-1}}(v) + c\sigma_{g^{-1}}(w)\right] &= \left[\tau_g(u), \sigma_{g^{-1}}(c + vw)\right] \\ &= [u, v + cw].\end{aligned}$$

This also shows that the bundle charts γ constructed in the proof of Theorem 18.3 are vector bundle charts in the sense of Definition 16.16. Note that the map ψ_u from part (iii) of Theorem 18.3 satisfies

$$\begin{aligned}\psi_u(v + cw) &= [u, v + cw] \\ &= [u, v] + c[u, w] \\ &= \psi_u(v) + c\psi_u(w),\end{aligned}$$

and hence ψ_u is linear. We have thus proved:

COROLLARY 18.5. *Let $\pi\colon P \to M$ be a principal G-bundle. Suppose σ is a faithful representation*[1] *of G on a vector space V. Then the associated bundle $P \times_G V$ is a vector bundle over M and the map ψ_u from part* (iii) *of Theorem* 18.3 *is a linear isomorphism.*

EXAMPLE 18.6. Let $V \to E \xrightarrow{\pi} M$ be a vector bundle, and let $\Pi\colon \mathrm{Fr}(E) \to M$ denote the frame bundle. The canonical representation of $\mathrm{GL}(V)$ on V produces a new vector bundle

$$\Pi_V\colon \mathrm{Fr}(E) \times_{\mathrm{GL}(V)} V \to M$$

It follows from Corollary 18.4 that this vector bundle is isomorphic to E. If $p \in M$ then a bundle chart (U, ε) about p provides an explicit isomorphism $E_p \cong \Pi_V^{-1}(p)$ via:

$$E_p \xrightarrow{\ \varepsilon_p\ } V \xrightarrow{\ \psi_{\varepsilon_p^{-1}}\ } \Pi_V^{-1}(p)$$

This may seem a bit silly: starting from a vector bundle E, we constructed its frame bundle, and then used the Associated Bundle Theorem to produce... E again.[2] But this is missing the point: the real power of Corollary 18.5 is that we are free to choose *any* faithful representation of $\mathrm{GL}(V)$.

DEFINITION 18.7. Let $V \to E \xrightarrow{\pi} M$ be a vector bundle. Let $\Pi\colon \mathrm{Fr}(E) \to M$ denote the frame bundle of E. The group $\mathrm{GL}(V)$ acts on the dual space $V^* = \mathrm{Hom}(V, \mathbb{R})$ by

$$\lambda \mapsto \lambda \circ A^{-1} \tag{18.4}$$

[1] A faithful representation is simply another word for an effective linear action, cf. part (iii) of Examples 12.13.

[2] Strictly speaking, all we have done is produce a new bundle which is isomorphic to E as a vector bundle.

for $A \in \mathrm{GL}(V)$ and $\lambda \in V^*$. This gives an associated vector bundle

$$\Pi_{V^*} : \mathrm{Fr}(E) \times_{\mathrm{GL}(V)} V^* \to M.$$

We call this vector bundle the **dual bundle** of E, and write it as E^*.

One can also give a very explicit construction of E^*. Namely, the total space E^* is given as the union

$$E^* := \bigsqcup_{p \in M} E_p^* \tag{18.5}$$

where $E_p^* = \mathrm{Hom}(E_p, \mathbb{R})$ is the dual vector space to E_p. A bundle chart (U, ε) on E gives rise to a bundle chart (U, ε^*) on E^* via

$$\varepsilon_p^*(\lambda)(v) := \lambda(\varepsilon_p^{-1}(v)), \qquad \lambda \in E_p^*,\ v \in V$$

together with an explicit isomorphism

$$E_p^* \xrightarrow{\varepsilon_p^*} V^* \xrightarrow{\psi_{(\varepsilon_p^*)^{-1}}} \Pi_{V^*}^{-1}(p)$$

Example 18.8. The cotangent bundle is the dual bundle to the tangent bundle.

The explicit construction of the dual bundle E^* in (18.5) is only isomorphic to the dual bundle from Definition 18.7. Nevertheless, we will suppress this in the discussion that follows, and regard them as being "the same". It is important to understand why this is harmless.

In general it is **not** alright to simply work with vector bundles up to isomorphism—at least, if one did then the whole dual bundle construction would be pointless, since any vector bundle is isomorphic to its dual bundle.[3]

However there is a stronger notion than isomorphism: **canonical isomorphism**.[4] Roughly speaking, to say two mathematical objects are *canonically* isomorphic is to say that the isomorphism does not involve making any choices. This really is a stronger property. Indeed, for any finite-dimensional vector space V, you hopefully remember from linear algebra that:

$$\begin{aligned} &V \cong V^* &&\text{(non-canonical isomorphism)} \\ &V \cong V^{**} &&\text{(canonical isomorphism)} \end{aligned} \tag{18.6}$$

The precise mathematical definition of canonical isomorphism will appear in the bonus section of the next chapter. For now it is only important to understand that:

[3] See Problem Sheet I.

[4] Also called **natural isomorphism**.

> If two mathematical objects are **canonically** isomorphic, it is **harmless** to regard them as actually being the same object; whereas when the isomorphism is non-canonical then regarding them as the same object "loses" information.

Returning to the situation at hand: the construction of the dual bundle in (18.5) is canonically isomorphic to Definition 18.7. This is not too hard to prove directly (namely, one just checks that the obvious isomorphism between the two doesn't depend on choices). The general statement (which covers all possible cases of interest) is given in Theorem 19.64 next chapter. For all subsequent vector bundle constructions, we will suppress canonical isomorphisms from our discussion without further comment.

DEFINITION 18.9. Let $G \to P \xrightarrow{\pi_1} M$ and $H \to Q \xrightarrow{\pi_2} M$ be principal bundles over the same base manifold M with corresponding right actions τ_1 and τ_2. Let

$$P \star Q := \bigsqcup_{p \in M} P_p \times Q_p.$$

Note that the total space of $P \star Q$ is *not* the product space $P \times Q$.

Define a right action τ of $G \times H$ on $P \star Q$ by

$$\tau_{(g,h)}(u_1, u_2) := \big(\tau_{1|g}(u_1), \tau_{2|h}(u_2)\big).$$

This is a free proper right action of $G \times H$ on $P \star Q$ whose orbits are exactly the fibres. Thus Remark 16.5 and Proposition 16.18 tells us that $P \star Q$ is a principal $G \times H$ bundle over M whose fibre over $p \in M$ is $P_p \times Q_p$. We call it the **product principal bundle**.

Let us apply this to vector bundles: if $V \to E \xrightarrow{\pi_1} M$ and $W \to F \xrightarrow{\pi_2} M$ are two vector bundles over the same manifold M, then $\mathrm{Fr}(E)$ is a principal $\mathrm{GL}(V)$-bundle over M and $\mathrm{Fr}(F)$ is a principal $\mathrm{GL}(W)$-bundle over M. Thus $\mathrm{Fr}(E) \star \mathrm{Fr}(F)$ is a principal $\mathrm{GL}(V) \times \mathrm{GL}(W)$ bundle over M.

DEFINITION 18.10. The group $\mathrm{GL}(V) \times \mathrm{GL}(W)$ acts on $V \times W$ by the canonical representation, and thus we can form the vector bundle

$$\big(\mathrm{Fr}(E) \star \mathrm{Fr}(F)\big) \times_{\mathrm{GL}(V) \times \mathrm{GL}(W)} (V \oplus W)$$

over M. This is a vector bundle with fibre $V \oplus W$ and we denote it by $E \oplus F$ and call it the **direct sum**[5] of E and F.

The direct sum bundle can also be explicitly constructed as

$$E \oplus F := \bigsqcup_{p \in M} E_p \oplus F_p.$$

[5] **Warning:** As vector spaces, the direct sum $V \oplus W$ is the same thing as the product $V \times W$, and we often use the notation interchangeably. For vector bundles, we always use the $\oplus$ notation. This is because $E \times F$ is used to denote a different bundle; see Problem H.5.

This discussion can be summed up by the following:

> METATHEOREM. *Anything you can do with vector spaces, you can also do with vector bundles.*

The "proof" of the Metatheorem is Corollary 18.5. Or more accurately: the statement of Corollary 18.5 is one way of turning the Metatheorem into an precise mathematical statement.[6]

In the next chapter, we will see three further applications of the Metatheorem:

(i) If V and W are vector spaces, then the set $\mathrm{Hom}(V, W)$ of linear maps from V to W is a vector space; thus if E and F are vector bundles then there is a well-defined **homomorphism bundle** $\mathrm{Hom}(E, F)$.
(ii) If V and W are vector spaces, their tensor product $V \otimes W$ is another vector space; thus if E and F are vector bundles then there is a well-defined **tensor bundle** $E \otimes F$.
(iii) If V is a vector space, its exterior algebra $\bigwedge V$ is another vector space; thus if E is a vector bundle there is a well-defined **exterior algebra bundle** $\bigwedge E$.

We conclude this chapter by returning to the difference between principal bundles and vector bundles. Corollaries 18.4 and 18.5 tell us that studying vector bundles over a given manifold M is essentially the same thing as studying principal G-bundles over M for G a *matrix* Lie group. However not all Lie groups are matrix Lie groups, and thus principal bundles are more general than vector bundles.

---

[6] In the bonus section of the next chapter, we will present an entirely different way of formulating and proving the Metatheorem, using category-theoretic tools instead of principal bundles.

Chapter 19
Tensor and Exterior Algebras

In this chapter we continue our theme of constructing new vector bundles from old, but this time we focus on two constructions you may be less familiar with on the linear algebra level.

Definition 19.1. Let V and W be two vector spaces. Their **tensor product** is the vector space $V \otimes W$ which is defined as follows. First, let $F(V \times W)$ denote the (infinite-dimensional) vector space which has as basis all pairs (v, w) where $v \in V$ and $w \in W$. Thus an element of $F(V \times W)$ consists of a finite linear combination of pairs (v, w) with $v \in V$ and $w \in W$. Now let $R(V, W)$ denote the linear subspace of $F(V \times W)$ generated by the set of all elements of the form

$$\begin{cases} (v_1 + v_2, w) - (v_1, w) - (v_2, w), & v_1, v_2 \in V, \ w \in W, \\ (v, w_1 + w_2) - (v, w_1) - (v, w_2), & v \in V, \ w_1, w_2 \in W, \\ c(v, w) - (cv, w), & v \in V, \ w \in W, \ c \in \mathbb{R}, \\ c(v, w) - (v, cw), & v \in V, \ w \in W, \ c \in \mathbb{R}. \end{cases}$$

Let $V \otimes W$ denote the quotient vector space $F(V \times W)/R(V, W)$. The coset in $V \otimes W$ containing (v, w) is denoted by $v \otimes w$. By construction one has

$$\begin{cases} (v_1 + v_2) \otimes w = v_1 \otimes w + v_2 \otimes w, & v_1, v_2 \in V, \ w \in W, \\ v \otimes (w_1 + w_2) = v \otimes w_1 + v \otimes w_2, & v \in V, \ w_1, w_2 \in W, \\ c(v \otimes w) = (cv) \otimes w, & v \in V, \ w \in W, \ c \in \mathbb{R}, \\ c(v \otimes w) = v \otimes (cw), & v \in V, \ w \in W, \ c \in \mathbb{R}. \end{cases}$$

A typical element in $V \otimes W$ is a finite sum $\sum_i c_i \, v_i \otimes w_i$ where the c_i are real numbers. An element of the form $v \otimes w$ is called **decomposable**.

There is a natural bilinear map $\otimes \colon V \times W \to V \otimes W$ that sends $(v, w) \mapsto v \otimes w$. Here is a useful property of the tensor product.

W. J. Merry, *Lectures on Differential Geometry I*,
https://doi.org/10.1007/978-3-032-03733-6_19

Lemma 19.2. *Let V, W and Z be vector spaces and suppose $b\colon V \times W \to Z$ is a bilinear map. Then there exists a unique linear map $\bar{b}\colon V \otimes W \to Z$ such that the following diagram commutes:*

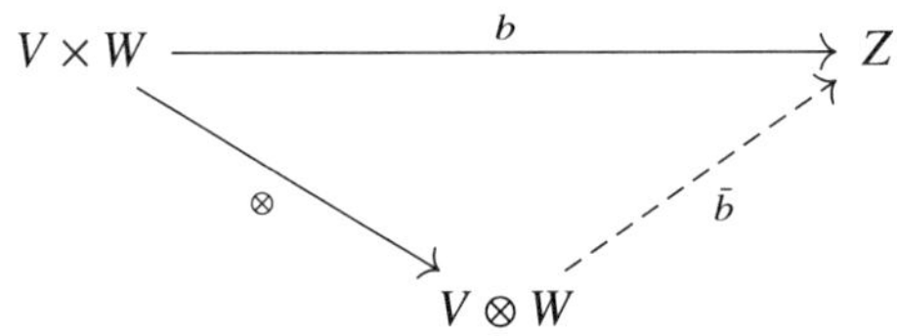

Moreover this property uniquely characterises $V \otimes W$.

Proof. Let $b\colon V \times W \to Z$ be a bilinear function. We extend b by linearity to a map $F(V \times W) \to Z$. Bilinearity then tells us that $R(V, W) \subset \ker b$, and hence b factors to define a homomorphism $\bar{b}\colon V \otimes W \to Z$ such that $\bar{b}(v \otimes w) = b(v, w)$ for all $(v, w) \in V \times W$. Moreover the map $\bar{b}$ is unique, since the decomposable elements generate $V \otimes W$.

Finally, to see why this property uniquely determines $\otimes$, suppose that X is another vector space equipped with a bilinear map $\beta\colon V \times W \to X$ with the property that if $B\colon V \times W \to Z$ is bilinear then there exists a *unique* linear map $\bar{B}\colon X \to Z$ such that the diagram commutes[1]:

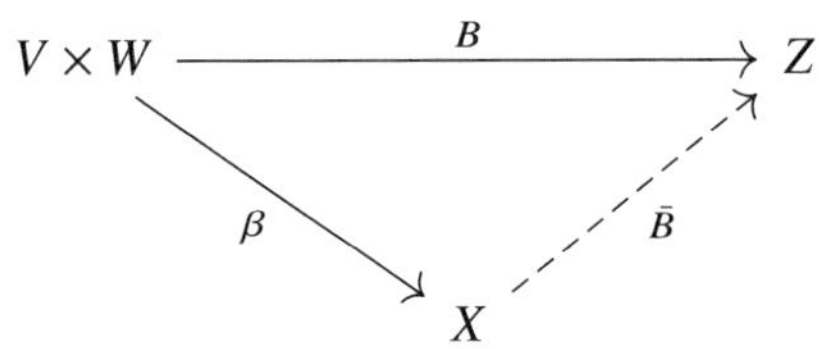

We apply this with $Z = V \otimes W$ and $B := \otimes$. This gives us a unique linear map $\bar{B}\colon X \to V \otimes W$ such that $\bar{B} \circ \beta = \otimes$. Now we go back to our original diagram and chose $Z = X$ and $b := \beta$. Thus we get a unique linear map $\bar{b}\colon V \otimes W \to X$ such that the following spliced diagram commutes.

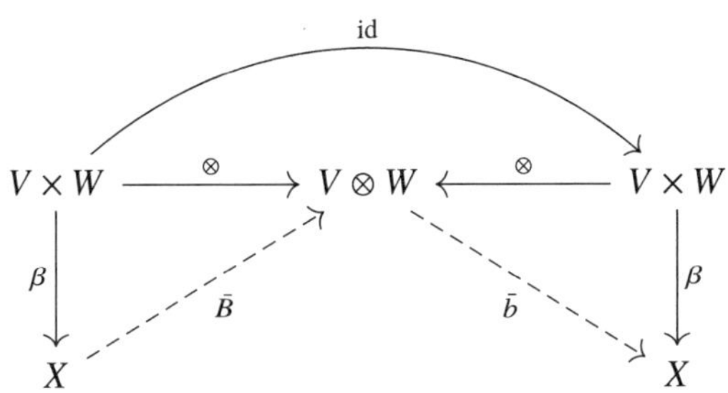

Thus the composition $\bar{b} \circ \bar{B}$ makes this diagram commute:

[1] The philosophy behind this style of proof is explained in the bonus section below.

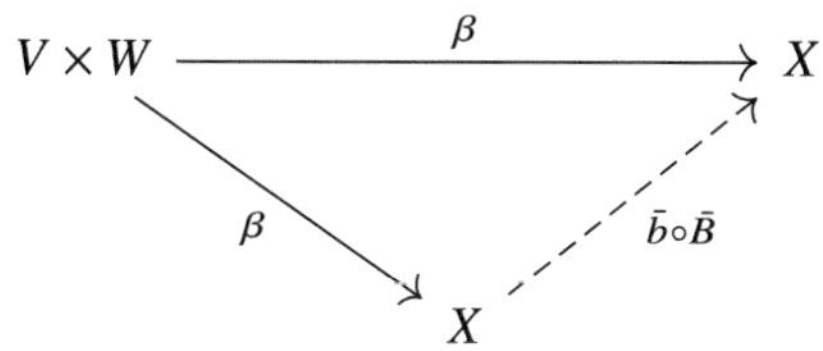

But there is meant to only be one map that makes this diagram commute, and another choice is the identity map $X \longrightarrow X$. Thus $\bar{b} \circ \bar{B} = \mathrm{id}_X$. Similarly $\bar{B} \circ \bar{b} = \mathrm{id}_{V \otimes W}$, and we conclude that X and $V \otimes W$ are isomorphic. This completes the proof. ■

Given two vector spaces V and W, let $\mathrm{Hom}(V, W)$ denote the set of linear maps from V to W. Thus $\mathrm{Mat}(m) = \mathrm{Hom}(\mathbb{R}^m, \mathbb{R}^m)$ and $V^* = \mathrm{Hom}(V, \mathbb{R})$.

Corollary 19.3. *Let V and W denote vector spaces, where V is finite-dimensional. There is a natural isomorphism* $\mathrm{Hom}(V, W) \cong V^* \otimes W$.

Proof. Define $b \colon V^* \times W \longrightarrow \mathrm{Hom}(V, W)$ by $b(\lambda, w)(v) := \lambda(v)w$. This gives us a linear map $\bar{b} \colon V^* \otimes W \longrightarrow \mathrm{Hom}(V, W)$ by Lemma 19.2. This map is an isomorphism, as an inverse $\mathrm{Hom}(V, W) \longrightarrow V^* \otimes W$ is given by $\ell \mapsto e^i \otimes \ell e_i$, where (e_i) is any basis of V and (e^i) is the dual basis of V^*. ■

Corollary 19.4. *If (e_i) is a basis for V and (e'_j) is a basis for W then $(e_i \otimes e'_j)$ is a basis for $V \otimes W$. Thus* $\dim(V \otimes W) = \dim V \cdot \dim W$.

The proof of the next result in on Problem Sheet H.

Lemma 19.5. *If V, W and Z are vector spaces then there are natural isomorphisms $V \otimes W \cong W \otimes V$ and $(V \otimes W) \otimes Z \cong V \otimes (W \otimes Z)$.*

Lemma 19.5 implies that we can unambiguously write $V \otimes W \otimes Z$.

Definition 19.6. Let V be a vector space and let h and k be non-negative integers. Define a new vector space

$$T^{h,k}V := \underbrace{V \otimes \cdots \otimes V}_{h \text{ copies}} \otimes \overbrace{V^* \otimes \cdots \otimes V^*}^{k \text{ copies}}.$$

One calls an element of $T^{h,k}V$ a **tensor of type** (h, k).
Warning: Do not confuse this notation with the tangent space of V!
The vector space $T^{h,k}V$ has dimension $(\dim V)^{h+k}$.

Note we are using Lemma 19.5 to write the right-hand side without brackets. Lemma 19.5 also shows us that it is unimportant in which order we present the factors: for convenience we write the V factors first and the V^* factors afterwards.

DEFINITION 19.7. Let $\mathrm{Mult}_{h,k}(V)$ denote the space of multilinear maps

$$\underbrace{V \times \cdots \times V}_{h \text{ copies}} \times \overbrace{V^* \times \cdots \times V^*}^{k \text{ copies}} \to \mathbb{R}.$$

Thus $\mathrm{Mult}_{1,0}(V) = V^*$ and $\mathrm{Mult}_{0,1}(V) = V^{**} \cong V$.

PROPOSITION 19.8. *There is a canonical isomorphism between the vector space $T^{h,k}V$ and the vector space $\mathrm{Mult}_{k,h}(V)$.*[2]

The proof of Proposition 19.8 uses the following piece of linear algebra.

DEFINITION 19.9. Let V and W be vector spaces. A **perfect pairing** of V with W is a bilinear map $\beta\colon V \times W \to \mathbb{R}$ such that $\beta(v, \cdot)$ is identically zero if and only if $v = 0$, and $\beta(\cdot, w)$ is identically zero if and only if w is zero.

PROPOSITION 19.10. *If there exists a perfect pairing β between V and W then the map $V \to W^*$ given by $v \mapsto \beta(v, \cdot)$ is a linear isomorphism*[3].

EXAMPLE 19.11. The natural isomorphism $V \cong V^{**}$ arises from the perfect pairing $V \times V^* \to \mathbb{R}$ given by $(v, \lambda) \mapsto \lambda(v)$.

Proof of Proposition 19.8. We prove the result in two steps.

1. In this step we define a perfect pairing of $T^{h,k}V$ with $T^{h,k}V^*$. Namely, if

$$v = v_1 \otimes \cdots \otimes v_h \otimes \lambda^1 \otimes \cdots \otimes \lambda^k \in T^{h,k}V$$

and

$$w = \eta^1 \otimes \cdots \otimes \eta^h \otimes w_1 \otimes \cdots \otimes w_k \in T^{h,k}V^*$$

then we can naturally feed them each other

$$\beta(v, w) := \prod_{i=1}^{h} \eta^i(v_i) \cdot \prod_{j=1}^{k} \lambda^j(w_j) \tag{19.1}$$

We are *not* using the Einstein Summation Convention in this formula—this is a product not a sum!

Now extend this bilinearly to all elements. It is immediate that this is a perfect pairing; thus by Proposition 19.10 we have

$$T^{h,k}V \cong \left(T^{h,k}V^*\right)^*$$

2. In this step we prove that

$$\mathrm{Mult}_{h,k}(V^*) \cong \left(T^{h,k}V^*\right)^* \tag{19.2}$$

[2] Note the h and the k swapped round—this is *not* a typo!

[3] This result fails if V is infinite-dimensional.

The proof is by induction on $h + k$. The case $h + k = 1$ is exactly Lemma 19.2. The inductive step follows Lemma 19.5 and Lemma 19.2 again. Explicitly, given a map $b \in \mathrm{Mult}_{h,k}(V^*)$ there exists a unique $\bar{b}\colon T^{h,k}V^* \to \mathbb{R}$, i.e. $T \in \left(T^{h,k}V^*\right)^*$ such that the following diagram commutes[4].

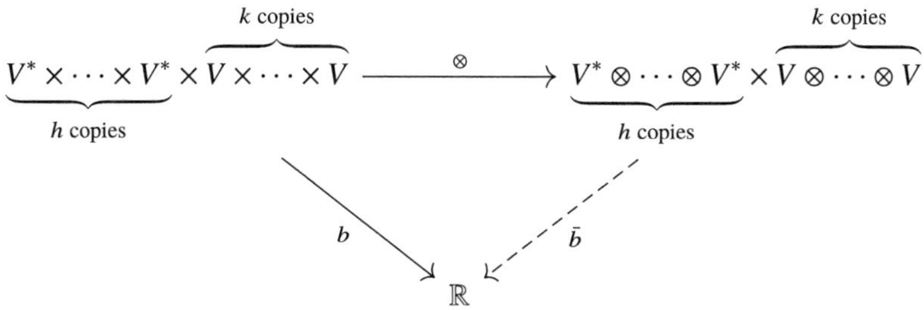

Since b is uniquely determined by $\bar{b}$, this sets up the desired isomorphism (19.2). Finally, by definition we have $\mathrm{Mult}_{h,k}(V^*) \cong \mathrm{Mult}_{k,h}(V)$, and so the proof is complete. ∎

Remark 19.12. On decomposable elements the isomorphism $T^{h,k}V \cong \mathrm{Mult}_{k,h}(V)$ is easier to describe. Suppose for simplicity $A \in T^{2,3}V$ is the decomposable element

$$A = v_1 \otimes v_2 \otimes \lambda^1 \otimes \lambda^2 \otimes \lambda^3.$$

If we use Proposition 19.8 to regard A as an element of $\mathrm{Mult}_{3,2}\,V$ then A is given by

$$(w_1, w_2, w_3, \eta^1, \eta^2) \mapsto \eta^1(v_1)\,\eta^2(v_2)\,\lambda^1(w_1)\,\lambda^2(w_2)\,\lambda^3(w_3).$$

We can use the Metatheorem to transfer these linear algebra constructions to vector bundles:

Corollary 19.13. *Let $V \to E \xrightarrow{\pi_1} M$ and $W \to F \xrightarrow{\pi_2} M$ be two vector bundles. Then there is a vector bundle $V \otimes W \to E \otimes F \xrightarrow{\pi} M$ whose fibre over p is $E_p \otimes F_p$.*

Proof. We form the principal $\mathrm{GL}(V) \times \mathrm{GL}(W)$ bundle $\mathrm{Fr}(E) \star \mathrm{Fr}(F)$ as in Definition 18.9. The canonical action of $\mathrm{GL}(V) \times \mathrm{GL}(W)$ on $V \times W$ induces by Lemma 19.2 an action of $\mathrm{GL}(V) \times \mathrm{GL}(W)$ on $V \otimes W$. Corollary 18.5 then produces our desired vector bundle

$$E \otimes F \overset{\text{def}}{=} \big(\mathrm{Fr}(E) \star \mathrm{Fr}(F)\big) \times_{\mathrm{GL}(V)\times\mathrm{GL}(W)} (V \otimes W).$$

This completes the proof. ∎

Corollary 19.14. *Let $V \to E \xrightarrow{\pi_1} M$ and $W \to F \xrightarrow{\pi_2} M$ be two vector bundles. Then there is a vector bundle $\mathrm{Hom}(V, W) \to \mathrm{Hom}(E, F) \xrightarrow{\pi} M$ whose fibre over p is $\mathrm{Hom}(E_p, F_p)$.*

We call $\mathrm{Hom}(E, F)$ the **homomorphism bundle**. In the special case $E = F$ we write $\mathrm{End}(E) := \mathrm{Hom}(E, E)$ and call it the **endomorphism bundle**.

[4] This diagram is rotated compared to the diagram in the statement of Lemma 19.2 so it fits on the page better.

Proof. The group $\mathrm{GL}(V) \times \mathrm{GL}(W)$ acts on $\mathrm{Hom}(V, W)$ via

$$\Phi \mapsto B \circ \Phi \circ A^{-1}, \tag{19.3}$$

for $A \in \mathrm{GL}(V)$ and $B \in \mathrm{GL}(W)$. Now argue as above. ■

COROLLARY 19.15. *Let E and F be two vector bundles over M. Then there is a natural vector bundle isomorphism*

$$\mathrm{Hom}(E, F) \cong E^* \otimes F.$$

Proof. Under the isomorphism from Corollary 19.3, the action of $\mathrm{GL}(V) \times \mathrm{GL}(W)$ on $\mathrm{Hom}(V, W)$ (19.3) is exactly the same as the action of $\mathrm{GL}(V) \times \mathrm{GL}(W)$ on $V^* \otimes W$ given by combining (18.4) with the action used in the proof of with Corollary 19.13. Corollary 18.4 then supplies the desired isomorphism. ■

Alternatively, one can argue directly: define

$$\Phi \colon \mathrm{Hom}(E, F) \to E^* \otimes F$$

fibrewise by declaring that the map $\Phi_p : \mathrm{Hom}(E_p, F_p) \to E_p^* \otimes F_p$ is the isomorphism from Corollary 19.3. This assignment $p \mapsto \Phi_p$ is smooth[5] and thus Φ is a vector bundle isomorphism.

Similarly Lemma 19.5 tells us that the tensor product of vector bundles is commutative and associative:

COROLLARY 19.16. *Let D, E and F be three vector bundles over M. Then the bundles $E \otimes F$ and $F \otimes E$ are isomorphic, and the bundles $D \otimes (E \otimes F)$ and $(D \otimes E) \otimes F$ are isomorphic.*

More generally:

COROLLARY 19.17. *Let E be a vector bundle over M. Then there is a vector bundle $T^{h,k}E$ over M whose fibre over p is the vector space $T^{h,k}E_p$.*[6]

Let us recall the formal definition of an algebra.

DEFINITION 19.18. A vector space V is said to be an **algebra** if there exists a bilinear map[7] $V \times V \to V$, which we call **multiplication** and write as $(v, w) \mapsto vw$. If V and W are algebras, an **algebra morphism** $\ell \colon V \to W$ is a linear map such that

$$\ell(vw) = \ell(v)\ell(w), \qquad \forall\, v, w \in V.$$

Whilst each $T^{h,k}V$ is not an algebra, if we sum them all together we obtain one.

[5] If you are worried about why $p \mapsto \Phi_p$ is smooth, you could use Proposition 20.23 from the next chapter.

[6] Again: do not confuse $T^{h,k}E$ with the tangent bundle TE. $T^{h,k}E$ is a bundle over M, whereas TE is a bundle over E.

[7] By Lemma 19.2 this is equivalent to a linear map $V \otimes V \to V$.

DEFINITION 19.19. The **tensor algebra** of V is defined to be

$$\widetilde{T}V := \bigoplus_{h,k\geq 0} T^{h,k}V,$$

where $T^{0,0}V := \mathbb{R}$. (The tilde is there to prevent confusion with the tangent bundle of V.) This is a *graded algebra*, in the sense that $\otimes$ gives a natural map

$$\otimes \colon T^{h,k}V \times T^{h_1,k_1}V \to T^{h+h_1,k+k_1}V. \tag{19.4}$$

The natural map is defined as one would guess: on decomposable elements it simply tensors everything together and then rearranges the factors so the V elements come first, so as to fit with our convention. We illustrate this with $(h,k) = (1,2)$ and $(h_1,k_1) = (2,1)$:

$$\Big((v_1 \otimes \lambda^1 \otimes \lambda^2), (w_1 \otimes w_2 \otimes \eta^1)\Big) \mapsto v_1 \otimes w_1 \otimes w_2 \otimes \lambda^1 \otimes \lambda^2 \otimes \eta^1. \tag{19.5}$$

If $(h,k) = (0,0)$ then tensor multiplication with a scalar is defined to be normal scalar multiplication, i.e.

$$c \otimes v := cv, \qquad c \in \mathbb{R},\ v \in V.$$

REMARK 19.20. The space $\widetilde{T}V$ is an *infinite-dimensional* vector space[8]. This means we use **cannot** use Corollary 18.5 to produce a "vector bundle" $\widetilde{T}V \to \widetilde{T}E \to M$ out of a vector bundle $V \to E \to M$.

We now introduce another linear algebra construction, called the **exterior algebra**. This will associate to a vector space V another (finite-dimensional) vector space $\bigwedge V$ which, like the tensor algebra $\widetilde{T}V$, admits an algebra structure.

Let V be a vector space. Let $\widetilde{T}^+V$ denote the subalgebra given by $\widetilde{T}^+V := \bigoplus_{h\geq 0} T^{h,0}V$. Let $\widetilde{I}V$ denote the two-sided ideal in $\widetilde{T}^+V$ generated by all elements of the form $v \otimes v$ for $v \in V$. Thus for instance $u \otimes v \otimes v \otimes w$ belongs to $\widetilde{I}V$.

DEFINITION 19.21. The **exterior algebra** is defined to be the quotient algebra $\bigwedge V := \widetilde{T}^+V/\widetilde{I}V$. We denote the image of $v_1 \otimes \cdots \otimes v_h$ in $\bigwedge V$ by $v_1 \wedge \cdots \wedge v_h$ and call $\wedge$ the **wedge product**.

Such an element $v_1 \wedge \cdots \wedge v_h$ is called **decomposable**. The space $\bigwedge V$ is the quotient of one infinite-dimensional vector space by another. As we will shortly see, $\bigwedge V$ is always finite-dimensional.

Let $\bigwedge^h V$ be the image of $T^{h,0}V$ in $\bigwedge V$ under the projection $\widetilde{T}^+V \to \bigwedge V$. There is a canonical vector space isomorphism

$$\textstyle\bigwedge^h V \cong T^{h,0}V \Big/ I^h V,$$

[8] Actually this is only half the story: $\widetilde{T}E$ is an example of an infinite-dimensional vector bundle over M. This goes beyond the scope of this course, however, and we won't need it.

where $I^h V := T^{h,0}V \cap \widetilde{I}V$. Note that $\bigwedge^1 V = V$ and $\bigwedge^0 V = \mathbb{R}$. This definition may seem a little abstract, so let us unpack things a bit.

Proposition 19.22 (Properties of the wedge product). *Let V be a vector space. Then:*

(i) *For all $v, w \in V$, $v \wedge w = -w \wedge v$.*
(ii) *Assume $h, k > 0$. If $v \in \bigwedge^h V$ and $w \in \bigwedge^k V$ then $v \wedge w \in \bigwedge^{h+k} V$ and*

$$v \wedge w = (-1)^{hk} w \wedge v,$$

This continues to hold if either $h = 0$ or $k = 0$ provided we use the convention that for a real number c and a vector v, one has $c \wedge v := cv$.
(iii) *If $v_1 \wedge \cdots \wedge v_h \in \bigwedge^h V$ is a decomposable element then transposing v_i with v_j acts as multiplication by -1:*

$$v_1 \wedge \cdots \wedge v_i \wedge \cdots \wedge v_j \wedge \cdots \wedge v_h = -v_1 \wedge \cdots \wedge v_j \wedge \cdots \wedge v_i \wedge \cdots \wedge v_h$$

(iv) *If $\varrho \in \mathfrak{S}_h$ is a permutation on h letters and $v_i \in V$ then*

$$v_{\varrho(1)} \wedge \cdots \wedge v_{\varrho(h)} = \operatorname{sgn}(\varrho) v_1 \wedge \cdots \wedge v_h.$$

Proof. To prove part (i), we note that for any $u \in V$, $u \otimes u$ belongs to $\widetilde{I}V$, and thus in $\bigwedge V$, $u \wedge u = 0$. Applying this with $u = v + w$ we have

$$\begin{aligned} 0 &= (v + w) \wedge (v + w) \\ &= v \wedge v + v \wedge w + w \wedge v + w \wedge w \\ &= v \wedge w + w \wedge v. \end{aligned}$$

To prove part (ii), as both sides are linear in v and w, it suffices to verify it for decomposable elements, and for such, the conclusion follows by repeated applications of part (i). Next, to prove part (iii), we may assume $i < j$. Set $u := v_{i+1} \wedge \cdots \wedge v_{j-1}$. Then by part (ii) one has

$$v_i \wedge u \wedge v_j = -v_j \wedge u \wedge v_i,$$

and thus part (iii) follows. Finally, part (iv) is immediate from the fact that any permutation may be written as a product of transpositions. ■

There is an analogous universal mapping property for the exterior algebra.

Definition 19.23. Let V and W be vector spaces. Let $\mathrm{Alt}_h(V, W)$ denote the space of **alternating h-linear maps**, i.e. multilinear maps $A \colon V \times \cdots \times V \to W$ (h times) that vanish whenever any two of the arguments are equal:

$$A(\cdots, v, \cdots, v, \cdots) = 0.$$

We abbreviate $\mathrm{Alt}_h(V) = \mathrm{Alt}_h(V, \mathbb{R})$.

The map $\wedge \colon V \times \cdots \times V \to \bigwedge^h V$ given by sending $(v_1, \ldots, v_h) \mapsto v_1 \wedge \cdots \wedge v_h$ is an example of such a map. We aim to prove the following alternating version of Proposition 19.8:

PROPOSITION 19.24. *There is a canonical isomorphism between* $\bigwedge^h V^*$ *and* $\mathrm{Alt}_h(V)$.

The proof strategy is similar to that of Proposition 19.8, and we will be brief. First, we need an analogue of Lemma 19.2.

LEMMA 19.25. *Let V and W be vector spaces. For any* $A \in \mathrm{Alt}_h(V, W)$ *there is a unique linear map* $a \colon \bigwedge^h V \to W$ *such that the following diagram commutes*:

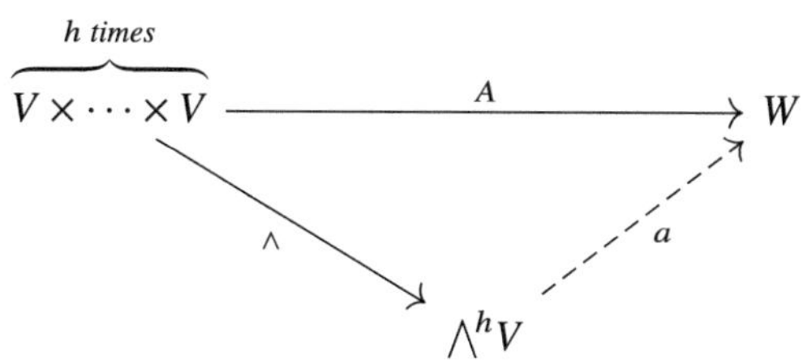

Moreover $\bigwedge^h V$ *is uniquely characterised by this property.*

The proof of Lemma 19.25 is on Problem Sheet H.

Proof of Proposition 19.24. Just as in the proof of step 2 of Proposition 19.8, an inductive argument based on Lemma 19.25 tells us that we can identify

$$\mathrm{Alt}_h(V) \cong \left(\bigwedge^h V\right)^*.$$

The next step is to exhibit a perfect pairing of $\bigwedge^h V^*$ with $\bigwedge^h V$. This formula is a little harder to guess than in (19.1), but once you know the formula it is easy to check. Namely, we define

$$\alpha \colon \bigwedge^h V^* \times \bigwedge^h V \to \mathbb{R}$$

by declaring on decomposable elements that

$$\alpha\big(\lambda^1 \wedge \cdots \wedge \lambda^h, v_1 \wedge \cdots \wedge v_h\big) := \det A,$$

where A is the $h \times h$ matrix whose (i, j)th entry is $\lambda^i(v_j)$. Then extend α by bilinearity to all of $\bigwedge^h V^* \times \bigwedge^h V^*$. We invite you to verify this is indeed a perfect pairing. ■

For later use, let us state part of the proof of Proposition 19.24 as a separate corollary.

COROLLARY 19.26. *Let* $\lambda^1, \ldots, \lambda^h \in V^*$ *and* $v_1, \ldots, v_h \in V$. *Then viewing* $\lambda^1 \wedge \cdots \wedge \lambda^h$ *as an element of* $\mathrm{Alt}_h(V)$, *one has*

$$\lambda^1 \wedge \cdots \wedge \lambda^h(v_1, \ldots, v_h) = \det A,$$

where A *is the* $h \times h$ *matrix whose* (i, j)*th entry is* $\lambda^i(v_j)$.

On Problem Sheet H you are also asked to show:

LEMMA 19.27. *Let V be a vector space of dimension n with basis $\{e_1, \dots, e_n\}$. Then*

$$\{e_{i_1} \wedge \cdots \wedge e_{i_h} \mid 1 \le i_1 < \cdots < i_h \le n\}$$

is a basis of $\bigwedge^h V$ and $\bigwedge^h V = 0$ for $h > n$. Thus $\dim \bigwedge^h V = \binom{n}{h}$ and $\dim \bigwedge V = 2^n$.

We end this chapter by applying the Metatheorem to the operations $\bigwedge^h$ and $\bigwedge$.

COROLLARY 19.28. *Let $V \to E \xrightarrow{\pi} M$ be a vector bundle of rank n. Then for any $0 \le h \le n$, there is a vector bundle $\bigwedge^h E \to M$ whose fibre over $p \in M$ is given by $\bigwedge^h E_p$. This vector bundle has rank $\binom{n}{h}$. Similarly there is a vector bundle $\bigwedge E \to M$ of rank 2^n whose fibre over $p \in M$ is given by $\bigwedge E_p$. It is the direct sum of the vector bundles $\bigwedge^h E$.*

The bundle $\bigwedge E$ inherits an algebra structure from the algebra structure on $\bigwedge V$.

DEFINITION 19.29. Let V be vector space which is also an algebra in the sense of Definition 19.18, and suppose that $V \to E \xrightarrow{\pi} M$ is a vector bundle. We say that E is an **algebra bundle** if each fibre E_p admits the structure of an algebra, and there exists a vector bundle atlas $\{(U_a, \varepsilon_a)\}$ such that for each $p \in U_a$ the map $(\varepsilon_a)_p : E_p \to V$ is not only a linear isomorphism but also an algebra isomorphism.

COROLLARY 19.30. *Let $V \to E \xrightarrow{\pi} M$ be a vector bundle. Then $\bigwedge V \to \bigwedge E \to M$ is an algebra bundle.*

The proof of Corollary 19.30 is left for you on Problem Sheet H.

REMARK 19.31. In Volume II Chapter 7 we will work with **Lie algebra bundles**, which are algebra bundles with the additional property that the algebra structure is actually a Lie algebra.

Bonus Material for Chapter 19

In the bonus material for this chapter, we are introducing elements of a field of mathematics called **category theory**. This material will not be *needed* at any point in this book. Our aim is to give a category-theoretic proof of the Metatheorem from the previous chapter—see Theorem 19.60 below—which entirely bypasses principal and associated bundles.

In a nutshell, category theory is the an attempt to make "proof by analogy" a valid proof tactic. That is, category theory is an interdisciplinary language that allows one to describe certain general phenomena that crop up in mathematical arguments across the board. The advantage of possessing such a language is clear—it allows one to isolate the essence of a given statement or proof technique, thus allowing for concise and clean proofs. It is also efficient: a single category-theoretic blueprint can simultaneously prove diverse statements in number theory, geometry, algebra, analysis, and so on. Category theory is also useful in theoretical computer science; indeed, many functional programming languages (eg. Haskel, Scala) are almost literal interpretations of categorical methods. The generality comes at a price, though: category theory is often (lovingly) referred to as **abstract nonsense** [5].

You probably have already met several "category-theoretic" arguments in your mathematical career so far. Roughly speaking, a category-theoretic argument is one that focuses on transformations between objects of a given type, rather than on the objects themselves. An example of this is Lemma 19.2 above, which characterised the tensor product via a *universal property*. In general, a universal property typically expresses a certain role that a given mathematical object plays in relation to other objects of its type, and the abstract categorical theorem is: *if an object can be described by a universal property, then it is unique up to a unique isomorphism*. If we were allowed to quote this result, the proof of Lemma 19.2 would have been over at the end of the first paragraph.

Before we get started on the definitions, let us list a few results which you have probably already met that can all be proved used categorical methods:

- If A and B are sets and $f\colon A \times B \to \mathbb{R}$ is a function, then
$$\sup_{a\in A}\inf_{b\in B} f(a,b) \le \inf_{b\in B}\sup_{a\in A} f(a,b)$$
whenever the infima and suprema exist.
- *Cayley's Theorem*: Any finite group is isomorphic to a subgroup of a permutation group.
- Every row operation on matrices with m rows is given by left multiplication by some $m \times m$ matrix.
- A continuous bijection between compact Hausdorff spaces is a homeomorphism.

And now the definitions:

Definitions 19.32. A **category** C consists of three ingredients. The first is a *class* $\mathrm{obj}(\mathsf{C})$ of **objects**. Secondly, for each ordered pair of objects (A, B) there is a *set* $\mathrm{Hom}(A, B)$ of **morphisms** from A to B. Sometimes instead of $f \in \mathrm{Hom}(A, B)$ we write $f : A \to B$ or $A \xrightarrow{f} B$. Finally, there is a rule, called **composition**, which associates to every ordered triple (A, B, C) of objects a map

$$\mathrm{Hom}(A, B) \times \mathrm{Hom}(B, C) \to \mathrm{Hom}(A, C),$$

written

$$(f, g) \mapsto g \circ f,$$

which satisfies the following three axioms:

(i) The Hom sets are pairwise disjoint; that is, each $f \in \mathrm{Hom}(A, B)$ has a unique **domain** A and a unique **target** B.

(ii) Composition is associative whenever defined, i.e. given

$$A \xrightarrow{f} B \xrightarrow{g} C \xrightarrow{h} D$$

one has

$$(h \circ g) \circ f = h \circ (g \circ f).$$

(iii) For each $A \in \text{obj}(\mathsf{C})$ there is a unique morphism $\text{id}_A \in \text{Hom}(A, A)$ called the **identity** which has the property that $f \circ \text{id}_A = f$ and $\text{id}_B \circ f = f$ for every $f : A \to B$.

Remark 19.33. Note that we said that $\text{obj}(\mathsf{C})$ was a *class* and $\text{Hom}(A, B)$ was a *set*. There is (an important, but technical) difference between a class and a set. If you've ever taken a class on logic/set theory, you'll know that not every "collection" of objects is formally a set. For instance, the collection of all sets is itself not a set[9]. A class is a more general concept (the collection of all sets is a class). Nevertheless, as far as we're concerned, the distinction is essentially irrelevant.

Here are six examples of categories. The first three are algebraic in nature.

Example 19.34. The category Sets of **sets**. The objects of Sets are all the sets, $\text{Hom}(A, B)$ is just the set $\text{Maps}(A, B)$ of all functions from A to B, and composition is just the usual composition of functions.

Example 19.35. The category Groups of **groups**. The objects of Groups are just groups, $\text{Hom}(G, H)$ is the set of all *group homomorphisms* from G to H, and composition is just the usual composition of homomorphisms.

Example 19.36. The category $\mathsf{Vect} = \mathsf{Vect}_{\mathbb{R}}$ of **finite-dimensional real vector spaces**. The objects of Vect are finite-dimensional real vector spaces, and $\text{Hom}(V, W)$ is the set $\text{Hom}(V, W)$ of all *linear maps* from V to W.

Here are three more examples more pertinent to this course.

Example 19.37. The category Top of **topological spaces**. The objects of Top are all the topological spaces, $\text{Hom}(X, Y)$ is just the set $C(X, Y)$ of all *continuous* functions from X to Y, and composition is just the usual composition of functions.

Example 19.38. The category Man of **smooth manifolds**. The objects of Man are smooth manifolds, and $\text{Hom}(M, N)$ is the set $C^\infty(M, N)$ of all *smooth* maps $\varphi \colon M \to N$. Composition is given by normal composition of maps; this is well defined by Proposition 1.21.

Example 19.39. The category $\mathsf{VectBundles}$ of **vector bundles**. The objects of $\mathsf{VectBundles}$ are vector bundles $\pi \colon E \to M$, and a morphism from $\pi_1 : E_1 \to M_1$ to $\pi_2 \colon E_2 \to M_2$ is a pair (Φ, φ), where $\varphi \colon M_1 \to M_2$ is a smooth map and $\Phi \colon E_1 \to E_2$ is a vector bundle morphism from E_1 to E_2 along φ.

[9] This is Russell's Paradox [12].

REMARK 19.40. The category Vect is rather special: its morphism sets are themselves objects of the category. That is, if V and W are vector spaces then $\mathrm{Hom}(V, W)$ is itself naturally a vector space. This is not true in the category of Groups—the set of all group homomorphisms from one group to another typically does not have a group structure. Similarly the set $C^\infty(M, N)$ of smooth maps between two smooth manifolds is never itself a (finite-dimensional) manifold when $\dim M > 0$.

REMARK 19.41. The fact that we require the morphism sets to be pairwise disjoint has several pedantic consequences. For example, suppose $A \subsetneq B$ are two sets. Then the inclusion $\iota\colon A \hookrightarrow B$ and the identity map $\mathrm{id}_A\colon A \to A$ are different morphisms, since they have different targets. One should be aware that we only allow the composition $g \circ f$ when the range of f is *exactly* the same as the domain of g. Suppose L, M, N and P are manifolds, and suppose M is an embedded submanifold of N. Let $\varphi\colon L \to M$ be smooth and let $\psi\colon N \to P$ be smooth. Then as we have seen, the composition $\psi \circ \varphi\colon L \to P$ is also smooth (since M is embedded). Nevertheless, from the point of view of category theory, the composition $\psi \circ \varphi$ *does not exist*! Rather, one must first take the inclusion $\iota\colon M \hookrightarrow N$ and then consider the composition $\psi \circ \iota \circ \varphi$, which is a well-defined element of the morphism space $C^\infty(L, P)$.

DEFINITION 19.42. Suppose C and D are two categories. We say that C is a **subcategory** of D if:

1. $\mathrm{obj}(\mathsf{C}) \subseteq \mathrm{obj}(\mathsf{D})$;
2. $\mathrm{Hom}_{\mathsf{C}}(A, B) \subseteq \mathrm{Hom}_{\mathsf{D}}(A, B)$ for all $A, B \in \mathrm{obj}(\mathsf{C})$;
3. if $f \in \mathrm{Hom}_{\mathsf{C}}(A, B)$ and $g \in \mathrm{Hom}_{\mathsf{C}}(B, C)$ then the composite $g \circ f \in \mathrm{Hom}_{\mathsf{C}}(A, C)$ is equal to the composite $g \circ f \in \mathrm{Hom}_{\mathsf{D}}(A, C)$;
4. if $C \in \mathrm{obj}(\mathsf{C})$ then $\mathrm{id}_C \in \mathrm{Hom}_{\mathsf{C}}(C, C)$ is equal to $\mathrm{id}_C \in \mathrm{Hom}_{\mathsf{D}}(C, C)$.

If for every pair $A, B \in \mathrm{obj}(\mathsf{C})$ one always has $\mathrm{Hom}_{\mathsf{C}}(A, B) = \mathrm{Hom}_{\mathsf{D}}(A, B)$ then we say that C is a **full subcategory** of D.

EXAMPLE 19.43. Here are two examples of subcategories:

(i) The category Ab of *abelian* groups is a full subcategory of the category Groups.
(ii) Let $\mathsf{Vect}^{\leq\infty}$ denote the category of *all* real vector spaces (finite-dimensional or infinite-dimensional). Then Vect is a full subcategory of $\mathsf{Vect}^{\leq\infty}$.

A **functor** is a map from one category to another. These come in two flavours: *covariant* and *contravariant*. We discuss the former first.

DEFINITION 19.44. Suppose C and D are two categories. A **covariant functor** $\mathsf{F}\colon \mathsf{C} \to \mathsf{D}$ associates to each $A \in \mathrm{obj}(\mathsf{C})$ an object $\mathsf{F}(A) \in \mathrm{obj}(\mathsf{D})$, and to each morphism $A \xrightarrow{f} B$ in C a morphism $\mathsf{F}(A) \xrightarrow{\mathsf{F}(f)} \mathsf{F}(B)$ in D which satisfies the following two axioms:

1. If $A \xrightarrow{f} B \xrightarrow{g} C$ in C then $\mathsf{F}(A) \xrightarrow{\mathsf{F}(f)} \mathsf{F}(B) \xrightarrow{\mathsf{F}(g)} \mathsf{F}(C)$ in D and
$$\mathsf{F}(g \circ f) = \mathsf{F}(g) \circ \mathsf{F}(f).$$
2. $\mathsf{F}(\mathrm{id}_A) = \mathrm{id}_{\mathsf{F}(A)}$ for every $A \in \mathrm{obj}(\mathsf{C})$.

The easiest example of a functor is a **forgetful functor**:

Example 19.45. The forgetful functor $\mathsf{Top} \longrightarrow \mathsf{Sets}$ simply "forgets" the topological structure. Thus it assigns to each topological space its underlying set, and to each continuous function it assigns the same function, considered now simply as a map between two *sets* (i.e. it "forgets" the function is continuous). The same thing works as a functor $\mathsf{Man} \longrightarrow \mathsf{Top}$, where one "forgets" the smooth manifold structure.

Example 19.46. There is slightly more interesting forgetful functor $\mathsf{VectBundles} \longrightarrow \mathsf{Man}$ that sends a vector bundle $\pi\colon E \longrightarrow M$ to its base space M (i.e. it "forgets" the vector bundle sitting over the base). On morphisms, this functor just "forgets" the vector bundle morphism: $(\varphi, \Phi) \mapsto \varphi$.

Here is a pertinent example of a functor from the category Vect to itself:

Examples 19.47. Let V be a fixed vector space. There is a covariant functor

$$\mathrm{Hom}(V, \cdot)\colon \mathsf{Vect} \longrightarrow \mathsf{Vect}$$

that assigns to a vector space W the vector space $\mathrm{Hom}(V, W)$. If $\ell\colon W_1 \longrightarrow W_2$ is a linear map then

$$\mathrm{Hom}(V, \cdot)(\ell)\colon \mathrm{Hom}(V, W_1) \longrightarrow \mathrm{Hom}(V, W_2)$$

is given by $\ell_1 \mapsto \ell_1 \circ \ell$.

In this chapter we constructed several functors:

Example 19.48. There is a covariant functor $\otimes\colon (\mathsf{Vect}, \mathsf{Vect}) \longrightarrow \mathsf{Vect}$ given by $(V, W) \mapsto V \otimes W$.

Example 19.49. There is a covariant functor $\bigwedge\colon \mathsf{Vect} \longrightarrow \mathsf{Vect}$

Algebraic topology is an excellent source of functors. For instance:

- the **fundamental group** π_1 is a covariant functor from the **pointed homotopy category** hTop_* to Groups;
- the **higher homotopy groups** are covariant functors $\pi_n\colon \mathsf{hTop}_* \longrightarrow \mathsf{Ab}$;
- **singular homology** (or indeed, any homology theory) is a covariant functor $\mathsf{hTop}^2 \longrightarrow \mathsf{Ab}$, where hTop^2 is the homotopy category of pairs.

One can also formulate the definition of a functor of more than one variable. This requires us to define the notion of a product category.

Definition 19.50. Let C and D be two categories. The **product category** (C, D) is the category whose objects are ordered pairs (C, D) where $C \in \mathrm{obj}(\mathsf{C})$ and $D \in \mathrm{obj}(\mathsf{D})$, and

$$\mathrm{Hom}((C, D), (C', D')) = \{(f, g) \mid f \in \mathrm{Hom}_{\mathsf{C}}(C, C'),\ g \in \mathrm{Hom}_{\mathsf{D}}(D, D')\}.$$

The composition $(f, g) \circ_{(\mathsf{C},\mathsf{D})} (f', g')$ is defined as you expect:

$$(f, g) \circ_{(\mathsf{C},\mathsf{D})} (f', g') := ((f \circ_{\mathsf{C}} f'), (g \circ_{\mathsf{D}} g')).$$

The identity element $\mathrm{id}_{(C,D)}$ is simply the pair $(\mathrm{id}_C, \mathrm{id}_D)$.

EXAMPLE 19.51. The category (Vect, Vect) has objects ordered pairs (V, W) of vector spaces, and morphisms pairs of linear maps.

DEFINITION 19.52. A **covariant functor of two variables** is a covariant functor defined on a product category: $\mathsf{F}\colon (\mathsf{C}, \mathsf{D}) \to \mathsf{E}$.

EXAMPLE 19.53. Let V and W be vector spaces. Then the **direct sum** $V \oplus W$ of V and W is another vector space[10]. Thus we get a functor $\oplus\colon (\mathsf{Vect}, \mathsf{Vect}) \to \mathsf{Vect}$ that assigns to (V, W) the vector space $V \oplus W$, and assigns to a pair (ℓ_1, ℓ_2) of linear maps $\ell_1\colon V_1 \to V_2$ and $\ell_2\colon W_1 \to W_2$ the linear map $\ell_1 \oplus \ell_2\colon V_1 \oplus W_1 \to V_2 \oplus W_2$.

In the same way, one can form a k**-fold product category** $(\mathsf{C}_1, \ldots, \mathsf{C}_k)$ of categories C_i, and a **covariant functor of** k **variables** is a covariant functor of the form $\mathsf{F}\colon (\mathsf{C}_1, \ldots, \mathsf{C}_k) \to \mathsf{D}$. For example, there is a functor

$$(\mathsf{Vect}, \ldots, \mathsf{Vect}) \to \mathsf{Vect}, \qquad (V_1, \ldots, V_k) \to V_1 \oplus \cdots \oplus V_k.$$

A contravariant functor is defined in almost the same way, but it reverses the arrows.

DEFINITION 19.54. Suppose C and D are two categories. A **contravariant functor** $\mathsf{G}\colon \mathsf{C} \to \mathsf{D}$ associates to each $A \in \mathrm{obj}(\mathsf{C})$ an object $\mathsf{G}(A) \in \mathrm{obj}(\mathsf{D})$, and to each morphism $A \xrightarrow{f} B$ in C a morphism $\mathsf{G}(B) \xrightarrow{\mathsf{G}(f)} \mathsf{G}(A)$ in D which satisfies the following two axioms:

1. If $A \xrightarrow{f} B \xrightarrow{g} C$ in C then $\mathsf{G}(C) \xrightarrow{\mathsf{G}(g)} \mathsf{G}(B) \xrightarrow{\mathsf{G}(f)} \mathsf{G}(A)$ in D and
$$\mathsf{G}(g \circ f) = \mathsf{G}(f) \circ \mathsf{G}(g).$$
2. $\mathsf{G}(\mathrm{id}_A) = \mathrm{id}_{\mathsf{G}(A)}$ for every $A \in \mathrm{obj}(\mathsf{C})$.

Here is a simple example of a contravariant functor on the category of vector spaces.

EXAMPLE 19.55. Let W be a fixed vector space. Then there is a contravariant functor

$$\mathrm{Hom}(\cdot, W)\colon \mathsf{Vect} \to \mathsf{Vect}$$

that assigns to a vector space V the vector space $\mathrm{Hom}(V, W)$. If $\ell\colon V_1 \to V_2$ is a linear map[11] then

$$\mathrm{Hom}(\cdot, W)(\ell)\colon \mathrm{Hom}(V_2, W) \to \mathrm{Hom}(V_1, W)$$

is given by $\ell_2 \mapsto \ell_2 \circ \ell$. It is important for you to understand why $\mathrm{Hom}(V, \cdot)$ is covariant but $\mathrm{Hom}(\cdot, W)$ is contravariant.

Taking $W = \mathbb{R}$ shows that $V \mapsto V^*$ is a contravariant functor.

REMARK 19.56. Going back to algebraic topology, **singular cohomology** is a contravariant functor $\mathsf{hTop}^2 \to \mathsf{Ab}$. Later in this book we will look at **de Rham cohomology**.

---

[10] We use the direct sum notation here as well (instead of the product) so as not to confuse you with the product category.

[11] Note the order of V_1 and V_2!

Similarly one can consider contravariant functors of more than one variable. In fact, one can even consider functors that are covariant in some variables and contravariant in others. This is easiest to see with an example.

EXAMPLE 19.57. Let $\mathrm{Hom}(\cdot,\cdot)\colon (\mathsf{Vect}, \mathsf{Vect}) \to \mathsf{Vect}$ denote the functor that sends a pair (V, W) to the vector space $\mathrm{Hom}(V, W)$. As Example 19.47 and 19.55 showed, this is contravariant in the first variable and covariant in the second variable. If $\ell_1 : V_1 \to V_2$ and $\ell_2 \colon W_1 \to W_2$ then

$$\mathrm{Hom}(\cdot,\cdot)(\ell_1,\ell_2)\colon \mathrm{Hom}(V_2, W_1) \to \mathrm{Hom}(V_1, W_2)$$

sends a linear map $a\colon V_2 \to W_1$ to the linear map $\ell_2 \circ a \circ \ell_1 \colon V_1 \to W_2$.

We have now almost arrived at the correct setting for which to prove the Metatheorem. The only thing that is left is to take into the account that we require our functors to be smooth.

DEFINITION 19.58. Let $\mathsf{F}\colon \mathsf{Vect} \to \mathsf{Vect}$ be a covariant functor. We say that F is **smooth** if for any two vector spaces V, W, the map

$$\mathrm{Hom}(V, W) \to \mathrm{Hom}(\mathsf{F}(V), \mathsf{F}(W)), \qquad \ell \mapsto \mathsf{F}(\ell)$$

is itself smooth in the normal sense.

A similar definition makes sense for functors of k variables which are covariant in some variables and contravariant in others, provided one remembers to flip the domain and target in each contravariant variable:

DEFINITION 19.59. Let $\mathsf{F}\colon (\mathsf{Vect}, \dots, \mathsf{Vect}) \to \mathsf{Vect}$ be a functor of k variables of either (or mixed) variance. We say that F is a **smooth functor** if for any vector spaces $V_1, \dots V_k$ and $W_1, \dots, W_k$, the induced map

$$\bigoplus_{i=1}^{r} \tilde{\mathrm{Hom}}(V_i, W_i) \to \mathrm{Hom}(\mathsf{F}(V_1, \dots, V_k), \mathsf{F}(W_1, \dots W_k)),$$
$$(\ell_1, \dots, \ell_k) \mapsto \mathsf{F}(\ell_1, \dots, \ell_k) \tag{19.6}$$

where

$$\tilde{\mathrm{Hom}}(V_i, W_i) := \begin{cases} \mathrm{Hom}(V_i, W_i), & \text{if } \mathsf{F} \text{ is covariant in the } i\text{th variable,} \\ \mathrm{Hom}(W_i, V_i), & \text{if } \mathsf{F} \text{ is contravariant in the } i\text{th variable,} \end{cases}$$

is a smooth map in the usual sense (note again each side is simply a vector space).

In fact, in all the examples we have seen, the map (19.6) is actually a linear map (and so is certainly smooth). We emphasise though that for a general functor this may not be the case. Here now is a precise statement of the Metatheorem.

Theorem 19.60 *Let* $\mathsf{F}\colon (\mathsf{Vect}, \dots, \mathsf{Vect}) \to \mathsf{Vect}$ *be a smooth functor of* k *variables of either variance in each variable. Let* $\pi_i \colon E_i \to M$ *be* k *vector bundles. Define*

$$\mathsf{F}(E_1, \ldots, E_k) := \bigsqcup_{p \in M} \mathsf{F}(E_{1|p}, \ldots, E_{k|p}),$$

with associated projection $\pi \colon \mathsf{F}(E_1, \ldots, E_k) \to M$. *Then* $\mathsf{F}(E_1, \ldots, E_k)$ *is a vector bundle.*

The proof is very easy, but it is notationally quite challenging. We recommend you write out for yourself the case $k = 2$ where F is say, contravariant in the first variable and covariant in the second (the $\mathrm{Hom}(\cdot, \cdot)$ functor from Example 19.57 is such an example). Once you understand this, the general case is just messier.

Proof. Choose an open set $U \subset M$ over which all the E_i are trivial, i.e. so that there exist vector bundle charts $\varepsilon_i \colon \pi_i^{-1}(U) \to \mathbb{R}^{n_i}$, where E_i has rank n_i. Then for each $p \in U$ and each i, we have a linear isomorphism $\varepsilon_{i|p} \colon E_{i|p} \to \mathbb{R}^{n_i}$. Set

$$\tilde{\varepsilon}_{i|p} := \begin{cases} \varepsilon_{i|p} \colon E_{i|p} \to \mathbb{R}^{n_i}, & \text{if } \mathsf{F} \text{ is covariant in the } i\text{th variable,} \\ \varepsilon_{i|p}^{-1} \colon \mathbb{R}^{n_i} \to E_{i|p}, & \text{if } \mathsf{F} \text{ is contravariant in the } i\text{th variable.} \end{cases}$$

Since F is a functor, we can feed it the morphisms $\tilde{\varepsilon}_{i,p}$ to get a map

$$\tilde{\varepsilon}_p = \mathsf{F}(\tilde{\varepsilon}_{1|p}, \ldots, \tilde{\varepsilon}_{k|p}) \in \mathrm{Hom}\Big(\mathsf{F}\big(E_{1|p}, \ldots, E_{k|p}\big), \mathsf{F}\big(\mathbb{R}^{n_1}, \ldots, \mathbb{R}^{n_k}\big)\Big)$$

By functoriality, $\tilde{\varepsilon}_p$ is a linear isomorphism. Define $\tilde{\varepsilon} \colon \pi^{-1}(U) \to \mathsf{F}(\mathbb{R}^{n_1}, \ldots, \mathbb{R}^{n_k})$ by letting $\tilde{\varepsilon}$ be equal to $\tilde{\varepsilon}_p$ on $\mathsf{F}(E_{1|p}, \ldots, E_{k|p})$. We now declare $\tilde{\varepsilon}$ to be a bundle chart for $\mathsf{F}(E_1, \ldots, E_k)$ over U.

To complete the proof, we need to show that the transition functions are smooth linear isomorphisms. For this, suppose $\gamma_i \colon \pi_i^{-1}(U) \to \mathbb{R}^{k_i}$ were different choices of vector bundle chart on each E_i, with corresponding chart $\tilde{\gamma}$ on $\mathsf{F}(E_1, \ldots, E_k)$. Let $g_i(p) = \varepsilon_{i|p} \circ \gamma_{i|p}^{-1}$ denote the transition functions on E_i from the ε_i to the γ_i. We must show that the transition function $g(p) := \tilde{\varepsilon}_p \circ \tilde{\gamma}_p^{-1}$ is smooth and linear. But this again follows almost immediately from functoriality. If $p \in U$ then

$$\begin{aligned} g(p) &= \tilde{\varepsilon}|_{\mathsf{F}(E_{1|p}, \ldots, E_{k|p})} \circ \tilde{\gamma}|^{-1}_{\mathsf{F}(E_{1|p}, \ldots, E_{k|p})} \\ &= \mathsf{F}(\tilde{\varepsilon}_{1|p}, \ldots, \tilde{\varepsilon}_{k|p}) \circ \mathsf{F}(\tilde{\gamma}_{1|p}, \ldots, \tilde{\gamma}_{k|p})^{-1} \\ &= \mathsf{F}(\tilde{\varepsilon}_{1|p}, \ldots, \tilde{\varepsilon}_{k|p}) \circ \mathsf{F}(\tilde{\gamma}_{1|p}^{-1}, \ldots, \tilde{\gamma}_{k|p}^{-1}) \\ &= \mathsf{F}\big(\tilde{\varepsilon}_{1|p} \circ \gamma_{1|p}^{-1}, \ldots \tilde{\varepsilon}_{k|p} \circ \gamma_{k|p}^{-1}\big) \\ &= \mathsf{F}\big(\tilde{g}_1(p), \ldots \tilde{g}_k(p)\big), \end{aligned}$$

where

$$\tilde{g}_i(p) := \begin{cases} g_i(p) & \text{if } \mathsf{F} \text{ is covariant in the } i\text{th variable,} \\ g_i(p)^{-1} & \text{if } \mathsf{F} \text{ is contravariant in the } i\text{th variable.} \end{cases}$$

Thus since F is a functor, $\mathsf{F}(\tilde{g}_1(p), \ldots \tilde{g}_k(p))$ is a linear isomorphism. Moreover since F is a smooth functor, $p \mapsto \mathsf{F}(\tilde{g}_1(p), \ldots \tilde{g}_k(p))$ depends smoothly on p. This completes the proof. ∎

Theorem 19.60 bypasses Corollary 18.5. But it is not yet enough by itself, as we have not yet proved the analogue of Corollary 18.4—namely, that the vector bundle given by Theorem 19.60 is unique up to isomorphism.

This however is easily rectified, but it requires introducing **natural transformations**, which, roughly speaking, are functors between functors.

Definition 19.61. Let C and D be two categories, and let $\mathsf{F}, \mathsf{G} \colon \mathsf{C} \to \mathsf{D}$ be two functors. A **natural transformation** $\tau \colon \mathsf{F} \to \mathsf{G}$ is a family of morphisms $\tau_C \colon \mathsf{F}(C) \to \mathsf{G}(C)$ for each $C \in \mathrm{obj}(\mathsf{C})$ such that for any morphism $f \colon A \to B$ in C the following diagram commutes:

$$\begin{array}{ccc} \mathsf{F}(A) & \xrightarrow{\tau_A} & \mathsf{G}(A) \\ {\scriptstyle \mathsf{F}(f)}\downarrow & & \downarrow{\scriptstyle \mathsf{G}(f)} \\ \mathsf{F}(B) & \xrightarrow[\tau_B]{} & \mathsf{G}(B) \end{array}$$

If each morphism τ_C is an isomorphism then we say that τ is a **natural isomorphism**.

In the previous chapter we briefly discussed the difference between canonical isomorphisms and non-canonical isomorphisms. To illustrate this, let us give a proof of equation (18.6) from the previous chapter.

Theorem 19.62 *A finite-dimensional vector space V is canonically isomorphic to its double dual V^{**}.*

Proof. Let $\mathsf{F} \colon \mathsf{Vect} \to \mathsf{Vect}$ denote the functor

$$\mathsf{F}(V) := V^{**} = \mathrm{Hom}(\mathsf{Hom}(V, \mathbb{R}), \mathbb{R}).$$

If $A\ell \colon V \to W$ is a linear map then $\mathsf{F}(\ell) \colon \mathsf{F}(V) \to \mathsf{F}(W)$ is the linear map usually written as $\ell^{**} \colon V^{**} \to W^{**}$ and defined by

$$\ell^{**}(\varphi)(\eta) = \varphi(\eta \circ \ell) \qquad \varphi \in V^{**}, \eta \in W^{*}.$$

Let $\mathsf{ev}_V \colon V \to V^{**}$ denote the map

$$\mathsf{ev}_V(v)(\eta) := \eta(v), \qquad \eta \in V^{*}.$$

We claim that ev is a natural isomorphism from the identity functor to F. This comes down to showing that the following diagram commutes for any pair of vector spaces V, W and any linear map $\ell \colon V \to W$:

$$\begin{array}{ccc} V & \xrightarrow{\mathsf{ev}_V} & V^{**} \\ {\scriptstyle \ell}\downarrow & & \downarrow{\scriptstyle \ell^{**}} \\ W & \xrightarrow[\mathsf{ev}_W]{} & W^{**} \end{array}$$

This is trivial: if $\eta \in W^*$ and observe:

$$\begin{aligned}\ell^{**}\mathsf{ev}_V(v)(\eta) &= \mathsf{ev}_V(v)(\eta \circ \ell)\\ &= \eta(\ell v)\\ &= \mathsf{ev}_{\ell v}(\eta).\end{aligned}$$

The proof is complete. ∎

REMARK 19.63. If we work on the larger category $\mathsf{Vect}^{\leq\infty}$ of all (not necessarily finite-dimensional) vector spaces, ev is still a natural transformation, but no longer a natural isomorphism.

Theorem 19.60 admits the following enhancement.

Theorem 19.64 *Let* F_1 *and* F_2 *be two functors as in the statement of Theorem* 19.60. *Assume there exists a smooth natural isomorphism* $\tau\colon \mathsf{F}_1 \to \mathsf{F}_2$. *Then the vector bundles obtained by applying Theorem* 19.60 *to* F_1 *and* F_2 *are naturally isomorphic.*

Corollary 19.3 and Corollary 19.16 are special cases of Theorem 19.64.

Problem Sheet H

A (♣) indicates the problem is non-examinable. This does not necessarily imply it is difficult.

Problem H.1 Let V, W and U be vector spaces. Prove there are natural isomorphisms $V \otimes W \cong W \otimes V$ and $(U \otimes V) \otimes W \cong U \otimes (V \otimes W)$.

Problem H.2 Let V and W be vector spaces. Prove that for any $A \in \mathrm{Alt}_h(V, W)$ there is a unique linear map $a \colon \bigwedge^h V \to W$ such that the following diagram commutes:

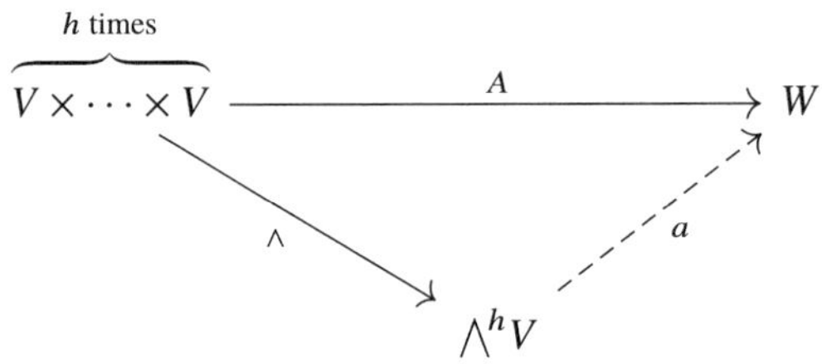

Prove moreover that $\bigwedge^h V$ is uniquely characterised by this property.

Problem H.3 Let V be a vector space of dimension n with basis $\{e_1, \dots, e_n\}$. Prove that

$$\{e_{i_1} \wedge \cdots \wedge e_{i_h} \mid 1 \le i_1 < \cdots < i_h \le n\}$$

is a basis of $\bigwedge^h V$ and $\bigwedge^h V = 0$ for $h > n$. Deduce that $\dim \bigwedge^h V = \binom{n}{h}$ and that $\dim \bigwedge V = 2^n$.

Problem H.4 Let $\varphi \colon M \to N$ be a smooth map, and suppose $L \to E \xrightarrow{\pi} N$ is a fibre bundle. Set

$$\varphi^* E := \{(p, u) \in M \times E \mid \varphi(p) = \pi(u)\},$$

with projection $\mathrm{pr}_1 : \varphi^* E \to M$.

(i) Prove that $\varphi^* E$ is a fibre bundle over M with fibre L such that (φ, pr_2) is a fibre bundle morphism:

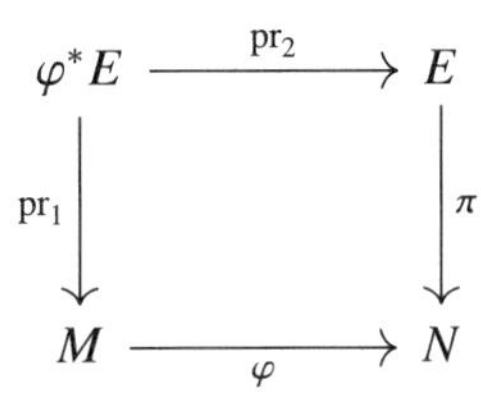

(ii) Prove that

$$T_{(p,u)}(\varphi^* E) = \{(\xi, \zeta) \in T_p M \times T_u E \mid D\varphi(p)\xi = D\pi(p)\zeta\}.$$

(iii) Suppose $\psi\colon K \to M$ is another smooth map. Prove that $\psi^*(\varphi^* E)$ and $(\varphi \circ \psi)^* E$ are isomorphic fibre bundles over K.

(iv) Assume now that E has structure group G[12]. Prove that $\varphi^* E$ has structure group G. Deduce that the pullback of a vector bundle is a vector bundle, and the pullback of a principal bundle is a principal bundle.

(v) Prove that the isomorphism from part (iii) can be taken to be an isomorphism of G-fibre bundles (or a subgroup thereof).

PROBLEM H.5 Let $L \to E \xrightarrow{\pi_1} M$ and $K \to F \xrightarrow{\pi_2} N$ be fibre bundles with structure groups G and H respectively. Prove that $(\pi_1, \pi_2)\colon E \times F \to M \times N$ is another fibre bundle with fibre $L \times K$ and structure group $G \times H$. We call this the **external product** of E and F.

(♣)PROBLEM H.6 Let $\iota\colon M \to M \times M$ denote the diagonal map $p \mapsto (p, p)$.

(i) Let E and F be two vector bundles over M. Prove that

$$E \oplus F = \iota^*(E \times F).$$

(ii) Let P and Q be two principal bundles over M. Prove that

$$P \star Q = \iota^*(P \times Q).$$

PROBLEM H.7 Let E be a vector bundle over M and F a vector bundle over N. Suppose $\Psi\colon E \to F$ is any smooth map that maps each fibre E_p for $p \in M$ linearly onto some fibre F_q for $q \in N$. Prove that $\Psi = \Phi_2 \circ \Phi_1$ where Φ_1 is a vector bundle homomorphism and Φ_2 is a vector bundle isomorphism along a map $M \to N$.

PROBLEM H.8 Let $V \to E \xrightarrow{\pi_1} M$ and $W \to F \xrightarrow{\pi_2} M$ be two vector bundles over M, and let $\Phi\colon E \to F$ be a vector bundle homomorphism.

(i) Assume Φ is injective on each fibre. Consider the quotient vector space

$$\overline{E}_p := F_p / \Phi_p(E_p).$$

Prove that $\overline{E} := \bigsqcup_{p \in M} \overline{E}_p$ is a vector bundle over M with fibre W/V. Deduce that $\operatorname{im} \Phi$ is a vector subbundle of F.

(ii) Assume that Φ is surjective on each fibre. Let

$$Z_p := \ker \Phi_p \subset E_p.$$

Prove that $Z := \bigsqcup_{p \in M} Z_p$ is a vector bundle over M. What is the fibre?

(♣)PROBLEM H.9 Let $\varphi\colon M \to N$ be a smooth map and suppose $\pi\colon E \to N$ is a vector bundle, which we illustrate pictorially as:

[12] i.e. E is a (G, σ)-fibre bundle for some effective action σ of G on L.

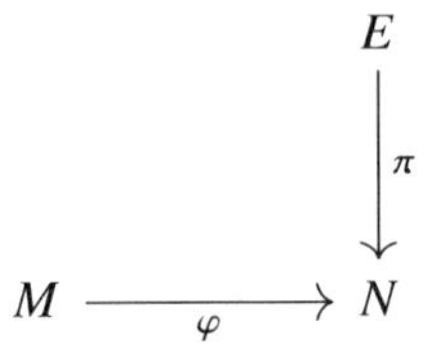

(δ)

A **solution** of the diagram (δ) is a vector bundle $\pi_1 : F \to M$ together with a vector bundle morphism $\Phi \colon F \to E$ along φ:

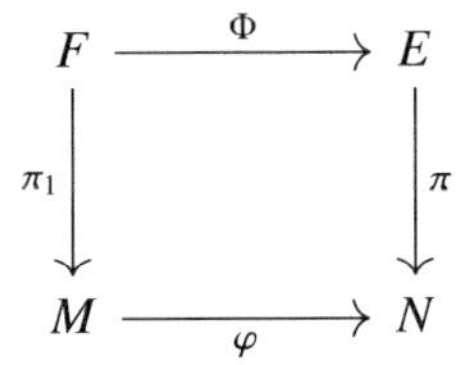

As we have seen, one possible solution is the pullback bundle $\varphi^* E$:

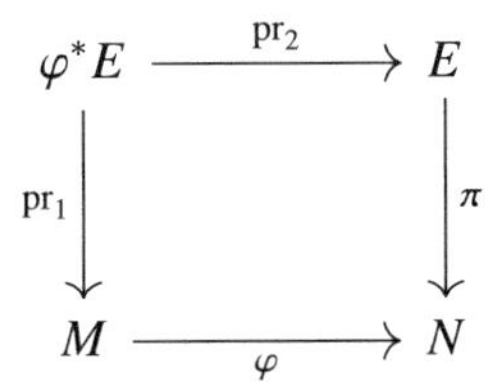

The aim of this problem is to prove that the pullback bundle can be characterised as a solution to a *universal mapping problem*: namely, that $\varphi^* E$ is the "most efficient" solution in the following sense:

Suppose $\pi_1 : F \to M$ and Φ is *any* solution to (δ). Prove there exists a *unique* vector bundle homomorphism $\Psi \colon F \to \varphi^* E$ such that the following diagram commutes:

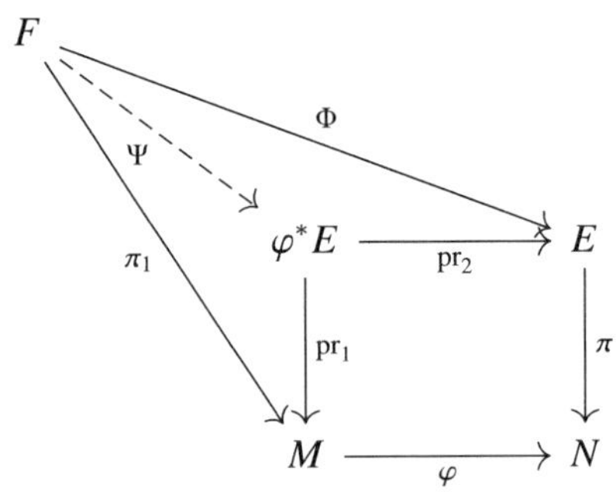

Prove moreover that $\varphi^* E$ is uniquely determined by this property. Explicitly this means that if $\widetilde{\pi} \colon \widetilde{E} \to M$ and $\widetilde{\Phi}$ is another solution to the diagram (δ) with the property that for *any* solution $\pi_1 : F \to M$ and Φ of (δ) there exists a *unique* vector bundle homomorphism $\widetilde{\Psi} \colon F \to \widetilde{E}$ such that the following commutes:

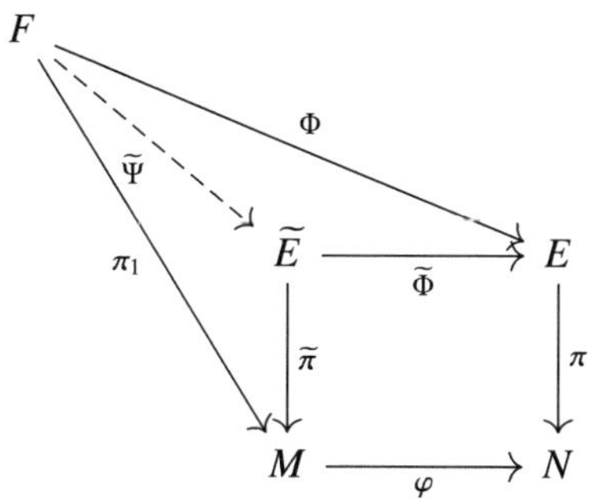

then in fact $\widetilde{E}$ is isomorphic as a vector bundle over M to $\varphi^* E$.

Bonus Problem(s) for Sheet H

These problem(s) are hard, and are included for enthusiasts only. Solutions will not be provided.

(♣)PROBLEM H.10 Let M be a smooth compact connected manifold of dimension 4. Let $G = \mathrm{SU}(2)$. How "many" isomorphism classes of principal G-bundles $P \to M$ are there[13]?

[13] This problem is relevant in gauge theory; a topic we return to in Volume II.

Chapter 20
Sections of Vector Bundles

A fibre bundle $\pi\colon E \to M$ is a surjective submersion between manifolds with the property that the domain E has extra structure. Smooth maps that go in the opposite direction are—from the point of view of fibre bundles—uninteresting unless they respect this extra structure.

DEFINITION 20.1. Let $L \to E \xrightarrow{\pi} M$ be a fibre bundle. A **section**[1] of E is a smooth map $s\colon M \to E$ such that $\pi \circ s = \mathrm{id}$, that is, a smooth map $s\colon M \to E$ such that

$$s(p) \in E_p, \qquad \forall\, p \in M. \tag{20.1}$$

The set of all sections is denoted by $\Gamma(E)$. A **local section** of E is a section of the bundle $\pi^{-1}(U) \to U$ of E over an open set $U \subset M$. We denote by $\Gamma(U, E)$ the set of all local sections with domain U.

EXAMPLES 20.2. Here are some examples of sections:

(i) Let M be a manifold. A vector field X on M is a section of the tangent bundle. Thus

$$\mathfrak{X}(M) = \Gamma(TM).$$

Similarly a vector field X defined on an open subset of M is a local section:

$$\mathfrak{X}(U) = \Gamma(U, TM).$$

In particular, if (U, x) is a chart on M with local coordinates (x^i) then $\frac{\partial}{\partial x^i}$ is an element of $\Gamma(U, TM)$.

(ii) In a similar vein, if $f \in C^\infty(M)$ then in Example 5.2 we defined a section df of T^*M. If $f \in C^\infty(U)$ then $df \in \Gamma(U, T^*M)$.

[1] We should really say "smooth section", but since we will never consider non-smooth sections, we omit the adjective.

W. J. Merry, *Lectures on Differential Geometry I*,
https://doi.org/10.1007/978-3-032-03733-6_20

(iii) A section of the trivial fibre bundle $M \times L \to M$ is the same thing as a smooth map $M \to L$. Thus for instance, a section of $M \times \mathbb{R} \to M$ is just a smooth function on M.

(iv) Let $\pi_1 : E \to M$ and $\pi_2 : F \to M$ denote two vector bundles, and consider the vector bundle $\operatorname{Hom}(E, F)$. A section $\Phi \in \Gamma(\operatorname{Hom}(E, F))$ is a smooth map $p \mapsto \Phi_p$ where $\Phi_p : E_p \to F_p$ is a linear map. Thus:

$$\Gamma(\operatorname{Hom}(E, F)) = \{\text{vector bundle homomorphisms } \Phi : E \to F\}.$$

Local sections always exist:

Lemma 20.3. *Let $L \to E \xrightarrow{\pi} M$ be a fibre bundle and let $p \in M$. Then there exists a neighbourhood U of p and a local section $s \in \Gamma(U, E)$.*

Proof. This follows directly from the definitions. ■

The existence of a global section is sometimes not automatic:

Proposition 20.4. *Let M be a smooth manifold.*

(i) *Let $\pi : E \to M$ be a vector bundle. Then $\Gamma(E) \neq \emptyset$.*
(ii) *Let $\pi : P \to M$ be a principal G-bundle. Then $\Gamma(P) \neq \emptyset$ if and only if $P = M \times G$ is trivial.*

Proof. We prove the two cases separately.

- (i): The map
$$o : M \to E, \qquad p \mapsto 0 \in E_p$$
is a global smooth section. We call o the **zero section**.
- (ii): If $P = M \times G$ is the trivial bundle, then for any $g \in G$ the map $s(p) := (p, g)$ is a section. Conversely, let τ denote the free right action and suppose $s : M \to P$ is a section. Then since p and $s(\pi(p))$ belong to the same fibre for each $p \in P$, there is a well-defined equivariant map $\varepsilon : P \to G$ such that
$$p = \tau_{\varepsilon(p)}(s(\pi(p))), \qquad \forall\, p \in P.$$
We claim that ε is a principal bundle chart, whence P is a trivial bundle. For this we need to prove that $(\pi, \varepsilon) : P \to M \times G$ is a diffeomorphism. But this follows from Lemma 17.15, since (π, ε) is a principal bundle morphism along the identity map on M. ■

For vector bundles we can do more than just ask for a single local section.

Definition 20.5. Let $\pi : E \to M$ be a vector bundle of rank n and let $U \subset M$ be open. A **local frame** for E over U is a collection $(e_1, \dots, e_n)$ of sections $e_i \in \Gamma(U, E)$ such that $\{e_1(p), \dots, e_n(p)\}$ form a basis of the vector space E_p for each $p \in U$.

Here are three equivalent ways to think of local frames:

LEMMA 20.6. *Let $\pi\colon E \to M$ be a vector bundle and suppose $U \subset M$ is an open set. The following are equivalent.*

(i) *There exists a local frame for E over U.*
(ii) *There exists a vector bundle chart $\varepsilon\colon \pi^{-1}(U) \to \mathbb{R}^n$.*
(iii) *There exists a section of the frame bundle over U: $\Gamma(U, \mathrm{Fr}(E)) \neq \emptyset$.*

Proof. Suppose (e_i) is a local frame over U. Then every point $v \in \pi^{-1}(U)$ can be written as uniquely as linear combination $v = a^i\, e_i(p)$. We define

$$\varepsilon\colon \pi^{-1}(U) \to \mathbb{R}^n, \qquad v \mapsto (a^1, \dots, a^n). \tag{20.2}$$

This is a vector bundle chart. This shows that (i) $\Rightarrow$ (ii).

Conversely if $\varepsilon\colon \pi^{-1}(U) \to \mathbb{R}^n$ is a vector bundle chart then if we define

$$e_i(p) := \varepsilon_p^{-1}(e_i),$$

where e_i is the standard basis vector in $\mathbb{R}^n$, then e_i is smooth (use the argument from the proof of Lemma 20.9) and the collection $\{e_i(p)\}$ is a basis of E_p since ε_p is a linear isomorphism. This shows (ii) $\Rightarrow$ (i).

A section s of the frame bundle $\mathrm{Fr}(E)$ over U is by definition a smooth map $s\colon U \to \mathrm{Fr}(E)$ such that $s(p) \in \mathrm{Fr}(E_p)$, that is, $s(p)$ is a linear isomorphism $\mathbb{R}^n \to E_p$. Define $e_i \in \Gamma(U, E)$ by $e_i(p) := s(p)e_i$. Then (e_i) is a local frame for E over U. Conversely starting with (e_i) and defining s by the same equation produces a local section of $\mathrm{Fr}(E)$ over U. This shows (i) $\Leftrightarrow$ (iii). ∎

A **global frame** of a vector bundle is a frame defined on $U = M$. The next statement is the generalisation to arbitrary vector bundles of Problem F.1.

COROLLARY 20.7. *A vector bundle $\pi\colon E \to M$ admits a global frame if and only if it is trivial.*

LEMMA 20.8. *Let $\pi\colon E \to M$ be a fibre bundle and let $s \in \Gamma(U, E)$. Then $s(U)$ is an embedded submanifold of E of dimension equal to the dimension of M.*

Proof. If (V, x) is a chart on M with $V \subset U$ then $x \circ \pi$ is a chart on $s(V)$. ∎

Applying this to the zero section of a vector bundle allows us to see $M \cong o(M)$ as an embedded submanifold of E.

The space of sections of a vector bundle has extra structure not present in normal fibre bundles. We already saw this for vector fields in Chapter 8, but let us go over it again here.

LEMMA 20.9. *Let $\pi\colon E \to M$ be a vector bundle.*[2] *Then for any non-empty open set $U \subset M$, the set $\Gamma(U, E)$ is an infinite-dimensional real vector space and a module over the ring $C^\infty(U)$.*

[2] Here and elsewhere one should implicitly assume that all vector bundles have strictly positive rank, and that M has a strictly positive dimension.

Proof. Suppose $s \in \Gamma(U, E)$. Let $x\colon V \to O$ be a chart on $V \subset U$ and let ε be a vector bundle chart on E defined on $\pi^{-1}(V)$. Then as in Remark 16.5, we may take $(x \circ \pi, \varepsilon)$ as a chart on E. The assumption that s is smooth means that the composition

$$(x \circ \pi, \varepsilon) \circ s \circ x^{-1}\colon O \to O \times \mathbb{R}^n$$

is smooth. Moreover the section property tells us that this local map is of the form

$$(x \circ \pi, \varepsilon) \circ s \circ x^{-1} = (\mathrm{id}, \tilde{s}) \tag{20.3}$$

where $\tilde{s}\colon O \to \mathbb{R}^n$ is some smooth map. Just as in the proof of Proposition 8.2, this argument reverses, and we see that a map s satisfying the section property is smooth if and only if each local map $\tilde{s}$ is smooth.

The map sending $s \mapsto \tilde{s}$ establishes a linear isomorphism between $\Gamma(V, E)$ and $C^\infty(O, \mathbb{R}^n) = C^\infty(O, \mathbb{R})^n = C^\infty(V, \mathbb{R})^n$.[3]

Since $n > 0$ and V is positive dimensional $C^\infty(V, \mathbb{R})^n$ is infinite-dimensional and thus so is $\Gamma(V, E)$. Pick now a nowhere zero $s_0 \in \Gamma(V, E)$, then $f \mapsto f s_0$ defines a linear injective map $C_c^\infty(V) \to \Gamma(U, E)$, and thus $\Gamma(U, E)$ is infinite dimensional since $C_c^\infty(V)$ is (see Remark 8.14).

To show $\Gamma(U, E)$ is a module over the ring $C^\infty(U)$ is now trivial: if r and s are two sections and $c \in \mathbb{R}$ then $p \mapsto cr(p) + s(p)$ certainly satisfies the section property (20.1), and its local expression is given by $c\tilde{r} + \tilde{s}$ which is smooth if $\tilde{r}$ and $\tilde{s}$ are. Moreover if $f \in C^\infty(U)$ then we define

$$(fs)(p) := f(p)s(p), \qquad p \in U.$$

The local expression of fs is $\tilde{f}\tilde{s}$ where $\tilde{f} = f \circ x^{-1}$. This is smooth.

Finally, the vector space $C^\infty(U)$ is infinite-dimensional by Lemma 2.13, and hence any non-zero module over it is *a fortiori* infinite-dimensional over $\mathbb{R}$. ■

The analogue of part (iii) of Proposition 8.2 also holds for general vector bundles.

Remark 20.10. If (e_i) is a local frame for E over U then any (not necessarily smooth) map $s\colon U \to E$ satisfying the section property (20.1) can be written as

$$s = f^i e_i, \qquad \text{for some functions } f^i\colon U \to \mathbb{R}.$$

If we take the vector bundle chart ε on E from (20.2) associated to the local frame (e_i) then for any chart x on M with appropriate domain, the function $\tilde{s}$ associated to s from (20.3) is given by

$$\tilde{s}(q) = \left(f^1(x^{-1}(q)), \ldots, f^n(x^{-1}(q))\right).$$

This tells us that s is smooth (and hence belongs to $\Gamma(U, E)$) if and only if the functions f^i are smooth functions on U.

The next lemma is analogous to Problem D.2 and is thus left as an exercise.

[3] In other words $\Gamma(V, E)$ is a free $C^\infty(V)$-module of rank n.

Lemma 20.11. *Let $\pi\colon E \to M$ be a vector bundle. Let $p \in M$ and $v \in E_p$. Then there exists a section $s \in \Gamma(E)$ with $s(p) = v$.*

An application of the Bump Functions Lemma 3.2 gives us the following result.

Lemma 20.12. *Let $\pi\colon E \to M$ be a vector bundle and let $s \in \Gamma(U, E)$. Fix $p \in U$. Then there exists a global section $r \in \Gamma(E)$ such that r agrees with s on a neighbourhood of p.*

Proof. Choose a neighbourhood V of p with $\overline{V} \subset U$. Choose a bump function $\chi\colon M \to \mathbb{R}$ such that $\chi(q) = 1$ for all $q \in V$ and such that $\operatorname{supp}(\chi) \subset U$. Define $r\colon M \to E$ by

$$r(q) := \begin{cases} \chi(q)s(q), & q \in U, \\ 0, & q \in M \setminus U. \end{cases}$$

Then r is smooth and agrees with s on the neighbourhood V of p. ∎

Definition 20.13. A local frame (e_i) of E over U determines a local frame (e^i) of the dual bundle E^* over U by requiring that

$$e^i(p)(e_j(p)) = \delta^i_j, \qquad \text{for all } p \in U.$$

(Exercise: Why is this smooth?)

We call (e^i) the **dual frame** to (e_i).

Convention. If $\lambda \in \Gamma(U, E^*)$ is a section of a *dual* bundle, we normally write λ_p instead of $\lambda(p)$ for the value of λ at p. This is consistent with our notation df_p for the differential of f at p, i.e. the value of the section $df \in \Gamma(T^*M)$ at p. Thus $\lambda_p\colon E_p \to \mathbb{R}$ is a linear map.

Remark 20.14. If $s \in \Gamma(U, E)$ then if we write $s = f^i e_i$ for smooth functions f^i as per Remark 20.10 then observe that

$$e^i_p(s(p)) = f^i(p).$$

Similarly if $\lambda \in \Gamma(U, E^*)$ is any section of the dual bundle then we can write $\lambda = g_i e^i$ where the function $g_i \in C^\infty(U)$ are given by

$$g_i(p) = \lambda_p(e_i(p)).$$

Example 20.15. Let M be a smooth manifold, and let (U, x) be a chart on M with local coordinates (x^i). Then

$$\left\{ \frac{\partial}{\partial x^i} \mid i = 1, \dots, m \right\}$$

is a local frame for TM over U. Similarly

$$\{dx^i \mid i = 1, \dots, m\}$$

is a local frame for T^*M over U. This is the dual frame. Taking this one step further,

$$\left\{\frac{\partial}{\partial x^i} \otimes dx^j \mid 1 \le i, j \le m\right\}$$

is a local frame for $TM \otimes T^*M$ over U.

We now introduce two key properties that a linear operator between spaces of sections may or may not have.

DEFINITION 20.16. Let $\pi_1 \colon E \to M$ and $\pi_2 \colon F \to M$ be two vector bundles over the same manifold M. Suppose $\zeta \colon \Gamma(E) \to \Gamma(F)$ is an $\mathbb{R}$-linear operator.

- We say that ζ is a **local operator** if whenever a section $s \in \Gamma(E)$ vanishes on an open set $U \subset M$, $\zeta(s) \in \Gamma(F)$ also vanishes on U.
- We call ζ a **point operator** if whenever a section $s \in \Gamma(E)$ vanishes at a point p, $\zeta(s)$ also vanishes at p.

Any point operator is clearly a local operator, but the converse is not true.

EXAMPLE 20.17. By part (iii) of Example 20.2, the space $C^\infty(\mathbb{R})$ can be identified with the space of all sections of the trivial bundle $\mathbb{R} \times \mathbb{R} \to \mathbb{R}$. Differentiation

$$C^\infty(\mathbb{R}) \to C^\infty(\mathbb{R}), \qquad \gamma \mapsto \gamma'$$

is a local operator (since if γ is constant on an open set its derivative is also constant on that open set) but it is not a point operator.

More generally:

EXAMPLE 20.18. Let M be a smooth manifold, and let $X \in \mathfrak{X}(M)$ denote a vector field. Regard X as a derivation of $C^\infty(M)$ as in Proposition 8.7, or equivalently, as a linear operator $\Gamma(M \times \mathbb{R}) \to \Gamma(M \times \mathbb{R})$ (as in part (iii) of Examples 20.1). Then $f \mapsto X(f)$ is a local operator by Corollary 3.4, but not a point operator.

Local operators behave well under restriction.

PROPOSITION 20.19. *Let $\pi_1 \colon E \to M$ and $\pi_2 \colon F \to M$ be two vector bundles and suppose $\zeta \colon \Gamma(E) \to \Gamma(F)$ is a local operator. Then for each open set $U \subset M$, there is a unique $\mathbb{R}$-linear map $\zeta^U \colon \Gamma(U, E) \to \Gamma(U, F)$, called the* **restriction** *of ζ to U, such that for any global section s, one has*

$$\zeta^U(s|_U) = \zeta(s)|_U. \tag{20.4}$$

Proof. Let $s \in \Gamma(U, E)$ and fix $p \in U$. By Lemma 20.12 there exists a global section r of E that agrees with s in some neighbourhood V of p. We set

$$\zeta^U(s)(p) := \zeta(r)(p).$$

This is well-defined, i.e. independent of the choice of r, since ζ is a local operator. Since $\zeta(r)$ is smooth by assumption, it follows $\zeta^U(s)$ is smooth at p, and since p was arbitrary, $\zeta^U(s)$ is smooth. Finally, if s is a global section then s is an extension of $s|_U$ for any open U, and thus (20.4) follows. ∎

As the proof of Proposition 20.19 shows, the operator ζ^U is itself a local operator.

So far we have looked at operators which are $\mathbb{R}$-linear. Since $\Gamma(E)$ and $\Gamma(F)$ are modules over $C^\infty(M)$, we could instead restrict our attention to operators that are $C^\infty(M)$-linear, i.e.

$$\zeta(fs) = f\zeta(s), \qquad \forall f \in C^\infty(M),\ s \in \Gamma(E).$$

Note that $C^\infty(M)$-linear implies $\mathbb{R}$-linear, since we may regard $\mathbb{R} \subset C^\infty(M)$ via the constant functions.

This is the same thing as working with point operators, as the next result shows.

Theorem 20.20 *Let $\pi_1 \colon E \to M$ and $\pi_2 \colon F \to M$ be two vector bundles over M. An $\mathbb{R}$-linear operator $\zeta \colon \Gamma(E) \to \Gamma(F)$ is a point operator if and only if ζ is $C^\infty(M)$-linear.*

Proof. We prove the result in four steps.

1. In this first step we show that a $C^\infty(M)$-linear operator $\zeta \colon \Gamma(E) \to \Gamma(F)$ is a local operator. Suppose $s \in \Gamma(E)$ vanishes on an open set U. Let $p \in U$, and choose a bump function $\chi \colon M \to \mathbb{R}$ such that $\chi(p) = 1$ and $\operatorname{supp}(\chi) \subset U$. Then χs is identically zero on M, and so $\zeta(\chi s)$ is identically zero. However evaluating at p and using $C^\infty(M)$-linearity,

$$0 = \zeta(\chi s)(p) = \chi(p)\zeta(s)(p) = \zeta(s)(p).$$

Since p was an arbitrary point of U, we have $\zeta(s)|_U \equiv 0$ as required.

2. Since ζ is a local operator, by Proposition 20.19 the local operators ζ^U are defined. In this step we show that such a local operator ζ^U is itself $C^\infty(U)$-linear.

Fix $s \in \Gamma(U, E)$, $f \in C^\infty(U)$. We want to show that $\zeta^U(fs) = f\zeta^U(s)$. Fix $p \in U$ and let $r \in \Gamma(E)$ denote a global section that agrees with s on a neighbourhood of p, and let g be a global smooth function that agrees with f on a neighbourhood of p. The existence of g is a special case of Lemma 20.12, cf. part (iii) of Examples 20.2, but it was also proved directly in Step 2 of Proposition 3.3. Then

$$\begin{aligned}\zeta^U(fs)(p) &= \zeta(gr)(p)\\ &= g(p)\zeta(r)(p)\\ &= f(p)\zeta^U(s)(p).\end{aligned}$$

Since p was arbitrary, we see that $\zeta^U(fs) = f\zeta^U(s)$, as required.

3. We now show that ζ is actually a point operator. Let $s \in \Gamma(E)$. Suppose $s(p) = 0$. Choose an neighbourhood U of p admitting a local frame (e_i). Then we can write

$$s|_U = f^i e_i, \qquad f^i \in C^\infty(U).$$

Since $s(p) = 0$ we have $f^i(p) = 0$ for each i. We now compute:

$$\begin{aligned}\zeta(s)(p) &= \zeta^U(s|_U)(p) \\ &= \zeta^U(f^i e_i)(p) \\ &= f^i(p)\zeta^U(e_i)(p) \\ &= 0,\end{aligned}$$

where the first equality used Proposition 20.19 and the penultimate equality used the previous step.

4. Finally we prove the converse: suppose $\zeta\colon \Gamma(E) \to \Gamma(F)$ is a point operator. Fix $f \in C^\infty(M)$, $s \in \Gamma(E)$ and $p \in M$. Let $c := f(p)$. Then $fs - cs$ vanishes at p, and thus $\zeta(fs - cs)(p) = 0$ as ζ is a point operator. Since ζ is $\mathbb{R}$-linear,

$$\begin{aligned}\zeta(fs)(p) &= \zeta(cs)(p) \\ &= c\zeta(s)(p) \\ &= f(p)\zeta(s)(p).\end{aligned}$$

Since p was arbitrary, $\zeta(fs) = f\zeta(s)$. This completes the proof. ∎

Let us now return to part (iv) of Examples 20.2: a vector bundle homomorphism $\Phi\colon E \to F$ is the same thing as a section of the homomorphism bundle $\mathrm{Hom}(E, F)$. The aim of the rest of this chapter is to give yet another alternative description of a vector bundle homomorphism.

Definition 20.21. Let $\pi_1 : E \to M$ and $\pi_2 : F \to M$ denote two vector bundles over the same manifold M. Let $\Phi\colon E \to F$ denote a vector bundle homomorphism. We define an operator

$$\Phi_* : \Gamma(E) \to \Gamma(F), \qquad s \mapsto \Phi \circ s.$$

Proposition 20.22. *Let $\pi_1 : E \to M$ and $\pi_2 : F \to M$ denote two vector bundles over the same manifold M. Let $\Phi\colon E \to F$ denote a vector bundle homomorphism. Then $\Phi_* : \Gamma(E) \to \Gamma(F)$ is $C^\infty(M)$-linear, and hence a point operator.*

Proof. The map Φ_* is clearly a linear map between the two vector spaces $\Gamma(E)$ and $\Gamma(F)$. More is true: Φ_* is actually a *module homomorphism*, i.e. it is linear over $C^\infty(M)$. Indeed, if $f \in C^\infty(M)$, $s \in \Gamma(E)$, and $p \in M$ then

$$\begin{aligned}\Phi_*(fs)(p) &= \Phi \circ (fs)(p) \\ &= \Phi|_p(f(p)s(p)) \\ &= f(p)\Phi_p(s(p)) \\ &= (f\Phi_*(s))(p),\end{aligned}$$

where the penultimate equality used that Φ_p is a linear map. ∎

The main result of this chapter, the Hom-Gamma Theorem 20.25, states that every point operator $\Gamma(E) \to \Gamma(F)$ is a of the form Φ_*. This requires some more preparation.

As we have seen in Example 20.18, a vector field on a manifold can be thought of an operator on the space of sections of the trivial bundle $M \times \mathbb{R}$ via $f \mapsto X(f)$. The next result generalises part (ii) of Proposition 8.2 to arbitrary vector bundles.

PROPOSITION 20.23. *Let $\pi_1 : E \to M$ and $\pi_2 : F \to M$ be two vector bundles. Suppose $\Phi : E \to F$ is a fibre-preserving map such that $\Phi_p : E_p \to F_p$ is linear for every $p \in M$. Then Φ is smooth (and hence a vector bundle homomorphism) if and only if $\Phi_*(s) := \Phi \circ s$ belongs to $\Gamma(U, F)$ for every $s \in \Gamma(U, E)$.*

Proof. If Φ is smooth then certainly $\Phi \circ s$ is smooth. For the converse, let $p \in M$ and suppose (U, x) is a chart on M with local coordinates (x^i). We may assume that both E and F admit local frames over U; call them (e_j) and (e'_i) respectively. Since Φ_* maps smooth sections to smooth sections, there are functions $f_j^i \in C^\infty(U)$ such that

$$\Phi_*(e_j) = f_j^i \, e'_i.$$

Let ε_1 and ε_2 denote the vector bundle charts on E and F respectively associated to (e_j) and (e'_i) from part (ii) of Lemma 20.6. Then $(x \circ \pi_1, \varepsilon)$ is a manifold chart on E on $\pi_1^{-1}(U)$, and $(x \circ \pi_2, \varepsilon_2)$ is a manifold chart on F on $\pi_2^{-1}(U)$. Let $a^j \in C^\infty(\pi_1^{-1}(U))$ denote the smooth functions defined implicitly by the requirement

$$v = a^j(v) e_j(\pi_1(v)), \qquad \forall\, v \in \pi_1^{-1}(U).$$

Then the local expression of Φ is of the form:

$$(x \circ \pi_2, \varepsilon_2) \circ \Phi \circ (x \circ \pi_1, \varepsilon_1)^{-1} = \big(\mathrm{id}, (a^j f_j^1, \dots, a^j f_j^n) \circ x^{-1}\big),$$

which is smooth. ■

PROPOSITION 20.24. *Let $\pi_1 : E \to M$ and $\pi_2 : F \to M$ be two vector bundles. Suppose $\zeta : \Gamma(E) \to \Gamma(F)$ is a $C^\infty(M)$-linear map. Then for each $p \in M$ there is a unique linear map $\Phi_p : E_p \to F_p$ such that for all $s \in \Gamma(E)$, one has*

$$\Phi_p(s(p)) = \zeta(s)(p).$$

Proof. Fix $p \in M$ and $v \in E_p$. By Lemma 20.11 there exists a section s such that $s(p) = v$. Define $\Phi_p(v) := \zeta(s)(p)$. This definition is independent of the choice of s, since if s_1 was another such section then $(s - s_1)(p) = 0$, and thus $\zeta(s)(p) - \zeta(s_1)(p) = \zeta(s - s_1)(p) = 0$ since ζ is a point operator by Theorem 20.20.

We claim that Φ_p is a linear map. Indeed, if $v_1, v_2 \in E_p$ and $c \in \mathbb{R}$, then if s_1 and s_2 are sections such that $s_i(p) = v_i$ then $cs_1 + s_2$ is a section whose value at p is $cv_1 + v_2$ and

$$\begin{aligned}\Phi_p(cv_1 + v_2) &= \zeta(cs_1 + s_2)(p) \\ &= c\zeta(s_1)(p) + \zeta(s_2)(p) \\ &= c\Phi_p(v_1) + \Phi_p(v_2).\end{aligned}$$

This completes the proof. ■

We now move onto our main result.

Theorem 20.25 (The Hom-Gamma Theorem) *Let $\pi_1\colon E \to M$ and $\pi_2\colon F \to M$ be two vector bundles. Then there is a one-to-one correspondence between*

$$\{\textit{vector bundle homomorphisms } \Phi\colon E \to F\}$$

and

$$\{\textit{point operators } \zeta\colon \Gamma(E) \to \Gamma(F)\}$$

given by

$$\Phi \mapsto \Phi_*.$$

The reason for the name will be explained after the proof.

Proof. We first prove surjectivity. If $\zeta\colon \Gamma(E) \to \Gamma(F)$ is a $C^\infty(M)$-linear map then by Proposition 20.24 there exists a linear map $\Phi_p\colon E_p \to F_p$ such that for any $s \in \Gamma(E)$, $\Phi_p(s(p)) = \zeta(s)(p)$. Define $\Phi\colon E \to F$ by declaring that $\Phi|_{E_p} = \Phi_p$. Then by Proposition 20.23, the map Φ is a vector bundle homomorphism, and clearly $\Phi_* = \zeta$.

To prove injectivity, suppose $\Phi_* = \Psi_*$. Let $p \in M$ and $v \in E_p$ and let $s \in \Gamma(E)$ be a section such that $s(p) = v$ (using Lemma 20.11). Then

$$\begin{aligned}\Phi(v) &= \Phi(s(p)) \\ &= \Phi_*(s)(p) \\ &= \Psi_*(s)(p) \\ &= \Psi(s(p)) \\ &= \Psi(v).\end{aligned}$$

Since p and v were arbitrary we conclude $\Phi = \Psi$ as required. ■

Remark 20.26. Why the name "Hom-Gamma Theorem"? Suppose R is a commutative ring and X, Y are two R-modules. Let us write

$$\operatorname{Hom}(X, Y) = \{f\colon X \to Y \text{ is an } R\text{-module homomorphism}\}.$$

In the case $R = \mathbb{R}$, modules are just vector spaces, and this coincides with our notation for $\operatorname{Hom}(V, W)$ as the set of linear maps from V to W. We now apply this with $R = C^\infty(M)$. By Theorem 20.20, an element

$$\zeta \in \operatorname{Hom}\big(\Gamma(E), \Gamma(F)\big)$$

is the same thing as a point operator. Thus the Hom-Gamma Theorem 20.25 tells us that

$$\Gamma\big(\operatorname{Hom}(E, F)\big) \cong \operatorname{Hom}\big(\Gamma(E), \Gamma(F)\big), \tag{20.5}$$

in other words,

The operations Hom *and* Γ *commute.*

Bonus Material for Chapter 20

In this bonus section, we make another algebraic interlude and introduce the notion of a **sheaf**. As with the category theory material from the previous chapter, this is for interest only, and is not needed to understand any of the subsequent chapters.

Roughly speaking, a presheaf is a way to assign data locally to open subsets of a topological space in such a way that it is compatible with restrictions. A sheaf is a presheaf for which it is possible to go backwards and reassemble global data from local data.

DEFINITION 20.27. Let X denote a topological space. A **presheaf $\mathcal{F}$ of sets on X** consists of:

(i) A set $\mathcal{F}(U)$ for every open set $U \subset X$.
(ii) For every pair $U \subset V$ of open sets a map $\operatorname{res}_U^V \colon \mathcal{F}(V) \to \mathcal{F}(U)$ called the **restriction map** such that $\operatorname{res}_U^U = \operatorname{id}_{\mathcal{F}(U)}$ for every U and such that

$$\operatorname{res}_U^W = \operatorname{res}_U^V \circ \operatorname{res}_V^W, \qquad \text{whenever } U \subset V \subset W.$$

DEFINITION 20.28. Let $\mathcal{F}$ and $\mathcal{G}$ be two presheaves on X. A **morphism of presheaves** $\zeta \colon \mathcal{F} \to \mathcal{G}$ is a family of maps $\zeta^U : \mathcal{F}(U) \to \mathcal{G}(U)$ such that for every pair of open sets $U \subset V$ the following diagram commutes:

$$\begin{array}{ccc} \mathcal{F}(V) & \xrightarrow{\ \zeta^V\ } & \mathcal{G}(V) \\ {\scriptstyle \operatorname{res}_U^V}\big\downarrow & & \big\downarrow{\scriptstyle \operatorname{res}_U^V} \\ \mathcal{F}(U) & \xrightarrow[\ \zeta^U\ ]{} & \mathcal{G}(U) \end{array}$$

If $\zeta \colon \mathcal{F} \to \mathcal{G}$ and $\xi \colon \mathcal{G} \to \mathcal{H}$ are two morphisms of presheaves over X then their composition $\xi \circ \zeta \colon \mathcal{F} \to \mathcal{H}$ is defined as one would guess:

$$(\xi \circ \zeta)_U := \xi^U \circ \zeta^U.$$

An **isomorphism** is a presheaf morphism such that each ζ^U is a bijection.

This gives us the category $\mathsf{PSh}(X; \mathsf{Sets})$ of presheaves on X whose objects are the presheaves on X and whose morphisms are presheaf morphisms.

DEFINITION 20.29. Let C be an arbitrary category. A **presheaf $\mathcal{F}$ on X with values in C** is defined in almost the same way, only now each $\mathcal{F}(U)$ must be an object of C, each restriction map res_U^V must be a morphism in C, and morphisms between two presheaves must also be morphisms in C.

To give a concrete example, let's take $\mathsf{C} = \mathsf{Vect}$. A presheaf of vector spaces is thus an assignment of a vector space $\mathcal{F}(U)$ for every open set $U \subset X$, and the restriction maps res_U^V must be linear transformations $\mathcal{F}(V) \to \mathcal{F}(U)$. Finally if $\zeta \colon \mathcal{F} \to \mathcal{G}$ is a morphism of presheaves of vector spaces then each ζ^U must be a linear transformation $\mathcal{F}(U) \to \mathcal{G}(U)$. In particular, an isomorphism of presheaves of vector spaces requires each ζ^U to be a linear isomorphism.

Remark 20.30. Here is an alternative more categorical definition of a presheaf. Let $\mathsf{Open}(X)$ denote the category whose objects are the open sets of X and, for two open sets U, V, the morphism space $\mathrm{Hom}(U, V)$ consists of the inclusion map $U \hookrightarrow V$ if $U \subset V$ and is empty otherwise. Then a presheaf on X with values in C is simply a contravariant functor $\mathsf{Open}(X) \to \mathsf{C}$. A morphism $\zeta \colon \mathcal{F} \to \mathcal{G}$ is a *natural transformation* ζ between the two functors.

If $\mathcal{F}$ is a presheaf on X and $s \in \mathcal{F}(V)$ then for $U \subset V$ we normally abbreviate

$$s|_U := \mathrm{res}_U^V(s)$$

This fits in with the idea that we are "restricting" s to U. In fact, every single presheaf we will care about in the course will be a presheaf of *functions*, which we now define, and in this case restriction really *is* restriction.

Definition 20.31. Let X be a topological space and let S be a fixed set. A **presheaf of S-valued functions** is a presheaf with the property that $\mathcal{F}(U) \subset \mathrm{Maps}(U, S)$ for all open sets $U \subset X$, where $\mathrm{Maps}(U, S)$ denotes the set of all functions from U to S (i.e. the morphism set in category Sets).

Definition 20.32. Let $\mathcal{F}$ and $\mathcal{G}$ be two presheaves on X. We say that $\mathcal{F}$ is a **subpresheaf** of $\mathcal{G}$ if for every open set $U \subset X$, $\mathcal{F}(U) \subset \mathcal{G}(U)$, and for all $U \subset V$ open sets the restriction maps $\mathcal{F}(V) \to \mathcal{F}(U)$ are induced by the restriction maps $\mathcal{G}(V) \to \mathcal{G}(U)$.

Thus if $\mathcal{F}$ is any presheaf of S-valued functions on X then $\mathcal{F}$ is a subpresheaf of the presheaf of *all* S-valued functions on X.

Examples 20.33. Let us see some standard examples of presheaves that will be relevant to this course.

(i) Let X be a topological space and take $S = \mathbb{R}$. Let C_X denote the presheaf that assigns to an open set $U \subset X$ the set of continuous real-valued functions on X:

$$C_X(U) := C(U, \mathbb{R}) = \{f \colon U \to \mathbb{R} \text{ continuous}\}.$$

C_X is not just a presheaf of sets, but a presheaf of $\mathbb{R}$-algebras (and thus also a presheaf of rings and (infinite-dimensional) vector spaces).

(ii) We can also consider differentiable functions. Take $X = \mathbb{R}$ and let $\mathcal{F}(U) = C^\infty(U)$ denote the set of all smooth functions $U \to \mathbb{R}$. This is a subpresheaf of $C_{\mathbb{R}}$. We can think of differentiation as a morphism $D \colon \mathcal{F} \to \mathcal{F}$. This is a morphism of presheaves of vector spaces, since

$$D(af + bg) = af' + bg' = aD(f) + bD(g)$$

but it is *not* a morphism of presheaves of algebras, since

$$D(fg) = fg' + f'g \neq D(f)D(g).$$

(iii) More generally, let M be a smooth manifold. Then the assignment $U \mapsto C^\infty(U)$ is a presheaf of $\mathbb{R}$-algebras on M. As before, differentiation is a morphism of presheaves of vector spaces, but not of algebras. We normally denote this presheaf by C^∞_M.

(iv) Let $\pi\colon E \to M$ be a vector bundle. Then $U \mapsto \Gamma(U, E)$ is a presheaf of (infinite-dimensional) vector spaces on M. It is not a presheaf of algebras,[4] since in general there is no way to multiply two sections together. We usually denote this presheaf by $\mathcal{E}_E$.

(v) Let X be any topological space and let S be any set. Let $\mathcal{F}(U)$ denote the set of all *constant* functions $U \to S$. Since a constant function $f\colon U \to S$ can be identified with its image $s := f(U)$, one can simply think of $\mathcal{F}(U)$ as being equal to S itself. In this case, all restriction maps are the identity map id_S. We call this the **constant presheaf on X with values in S**.

Let us now introduce a sheaf, which is a presheaf with an additional property.

Definition 20.34. Let $\mathcal{F}$ be a presheaf on X (of sets, rings, groups, etc.). We say that $\mathcal{F}$ is a **sheaf** if the following condition is satisfied: for any open set $U \subset X$ and any open cover $\{U_a \mid a \in A\}$ of U, if we are given a collection $s_a \in \mathcal{F}(U_a)$ such that

$$s_a|_{U_a \cap U_b} = s_b|_{U_a \cap U_b}, \qquad \forall\, a, b \in A \text{ such that } U_a \cap U_b \neq \emptyset, \tag{20.6}$$

then there exists a *unique* $s \in \mathcal{F}(U)$ such that $s|_{U_a} = s_a$ for all $a \in A$.

Remark 20.35. Taking $U = \emptyset$ and choosing the covering with empty index set $A = \emptyset$ shows that if $\mathcal{F}$ is a sheaf then $\mathcal{F}(\emptyset)$ is a set consisting of one element.

A morphism $\zeta\colon \mathcal{F} \to \mathcal{G}$ of sheaves is simply a morphism of the underlying presheaves, and we denote by $\mathsf{Sh}(X; \mathsf{C})$ the category of sheaves on X with values in C.

Remark 20.36. If we start with a presheaf of functions, as in Definition 20.31, the condition (20.6) can be phrased in a slightly simpler fashion: if $\mathcal{F}$ is a presheaf of S-valued functions on X then $\mathcal{F}$ is a sheaf if and only if for any open set $U \subset X$ and any open cover $\{U_a \mid a \in A\}$ of U, if $f\colon U \to S$ is any function such that $f|_{U_a} \in \mathcal{F}(U_a)$ for each $a \in A$, then $f \in \mathcal{F}(U)$.

This reformulation makes it clear that the presheaf C_X of continuous functions on a topological space is actually a sheaf.

Proposition 20.37. *Let M be a smooth manifold. Then C^∞_M is a sheaf. More generally, if $\pi\colon E \to M$ is any vector bundle over M then $\mathcal{E}_E$ is a sheaf.*

---

[4] Unless E is an algebra bundle, cf. Definition 19.29.

Not everything is a sheaf however: the presheaf of constant functions from part (v) of Example 20.33 is not a sheaf if X contains two disjoint non-empty open subsets and S has more than one element.[5]

There is a natural way to turn a presheaf into a sheaf. This procedure is called the **sheafification** of a presheaf. The definition is rather complicated, and for our purposes unimportant (since the relevant presheaves in this course are already sheaves thanks to Proposition 20.37). Thus we will content ourselves with giving the definition only in the special case of a presheaf of functions.

PROPOSITION 20.38. *Let X be a topological space and let S be a set. Suppose $\mathcal{F}$ is a presheaf of S-valued functions on X. Let*

$$\tilde{\mathcal{F}}(U) := \Big\{ f : U \to S \mid \textit{there exists an open covering } \{U_a \mid a \in A\} \\ \textit{of } U \textit{ such that } f|_{U_a} \in \mathcal{F}(U_a) \textit{ for all } a \in A. \Big\}$$

Then $\tilde{\mathcal{F}}$ is a sheaf and the inclusion $\mathcal{F}(U) \hookrightarrow \tilde{\mathcal{F}}(U)$ induces a morphism of presheaves $\iota : \mathcal{F} \to \tilde{\mathcal{F}}$.

Proof. This is clear from the reformulation of the sheaf condition given in Remark 20.36—we simply added in all the functions that were needed in order for $\mathcal{F}$ to be a sheaf. ■

REMARK 20.39. If $\mathcal{F}$ already was a sheaf, then clearly $\mathcal{F} = \tilde{\mathcal{F}}$. Indeed, in this case all the functions we added were already in $\mathcal{F}$.

EXAMPLE 20.40. Let $\mathcal{F}$ be the presheaf of constant S-valued functions on X. As we have remarked before, this is typically not a sheaf. However it is very easy to describe the sheaf obtained from $\mathcal{F}$ via Proposition 20.38. Indeed, a little thought shows that the sheaf $\tilde{\mathcal{F}}$ is exactly the *locally constant* functions on S:

$$\tilde{\mathcal{F}}(U) = \{ f : U \to S \mid f \text{ is locally constant} \} .$$

REMARK 20.41. The sheafification can be defined via a universal property (compare Lemma 19.2): Let $\mathcal{F}$ be a presheaf on X. The sheafification $\tilde{\mathcal{F}}$ and the morphism $\iota : \mathcal{F} \to \tilde{\mathcal{F}}$ of presheaves has the property that if $\mathcal{G}$ is any sheaf on X and $\zeta : \mathcal{F} \to \mathcal{G}$ is any morphism of presheaves, then there exist a *unique* morphism of sheaves $\tilde{\zeta} : \tilde{\mathcal{F}} \to \mathcal{G}$ such that the following diagram commutes:

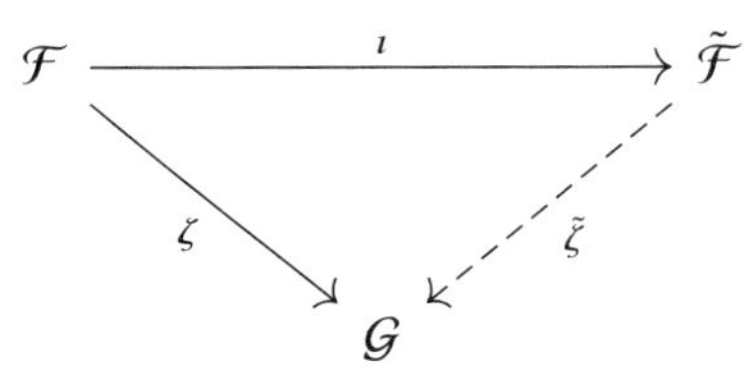

[5] Exercise: Why?

As such, via abstract nonsense, the sheafification is unique up to isomorphism.

We now move onto discussing the **stalk** of a presheaf. This generalises the notation of a germ of a function that we discussed in Chapter 2.

Definition 20.42. Let $\mathcal{F}$ be a presheaf on X, and let $x \in X$. We define the **stalk** of $\mathcal{F}$ at x to be:

$$\mathcal{F}_p := \{(U, s) \mid U \text{ is a neighbourhood of } p,\ s \in \mathcal{F}(U)\} \Big/ \sim$$

where $(U, s) \sim (V, t)$ if there exists a neighbourhood $W \subset U \cap V$ such that $s|_W \equiv t|_W$.

Thus for any neighbourhood U of p there exists a canonical map $\mathcal{F}(U) \to \mathcal{F}_p$ that sends s to the equivalence class of (U, s) in $\mathcal{F}_p$, which we denote by $\underline{s}$.

Lemma 20.43. *Let $\zeta : \mathcal{F} \to \mathcal{G}$ be a morphism of presheaves. Then for each $x \in X$ there is a well-defined map $\zeta^p : \mathcal{F}_p \to \mathcal{G}_p$ defined as follows: if $\underline{s} \in \mathcal{F}_p$ is represented by (U, s), then we declare that $(U, \zeta^U(s))$ is a representative of $\zeta^p(\underline{s})$. Thus the following diagram commutes:*

$$\begin{array}{ccc} \mathcal{F}(U) & \xrightarrow{s \mapsto \underline{s}} & \mathcal{F}_p \\ \Big\downarrow{\scriptstyle \zeta^U} & & \Big\downarrow{\scriptstyle \zeta^p} \\ \mathcal{G}(U) & \xrightarrow[t \mapsto \underline{t}]{} & \mathcal{G}_p \end{array}$$

Proof. We need only check this is well-defined. Suppose $(U, s) \sim (V, t)$. Then there exists $W \subset U \cap V$ such that $s|_W \equiv t|_W$. Since ζ is a presheaf morphism, one has that

$$\zeta^U(s)|_W = \zeta^W(s|_W) = \zeta^W(t|_W) = \zeta^V(t)|_W.$$

Thus $(U, \zeta^U(s)) \sim (V, \zeta^V(t))$. ■

Remark 20.44. A more categorical way to define stalks is the following: given $p \in X$, let $\mathsf{Open}_p(X)$ denote the full subcategory of $\mathsf{Open}(X)$ (cf. Remark 20.30) consisting of neighbourhoods of p. Then if $\mathcal{F}$ is a presheaf on X, one has

$$\mathcal{F}_p = \varinjlim \mathcal{F}(U)$$

where the filtered colimit runs over $\mathsf{Open}_p(X)$. Similarly if $\zeta : \mathcal{F} \to \mathcal{G}$ is a morphism of presheaves then

$$\zeta^p = \varinjlim \zeta^U.$$

If $\mathcal{F}$ is a presheaf of groups, or rings, or modules, etc, then the stalks also inherit that structure. We saw this in the concrete example where $\mathcal{F} = C^\infty_M$ just after Definition 2.7. As another example, suppose $\mathcal{F}$ is a sheaf of groups. Then $\mathcal{F}_p$ is also a group, where we define the group law as follows: if $\underline{s}$ is represented by (U, s) and $\underline{t}$ is represented by (V, t), then we declare $\underline{s} \cdot \underline{t}$ to be the element represented by $(U \cap V, s|_{U\cap V} \cdot t|_{U\cap V})$.

REMARK 20.45. More generally, if C is a category in which filtered colimits exist then for any $p \in X$ there is a functor $\mathsf{PSh}(X; \mathsf{C}) \to \mathsf{C}$ given by $\mathcal{F} \mapsto \mathcal{F}_p$.

Let us now look at some operations on sheaves.

DEFINITION 20.46. Let U be an open set of X. Then if $\mathcal{F}$ is any presheaf on X then we can define a presheaf $\mathcal{F}|_U$ on U by setting $\mathcal{F}|_U(V) := \mathcal{F}(V)$ for $V \subset U$ open. If $\mathcal{F}$ is a sheaf then so is $\mathcal{F}|_U$.

DEFINITION 20.47. Let $\varphi\colon X \to Y$ be a continuous map from one topological space to another. Suppose $\mathcal{F}$ is a presheaf on X. We define a presheaf $\varphi_*(\mathcal{F})$ on Y by declaring that

$$\varphi_*(\mathcal{F})(U) := \mathcal{F}(\varphi^{-1}(U)), \qquad U \subset Y \text{ open}.$$

We call $\varphi_*(\mathcal{F})$ the **direct image** of $\mathcal{F}$ under φ. If $\zeta\colon \mathcal{F} \to \mathcal{G}$ is a morphism of presheaves on X then $\varphi_*(\zeta)\colon \varphi_*(\mathcal{F}) \to \varphi_*(\mathcal{G})$ is a morphism on presheaves on Y, where

$$\varphi_*(\zeta)_U := \zeta^{\varphi^{-1}(U)}\colon \varphi_*(\mathcal{F})(U) = \mathcal{F}(\varphi^{-1}(U)) \to \mathcal{G}(\varphi^{-1}(U)) = \varphi_*(\mathcal{G})(U).$$

In this way we get a functor from presheaves on X to presheaves on Y. If $\mathcal{F}$ is a sheaf on X then it is clear that $\varphi_*(\mathcal{F})$ is a sheaf on Y.

DEFINITION 20.48. A **continuous ringed space** consists of a pair $(X, \mathcal{F})$ where X is a topological space and $\mathcal{F}$ is a subsheaf of the sheaf C_X of $\mathbb{R}$-algebras from part (i) of Example 20.33. Explicitly, this means:

- $\mathcal{F}$ is a sheaf and $\mathcal{F}(U) \subset C(U, \mathbb{R})$ for each open set $U \subset X$.
- If $f, g \in \mathcal{F}(U)$ and $a, b \in \mathbb{R}$ then $af + bg$ and fg both belong to $\mathcal{F}(U)$.

REMARK 20.49. The name "continuous ringed space" is not quite standard. In algebraic geometry, given a commutative ring R, one studies the more general notion of a **ringed space**, which is defined to be a pair $(X, \mathcal{F})$, where X is a topological space and $\mathcal{F}$ is a sheaf of commutative, associative and unital R-algebras on X. Thus what I call a "continuous ringed space" is the special case where $R = \mathbb{R}$ and $\mathcal{F}$ is a subalgebra of the sheaf of continuous functions on X.

Algebraic geometers often restrict to a special class of ringed spaces, called **locally ringed spaces**, which are ringed spaces $(X, \mathcal{F})$ with the additional property that the stalk $\mathcal{F}_p$ is a *local ring* for every point $p \in X$ (i.e. it has a unique maximal ideal). All continuous ringed spaces in the sense of Definition 20.48 are locally ringed spaces; see Lemma 2.15.

DEFINITION 20.50. Let $(X, \mathcal{F})$ and $(Y, \mathcal{G})$ be two continuous ringed spaces. A **morphism of continuous ringed spaces** is a continuous map $\varphi\colon X \to Y$ with the following property:

$$f \in \mathcal{G}(U) \quad \Rightarrow \quad f \circ \varphi \in \mathcal{F}(\varphi^{-1}(U)), \qquad \text{for all open } U \subset Y. \tag{20.7}$$

Property (20.7) implies there is a well-defined sheaf morphism $\mathcal{G} \to \varphi_*(\mathcal{F})$ given by

$$f \in \mathcal{G}(U) \mapsto f \circ \varphi \in \varphi_*(\mathcal{F})(U).$$

An **isomorphism of continuous ringed spaces** is a homeomorphism φ such that both φ and φ^{-1} are morphisms of continuous ringed spaces.

We will now use the notion of a continuous ringed space to give an equivalent definition of a manifold. This definition is more in the spirit of algebraic geometry, and it has several advantages over the standard one, as we will shortly explain.

Definition 20.51. Let $(M, \mathcal{F})$ be a continuous ringed space. We say $(M, \mathcal{F})$ is a **smooth ringed space of dimension** n if for every point $p \in M$ there exists a neighbourhood U of p and a homeomorphism $x \colon U \to O$, where O is some open subset of $\mathbb{R}^n$, such that x defines an isomorphism of continuous ringed spaces

$$(U, \mathcal{F}|_U) \cong (O, C^\infty_O).$$

The next theorem tells us that this really is an alternative way to define a manifold.

Theorem 20.52 *Let M be a smooth manifold of dimension n. Then (M, C^∞_M) is a smooth ringed space of dimension n. Conversely, assume that $(M, \mathcal{F})$ is a smooth ringed space, and assume in addition that M is Hausdorff and second countable. Then there exists a unique smooth structure on M such that $\mathcal{F}$ becomes the sheaf C^∞_M.*

The proof is easy: one direction is clear from the definition of a smooth function on a manifold (Definition 2.1), and for the other direction we (work a bit and then) apply Proposition 1.17.

Remark 20.53. In many ways, starting Chapter 1 by defining a manifold via Definition 20.51 would have been more efficient. Here are some reasons why:

(i) There is no need to worry about equivalence classes of smooth atlases (cf. Remark 1.12).
(ii) The definition of what it means for a continuous map $\varphi \colon (M, C^\infty_M) \to (N, C^\infty_N)$ between two smooth manifolds to be smooth is much cleaner: it simply has to be a morphism of continuous ringed spaces.
(iii) The definition of a tangent vector as a derivation on the space of germs (i.e. the stalks of the sheaf C^∞_M) is far more natural.
(iv) This algebraic approach dramatically reduces the need to use local coordinates, which are messy and irritating.

Nevertheless, the best part of differential geometry is the "geometry", and this algebraic approach deletes most of said geometry. So we will not pursue it beyond this chapter.

We conclude this chapter by giving a sheaf-theoretic definition of a vector bundle. This will also allow us to reinterpret the Hom-Gamma Theorem 20.25 in a more algebraic way. First, some preliminary definitions.

Definition 20.54. Let $(X, \mathcal{F})$ be a continuous ringed space. Let $\mathcal{M}$ be a sheaf of abelian groups on X, and assume in addition that for every open set $U \subset X$, the abelian group $\mathcal{M}(U)$ has the structure of an $\mathcal{F}(U)$-module, and moreover the restriction morphisms respect this structure, i.e.

$$\operatorname{res}^V_U(fs) = \operatorname{res}^V_U(f)\operatorname{res}^V_U(s), \qquad \forall\, f \in \mathcal{F}(V),\ s \in \mathcal{M}(V).$$

Then we say that $\mathcal{M}$ is a **sheaf of $\mathcal{F}$-modules**. A morphism ζ from one sheaf $\mathcal{M}$ of $\mathcal{F}$-modules to another sheaf $\mathcal{N}$ of $\mathcal{F}$-modules is one such that each map $\zeta^U : \mathcal{M}(U) \to \mathcal{N}(U)$ is $\mathcal{F}(U)$-linear. We call such a ζ an **$\mathcal{F}$-morphism** of sheaves.

Here is an example.

Example 20.55. Let $\pi : E \to M$ be a vector bundle. Then the sheaf $\mathcal{E}_E$ of sections of E is a sheaf of C^∞_M-modules. Indeed, this is just a fancy way of rephrasing Lemma 20.9.

We can also rephrase some of the results from the previous chapter.

Corollary 20.56. *Let $\pi_1 : E_1 \to M$ and $\pi_2 : E_2 \to M$, and suppose $\zeta : \Gamma(E_1) \to \Gamma(E_2)$ is an $\mathbb{R}$-linear operator. Then ζ is a local operator in the sense of Definition* 20.16 *if and only if $\zeta = \zeta^M$ for a morphism of sheaves $\zeta : \mathcal{E}_{E_1} \to \mathcal{E}_{E_2}$.*

Proof. This is Proposition 20.19. ∎

Corollary 20.57. *Let $\pi_1 : E_1 \to M$ and $\pi_2 : E_2 \to M$ be two vector bundles over M. Suppose $\zeta : \mathcal{E}_{E_1} \to \mathcal{E}_{E_2}$ is a C^∞_M-morphism of sheaves. Then $\zeta^M : \Gamma(E_1) \to \Gamma(E_2)$ is a point operator in the sense of Definition* 20.16.

Proof. This is Theorem 20.20. ∎

Here is another more abstract example of an $\mathcal{F}$-module.

Example 20.58. Let $(X, \mathcal{F})$ be a continuous ringed space. Let $n \in \mathbb{N}$. Then the sum

$$\mathcal{F}^k(U) := \underbrace{\mathcal{F}(U) \oplus \cdots \oplus \mathcal{F}(U)}_{n \text{ copies}}$$

is a *free* $\mathcal{F}$-module of rank k.

More generally, if $\mathcal{M}$ is any $\mathcal{F}$-module over X then we say that $\mathcal{M}$ is **locally free of rank** n if for any $p \in X$ there exists a neighbourhood U of p and an $\mathcal{F}|_U$-isomorphism of sheaves $\mathcal{M}|_U \cong \mathcal{F}^k$. If $\mathcal{M}$ is locally free of rank n then with a little work one can show that the stalk $\mathcal{M}_p$ is a free $\mathcal{F}_p$-module of rank n.

Example 20.59. Let $\pi : E \to M$ be a vector bundle of rank n. Then the sheaf $\mathcal{E}_E$ is locally free of rank n. Indeed, this follows from the fact that for any $p \in M$, there exists a neighbourhood U of p such that E admits a local frame (e_i). Then any $s \in \Gamma(U, E)$ can be written as a

$$s = f^i e_i, \qquad a^i \in C^\infty(U),$$

and the correspondence $s \mapsto (f^1, \ldots, f^n)$ sets up an isomorphism $\mathcal{E}_E|_U$ with $(C^\infty_M|_U)^n$.

Just as in Theorem 20.52, it is actually possible to work backwards and *define* a vector bundle this way.

Theorem 20.60 *Let M be a smooth manifold and let $\mathcal{M}$ be a sheaf of locally free C^∞_M-modules of rank n. Then there exists a vector bundle $\pi : E \to M$ and a C^∞_M-isomorphism of sheaves from $\mathcal{M}$ to $\mathcal{E}_E$. Moreover E is unique up to vector-bundle isomorphism.*

Proof (Sketch). The stalk $\mathcal{F}_p M$ of C^∞_M is a local ring with maximal ideal $\mathfrak{m}_p$ equal to the kernel of the evaluation map.[6] The stalk $\mathcal{M}_p$ is a free $\mathcal{F}_p$-module of rank n. Thus if we set

$$E_p := \mathcal{M}_p / \mathfrak{m}_p \mathcal{M}_p$$

then E_p is a vector space of dimension n. Now set $E = \bigsqcup_{p \in M} E_p$. If $p \in M$ and $U \subset M$ is a neighbourhood such that $\mathcal{M}|_U \cong (C^\infty_M|_U)^n$ then this gives us a basis $\{e_1(x), \ldots, e_k(x)\}$ of E_p for every $p \in U$, and thus a local frame for E. This gives us a bundle chart via (20.2). We use this to define a fibre bundle structure on E via Remark 16.5. The transition functions arising from a different choice of local frame near p are linear by assumption, and thus we have built a vector bundle. ∎

Remark 20.61. Theorem 20.60 tells us that there is a one-to-one correspondence (up to isomorphism) between vector bundles and locally free sheaves. From the point of view of categories, this gives us a way to go from an object of the category of vector bundles to an object of the category of finite rank locally free sheaves. A souped-up version of the Hom-Gamma Theorem allows us to extend this to morphisms too: i.e. a vector bundle homomorphism $E \to F$ is equivalent to an C^∞_M-morphism of sheaves. This allows us to conclude the following result: *there is an equivalence of categories between the category of vector bundles over M and the category of finite rank locally free C^∞_M-modules.*

[6] See Lemma 2.15.

Chapter 21
Tensor Fields

In this chapter we investigate sections of the tensor bundles $T^{h,k}(TM)$. Such sections play a special role in differential geometry, and they are classically referred to as *tensor fields*. Tensor fields can be thought of as a generalisation of vector fields—indeed, a tensor field of type $(h, k) = (1, 0)$ is exactly a vector field. We already briefly met tensor fields in Remark 8.17—one of the goals of this chapter is to fill in the details from this remark.

DEFINITION 21.1. A **tensor field** of type (h, k) on M is a section of $T^{h,k}(TM)$. We normally use the special notation $\mathcal{T}^{h,k}(M)$ for tensor fields. The space of sections $\mathcal{T}^{h,k}(U) := \Gamma(U, T^{h,k}(TM))$ is defined similarly; these are the tensor fields of type (h, k) over U. Let us unpack this a bit. The bundle $T^{h,k}(TM)$ is the bundle whose fibre over $p \in M$ is

$$T^{h,k}(T_pM) := \underbrace{T_pM \otimes \cdots \otimes T_pM}_{h \text{ copies}} \otimes \overbrace{T_p^*M \otimes \cdots \otimes T_p^*M}^{k \text{ copies}}.$$

If $A \in \mathcal{T}^{h,k}(M)$ then we can think of the value of A at p, which we write either as $A(p)$ or A_p (the latter is preferred if there are many variables) as a multilinear map

$$A_p : \underbrace{T_p^*M \times \cdots \times T_p^*M}_{h \text{ copies}} \times \overbrace{T_pM \times \cdots \times T_pM}^{k \text{ copies}} \to \mathbb{R}$$

thanks to Proposition 19.8.

A tensor field of type $(1, 0)$ is just a vector field: in this case we think of $X(p) : T_p^*M \to \mathbb{R}$ as the linear map given by $X(p)(\lambda) := \lambda(X(p))$. If $A \in \mathcal{T}^{h,k}(M)$ and (U, x) is a chart on M then locally we can write

W. J. Merry, *Lectures on Differential Geometry I*,
https://doi.org/10.1007/978-3-032-03733-6_21

$$A = A^{i_1 \cdots i_h}_{j_1 \cdots j_k} \frac{\partial}{\partial x^{i_1}} \otimes \cdots \otimes \frac{\partial}{\partial x^{i_h}} \otimes dx^{j_1} \otimes \cdots \otimes dx^{j_k}$$

where the function $A^{i_1 \cdots i_h}_{j_1 \cdots j_k} \in C^\infty(U)$ is defined by

$$A^{i_1 \dots i_h}_{j_1 \dots j_k}(p) = A_p \left(dx^{i_1}_p, \dots, dx^{i_h}_p, \frac{\partial}{\partial x^{j_1}}\bigg|_p, \dots, \frac{\partial}{\partial x^{j_k}}\bigg|_p \right).$$

Expressions of this form are the main reason we introduced the Einstein Summation Convention!

If $A \in \mathcal{T}^{h,k}(U)$ and $B \in \mathcal{T}^{h_1,k_1}(U)$ then we can form their tensor product

$$A \otimes B \in \mathcal{T}^{h+h_1,k+k_1}(U)$$

which is defined pointwise via Definition 19.19, see (19.4) and (19.5). This means that if we set

$$\mathcal{T}(U) := \bigoplus_{h,k \geq 0} \mathcal{T}^{h,k}(U),$$

then $\mathcal{T}(U)$ is not just a module over the ring $C^\infty(U)$ but actually an algebra. By a slight abuse of language, we call $\mathcal{T}(M)$ the **tensor algebra** of M.

Warning: $\mathcal{T}(M)$ is not the space of sections of a vector bundle, as the tensor algebra $\widetilde{T}E$ of a vector bundle is not a vector bundle, cf. Remark 19.20.

Definition 21.2. Let $\zeta \colon \mathcal{T}(M) \to \mathcal{T}(N)$ be an $\mathbb{R}$-linear map. We say that ζ has **degree** $(i, j) \in \mathbb{Z}^2$ if

$$\zeta\big(\mathcal{T}^{h,k}(M)\big) \subset \mathcal{T}^{h+i,k+j}(N)$$

Here i and j can be negative; we use the convention that $T^{h,k}(TM) = \{0\}$ if either $h < 0$ or $k < 0$.

In the case $M = N$, we say that such an ζ is a **local operator**, respectively a **point operator**, if the restriction of ζ to each $\mathcal{T}^{h,k}(M)$ is a local, respectively a point, operator.

Remark 21.3. As in (19.5), strictly speaking the tensor $A \otimes B$ defined above needs its factors rearranging. If for instance $A = X_1 \otimes \omega_1$ and $B = X_2 \otimes \omega_2$ for vector fields X_i and 1-forms ω_i (i.e. ω_i is a tensor of type $(0, 1)$), then $A \otimes B$ should really be written as $X_1 \otimes X_2 \otimes \omega_1 \otimes \omega_2$ (so that the vector field factors come first). In practice, this is inconvenient, and so we will often not bother and just keep the factors unchanged, thus writing $A \otimes B = X_1 \otimes \omega_1 \otimes X_2 \otimes \omega_2$. This is harmless, since it was merely a convention to put the vector fields first (cf. Corollary 19.16).

A differential form is a section of the exterior algebra bundle of the cotangent bundle.

Definition 21.4. A **differential k-form** (often simply called "a k-form") on M is a section of $\bigwedge^k T^*M$. We use the special notation $\Omega^k(M)$ for the space of differential k-forms. If $\omega \in \Omega^k(M)$ and $p \in M$ then we can think of ω_p as an alternating map

$$\omega_p \colon \underbrace{T_pM \times \cdots \times T_pM}_{k \text{ copies}} \to \mathbb{R},$$

thanks to Proposition 19.24.

If $\omega \in \Omega^k(M)$ and (U, x) is a chart on M then locally we can write

$$\omega = \omega_{i_1 \cdots i_k} dx^{i_1} \wedge \cdots \wedge dx^{i_k} \tag{21.1}$$

where $\omega_{i_1 \dots i_k} \in C^\infty(U)$. We define

$$\Omega(M) = \bigoplus_{0 \le k \le m} \Omega^k(M),$$

with $\Omega(U)$ defined similarly. Thus an element of $\Omega(M)$ is a sum $\sum_{i=0}^m \omega_i$ where $\omega_i \in \Omega^i(M)$. Since an alternating multilinear map is (in particular) a multilinear map, we see that any differential k-form may be regarded as a tensor of type $(0, k)$. For $k = 1$ this is an equality (since in dimension one every linear map is alternating):

$$\Omega^1(M) = \mathcal{T}^{0,1}(M)$$

But for $k \geq 2$, there are (many) multilinear maps that are not alternating, and thus not every tensor of type $(0, k)$ is a differential form.

$$\Omega^k(M) \subsetneq \mathcal{T}^{0,k}(M).$$

As above, an $\mathbb{R}$-linear map $\zeta \colon \Omega(M) \to \Omega(N)$ is said to have **degree** h if $\zeta(\Omega^k(M)) \subset \Omega^{h+k}(M)$.

Identifying tensors: For low values of (h, k) we have various different notations for $\mathcal{T}^{h,k}(M)$.

- $\mathcal{T}^{0,0}(M) = C^\infty(M)$.
- $\mathcal{T}^{1,0}(M) = \mathfrak{X}(M)$.
- $\mathcal{T}^{0,1}(M) = \Omega^1(M)$.
- $\mathcal{T}^{1,1}(M) = \Gamma(\mathrm{End}(TM))$.

The following theorem tells us how to recognise tensor fields and differential forms "in the wild".

Theorem 21.5 (The Tensor and Differential Form Criterion) *Let M be a smooth manifold and let $U \subset M$ be a non-empty open set.*

(i) *There is a canonical identification between $\mathcal{T}^{h,k}(U)$ and $C^\infty(U)$-multilinear functions*

$$\underbrace{\Omega^1(U) \times \cdots \times \Omega^1(U)}_{h \text{ copies}} \times \overbrace{\mathfrak{X}(U) \times \cdots \times \mathfrak{X}(U)}^{k \text{ copies}} \to C^\infty(U).$$

(ii) *There is a canonical identification between $\Omega^k(U)$ and alternating $C^\infty(U)$-multilinear functions*

$$\underbrace{\mathfrak{X}(U) \times \cdots \times \mathfrak{X}(U)}_{k \text{ copies}} \to C^\infty(U).$$

Proof. The case $(h,k) = (1,0)$ or $(h,k) = (0,1)$ of part (i) follows from the Hom-Gamma Theorem 21.5. Indeed, suppose for definiteness $(h,k) = (0,1)$. We apply the Hom-Gamma Theorem with $E = TM$ and $F = M \times \mathbb{R}$ the trivial bundle. Then in the notation of (20.5),

$$\operatorname{Hom}\big(\Gamma(E), \Gamma(F)\big) = \big\{C^\infty(M)\text{-linear maps } \mathfrak{X}(M) \to C^\infty(M)\big\}.$$

and

$$\begin{aligned} \Gamma\big(\operatorname{Hom}(E,F)\big) &= \Gamma(E^* \otimes \mathbb{R}) \\ &= \Gamma(E^*) \\ &= \mathcal{T}^{0,1}(M), \end{aligned}$$

where the first equality used Corollary 19.14 and the second equality uses the fact that for any real vector space V there is a canonical isomorphism $V \cong V \otimes \mathbb{R}$.

The general case for part (i) can be proved by induction on $h + k$; the details of this argument are left to you on Problem Sheet I. The proof of part (ii) proceeds along similar lines, and is also omitted. ■

We will typically suppress this isomorphism from our notation, and thus interchangeably regard a tensor field A over U either as an element of $\mathcal{T}^{h,k}(U)$, or as an appropriate multilinear map, and similarly with differential forms. In the bonus section below we explain why Theorem 21.5 could also be called a "Tensor-Gamma Theorem".

Remark 21.6. Every differential geometer should at one point in their life compute tensor fields in local coordinates.[1] This is left as a wholesome exercise for you to enjoy on Problem Sheet I.

We will return to differential forms in the next chapter. For now we concentrate on tensors. Let us now think all the way back to Chapter 8. If $\varphi\colon M \to N$ is a diffeomorphism, we defined two isomorphisms (Definition 8.15)

$$\varphi_*\colon C^\infty(M) \to C^\infty(N), \qquad \varphi_*\colon \mathfrak{X}(M) \to \mathfrak{X}(N).$$

The vector field $\varphi_* X$ is sometimes called the **push forward** of X by φ, and the map φ_* is called a **push forward** map.

Since $\mathcal{T}^{0,0}(M) = C^\infty(M)$ and $\mathcal{T}^{1,0}(M) = \mathfrak{X}(M)$, we can think of both of these maps φ_* as being defined on tensors of type $(0,0)$ or $(1,0)$. As hinted at in Remark 8.17, these two maps φ_* are both special cases of an $\mathbb{R}$-linear map between the tensor algebras:

$$\varphi_*\colon \mathcal{T}(M) \to \mathcal{T}(N)$$

[1] Once is enough.

which **preserves type**.[2]

REMARK 21.7. The same is true of the Lie derivative. We will come back to this next chapter.

We shall take a slightly circuitous route to defining the map φ_* and first define a type-preserving map going the other way

$$\varphi^*\colon \mathcal{T}(N) \to \mathcal{T}(M).$$

The upper star indicates that $\varphi \mapsto \varphi^*$ is contravariant, i.e. it reverses the direction of the arrows.

In fact, some of this construction works for arbitrary smooth maps (not necessarily diffeomorphisms), and we cover this first as a special case. This will be useful next chapter when we talk about pullbacks of differential forms.

DEFINITION 21.8. Let $\varphi\colon M \to N$ be smooth (not necessarily a diffeomorphism). Let $k \geq 1$ and suppose $A \in \mathcal{T}^{0,k}(N)$ is a $(0, k)$ tensor field on N. We define the **pullback** of A by φ, written $\varphi^* A$, to be the tensor field on M defined pointwise as follows: for $p \in M$ and $\xi_1, \dots \xi_k \in T_p M$, set

$$(\varphi^* A)_p(\xi_1, \dots, \xi_k) := A_{\varphi(p)}\big(D\varphi(p)\xi_1, \dots, D\varphi(p)\xi_k\big),$$

We extend this to cover the case $k = 0$ by setting

$$\varphi^* f := f \circ \varphi, \qquad f \in C^\infty(N).$$

It is immediate that φ^* is $\mathbb{R}$-linear. Moreover if $f \in C^\infty(N)$ then

$$\varphi^*(fA) = (\varphi^* f)(\varphi^* A),$$

Note also that if $\psi\colon L \to M$ then $(\varphi \circ \psi)^* = \psi^* \circ \varphi^*$. If we consider all these maps φ_* at once we get a type-preserving operator

$$\varphi^*\colon \bigoplus_{k \geq 0} \mathcal{T}^{0,k}(N) \to \bigoplus_{k \geq 0} \mathcal{T}^{0,k}(M)$$

which moreover is an algebra morphism, i.e.

$$\varphi^*(A \otimes B) = \varphi^* A \otimes \varphi^* B.$$

To extend this definition to tensors of arbitrary type, we need to assume that φ is a diffeomorphism. First let us introduce the notion of a **cotangent lift**.

DEFINITION 21.9. Let $\varphi\colon M \to N$ be a diffeomorphism. Then $D\varphi(p)\colon T_p M \to T_{\varphi(p)} N$ is a linear isomorphism for each p, and thus we can speak of its inverse $D\varphi(p)^{-1}\colon T_{\varphi(p)} N \to T_p M$. This coincides with the differential of φ^{-1} at the point $\varphi(p)$.

[2] i.e. is of type $(0, 0)$.

Thus there is a well-defined map

$$D\varphi^{\dagger} : T^*M \to T^*N,$$

called the **cotangent lift** defined for $p \in M$, $\lambda \in T_p^*M$ and $\xi \in T_{\varphi(p)}N$ by

$$\big(D\varphi^{\dagger}(p)\lambda\big)(\xi) := \lambda\big(D\varphi(p)^{-1}\xi\big)$$

The cotangent lift makes the following diagram commute:

$$\begin{array}{ccc} T^*M & \xrightarrow{\;D\varphi^{\dagger}\;} & T^*N \\ \Big\downarrow{\scriptstyle \pi} & & \Big\downarrow{\scriptstyle \pi} \\ M & \xrightarrow[\;\varphi\;]{} & N \end{array}$$

Thus $D\varphi^{\dagger}$ is a vector bundle morphism along φ.

We emphasise $D\varphi^{\dagger}$ is only defined for diffeomorphisms.

Definition 21.10. Now suppose $\varphi\colon M \to N$ is a diffeomorphism. Then we can define the pullback tensor $\varphi^* A \in \mathcal{T}^{h,k}(M)$ for a tensor $A \in \mathcal{T}^{h,k}(N)$ of arbitrary type (h, k) by setting

$$\begin{aligned} &\varphi^* A_p(\lambda_1, \dots, \lambda_h, \xi_1, \dots, \xi_k) \\ &\quad := A_{\varphi(p)}\Big(D\varphi^{\dagger}(p)\lambda_1, \dots, D\varphi^{\dagger}(p)\lambda_h, D\varphi(p)\xi_1, \dots, D\varphi(p)\xi_k\Big) \end{aligned}$$

for $p \in M$, $\lambda_1, \dots, \lambda_h \in T_p^*M$ and $\xi_1, \dots, \xi_k \in T_pM$.

Remark 21.11. More generally, if $\varphi\colon U \subset M \to N$ is a locally defined map which is a diffeomorphism onto its image $V = \varphi(U)$ then we can still use φ to pull back tensor fields of arbitrary type from V to U.

Thus when φ is a diffeomorphism there is a well-defined operator $\varphi^* : \mathcal{T}(N) \to \mathcal{T}(M)$ which is an algebra morphism, that is,

$$\varphi^*(A \otimes B) = \varphi^* A \otimes \varphi^* B. \tag{21.2}$$

Summary:

- A tensor of type $(0, k)$ can be pulled back by an arbitrary smooth map.
- A tensor of type (h, k) for $h > 0$ can only be pulled back by a diffeomorphism.

So much for pullbacks. What about push forwards? The answer is simple:

Definition 21.12. Let $\varphi\colon M \to N$ be a diffeomorphism. We define the **push forward** map

$$\varphi_*\colon \mathcal{T}(M) \to \mathcal{T}(N)$$

by

$$\varphi_* := (\varphi^{-1})^*.$$

Thus the push forward is the pullback of the inverse.

In the special case $h = k = 0$, the map φ_* sends a function f to $f \circ \varphi^{-1}$. In the special case $h = 1$ and $k = 0$, one has

$$(\varphi_* X)(p) = D\varphi(\varphi^{-1}(p))X(\varphi^{-1}(p)),$$

and thus in both cases these extend the definitions from Chapter 8.

Bonus Material for Chapter 21

In Chapter 19 we defined tensor products for finite-dimensional real vector spaces. However everything would have worked (without any changes at all) if we worked with finite rank modules over a fixed commutative ring R. A more interesting question is to what extent the finite rank hypothesis was needed. Indeed, suppose V is a module over a commutative ring R. The dual module is defined $V^* = \mathrm{Hom}_R(V, R)$, and the space $\mathrm{Mult}_{h,k}(V)$ is then defined to be the set of multilinear maps

$$\underbrace{V \times \cdots \times V}_{h \text{ copies}} \times \overbrace{V^* \times \cdots \times V^*}^{k \text{ copies}} \to R.$$

One can then ask the question: *is it true that* $T^{h,k}V$ *and* $\mathrm{Mult}_{s,r}(V)$ *are isomorphic modules*?

$$T^{h,k}V \overset{?}{\cong} \mathrm{Mult}_{k,h}(V). \tag{21.3}$$

The answer in general is **no**. Nevertheless (21.3) it is true for *some* infinite-rank modules.

Definition 21.13. Let R be a commutative ring[3] and V an R-module. We say that V is **projective** if for any R-module Z and any module homomorphism $f\colon V \to Z$, then for any surjective module homomorphism $g\colon W \to Z$ from some other R-module W, there exists a module homomorphism $h\colon V \to W$ such that the following commutes:

[3] All rings are assumed to be unital.

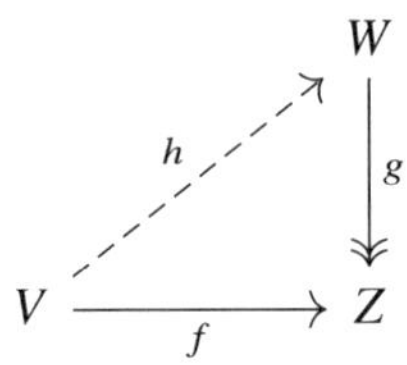

Warning: We do **not** require h to be unique. This is **not** a universal property!

On Problem Sheet I you will prove the following generalisation of Proposition 19.8.

PROPOSITION 21.14. *Let R be a commutative ring and let V be a finitely generated projective R-module. Then for all $h, k \geq 0$,*

$$T^{h,k}V \cong \mathrm{Mult}_{k,h}(V).$$

Moreover, as you will also prove on Problem Sheet I, the space of sections of a vector bundle satisfy the hypotheses of Proposition 21.14:

PROPOSITION 21.15. *Let $\pi \colon E \to M$ be a vector bundle. For any open set $U \subset M$, the space $\Gamma(U, E)$ is a finitely generated projective $C^\infty(U)$-module.*

This gives us the following cute corollary.

COROLLARY 21.16. *The operators Γ and $T^{h,k}$ also "commute". That is, there is an isomorphism of $C^\infty(M)$-modules*

$$\Gamma\big(T^{h,k}(TM)\big) \cong T^{h,k}\big(\Gamma(TM)\big),$$

Proof. The Tensor Criterion Theorem 21.5 tells us that

$$\mathcal{T}^{h,k}\big(\mathfrak{X}(U)\big) \cong \mathrm{Mult}_{k,h}\big(\mathfrak{X}(U)\big).$$

Combining this with Propositions 21.14 and 21.15 we obtain

$$\mathcal{T}^{h,k}(U) \cong T^{h,k}\big(\mathfrak{X}(U)\big),$$

which is just compact notation for

$$\Gamma\big(U, T^{h,k}(TM)\big) \cong T^{h,k}\big(\Gamma(U, TM)\big).$$

■

Chapter 22
The Lie Derivative Revisited

In this chapter we introduce the notion of a **tensor derivation** on a manifold. We show that the Lie derivative is a tensor derivation, and thus unify the two definitions of the Lie derivative from Definition 10.1.

DEFINITION 22.1. Let V be a vector space and fix $h, k \geq 0$. Choose $i \leq h$ and $j \leq k$. The (i, j)**th contraction**, written $C^{i,j}$ is the linear operator

$$C^{i,j} : T^{h,k}V \to T^{h-1,k-1}V$$

defined on decomposable elements by feeding the ith V-factor to the jth V^* factor:

$$\begin{aligned} &C^{i,j}(v_1 \otimes \cdots \otimes v_i \otimes \cdots \otimes v_h \otimes \lambda^1 \otimes \cdots \otimes \lambda^j \otimes \cdots \otimes \lambda^k) := \\ &\lambda^j(v_i) \cdot v_1 \otimes \cdots \otimes v_{i-1} \otimes v_{i+1} \otimes \cdots \otimes \lambda^{j-1} \otimes \lambda^{j+1} \otimes \cdots \otimes \lambda^k \end{aligned}$$

and then extending by linearity.

LEMMA 22.2. *Assume* $\dim V = n$. *Let* $A \in T^{h,k}V$*, and regard* A *as defining an element of* $\mathrm{Mult}_{k,h}(V)$ *as in Proposition* 19.8. *Let* $\{e_1, \ldots, e_n\}$ *be a basis of* V *with dual basis* $\{e^1, \ldots, e^n\}$ *of* V^*. *Then if we regard* $C^{i,j}A$ *also as an element of* $\mathrm{Mult}_{k-1,h-1}(V)$*, one has*

$$\begin{aligned} &(C^{i,j}A)\big(w_1, \ldots, w_{k-1}, \eta^1, \ldots, \eta^{h-1}\big) \\ &\quad = \sum_{l=1}^{n} A\big(w_1, \ldots, \underbrace{e_l}_{j\text{th position}}, \ldots, w_{k-1}, \eta^1, \ldots, \underbrace{e^l}_{i\text{th position}}, \ldots \eta^{h-1}\big). \end{aligned}$$

W. J. Merry, *Lectures on Differential Geometry I*,
https://doi.org/10.1007/978-3-032-03733-6_22

Proof. It suffices to prove the equality for decomposable elements. Let us temporarily add a tilde to denote the multilinear map corresponding to a given tensor. For simplicity[1] assume $(h, k) = (2, 3)$, and assume $A = v_1 \otimes v_2 \otimes \lambda^1 \otimes \lambda^2 \otimes \lambda^3$. Then as in Remark 19.12, the corresponding map $\widetilde{A} \in \mathrm{Mult}_{3,2}(V)$ is given by

$$\widetilde{A}(w_1, w_2, w_3, \eta^1, \eta^2) = \eta^1(v_1)\,\eta^2(v_2)\,\lambda^1(w_1)\,\lambda^2(w_2)\,\lambda^3(w_3).$$

Take $(i, j) = (1, 2)$. Then $C^{1,2}A = \lambda^2(v_1) \cdot v_2 \otimes \lambda^1 \otimes \lambda^3$. We compute[2]

$$\begin{aligned}(C^{1,2}\widetilde{A})(w_1, w_2, \eta^1) &\overset{\text{def}}{=} \sum_{l=1}^{n} \widetilde{A}(w_1, e_l, w_2, e^l, \eta^1)\\ &= \sum_{l=1}^{n} e^l(v_1)\,\eta^1(v_2)\,\lambda^1(w_1)\,\lambda^2(e_l)\,\lambda^3(w_2)\\ &= \left(\sum_{l=1}^{n} e^l(v_1)\,\lambda^2(e_i)\right)\eta^1(v_2)\,\lambda^1(w_1)\,\lambda^3(w_2).\end{aligned}$$

But

$$\sum_{l=1}^{n} e^l(v_1)\,\lambda^2(e_l) = \lambda^2(v_1),$$

and thus

$$C^{1,2}\widetilde{A} = \widetilde{C^{1,2}A}$$

which is what we wanted to prove. ∎

A contraction $C^{i,j}$ extends to define an operator on tensor fields in an obvious fashion. For instance, if $A \in \mathcal{T}^{2,1}(M)$ is the decomposable tensor $X \otimes Y \otimes \omega$ then $C^{1,1}A = \omega(X)Y$.

LEMMA 22.3. *For any $1 \le i \le h$ and $1 \le j \le k$, the contraction $C^{i,j}$ defines a point operator $\mathcal{T}^{h,k}(M) \to \mathcal{T}^{h-1,k-1}(M)$.*

Proof. We need only check the point property, but this is immediate from the definition: if $A_p = 0$ then $(C^{i,j}A)_p = 0$. ∎

DEFINITION 22.4. Let $\zeta\colon \mathcal{T}(M) \to \mathcal{T}(N)$ be an operator which preserves type. We say that ζ **commutes with all contractions** if for any any $1 \le i \le h$ and $1 \le j \le k$ one has

$$\zeta \circ C^{i,j} = C^{i,j} \circ \zeta$$

as maps $\mathcal{T}^{h,k}(M) \to \mathcal{T}^{h-1,k-1}(N)$.

EXAMPLE 22.5. Let $\varphi\colon M \to N$ be a diffeomorphism. Then φ^* and φ_* both commute with all contractions.

Here is the main definition for this chapter.

[1] The general case is only notationally more complicated.

[2] We include the summation signs to minimise the risk of confusion.

DEFINITION 22.6. A **tensor derivation** is an $\mathbb{R}$-linear operator $\zeta\colon \mathcal{T}(M) \to \mathcal{T}(M)$ that preserves type, commutes with all contractions, and satisfies

$$\zeta(A \otimes B) = \zeta A \otimes B + A \otimes \zeta B \tag{22.1}$$

for all $A, B \in \mathcal{T}(M)$.

We refer to (22.1) as the **derivation property**. Operators that satisfy this property are automatically local operators.

LEMMA 22.7. *Suppose $\zeta\colon \mathcal{T}(M) \to \mathcal{T}(M)$ is an $\mathbb{R}$-linear map that satisfies the derivation property. Then ζ is a local operator. Thus in particular tensor derivations are local operators.*

Proof. This is a standard argument which we have seen many times by now. Suppose $A \in \mathcal{T}^{h,k}(M)$ vanishes on $U \subset M$. Let $p \in U$ and let $\chi\colon M \to \mathbb{R}$ denote a bump function with $\chi(p) = 1$ and $\operatorname{supp}(\chi) \subset U$. Then $\eta A = \chi \otimes A$ vanishes identically on M, and hence by $\mathbb{R}$-linearity $\zeta(\chi \otimes A) = 0$. Then by the derivation property

$$\begin{aligned} 0 &= \zeta(\chi \otimes A)(p) \\ &= \chi(p)(\zeta A)(p) + (\zeta\chi)(p)A(p) \\ &= (\zeta A)(p). \end{aligned}$$

As p was an arbitrary point of U, it follows that ζA vanishes on U. ∎

REMARK 22.8. Combining Lemma 22.7 and Proposition 20.19 shows that tensor derivations induce operators $\zeta^U : \mathcal{T}(U) \to \mathcal{T}(U)$ for each open set $U \subset M$. These operators are actually tensor derivations in their own right: the fact that ζ^U commutes with contractions follows from the fact that contractions are local operators, and the fact that ζ^U satisfies the derivation property is immediate.

LEMMA 22.9. *Suppose ζ is a tensor derivation on M, and let $U \subset M$ be open. Fix $A \in \mathcal{T}^{h,k}(U)$, and suppose $X_1, \dots, X_k \in \mathfrak{X}(U)$, and $\omega_1, \dots, \omega_h \in \Omega^1(U)$. Then:*

$$\begin{aligned} \zeta^U(A)(\omega_1, \dots, \omega_h, X_1, \dots, X_k) = \zeta^U\big(A(\omega_1, \dots, \omega_h, X_1, \dots, X_k)\big) \\ - \sum_{i=1}^{h} A\big(\omega_1, \dots, \zeta^U(\omega_i), \dots, \omega_h, X_1, \dots X_k\big) \\ - \sum_{i=1}^{k} A\big(\omega_1, \dots, \omega_h, X_1, \dots, \zeta^U(X_i), \dots X_k\big). \end{aligned}$$

Proof. The (0, 0)-tensor $A(\omega_1, \dots, \omega_h, X_1, \dots, X_k)$ can be thought of as being obtained from the $(h+k, h+k)$ tensor $A \otimes \omega_1 \otimes \dots \otimes \omega_h \otimes X_1 \otimes \dots \otimes X_k$ by repeated contractions.[3] We write this symbolically as

[3] As in Remark 21.3, we don't bother to reorder the factors in this expression.

$$A(\omega_1, \ldots, \omega_h, X_1, \ldots, X_k) = C(A \otimes \omega_1 \otimes \cdots \otimes \omega_h \otimes X_1 \otimes \cdots \otimes X_k),$$

where C stands for repeated contractions. The claim now follows by repeatedly using the fact that ζ commutes with contractions and satisfies the derivation property. ■

COROLLARY 22.10. *Suppose ζ and ξ are two tensor derivations that agree on functions and vector fields. Then they are identical.*

Proof. Let ω be a 1-form. Then by Lemma 22.9 with $A = \omega$ we see for an arbitrary vector field X that

$$\begin{aligned}\zeta(\omega)(X) &= \zeta(\omega(X)) - \omega(\zeta(X)) \\ &= \xi(\omega(X)) - \omega(\xi(X)) \\ &= \xi(\omega)(X).\end{aligned}$$

Since X was arbitrary, this shows that $\zeta(\omega) = \xi(\omega)$, and since ω was arbitrary this shows that ζ and ξ coincide on tensors of type $(0, 1)$. Now for an arbitrary A, observe that Lemma 22.9 expands $\zeta(A)$ in such a way that all the other terms are of the form ζ eating a function, a vector field, or a 1-form. Thus $\zeta(A) = \xi(A)$ for arbitrary A. ■

The next result shows how one can work backwards and build a tensor derivation if we have something defined on functions and vector fields with the appropriate property.

PROPOSITION 22.11. *Suppose we have a type-preserving local operator ζ defined on smooth functions and vector fields which satisfies*

$$\begin{aligned}\zeta(fg) &= \zeta(f)g + f\zeta(g), \\ \zeta(fX) &= \zeta(f)X + f\zeta(X)\end{aligned} \tag{22.2}$$

for all $f, g \in C^\infty(M)$ and $X \in \mathfrak{X}(M)$.[4] *Then ζ extends uniquely to a tensor derivation.*

Note the two conditions in (22.2) are forced if we want ζ to be a tensor derivation, since $f \otimes g = fg$ and $f \otimes X = fX$.

Proof. Uniqueness is immediate from the previous corollary, since we have prescribed what ζ must do to vector fields and functions. The proof proceeds very similarly to that of Corollary 22.10: the derivation property coupled with our requirement that ζ commutes with contractions means that at every stage we have no choice how to proceed.

Namely, we define ζ on 1-forms by setting

$$\zeta(\omega)(X) = \zeta(\omega(X)) - \omega(\zeta(X)). \tag{22.3}$$

The hypotheses imply that $\zeta\colon \mathcal{T}^{0,1}(M) \to \mathcal{T}^{0,1}(M)$ is a local operator. Next to define ζ on $\mathcal{T}^{1,1}(M)$ we start with a tensor of the form $X \otimes \omega$. The derivation property requires us define

---

[4] i.e. we have two local operators $\zeta\colon \mathcal{T}^{0,0}(M) \to \mathcal{T}^{0,0}(M)$ and $\zeta\colon \mathcal{T}^{1,0}(M) \to \mathcal{T}^{1,0}(M)$.

$$\zeta(X \otimes \omega) := X \otimes \zeta(\omega) + \zeta(X) \otimes \omega.$$

If ζ commutes with the contraction $C^{1,1}\colon \mathcal{T}^{1,1}(M) \to C^\infty(M)$ then we need

$$\begin{aligned}\zeta(\omega(X)) &= C^{1,1}\big(\zeta(X \otimes \omega)\big)\\ &= C^{1,1}\big(X \otimes \zeta(\omega) + \zeta(X) \otimes \omega\big)\\ &= \zeta(\omega)(X) + \omega(\zeta(X)),\end{aligned}$$

and this is true by (22.3). This also shows that (22.3) was forced—no other choice would have worked. Now we use the formula from Lemma 22.9 to define ζ on all higher tensors. A check similar to the one we just did shows that the resulting object is a derivation that commutes with all contractions. ■

We now obtain our promised extension of the Lie derivative.

Theorem 22.12 *Let $X \in \mathfrak{X}(M)$. There exists a unique tensor derivation*

$$\mathcal{L}_X\colon \mathcal{T}(M) \to \mathcal{T}(M)$$

that extends the Lie derivative defined on functions and vector fields from Chapter 8.

Lemma 22.9 tells us how to compute $\mathcal{L}_X A$. For instance, if $A \in \mathcal{T}^{0,k}(M)$ then

$$\begin{aligned}(\mathcal{L}_X A)(Y_1, \dots, Y_k) &= X\big(A(Y_1, \dots, Y_k)\big)\\ &\quad - \sum_{i=1}^{k} A\big(Y_1, \dots, Y_{i-1}, \mathcal{L}_X Y_i, Y_{i+1}, \dots, Y_k\big).\end{aligned} \tag{22.4}$$

The next result, whose proof is deferred to Problem Sheet I, classifies all tensor derivations.

PROPOSITION 22.13. *Let M be a smooth manifold.*

(i) *Suppose $A \in \mathcal{T}^{1,1}(M) \cong \Gamma(\mathrm{End}(TM))$. Then there exists a unique tensor derivation ζ_A on M with the property that $\zeta_A(Y)(p) = A_p(Y(p))$ for any vector field Y and satisfies $\zeta_A(f) = 0$ for any function f.*
(ii) *Let ξ be an arbitrary tensor derivation. Then there exists a vector field X on M and $A \in \mathcal{T}^{1,1}(M)$ such that $\xi = \mathcal{L}_X + \zeta_A$. Thus the space of tensor derivations on M can be identified with $\mathfrak{X}(M) \times \Gamma(\mathrm{End}(TM))$.*

An explicit formula for the Lie derivative is given by the following result.

PROPOSITION 22.14. *Let $X \in \mathfrak{X}(M)$ with flow Φ_t. Then for any tensor field A, the Lie derivative $\mathcal{L}_X A$ is given by:*

$$\mathcal{L}_X A := \frac{d}{dt}\Big|_{t=0} \Phi_t^* A = \lim_{t\to 0} \frac{\Phi_t^* A - A}{t}. \tag{22.5}$$

The proof of Proposition 22.14 is given in the bonus section below.

We now switch our focus back to differential forms. The following statement is a minor variation of the Tensor Criterion 21.5, and the proof is left as an exercise.

Theorem 22.15 (The Differential Form Criterion) *Let M be a smooth manifold and let $U \subset M$ be a non-empty open set. Then there is a canonical identification between $\Omega^k(U)$ and* **alternating** *$C^\infty(U)$-multilinear functions*

$$\underbrace{\mathfrak{X}(U) \times \cdots \times \mathfrak{X}(U)}_{k \text{ copies}} \to C^\infty(U).$$

Since an alternating multilinear map is (in particular) a multilinear map, we see that any differential k-form may be regarded as a tensor of type $(0, k)$. But for $k \geq 2$, there are (many) multilinear maps that are not alternating, and thus not every tensor of type $(0, k)$ is a differential form. We define

$$\Omega(M) = \bigoplus_{0 \leq k \leq m} \Omega^k(M),$$

with $\Omega(U)$ defined similarly. Thus an element of $\Omega(M)$ is a sum $\sum_{i=0}^m \omega_i$ where $\omega_i \in \Omega^i(M)$.

Definition 22.16. If $\omega \in \Omega^h(M)$ and $\theta \in \Omega^k(M)$ then the **wedge product** is the differential form $\omega \wedge \theta \in \Omega^{h+k}(M)$ defined pointwise by

$$(\omega \wedge \theta)_p = \omega_p \wedge \theta_p$$

Since $\bigwedge^k V = 0$ if $k > \dim V$, the wedge product $\omega \wedge \theta$ is zero if $h + k > \dim M$. Note that by part (ii) of Proposition 19.22, one has

$$\omega \wedge \theta = (-1)^{hk} \theta \wedge \omega, \qquad \omega \in \Omega^h(M), \theta \in \Omega^k(M)$$

The wedge product gives $\Omega(M)$ the structure of graded ring, and in fact, also a $C^\infty(M)$-graded skew-commutative[5] algebra.

Here is a useful piece of linear algebra, whose proof is on Problem Sheet J. We let $\mathfrak{S}_k$ denote the group of all permutations on k letters.

Definition 22.17. Let $h, k \geq 0$. An (h, k)**-shuffle** is a permutation $\varrho \in \mathfrak{S}_{h+k}$ such that

$$\varrho(1) < \cdots < \varrho(h) \qquad \text{and} \qquad \varrho(h+1) < \cdots < \varrho(h+k).$$

We let $\operatorname{Shuffle}(h, k) \subset \mathfrak{S}_{h+k}$ denote the set of all (h, k)-shuffles.

Lemma 22.18. *Let V be a vector space and suppose $\omega \in \bigwedge^h V^*$ and $\theta \in \bigwedge^k V^*$. Let $v_i \in V$ for $i = 1, \ldots, h+k$. Then if we identify ω with an element of $\operatorname{Alt}_h(V)$, θ with an element of $\operatorname{Alt}_k(V)$, and $\omega \wedge \theta$ with an element of $\operatorname{Alt}_{h+k}(V)$, one has:*

[5] The "skew-commutative" refers to the sign $(-1)^{hk}$.

$$(\omega \wedge \theta)(v_1, \ldots, v_{h+k}) \\ = \frac{1}{h!k!} \sum_{\varrho \in \mathfrak{S}_{h+k}} \operatorname{sgn}(\varrho)\omega\big(v_{\varrho(1)}, \ldots, v_{\varrho(h)}\big)\, \theta\big(v_{\varrho(h+1)}, \ldots, v_{\varrho(h+k)}\big)$$

or equivalently

$$(\omega \wedge \theta)(v_1, \ldots, v_{h+k}) \\ = \sum_{\varrho \in \mathrm{Shuffle}(h,k)} \operatorname{sgn}(\varrho)\omega\big(v_{\varrho(1)}, \ldots, v_{\varrho(h)}\big)\, \theta\big(v_{\varrho(h+1)}, \ldots, v_{\varrho(h+k)}\big).$$

For low values of h and k this gives an easy method to compute the wedge product of two differential forms. For instance, we have:

COROLLARY 22.19. *Let $\omega, \theta \in \Omega^1(M)$ denote two 1-forms. Then*

$$(\omega \wedge \theta)_p(\xi, \zeta) = \omega_p(\xi)\theta_p(\zeta) - \omega_p(\zeta)\theta_p(\xi), \qquad \forall\, p \in M,\ \forall\, \xi, \zeta \in T_pM.$$

If $\varphi \colon M \to N$ is smooth we showed in Definition 21.8 how to pullback a tensor $A \in \mathcal{T}^{0,k}(N)$ to obtain a tensor $\varphi^*A \in \mathcal{T}^{0,k}(M)$. It is clear from the definition that if A is alternating then so is φ^*A. Thus φ^* restricts to define a map $\Omega(N) \to \Omega(M)$. For convenience, we state this again here:

DEFINITION 22.20. Let $\varphi \colon M \to N$ denote a smooth map. Given $\omega \in \Omega^k(N)$, we define the **pullback** form $\varphi^*\omega \in \Omega^k(M)$ by

$$(\varphi^*\omega)_p(\xi_1, \ldots, \xi_k) := \omega_{\varphi(p)}\big(D\varphi(p)\xi_1, \ldots, D\varphi(p)\xi_k\big).$$

The next lemma tells us that φ^* is an algebra homomorphism.

LEMMA 22.21. *If $\varphi \colon M \to N$ is a smooth map and $\omega, \theta \in \Omega(N)$ then*

$$\varphi^*(\omega \wedge \theta) = \varphi^*\omega \wedge \varphi^*\theta.$$

Proof. Immediate from Lemma 22.18 and the definition. ∎

Note also that (just as with tensors), the pullback operation is functorial: if $\varphi \colon M \to N$ and $\psi \colon L \to M$ then

$$(\varphi \circ \psi)^* = \psi^* \circ \varphi^* \tag{22.6}$$

as maps $\Omega(N) \to \Omega(L)$.

Since any differential form $\omega \in \Omega^k(M)$ can be thought of as a tensor of type $(0, k)$, we can apply the Lie derivative $\mathcal{L}_X$ to it to obtain another tensor of type $(0, k)$, denoted by $\mathcal{L}_X\omega$. In fact, from Lemma 22.9 the tensor $\mathcal{L}_X\omega$ is easily seen to be alternating, and hence $\mathcal{L}_X\omega$ is another differential k-form. Explicitly, by (22.4) one has for $\omega \in \Omega^1(M)$ and $X, Y \in \mathfrak{X}(M)$ that

$$(\mathcal{L}_X\omega)(Y) = X(\omega(Y)) - \omega([X, Y]), \tag{22.7}$$

and more generally

$$(\mathcal{L}_X\omega)(X_1,\dots,X_k) = X(\omega(X_1,\dots,X_k)) - \sum_{i=1}^{k} \omega(X_1,\dots,[X,X_i],\dots X_k) \tag{22.8}$$

for $\omega \in \Omega^k(M)$ and $X, X_1, \dots, X_k \in \mathfrak{X}(M)$.

Here is how the Lie derivative behaves with respect to the wedge product.

LEMMA 22.22. *Let $\omega, \theta \in \Omega(M)$. Then*

$$\mathcal{L}_X(\omega \wedge \theta) = \mathcal{L}_X(\omega) \wedge \theta + \omega \wedge \mathcal{L}_X(\theta).$$

Proof. Apply Proposition 22.29[6] with $\mathcal{A}(\omega,\theta) = \omega \wedge \theta$. ■

REMARK 22.23. The Lie derivative $\mathcal{L}_X$ gives us a way to "differentiate" a tensor field (or a differential form) with respect to a vector field, but it does *not* allow us to differentiate a tensor field (or differential form) with respect to a single tangent vector. Indeed, the value of $\mathcal{L}_X A$ at a point p depends on the values of X on a whole neighbourhood of p, not just on $X(p)$. This is because $X \mapsto \mathcal{L}_X$ is *not* $C^\infty(M)$-linear. For instance, if A is a 1-form ω then for $X, Y \in \mathfrak{X}(M)$ and $f \in C^\infty(M)$,

$$\begin{aligned}(\mathcal{L}_{fX}\omega)(Y) &= (fX)\omega(Y) - \omega([fX,Y]) \\ &= fX(\omega(Y)) - \omega(f[X,Y] - Y(f)X) \\ &= f(\mathcal{L}_X\omega)(Y) + Y(f)\omega(X),\end{aligned}$$

where the second line used Problem D.5. The presence of the "error term" $Y(f)\omega(X)$ shows that $\mathcal{L}_{fX} \neq f\mathcal{L}_X$.

The first topic we will cover in Volume II will be *connections* in vector bundles and principal bundles. A connection on the tangent bundle TM induces a *covariant derivative* ∇_X on tensor fields associated to every vector field X. As we will see, a covariant derivative ∇_X will have the nice property that $\nabla_{fX} = f\nabla_X$. The downside is that it requires a choice of extra structure (namely, a connection). Meanwhile the Lie derivative is canonical. See Volume II, Remark 4.7.

We conclude this chapter with the analogue of a tensor derivation on differential forms.

DEFINITION 22.24. Let M be a smooth manifold and let $h \in \mathbb{Z}$. A **graded derivation of degree** h on M is a local operator $\zeta \colon \Omega(M) \to \Omega(M)$ of degree h such that if $\omega \in \Omega^k(M)$ and $\theta \in \Omega(M)$ then

$$\zeta(\omega \wedge \theta) = \zeta\omega \wedge \theta + (-1)^{hk}\omega \wedge \zeta\theta. \tag{22.9}$$

[6] Strictly speaking, this is a slight modification of Proposition 22.29 for differential forms instead of tensors, but the proof is exactly the same.

We refer to (22.9) as the **graded derivation property**. As in Remark 22.8, if ζ is a graded derivation and $U \subset M$ is open then the induced map $\zeta^U : \Omega(U) \to \Omega(U)$ is another graded derivation.

EXAMPLE 22.25. The Lie derivative $\mathcal{L}_X$ is a graded derivation of degree 0 by Lemma 22.22.

DEFINITION 22.26. A 1-form $\omega \in \Omega^1(U)$ is called **exact** if $\omega = df$ for some $f \in C^\infty(U)$.

Just as a tensor derivation is entirely determined by what it does to functions and vector fields, a graded derivation is entirely determined by what it does to functions and exact 1-forms.

PROPOSITION 22.27. *Suppose ζ and ξ are two graded derivations of the same degree h. If ζ and ξ agree on functions and exact 1-forms then $\zeta = \xi$.*

Proof. Since a graded derivation is a local operator, by Problem J.1 it is entirely determined by all its restrictions ζ^U where $U \subset M$ is the domain of a chart (U, x). By (21.1), any $\omega \in \Omega^k(U)$ can be written as a sum of elements of the form

$$f\, dx^{i_1} \wedge \cdots \wedge dx^{i_k}.$$

Since ζ is $\mathbb{R}$-linear, ζ^U is determined by what it does to such a term. But by repeatedly applying (22.9), we see that $\zeta^U(f\, dx^{i_1} \wedge \cdots \wedge dx^{i_k})$ is determined by $\zeta^U(f)$ and $\zeta^U(dx^{i_j})$. Thus if two graded derivations agree on functions and exact 1-forms then they are identical. ■

REMARK 22.28. It follows directly from the definition that the space of graded derivations of a given degree h forms a vector space (with the caveat that the zero operator has to be regarded as a graded derivation of all degrees). Moreover if ζ and ξ are two graded derivations of degrees h and k respectively then

$$\zeta \circ \xi - (-1)^{hk} \xi \circ \zeta$$

is another graded derivation of degree $h + k$.[7]

In the next chapter we will investigate a particularly important graded derivation of degree $+1$: the *exterior differential*. This will lead us to the *de Rham cohomology* of a manifold.

[7] Exercise: Check this!

Bonus Material for Chapter 22

In this bonus section we prove Proposition 22.14. Let us temporarily denote the right-hand side of (22.5) by $\widetilde{\mathcal{L}}_X A$. Thus to prove Proposition 22.14 we must show that for any tensor A, one has:

$$\mathcal{L}_X A = \widetilde{\mathcal{L}}_X A. \tag{22.10}$$

The proof of (22.10) uses the following result, which will be useful elsewhere.

PROPOSITION 22.29. *Let (h_0, k_0), (h_1, k_1) and (h_2, k_2) be three pairs of non-negative integers. Suppose we are given a $C^\infty(M)$-bilinear operator*

$$\mathcal{A} \colon \mathcal{T}^{h_0,k_0}(M) \times \mathcal{T}^{h_1,k_1}(M) \to \mathcal{T}^{h_2,k_2}(M).$$

Assume in addition that $\mathcal{A}$ has the property that if $\varphi \colon U \to V$ is a local diffeomorphism between open sets of M then the corresponding local operators

$$\varphi^*\big(\mathcal{A}^V(A, B)\big) = \mathcal{A}^U(\varphi^* A, \varphi^* B). \tag{22.11}$$

Then for every vector field X on M, one has

$$\widetilde{\mathcal{L}}_X\big(\mathcal{A}(A, B)\big) = \mathcal{A}\big(\widetilde{\mathcal{L}}_X A, B\big) + \mathcal{A}\big(A, \widetilde{\mathcal{L}}_X B\big).$$

The proof of Proposition 22.29 is on Problem Sheet I.

Proof of Proposition 22.14. Since $\Phi_t^* = ((\Phi_t)^{-1})_* = (\Phi_{-t})_*$ it follows from the definitions that $\widetilde{\mathcal{L}}_X = \mathcal{L}_X$ on functions and vector fields. Thus if we can show that $\widetilde{\mathcal{L}}_X$ is a tensor derivation, it will follow from the uniqueness part of Theorem 22.12 that $\mathcal{L}_X = \widetilde{\mathcal{L}}_X$.

Thus we must show that $\widetilde{\mathcal{L}}_X$ is a derivation that commutes with contractions. For this we use Proposition 22.29. Taking $\mathcal{A}(A, B) := A \otimes B$ shows that $\widetilde{\mathcal{L}}_X$ is a derivation (note (22.11) is satisfied by (21.2)). Similarly taking for instance $\mathcal{A}(A, B) = C^{1,1}(A \otimes B)$ shows that

$$X(\omega(Y)) = \big(\widetilde{\mathcal{L}}_X \omega\big)(Y) + \omega\big(\widetilde{\mathcal{L}}_X Y\big).$$

More generally, taking $\mathcal{A}(A, B) = C^{h,k}(A \otimes B)$ shows that $\widetilde{\mathcal{L}}_X$ commutes with $C^{h,k}$. This completes the proof. ∎

From now on, we will just write $\mathcal{L}_X$ for both the operator $\mathcal{L}_X$ from Theorem 22.12 and the operator defined in (22.5).

Problem Sheet I

A (♣) indicates the problem is non-examinable. This does not necessarily imply it is difficult.

(♣)PROBLEM I.1 Let $\varphi\colon M \to N$ be a smooth map.

(i) Let $A \in \mathcal{T}^{h,k}(M)$ denote a tensor of type (h,k). Let (U,x) and (V,y) denote two charts on M with $U \cap V \neq \emptyset$. Then one can write

$$A = f^{i_1\cdots i_h}_{j_1\cdots j_k}\frac{\partial}{\partial x^{i_1}} \otimes \cdots \otimes \frac{\partial}{\partial x^{i_h}} \otimes dx^{j_1} \otimes \cdots \otimes dx^{j_k}$$

and

$$A = g^{i_1\cdots i_h}_{j_1\cdots j_k}\frac{\partial}{\partial y^{i_1}} \otimes \cdots \otimes \frac{\partial}{\partial y^{i_h}} \otimes dy^{j_1} \otimes \cdots \otimes dy^{j_k}$$

for smooth functions $f^{i_1\cdots i_h}_{j_1\cdots j_k} \in C^\infty(U)$ and $g^{i_1\cdots i_h}_{j_1\cdots j_k} \in C^\infty(V)$. Investigate the relationship between

$$f^{i_1\cdots i_h}_{j_1\cdots j_k}|_{U\cap V} \qquad \text{and} \qquad g^{i_1\cdots i_h}_{j_1\cdots j_k}|_{U\cap V}.$$

(ii) Let $\omega \in \Omega^k(M)$ denote a differential k-form. Let (U,x) and (V,y) denote two charts on M with $U \cap V \neq \emptyset$. Then one can write

$$\omega = f_{i_1\cdots i_k}\, dx^{i_1} \wedge \cdots \wedge dx^{i_k}$$

and

$$\omega = g_{i_1\cdots i_k}\, dy^{i_1} \wedge \cdots \wedge dy^{i_k}$$

for smooth functions $f_{i_1\cdots i_k} \in C^\infty(U)$ and $g_{i_1\cdots i_k} \in C^\infty(V)$. Investigate the relationship between

$$f_{i_1\cdots i_k}|_{U\cap V} \qquad \text{and} \qquad g_{i_1\cdots i_k}|_{U\cap V}.$$

(iii) Assume φ is a diffeomorphism, and let $A \in \mathcal{T}^{h,k}(N)$. Let (U,x) be a chart on M and (V,y) a chart on N with $\varphi(U) \subset V$. Write

$$\varphi^* A = f^{i_1\cdots i_h}_{j_1\cdots j_k}\frac{\partial}{\partial x^{i_1}} \otimes \cdots \otimes \frac{\partial}{\partial x^{i_h}} \otimes dx^{j_1} \otimes \cdots \otimes dx^{j_k}$$

and

$$A = g^{i_1\cdots i_h}_{j_1\cdots j_k}\frac{\partial}{\partial y^{i_1}} \otimes \cdots \otimes \frac{\partial}{\partial y^{i_h}} \otimes dy^{j_1} \otimes \cdots \otimes dy^{j_k}$$

for smooth functions $f^{i_1\cdots i_h}_{j_1\cdots j_k} \in C^\infty(U \cap \varphi^{-1}(V))$ and $g^{i_1\cdots i_h}_{j_1\cdots j_k} \in C^\infty(V)$. Investigate the relationship between

$$f^{i_1\cdots i_h}_{j_1\cdots j_k} \qquad \text{and} \qquad g^{i_1\cdots i_h}_{j_1\cdots j_k}.$$

(iv) Let $\omega \in \Omega^k(N)$. Let (U,x) be a chart on M and (V,y) a chart on N such that $\varphi(U) \subset V$. Then one can write

$$\varphi^*\omega = f_{i_1\cdots i_k}\, dx^{i_1} \wedge \cdots \wedge dx^{i_k}$$

and

$$\omega = g_{i_1\cdots i_k}\, dy^{i_1} \wedge \cdots \wedge dy^{i_k}$$

for smooth functions $f_{i_1\cdots i_k} \in C^\infty(U \cap \varphi^{-1}(V))$ and $g_{i_1\cdots i_k} \in C^\infty(V)$. Investigate the relationship between

$$f_{i_1\cdots i_k} \qquad \text{and} \qquad g_{i_1\cdots i_k}.$$

(v) Conclude that local coordinates are horrible.

Problem I.2 Let $\pi\colon E \to M$ be a vector bundle. An operator $\zeta\colon \Gamma(E) \to \Gamma(E)$ is said to satisfy the **Leibniz rule** if there exists a vector field X on M such that for any $f \in C^\infty(M)$ and $s \in \Gamma(E)$ one has

$$\zeta(fs) = (Xf)s + f\zeta(s).$$

Prove that an operator satisfying the Leibniz rule is a local operator but—provided $X \neq 0$—is not a point operator.

Problem I.3 Let M be a smooth manifold and let $E_1, \ldots, E_k$ and E be vector bundles over M. Let $\zeta\colon \Gamma(E_1) \times \cdots \times \Gamma(E_k) \to \Gamma(E)$ be a $C^\infty(M)$-multilinear operator. Prove that for each $p \in M$ there is a unique $\mathbb{R}$-multilinear map

$$\Phi_p\colon E_{1|p} \times \cdots \times E_{k|p} \to E_p$$

such that for all $s_i \in \Gamma(E_i)$ one has

$$\Phi_p(s_1(p), \ldots s_k(p)) = \zeta(s_1, \ldots, s_k)(p).$$

Problem I.4 Let M be a smooth manifold and let $U \subset M$ be a non-empty open set. Prove that there is a canonical identification between $\mathcal{T}^{h,k}(U)$ and $C^\infty(U)$-multilinear functions

$$\underbrace{\Omega^1(U) \times \cdots \times \Omega^1(U)}_{h \text{ copies}} \times \overbrace{\mathfrak{X}(U) \times \cdots \times \mathfrak{X}(U)}^{k \text{ copies}} \to C^\infty(U).$$

(♣)Problem I.5 This problem introduces the **vertical bundle** of a fibre bundle.[8]

(i) Let $\pi\colon E \to M$ be a fibre bundle with fibre L. Let

$$VE := \bigsqcup_{u \in E} \left\{\ker D\pi(u)\colon T_uE \to T_{\pi(u)}M\right\}$$

with projection map $\pi_V\colon VE \to E$. Prove that VE is a vector bundle over E of rank $l = \dim L$.

[8] Vertical bundles will play a major role in Volume II.

(ii) Assume now that $\pi\colon E \to M$ is a vector bundle. Prove that the vertical bundle VE is isomorphic as a vector bundle to the pullback bundle $\pi^*E \to E$.
(iii) Continue to assume that $\pi\colon E \to M$ is a vector bundle. Prove that the composition $\pi \circ \pi_V\colon VE \to M$ is another vector bundle over M which is isomorphic to the direct sum bundle $E \oplus E$.
(iv) Continue to assume that $\pi\colon E \to M$ is a vector bundle. View M as an embedded submanifold of TM via the zero section. Consider the composite bundle $\pi \circ \pi_V\colon VE \to M$. Prove that there is a vector bundle morphism $VE \to TE$ over the inclusion of M into TM.

Problem I.6 Let M be a smooth manifold.

(i) Suppose $A \in \mathcal{T}^{1,1}(M) \cong \Gamma(\mathrm{End}(TM))$. Prove there exists a unique tensor derivation ζ_A on M with the property that $\zeta_A(Y)(p) = A_p(Y(p))$ for any vector field Y and satisfies $\zeta_A(f) = 0$ for any function f.
(ii) Let ξ be an arbitrary tensor derivation. Prove that there exists a vector field X on M and $A \in \mathcal{T}^{1,1}(M)$ such that $\xi = \mathcal{L}_X + \zeta_A$. Deduce that the space of tensor derivations on M can be identified with $\mathfrak{X}(M) \times \Gamma(\mathrm{End}(TM))$.

Definition. Let $X \in \mathfrak{X}(M)$ with flow Φ_t. Define an operator $\widetilde{\mathcal{L}}_X$ on $\mathcal{T}(M)$ by

$$\widetilde{\mathcal{L}}_X A := \frac{d}{dt}\bigg|_{t=0} \Phi_t^* A = \lim_{t\to 0} \frac{\Phi_t^* A - A}{t}.$$

Note that Proposition 22.14 states that $\widetilde{\mathcal{L}}_X$ coincides with the Lie derivative $\mathcal{L}_X$. The proof of Proposition 22.14 uses Problem I.7. Thus to avoid a circular argument, you **cannot** use the fact that $\widetilde{\mathcal{L}}_X = \mathcal{L}_X$ while solving Problem I.7!

(♣)Problem I.7 Let (h_0, k_0), (h_1, k_1) and (h_2, k_2) be three pairs of non-negative integers. Suppose we are given a $C^\infty(M)$-bilinear operator

$$\mathcal{A}\colon \mathcal{T}^{h_0,k_0}(M) \times \mathcal{T}^{h_1,k_1}(M) \to \mathcal{T}^{h_2,k_2}(M).$$

Assume in addition that $\mathcal{A}$ has the property that if $\varphi\colon U \to V$ is a local diffeomorphism between open sets of M then the corresponding local operators

$$\varphi^*\big(\mathcal{A}^V(A, B)\big) = \mathcal{A}^U(\varphi^* A, \varphi^* B).$$

Prove that for every vector field X on M, one has

$$\widetilde{\mathcal{L}}_X\big(\mathcal{A}(A, B)\big) = \mathcal{A}\big(\widetilde{\mathcal{L}}_X A, B\big) + \mathcal{A}\big(A, \widetilde{\mathcal{L}}_X B\big).$$

Bonus Problem(s) for Sheet I

These problem(s) are hard, and are included for enthusiasts only. Solutions will not be provided.

(♣)PROBLEM I.8 Let R be a commutative ring and let V be a finitely generated projective R-module. Prove that for all $h, k \geq 0$,

$$T^{h,k}V \cong \mathrm{Mult}_{k,h}(V).$$

(♣)PROBLEM I.9 Let $\pi\colon E \to M$ be a vector bundle and let $U \subset M$ be an arbitrary open set (possibly equal to M). Prove that the space $\Gamma(U, E)$ is a finitely generated projective $C^\infty(U)$-module.

Chapter 23
The Exterior Differential

Differential forms are typically more important than tensors in geometry for two key reasons:

- We can differentiate them.
- We can integrate them.

We will discuss differentiation in this chapter. Integration will be covered starting in Chapter 25. Let us motivate this by considering the special case of a 0-form, i.e. a smooth function. If $f \in C^\infty(M)$ then we have already defined the differential of f as the exact 1-form $df \in \Omega^1(M)$. Generalising this, we will define the differential of a k-form ω to be a $(k+1)$-form $d\omega$.

Here is the main result of this chapter.

Theorem 23.1 (The Exterior Differential) *Let M be a smooth manifold. There is a unique graded derivation $d\colon \Omega(M) \to \Omega(M)$ of degree 1 that extends the operator $f \mapsto df$ on $\Omega^0(M)$ and satisfies*

$$d \circ d = 0. \tag{23.1}$$

We call d the **exterior differential operator** and refer to $d\omega$ as the **exterior differential** of ω (often shortened to the just "the **differential** of ω"). The importance of the condition (23.1) will be explained shortly.

Warning! The words "derivative" and "differential" are often used synonymously, but—at least as far as this course is concerned—there is an important difference. As explained in Remark 5.3, if f is a smooth function then the derivative Df and the differential df differ only by the dash-to-dot map. However if $\omega \in \Omega^k(M)$ for $k \geq 1$ then the derivative $D\omega$ and the differential $d\omega$ are very different objects. The latter is an element of $\Omega^{k+1}(M)$, whereas the former is a linear map $D\omega\colon TM \to T \bigwedge T^*M$!

Our proof of Theorem 23.1 will construct d in coordinates. We will give a coordinate-free expression for d at the end of the chapter in Theorem 23.13.

W. J. Merry, *Lectures on Differential Geometry I*,
https://doi.org/10.1007/978-3-032-03733-6_23

Proof of Theorem 23.1. We prove the result in three steps.

1. We first deal with uniqueness. This is immediate from Proposition 22.27, since we have specified what d does to functions and to exact 1-forms (namely, $d(df) = 0$).

2. To construct d it suffices by Problem J.1 to show that for any chart (U, x) there is a an operator $d^U : \Omega^k(U) \to \Omega^{k+1}(U)$ satisfying the requirements of the theorem, which moreover has the property that if (V, y) is another chart with $U \cap V \neq \emptyset$ then

$$d^U = d^V, \qquad \text{on } U \cap V. \tag{23.2}$$

In this step we define an operator $d^U : \Omega^k(U) \to \Omega^{k+1}(U)$ which satisfies the requirements of the theorem. In the last step we will show that (23.2) is satisfied.

To ease the notation we adopt the following shorthand: if $I = (i_1, \dots, i_k)$ is a subset of $\{1, \dots, m\}$ with $i_j < i_{j+1}$ for each $j = 1, \dots, k-1$ then we set:

$$dx^I := dx^{i_1} \wedge \cdots \wedge dx^{i_k}.$$

We also define $dx^I := 1$ if $I = \emptyset$. Thus any $\omega \in \Omega(U)$ can be written as a sum

$$\omega = \sum_I f_I \, dx^I.$$

We define

$$d^U \omega := \sum_I df_I \wedge dx^I.$$

If $\omega \in \Omega^k(U)$ then $d^U \omega \in \Omega^{k+1}(U)$. Moreover d^U is obviously $\mathbb{R}$-linear and satisfies the first bullet point by definition. Thus we need only check that $d^U(d^U \omega) = 0$ and that (22.9) holds.

To establish (22.9), we may assume $\omega = f \, dx^I$ and $\theta = g \, dx^J$. If either I or J are empty then (22.9) follows from the Leibniz rule $d(fg) = f dg + g df$. In the general case we argue as follows. Assume ω has degree k. Then:

$$\begin{aligned}
d^U(\omega \wedge \theta) &= d^U(fg \, dx^I \wedge dx^J) \\
&= d(fg) \wedge dx^I \wedge dx^J \\
&= (f dg + g df) \wedge dx^I \wedge dx^J \\
&= \left(df \wedge dx^I\right) \wedge (g \, dx^J) + (-1)^k (f \, dx^I) \wedge \left(dg \wedge dx^J\right) \\
&= d^U \omega \wedge \theta + (-1)^k \omega \wedge d^U \theta.
\end{aligned}$$

To see that $d^U(d^U \omega) = 0$ we first show that $d^U(df) = 0$ for any function $f \in C^\infty(U)$. For this write $df = \frac{\partial f}{\partial x^i} dx^i$ (cf. Definition 8.4). Then

$$d^U(df) = \frac{\partial^2 f}{\partial x^i \partial x^j} dx^i \wedge dx^j,$$

where we abbreviate $\frac{\partial^2 f}{\partial x^i \partial x^j} = \frac{\partial}{\partial x^j}\left(\frac{\partial f}{\partial x^i}\right)$. But by elementary calculus, $\frac{\partial^2 f}{\partial x^i \partial x^j}$ is symmetric in i and j, whereas $dx^i \wedge dx^j$ is anti-symmetric. Thus the sum cancels.

Next, if $f, g \in C^\infty(U)$ then using the graded derivation property we have

$$d^U(df \wedge dg) = d^U(df) \wedge dg - df \wedge d^U(dg) = 0,$$

and more generally if $f, f_1, \dots f_k \in C^\infty(U)$ are any functions then

$$d^U(df \wedge df_1 \wedge \cdots \wedge df_k) = 0.$$

Applying this with $f_j = x^{i_j}$ and using $\mathbb{R}$-linearity shows that $d^U(d^U\omega) = 0$ for any $\omega \in \Omega^k(U)$.

3. Suppose (V, y) is another chart defined on an open set such that $U \cap V \neq \emptyset$. By Proposition 22.27 applied to the graded derivations d^U and d^V restricted $\Omega(U \cap V)$, to prove (23.2) it suffices to show they agree on functions and on exact 1-forms. But this is clear: if $f \in C^\infty(U \cap V)$ then one has

$$d^U f = d^V f = df,$$

and the argument we just gave above showed that

$$d^U(df) = d^V(df) = 0.$$

This completes the proof. ∎

Definition 23.2. A differential form ω is said to be **closed** if $d\omega = 0$. A differential form ω is said to be **exact** if $\omega = d\theta$ for some θ (this extends Definition 22.26 to k-forms for $k > 1$. Since $d \circ d = 0$, any exact form is closed, but the converse is typically false. One denotes the quotient vector space by

$$H^k_{\mathrm{dR}}(M) := \frac{\{\text{closed } k\text{-forms}\}}{\{\text{exact } k\text{-forms}\}},$$

where the "dR" stands for "de Rham". An element of $H^k_{\mathrm{dR}}(M)$ is written as $[\omega]$, where ω is a closed k-form. Thus by definition

$$[\omega] = [\omega + d\theta].$$

We call $H^k_{\mathrm{dR}}(M)$ the ***k*****th de Rham cohomology group**[1] of M.

Do not be scared by the word "cohomology" if you are not familiar with algebraic topology. As far as this course is concerned, all that is important is that $H^k_{\mathrm{dR}}(M)$ is a quotient vector space.

[1] Even though the de Rham groups are actually vector spaces (and not just abelian groups), it is still common to refer to them as the "de Rham cohomology groups" of M.

In Chapter 27 we will see that the de Rham groups are a topological invariant of M. For now we content ourselves with computing the zeroth groups.

LEMMA 23.3. *If M is connected then $H^0_{\mathrm{dR}}(M) \cong \mathbb{R}$.*

Proof. Since there are no differential forms of negative degree, there are in particular no exact 0-forms. Thus $H^0_{\mathrm{dR}}(M)$ is simply the space of closed 0-forms. A function f satisfies $df = 0$ if and only if f is locally constant.[2] ■

LEMMA 23.4. *Let $\varphi\colon M \to N$ be a smooth map and let $\omega \in \Omega(N)$. Then*

$$\varphi^*(d\omega) = d(\varphi^*\omega).$$

Thus φ^* commutes with the exterior differential operators:

$$\begin{array}{ccc} \Omega(M) & \xrightarrow{\ d\ } & \Omega(M) \\ {\scriptstyle \varphi^*}\uparrow & & \uparrow{\scriptstyle \varphi^*} \\ \Omega(N) & \xrightarrow[\ d\ ]{} & \Omega(N) \end{array}$$

Proof. We first prove the lemma for the special case of 0-forms, i.e functions. Let $f \in C^\infty(N)$ and $X \in \mathfrak{X}(M)$. Then manipulating the definitions gives us

$$\begin{aligned} \varphi^*(df)(X) &= df(D\varphi(X)) \\ &= D\varphi(X)(f) \\ &= X(f \circ \varphi) \\ &= d(f \circ \varphi)(X) \\ &= d(\varphi^* f)(X). \end{aligned}$$

where each line in this block of equations is a function on M.

For the general case, by arguing as in the proof of Theorem 23.1 it suffices to work in a chart domain and assume ω is of the form $f\, dx^{i_1} \wedge \cdots \wedge dx^{i_k}$. Then since we already know that φ^* is an algebra homomorphism (Lemma 22.21) and we already proved the result for functions:

$$\begin{aligned} \varphi^*(d\omega) &= \varphi^*\big(df \wedge dx^{i_1} \wedge \cdots \wedge dx^{i_k}\big) \\ &= \varphi^*(df) \wedge \varphi^*(dx^{i_1}) \wedge \cdots \wedge \varphi^*(dx^{i_k}) \\ &= d(\varphi^*(f)) \wedge d(\varphi^* x^{i_1}) \wedge \cdots \wedge d(\varphi^* x^{i_k}) \\ &= d\big(\varphi^*(f) \wedge d(\varphi^* x^{i_1}) \wedge \cdots \wedge d(\varphi^* x^{i_k})\big) \\ &= d(\varphi^*\omega). \end{aligned}$$

[2] The same argument shows that if M has n connected components then $H^0_{\mathrm{dR}}(M) \cong \mathbb{R}^n$.

This completes the proof. ■

Corollary 23.5. *If $\varphi\colon M \to N$ is a smooth map then φ^* induces a well-defined map (also denoted by) $\varphi^*: H^k_{\mathrm{dR}}(N) \to H^k_{\mathrm{dR}}(M)$ via:*

$$[\omega] \mapsto [\varphi^*\omega].$$

Proof. By Lemma 23.4, φ^* maps closed forms to closed forms and exact forms to exact forms. ■

We now relate the Lie derivative to the exterior differential.

Proposition 23.6. *Let M be a smooth manifold and fix $X \in \mathfrak{X}(M)$. Then d commutes with $\mathcal{L}_X$:*

$$\begin{array}{ccc} \Omega(M) & \xrightarrow{\ d\ } & \Omega(M) \\ {\scriptstyle \mathcal{L}_X}\downarrow & & \downarrow{\scriptstyle \mathcal{L}_X} \\ \Omega(M) & \xrightarrow[\ d\ ]{} & \Omega(M) \end{array}$$

Proof. We first prove the result for functions. For a function f and a vector field Y, one has by (22.7) that

$$\begin{aligned} \mathcal{L}_X(df)(Y) &= X(Y(f)) - [X,Y](f) \\ &= Y(X(f)) \\ &= d(X(f))(Y) \\ &= d(\mathcal{L}_X f)(Y). \end{aligned}$$

Since Y was arbitrary, this shows that $\mathcal{L}_X(df) = d(\mathcal{L}_X f)$. For the general case, by Remark 22.28, $d\circ\mathcal{L}_X - \mathcal{L}_X\circ d$ is a graded derivation of degree $+1$. By Proposition 22.27, if we can show it vanishes on functions and exact 1-forms then it is identically zero. We just did the case for functions, and for an exact 1-form we have

$$\begin{aligned} d(\mathcal{L}_X(df)) - \mathcal{L}_X(d(df)) &= d(\mathcal{L}_X(df)) - 0 \\ &= d(d(\mathcal{L}_X f) \\ &= 0. \end{aligned}$$

This completes the proof. ■

We now move onto defining a third operator on $\Omega(M)$, called the interior product. As usual, this begins at the linear algebra level.

Definition 23.7. Let V be a vector space, and fix $v \in V$. Define $i_v : \bigwedge^k V^* \to \bigwedge^{k-1} V^*$ by declaring that on decomposable elements $\lambda^1 \wedge \cdots \wedge \lambda^k$ that

$$i_v(\lambda^1 \wedge \cdots \wedge \lambda^k) = \sum_{i=1}^{k} (-1)^{i+1} \lambda^i(v) \cdot \lambda^1 \wedge \cdots \wedge \lambda^{i-1} \wedge \lambda^{i+1} \wedge \cdots \wedge \lambda^k,$$

and then extending by linearity.

Straight from the definition, we see that:

Lemma 23.8. *Let $\omega \in \bigwedge^h V^*$ and $\theta \in \bigwedge^k V^*$. Then*

$$i_v(\omega \wedge \theta) = i_v\omega \wedge \theta + (-1)^h \omega \wedge i_v\theta.$$

An alternative characterisation of the interior product is given by the following statement, whose proof is on Problem Sheet J.

Lemma 23.9. *Let $v \in V$ and let $\omega \in \bigwedge^k V^*$. If we regard both ω and $i_v\omega$ as elements of $\mathrm{Alt}_k(V)$ and $\mathrm{Alt}_{k-1}(V)$ respectively (via Proposition 19.24), then*

$$(i_v\omega)(v_1, \ldots, v_{k-1}) = \omega(v, v_1, \ldots, v_{k-1}).$$

Note this shows that $i_v \circ i_v = 0$. We now transfer this to manifolds.

Proposition 23.10. *Let M be a smooth manifold and let $X \in \mathfrak{X}(M)$. There is a graded derivation $i_X : \Omega(M) \to \Omega(M)$ of degree -1 defined by declaring that if $\omega \in \Omega^k(M)$ for $k \geq 1$ then*

$$(i_X\omega)(X_1, \ldots, X_{k-1}) := \omega(X, X_1, \ldots, X_{k-1}),$$

for $X_1, \ldots, X_{k-1} \in \mathfrak{X}(M)$. Meanwhile for $k = 0$ we set $i_X f := 0$. One has $i_X \circ i_X = 0$. This is the unique graded derivation of degree -1 such that $i_X\omega = \omega(X)$ for ω a 1-form and $i_X f = 0$ for f a function.

The proof is immediate; uniqueness follows from Proposition 22.27.

Corollary 23.11. *Let $X, Y \in \mathfrak{X}(M)$. Then*

$$i_{[X,Y]} = \mathcal{L}_X \circ i_Y - i_Y \circ \mathcal{L}_X$$

as operators on $\Omega(M)$.

Proof. Both sides are graded derivations of degree -1 by Remark 22.28. Thus it suffices to check on functions and exact 1-forms. For functions both sides are zero. For an exact 1-form df this follows from (22.7) applied with $\omega = df$. ∎

The exterior differential, Lie derivative, and interior product are related by the following useful formula that goes by the somewhat flamboyant name of **Cartan's Magic Formula**.

Theorem 23.12 (Cartan's Magic Formula) *Let $X \in \mathfrak{X}(M)$. Then*

$$\mathcal{L}_X = d \circ i_X + i_X \circ d.$$

Proof. Again, both sides are graded derivations of degree 0 by Remark 22.28. Thus by Proposition 22.27 we need only check they agree on functions and exact 1-forms. On functions this follows from Lemma 10.2 and on exact 1-forms this was Proposition 23.6. ∎

We conclude this chapter by using Cartan's Magic Formula to give a coordinate free definition of the exterior differential d.

Theorem 23.13 *Let M be a smooth manifold, $\omega \in \Omega^k(M)$ and $X_0, \dots X_k \in \mathfrak{X}(M)$. Then:*

$$\begin{aligned} d\omega(X_0, \dots, X_k) = &\sum_{i=0}^{k} (-1)^i X_i\big(\omega(X_0, \dots, \widehat{X_i}, \dots, X_k)\big) \\ &+ \sum_{0 \le i < j \le k} (-1)^{i+j} \omega([X_i, X_j], X_0, \dots \widehat{X_i}, \dots, \widehat{X_j}, \dots, X_k). \end{aligned}$$

Here and elsewhere, the caret $\widehat{X_i}$ means that the X_i term should be omitted.

Proof. One has $d\omega(X_0, \dots, X_k) = i_{X_0}(d\omega)(X_1, \dots, X_k)$, which by Cartan's Magic Formula is equal to

$$(\mathcal{L}_{X_0}\omega)(X_1, \dots, X_k) - d(i_{X_0}\omega)(X_1, \dots, X_k). \tag{23.3}$$

We now argue by induction on k. If $k = 1$ then by (22.7) this becomes

$$\begin{aligned} &\mathcal{L}_{X_0}(\omega(X_1)) - \omega([X_0, X_1]) - d(\omega(X_0))(X_1) \\ &\qquad = X_0(\omega(X_1)) - X_1(\omega(X_0)) - \omega([X_0, X_1]), \end{aligned}$$

which is what we want. Now assume $k \ge 2$ and that the result is true for all forms of degree $k - 1$. By (22.8) the first term of (23.3) is equal to

$$X_0(\omega(X_1, \dots, X_k)) - \sum_{i=1}^{k} \omega(X_1, \dots, [X_0, X_i], \dots, X_k).$$

By induction, we have that $d(i_{X_0}\omega)(X_1, \dots, X_k)$ is equal to

$$\begin{aligned} &\sum_{i=1}^{k} (-1)^{i-1} X_i\big((i_{X_0}\omega)(X_1, \dots, \widehat{X_i}, \dots, X_k)\big) \\ &\qquad + \sum_{1 \le i < j \le k} (-1)^{i+j} \big(i_{X_0}(\omega)([X_i, X_j], X_1, \dots, \widehat{X_i}, \dots, \widehat{X_j}, \dots, X_k)\big). \end{aligned}$$

Putting this into (23.3) and checking the signs carefully gives the desired result. ∎

Chapter 24
Orientations and Manifolds with Boundary

We now move onto somewhat different topics and discuss *orientations* of vector bundles and *manifolds with boundaries*. These are preliminary topics we need to cover before we can state and prove *Stokes' Theorem*, which is about integrating differential forms on oriented manifolds with boundary.

As usual, let us start at the level of linear algebra. Of course, you have known since kindergarten what an orientation of a vector space is, but perhaps you haven't seen it in the "right" language yet.

Definition 24.1. Let V be a one-dimensional vector space. Then $V \setminus \{0\}$ has two components. An **orientation** of V is a choice of one of these components, which one then labels as "positive". The other component is then labelled "negative". A **positive basis** of V is a choice of any non-zero vector belonging to the positive component. A **negative basis** of V is a choice of any non-zero vector belonging to the negative component.

Example 24.2. The **standard orientation** of $\mathbb{R}$ is given by declaring that the positive numbers are (surprise!) the positive component of $\mathbb{R} \setminus \{0\}$.

Definition 24.3. Let V be a vector space. We will use the notation $\det V$ to mean $\bigwedge^n V$ where $n = \dim V$. One calls $\det V$ the **determinant** of V. From Lemma 19.27, the space $\det V$ is a one-dimensional vector space. Moreover if (e_i) is a basis for V then $e_1 \wedge \cdots \wedge e_n$ is a basis of $\det V$.

Definitions 24.4. Let V be a vector space of positive dimension. An **orientation** on V is a choice of orientation on $\det V$. Thus there are exactly two orientations. An **oriented vector space** is a vector space together with a choice of orientation. A basis (e_i) of an oriented vector space V is said to be **positive** if $e_1 \wedge \cdots \wedge e_n$ is a positive basis of $\det V$. If instead $e_1 \wedge \cdots \wedge e_n$ is a negative basis of $\det V$, then (e_i) is a negative basis of V.

W. J. Merry, *Lectures on Differential Geometry I*,
https://doi.org/10.1007/978-3-032-03733-6_24

EXAMPLE 24.5. If e_i denotes the standard ith basis vector in $\mathbb{R}^m$ then the **standard orientation** of $\mathbb{R}^m$ is given by declaring that $e_1 \wedge \cdots \wedge e_n$ is a positive basis of $\det \mathbb{R}^m$. Thus (e_i) is a positive basis of $\mathbb{R}^m$.

You are probably more used to thinking of the determinant of a linear transformation, rather than the determinant of a vector space itself. The motivation for this terminology goes as follows. Suppose that V and W are vector spaces of the same dimension n. A linear map $\ell\colon V \to W$ induces a linear map $\Phi_\ell\colon \det V \to \det W$, defined explicitly by

$$\Phi_\ell(v_1 \wedge \cdots \wedge v_n) := \ell v_1 \wedge \cdots \wedge \ell v_n.$$

This is a linear map between two one-dimensional vector spaces, and hence is multiplication by a scalar. This scalar is non-zero if and only if ℓ is an isomorphism. In general the precise value of this scalar depends on a choice of basis of V and W, but the linear map Φ_ℓ itself clearly does not. If ℓ is an isomorphism and V and W are oriented, then we say that ℓ is **orientation-preserving** if Φ_ℓ maps the positive component of $\det V$ to the positive component of $\det W$. Otherwise ℓ is **orientation-reversing**.

If $V = W$ and we choose the same basis for both the domain V and the target V then the value of the scalar is independent of the basis. In this case, it is common to call the scalar the **determinant of** ℓ. Explicitly, if (e_i) is our chosen basis then

$$\Phi_\ell(e_1 \wedge \cdots \wedge e_n) = (\det \ell) \cdot e_1 \wedge \cdots \wedge e_n.$$

It is convenient to extend the definition of Φ_ℓ to all linear maps by declaring that if $\ell\colon V \to W$ is a linear map with $\dim V \neq \dim W$ then $\Phi_\ell\colon \det V \to \det W$ is the zero map.

Exercise: Check this new definition of determinant coincides with the one you are used to from linear algebra. Use this to give slicker proofs of everything you learnt in your linear algebra course.

REMARK 24.6. If V is a vector space then an orientation on V canonically determines an orientation on the dual space V^* by declaring that the dual basis to a positive basis is positive.

Now we move to vector bundles. A vector bundle of rank one is often called a **line bundle**.

WARNING 24.7. This terminology is also popular in complex geometry and algebraic geometry too. But typically there people are working with *complex* vector bundles, not real vector bundles. A complex line bundle is (in particular) a two-dimensional real vector bundle. So when taken out of context, beware that the phrase "line bundle" may either refer to a one-dimensional real bundle or a one-dimensional complex bundle.

DEFINITION 24.8. Let E be a vector bundle over M. The **determinant line bundle** associated to E is the vector bundle $\det E \to M$ of rank one whose fibre over $p \in M$ is $\det E_p$.

Roughly speaking, a vector bundle $\pi\colon E \to M$ is oriented if each fibre E_p is given an orientation which depends smoothly on p. To make this precise, we prove the following result.

PROPOSITION 24.9. *Let $\pi\colon E \to M$ be a vector bundle of rank n over M. The following are equivalent:*

(i) *There is a smooth nowhere vanishing section $\mu \in \Gamma(\det E^*)$.*
(ii) *It is possible to reduce the structure group of E from $\mathrm{GL}(n)$ to $\mathrm{GL}^+(n)$.*
(iii) *The bundle $\det E^* \to M$ is a trivial bundle.*

Proof. We prove the result in three steps.

1. We first prove (i) implies (ii). Let $\{(U_a, \varepsilon_a) \mid a \in A\}$ be a vector bundle atlas for E. We may assume each U_a is connected. For each $a \in A$, we obtain a local frame $(e_1^a, \dots, e_n^a)$ for E over U_a via Lemma 20.6. Since μ is non-vanishing, for each $a \in A$ the function

$$\mu(e_1^a, \dots, e_n^a)\colon U_a \to \mathbb{R} \tag{24.1}$$

is either everywhere positive or everywhere negative.[1] If for a given a one has that (24.1) is positive, we do nothing. If instead for a given a one has that (24.1) is negative, we replace the local frame $(e_1^a, \dots, e_n^a)$ with the new one $(-e_1^a, \dots, e_n^a)$, and then, using Lemma 20.6 again, replace ε_a with the vector bundle chart corresponding to this new frame. Having done this, we may assume that (24.1) is a positive function for every $a \in A$.

We claim that our new bundle atlas, which we still denote by $\{(U_a, \varepsilon_a) \mid a \in A\}$, has its structure group contained in $\mathrm{GL}^+(n)$. Indeed, if $U_a \cap U_b \neq \emptyset$ then we can write

$$e_j^b(p) = f_j^i(p) e_i^b(p), \qquad \forall\, p \in U_a \cap U_b,$$

for smooth functions $f_j^i\colon U_a \cap U_b \to \mathbb{R}$. In fact, unravelling the definition shows that $f_j^i(p)$ is the (i, j)th entry of the transition matrix $\varepsilon_{ba}(p)$. Thus for any $p \in U_a \cap U_b$, by Corollary 19.26 we have

$$\mu(e_1^a, \dots, e_n^b)(p) = \det \varepsilon_{ba}(p) \cdot \mu(e_1^a, \dots, e_n^a)(p).$$

Thus $\det \varepsilon_{ba}(p) > 0$ for all $p \in U_a \cap U_b$, which is what we wanted to prove.

2. Now we show that (ii) implies (i). For this we start with a vector bundle atlas $\{(U_a, \varepsilon_a) \mid a \in A\}$ with structure group in $\mathrm{GL}^+(n)$ and we have to build a section μ. Let $(e_a^1, \dots, e_a^n)$ denote the dual frame to the local frame $(e_1^a, \dots, e_n^a)$ associated to ε_a, and let $\{\kappa_a \mid a \in A\}$ denote a partition of unity subordinate to the open cover $\{U_a \mid a \in A\}$. We now define

$$\mu\colon M \to \det E^*, \qquad \mu := \sum_{a \in A} \kappa_a\, e_a^1 \wedge \dots \wedge e_a^n.$$

[1] Explicitly, this is the function given by $p \mapsto \mu_p(e_1(p), \dots, e_n(p))$, where we view $\mu_p \in \bigwedge^n E_p^*$ as an element of $\mathrm{Alt}_n(E_p)$ via Proposition 19.24.

We need only check that μ is nowhere vanishing. Fix $p \in M$. Then there exists $b \in A$ such that $p \in U_b$. We evaluate μ on $e_1^b \wedge \cdots \wedge e_n^b$ at p to discover

$$\mu_p\big(e_1^b(p), \ldots, e_n^b(p)\big) = \sum_{a \in A} \Big(\det \varepsilon_{ba}(p) \Big) \kappa_a(p),$$

which is positive as desired.

3. Finally, since $\det E^*$ is a one-dimensional vector bundle, it is trivial if and only if it admits a nowhere vanishing section (this is a special case of Corollary 20.7). Thus (i) and (iii) are equivalent by Lemma 20.6. This completes the proof. ■

We now use Proposition 24.9 to define precisely what it means for a vector bundle to be orientable.

DEFINITION 24.10. Let $\pi: E \to M$ be a vector bundle. We say that E is **orientable** if either of the three conditions from Proposition 24.9 is satisfied.

Assume E is orientable. Define an equivalence relation on the space of non-vanishing sections of $\det E^*$ by declaring that μ_1 and μ_2 are equivalent if there exists a strictly positive smooth function f on M such that $\mu_2 = f\mu_1$. We call an equivalence class an **orientation** of E. An **oriented vector bundle** is an orientable vector bundle E together with a choice of orientation. If M is connected then there are exactly two possible orientations.

NOTATION. We typically denote an orientation with the symbol $\mathfrak{o}$; thus saying $\mu \in \mathfrak{o}$ means that μ is a non-vanishing section of $\det E^*$ belonging to the orientation $\mathfrak{o}$.

DEFINITION 24.11. Suppose $(E, \mathfrak{o})$ is an oriented vector bundle. We assign an orientation to each vector space E_p as follows: a basis (v_i) of E_p is positive if and only if $\mu_p(v_1, \ldots, v_n) > 0$ for some (and hence any) $\mu \in \mathfrak{o}$.

This clarifies the intuitive idea that an orientation of a vector bundle is an orientation of each fibre that depends smoothly on p. Similarly we say a local frame (e_i) is **positively oriented** if the function $\mu(e_1, \ldots, e_n)$ is positive.

Specialising this to our favourite type of vector bundle tells what it means for a manifold to be orientable.

DEFINITION 24.12. A manifold M is said to be **orientable** if $TM \to M$ is an orientable vector bundle.

In this case since $\Gamma(\det T^*M) = \Omega^m(M)$ is just the top-dimensional differential forms, an non-vanishing section of $\det T^*M$ is simply a nowhere vanishing differential m-form. This has its own special name:

DEFINITION 24.13. A **volume form** on a smooth manifold M is a nowhere-vanishing differential m-form.

A manifold together with a choice of orientation $\mathfrak{o}$ is called an **oriented manifold**. By a slight abuse of notation we often refer to $\mathfrak{o}$ as an orientation of M itself, rather than an orientation of TM.

Definition 24.14. Let $(M, \mathfrak{o}_M)$ and $(N, \mathfrak{o}_N)$ be two oriented manifolds of the same dimension m. Suppose $\varphi\colon M \to N$ is a diffeomorphism. Let $\mu_M \in \mathfrak{o}_M$ and $\mu_N \in \mathfrak{o}_N$. Then $\varphi^*\mu_N = f\mu_M$ for a smooth nowhere vanishing function $f \in C^\infty(M)$. We say that φ is **orientation preserving** if f is everywhere positive and **orientation reversing** if f is everywhere negative.[2] This does not depend on the choice of μ_M and μ_N.

Note that if M and N are not connected, it may be the case that φ is neither orientation preserving or reversing.

Definition 24.15. As a special case of this, a chart $x\colon U \to O$ on an oriented manifold M is said to be **positively oriented** if x is an orientation preserving diffeomorphism between manifolds U and O (here U inherits the orientation from M and O inherits the standard orientation from $\mathbb{R}^m$).

We conclude this discussion by restating Proposition 24.9 in the special case of a tangent bundle, since this will more convenient to refer back to in the future.

Corollary 24.16 (Orientability of manifolds). *Let M be a smooth manifold. The following are equivalent:*

(i) *M admits a volume form.*
(ii) *There exists a smooth atlas $\{x_a\colon U_a \to O_a \mid a \in A\}$ for M such that whenever $U_a \cap U_b \neq \emptyset$,*

$$\det D(x_a \circ x_b^{-1})(x_b(p)) > 0, \qquad \forall\, p \in U_a \cap U_b. \tag{24.2}$$

We call such an atlas a **positively oriented** *smooth atlas. Note that every chart x_a is then positively oriented.*
(iii) *The determinant line bundle of the cotangent bundle T^*M is a trivial bundle.*

On Problem Sheet J you will see some examples of orientable and non-orientable manifolds.

Let us now move on to defining manifolds with *boundary*. A serious defect of differential geometry so far (at least as we have defined it) is that many natural and interesting compact subsets of Euclidean space fail to be manifolds, and thus none of our results are applicable to them.

Two key examples are the *closed* unit ball D^m, or a closed interval $[a, b] \subset \mathbb{R}$. Neither of these are locally Euclidean spaces (of dimension m and 1 respectively), since points on their boundary do not have neighbourhoods that are homeomorphic to open subsets of $\mathbb{R}^m$ (or $\mathbb{R}$). But note in both cases their interior *is* a smooth manifold of the desired dimension. For the closed ball D^m, the interior is B^m which is an m-dimensional manifold, and for the interval $[a, b]$, the interior (a, b) is a one-dimensional manifold. Moreover the boundary in both cases is an $(m-1)$-dimensional manifold: for the closed ball, $\partial D^m = S^{m-1}$, and $\partial[a, b] = \{a, b\}$.

Warning 24.17. In Chapter 1 (cf. Remark 1.18) we noted that manifold theory had re-purposed the words "open" and "closed" and given them their own meanings, which

[2] In symbols we write this as $\varphi^*\mathfrak{o}_N = \pm\mathfrak{o}_M$.

in many cases were *not* the same as the topological definitions of open and closed. In these notes we elected not to use the "manifold" meanings, and thus for us the words "open" and "closed" should always be taken to have their standard topological meaning.

Unfortunately the same is true of the word "boundary". As we will shortly see, the "boundary" of a manifold does *not* necessarily coincide with the topological definition of the word boundary. This time we will favour the manifold definition of the word, and thus when we write ∂M this is always taken to mean the "manifold" definition of the boundary (which we will shortly introduce). We will use the phrase *topological boundary* to denote the boundary in the sense of point-set topology, and use the notation ∂^{top}. Thus for any subset Y of a topological space X,

$$\partial^{\mathrm{top}} Y := \overline{Y} \setminus \mathrm{int}(Y).$$

We will see several examples below where $\partial M \neq \partial^{\mathrm{top}} M$ for M a manifold with boundary.

DEFINITIONS 24.18. A pair of **half-spaces** of $\mathbb{R}^m$ is specified by two things: a linear functional $\lambda \in (\mathbb{R}^m)^*$, and a real number c, which gives us the

$$\mathbb{R}^m_{\lambda \geq c} := \{p \in \mathbb{R}^m \mid \lambda(p) \geq c\},$$
$$\mathbb{R}^m_{\lambda \leq c} := \{p \in \mathbb{R}^m \mid \lambda(p) \leq c\}.$$

In a similar way we have **open half-spaces**

$$\mathbb{R}^m_{\lambda > c} := \{p \in \mathbb{R}^m \mid \lambda(p) > c\},$$
$$\mathbb{R}^m_{\lambda < c} := \{p \in \mathbb{R}^m \mid \lambda(p) < c\}.$$

The intersection

$$\mathbb{R}^m_{\lambda = c} = \mathbb{R}^m_{\lambda \geq c} \cap \mathbb{R}^m_{\lambda \leq c} = \{p \in \mathbb{R}^m \mid \lambda(p) = c\}$$

is called a **hyperplane**.

EXAMPLE 24.19. Take $\lambda \colon \mathbb{R}^m \to \mathbb{R}$ denote the linear functional u^1, i.e.

$$\lambda(u^1, \dots, u^m) = u^1.$$

We define the **standard half-spaces** to be

$$\mathbb{R}^m_{u^1 \geq 0} := \left\{(u^1, \dots, u^m) \in \mathbb{R}^m \mid u^1 \geq 0\right\},$$
$$\mathbb{R}^m_{u^1 \leq 0} := \left\{(u^1, \dots, u^m) \in \mathbb{R}^m \mid u^1 \leq 0\right\},$$

which we will typically abbreviate by $\mathbb{R}^m_+$ and $\mathbb{R}^m_-$ respectively.

WARNING 24.20. It is more common in the literature to define the "standard" half-spaces using $\lambda = u^m$ instead. For instance, $\mathbb{R}^m_{u^m > 0}$ is the "upper half-plane" $\mathbb{H}^m$ usually used in hyperbolic geometry. We prefer to use the standard half-spaces from Example 24.19 for two reasons:

(*i*) As we will see in the next chapter, using $\mathbb{R}^m_-$ as our "model" half-space leads to simpler formulae when discussing integration. The reason for this is explained in Problem J.7.

(*ii*) The symbol $\mathbb{H}^m$ is usually understood to denote the half-space $\mathbb{R}^m_{u^m>0}$ which *in addition* has been endowed with its standard *hyperbolic metric* (a topic we will come back to in Volume II, Chapter 22). Since we are not making any statements about metrics here, to avoid confusion we prefer not to use the symbol $\mathbb{H}^m$.

Of course, at the end of the day it is essentially irrelevant which half-space we choose as our "standard" one; they all give rise to the same notion. We could equally as well set the entire theory up with our "standard" half-space being $\mathbb{R}^m_{\lambda\geq\pi}$, where

$$\lambda(u^1, \dots, u^m) := \sum_{i=1}^m (-1)^i u^i - \log\Gamma(m).$$

(This choice would be somewhat inconvenient when it came to computations though!)

With these considerations in mind, let us now define a topological manifold with boundary.

DEFINITION 24.21. A separable metrisable[3] space M is called a **topological manifold with boundary of dimension** m if every point $p \in M$ has a neighbourhood homeomorphic to an open subset of the standard half-space $\mathbb{R}^m_-$.

As with normal manifolds, by convention the dimension is usually understood to be the corresponding small letter.

Any topological manifold of dimension m is also a topological manifold with boundary of dimension m. This is because the intersection of an open set in $\mathbb{R}^m$ with $\mathbb{R}^m_-$ is open in $\mathbb{R}^m_-$. The converse is not necessarily true, however, since an open subset of $\mathbb{R}^m_-$ that intersects the hyperplane $\mathbb{R}^m_{u^1=0}$ is *not* an open subset of $\mathbb{R}^m$.

DEFINITION 24.22. Let M be a topological manifold with boundary. We say a point $p \in M$ is an **interior point**[4] if p admits a neighbourhood that is homeomorphic to an open subset of $\mathbb{R}^m$. We denote by $\operatorname{int}(M)$ the set of interior points. If p is not an interior point then we say p is a **boundary point**. We denote by ∂M the collection of boundary points.

The fact that the dimension is well-defined again requires us to invoke Brouwer's Invariance of Domain Theorem (cf. Remark 1.5). In the smooth case however this will be much easier.

EXAMPLE 24.23. Here are some examples of topological manifolds with boundary:

(i) A topological space M is a topological manifold of dimension m if and only if it is a topological manifold with boundary of dimension m such that $\partial M = \emptyset$.

[3] The assumptions "separable and metrisable" can be replaced with Hausdorff and second countable; cf. Proposition 1.32.

[4] This notion coincides with the topological one, see Proposition 24.24 below and Problem Sheet J.

(ii) Any half-space $\mathbb{R}^m_{\lambda\geq c}$ is a topological manifold with boundary of dimension m. The boundary $\partial\mathbb{R}^m_{\lambda\geq c}$ is $\mathbb{R}^m_{\lambda=c}$. More generally any open subset Q of $\mathbb{R}^m_{\lambda\geq c}$ is a topological manifold with boundary, with $\partial Q = Q \cap \mathbb{R}^m_{\lambda=c}$.

(iii) The closed unit ball D^m is a topological manifold with boundary of dimension m. (Exercise: Prove this!) One has $\partial D^m = S^{m-1}$.

(iv) The closed m-dimensional cube $\overline{\mathbb{I}}^m = [-1, 1]^m$ that we used in Chapter 14 is a topological manifold with boundary of dimension m. In this case $\partial\overline{\mathbb{I}}^m$ is homeomorphic to S^{m-1}.

(v) The punctured closed unit ball $D^m \setminus \{0\}$ is a topological manifold with boundary, since it is an open subset of the topological manifold with boundary D^m. This is an example where the manifold boundary is *not* the same as the topological boundary, since:

$$\partial(D^m \setminus \{0\}) = S^{m-1}, \qquad \partial^{\text{top}}(D^m \setminus \{0\}) = S^{m-1} \cup \{0\}.$$

(vi) More generally, any annulus which is half-open and half-closed, e.g.

$$A^{\leq R}_{>r} := \{p \in \mathbb{R}^m \mid r < \|p\| \leq R\},$$
$$A^{<R}_{\geq r} := \{p \in \mathbb{R}^m \mid r \leq \|p\| < R\},$$

is a topological manifold with boundary whose boundary consists of the boundary circle for which one has the non-strict equality:

$$\partial A^{\leq R}_{>r} = \{\|p\| = R\}, \qquad \partial A^{<R}_{\geq r} = \{\|p\| = r\},$$

meanwhile

$$\partial^{\text{top}} A^{\leq R}_{>r} = \partial^{\text{top}} A^{<R}_{\geq r} = \{\|p\| = r\} \cup \{\|p\| = R\}.$$

Proposition 24.24. *Let M be a topological manifold with boundary. Then* $\text{int}(M) \cap \partial M = \emptyset$. *Moreover* $\text{int}(M)$ *is a topological manifold without boundary of dimension m and ∂M is a topological manifold without boundary of dimension $m - 1$.*

Proof. The fact that $\text{int}(M) \cap \partial M = \emptyset$ uses Brouwer's Theorem as mentioned above (since $\mathbb{R}^m$ is not homeomorphic to $\mathbb{R}^{m-1}$). The rest is clear, since an open subset Q of $\mathbb{R}^m_{\lambda\geq c}$ that does not intersect $\mathbb{R}^m_{\lambda=c}$ is also open in $\mathbb{R}^m$, and if Q is open in $\mathbb{R}^m_{\lambda\geq c}$ then $Q \cap \mathbb{R}^m_{\lambda=c}$ is open in $\mathbb{R}^m_{\lambda=c} \cong \mathbb{R}^{m-1}$. ∎

Corollary 24.25. *If M is a topological manifold with boundary and $U \subset M$ is an open set then U is a topological manifold with boundary, and $\partial U = U \cap \partial M$.*

We now define smooth manifolds with boundary. We begin by extending by the definition of a diffeomorphism between open subsets of half-spaces. We already know (Definition 7.14) how to define what it means for a map to be smooth whose domain is not open, so it remains to extend this to the case when the range is also not open.

Definition 24.26. Let $Q \subset \mathbb{R}^m_{\lambda\geq c}$ denote an open set and $f : Q \to \mathbb{R}^n_{\eta\geq d}$ a continuous map. We say that f is **smooth** if the composition $\iota \circ f : Q \to \mathbb{R}^n$ is smooth in the sense of Definition 7.14, where $\iota\colon \mathbb{R}^n_{\eta \geq d} \hookrightarrow \mathbb{R}^n$ is the inclusion. If both $f\colon Q \to f(Q)$ and $f^{-1}\colon f(Q) \to Q$ are homeomorphisms between open sets of half-spaces that are smooth in this sense, then we say that f is a **diffeomorphism**.

The next result is standard calculus; the proof is omitted.

PROPOSITION 24.27. *Here are some properties of smooth maps between open sets of half-spaces:*

(i) *Let O be an open subset of $\mathbb{R}^m$ with non-empty intersection with $\mathbb{R}^m_{\lambda \geq c}$. Suppose $f, g\colon O \to \mathbb{R}^n$ are smooth maps in the usual sense.*[5] *If $f = g$ on $O \cap \mathbb{R}^m_{\lambda \geq c}$ then $Df(p) = Dg(p)$ for all $p \in O \cap \mathbb{R}^m_{\lambda \geq c}$.*
(ii) *Let $O \subset \mathbb{R}^m$ be open and $f\colon O \to \mathbb{R}^m_{\eta \geq d}$ be smooth.*[6] *If $f(p) \in \mathbb{R}^m_{\eta = d}$ for all $p \in O$ then $Df(p)$ has image in $\mathcal{T}_p\left(\mathbb{R}^m_{\eta=d}\right) \cong \mathbb{R}^m_{\eta=0}$ for all $p \in O$.*
(iii) *Suppose $Q_1 \subset \mathbb{R}^m_{\lambda \geq a}$ and $Q_2 \subset \mathbb{R}^n_{\eta \geq d}$ are open sets, and suppose $f\colon Q_1 \to Q_2$ is a diffeomorphism.*[7] *Assume $\partial Q_1 = Q_1 \cap \mathbb{R}^m_{\lambda = a}$ and $\partial Q_2 = Q_2 \cap \mathbb{R}^n_{\eta = d}$ are both non-empty. Then f induces diffeomorphisms $\partial Q_1 \to \partial Q_2$ and* $\operatorname{int}(Q_1) \to \operatorname{int}(Q_2)$ *in the sense of Definition* 1.8, *where we think of ∂Q_1 and ∂Q_2 as open subsets of $\mathbb{R}^{m-1}$ and $\mathbb{R}^{n-1}$ respectively.*

We then have:

DEFINITION 24.28. Let M be a topological manifold with boundary. A **smooth atlas** on M is a collection $\mathcal{X} = \{x_a\colon U_a \to Q_a \mid a \in A\}$, where $\{U_a \mid a \in A\}$ is an open cover of M, each Q_a is an open subset of some m-dimensional half-space $\mathbb{R}^m_{\lambda_a \geq c_a}$ (the precise half-space may depend on a), and each $x_a\colon U_a \to Q_a$ is a homeomorphism such that the usual compatibility condition is satisfied: if $a, b \in A$ are such that $U_a \cap U_b \neq \emptyset$ then the composition

$$x_b \circ x_a^{-1}\colon x_a(U_a \cap U_b) \to x_b(U_a \cap U_b)$$

should be a diffeomorphism in the sense of Definition 24.26.

We call each such x_a a **half-space chart**. One then defines a smooth structure in exactly the same way as in Definition 1.11, and this gives us the definition of a smooth manifold with boundary.

DEFINITION 24.29. A **smooth manifold with boundary of dimension** m is a pair $(M, \mathcal{X})$ where M is a topological manifold with boundary of dimension m, and $\mathcal{X}$ is a smooth structure on M in the sense of Definition 24.28.

Just as with Proposition 24.24 we have:

PROPOSITION 24.30. *Let M be a smooth manifold with boundary. Then* $\operatorname{int}(M) \cap \partial M = \emptyset$. *Moreover* $\operatorname{int}(M)$ *naturally inherits the structure of a smooth manifold without boundary of dimension m, and ∂M naturally inherits the structure of a smooth manifold without boundary of dimension $m - 1$.*

---

[5] I.e. in the sense of Definition 1.8.

[6] I.e. in the sense of Definition 24.26.

[7] I.e. in the sense of Definition 24.26.

Proof. This follows from part (iii) of Proposition 24.27. ■

EXAMPLE 24.31. All the examples from Example 24.23 are naturally smooth manifolds with boundary, except for the unit cube $\bar{\mathbb{I}}^m$, which is *not* a smooth manifold with boundary when $m \geq 2$. (See Problem Sheet K.)

Although the definition of a smooth atlas does not require all the half-space charts to take values in the same half-space, it is often convenient to assume they do.

DEFINITION 24.32. A **standard half-space chart** is a half-space chart $x \colon U \to Q$ with the property that Q is an open subset of our preferred standard half-space $\mathbb{R}^m_-$. A **standard smooth atlas** on a smooth manifold with boundary M is a smooth atlas as in Definition 24.28 all of whose charts are standard half-space charts.

It is easy to see that we may always assume this:

LEMMA 24.33. *Every smooth manifold with boundary admits a standard smooth atlas.*

REMARK 24.34. You might therefore ask what the point was in the more general definition. This is two-fold: firstly it is convenient when proving certain standard spaces are topological (resp. smooth) manifolds with boundary to be allowed more flexibility. Secondly, the distinction between good smooth atlases and normal smooth atlases is meaningful in dimension $m = 1$ when one in addition insists on orientability, as we will see in Proposition 24.40 below.

Many of the concepts we have covered so far in this course make sense for manifolds with boundary, and we don't have the time (or energy) to fill in the details, so let us just briefly summarise some of the important points:

- Partitions of unity still make sense for smooth manifolds with boundary, and they always exist.
- If M is a smooth manifold with boundary then T_pM is still an m-dimensional vector space for all $p \in M$. This is clear for $p \in \operatorname{int}(M)$, so suppose $p \in \partial M$. Let $x \colon U \to Q$ denote a half-space chart about p, where Q is an open set in some half-space $\mathbb{R}^m_{\lambda \geq c}$ and $x(p)$ lies in the hyperplane $\mathbb{R}^m_{\lambda = c}$. As before, a function f defined near p on M is smooth at p if and only if $f \circ x^{-1}$ is smooth near $z := x(p)$. Now recall by definition a function is smooth if and only if it admits a smooth extension to some open neighbourhood of z in $\mathbb{R}^m$. If g and h are any two such extensions of $f \circ x^{-1}$ then by part (i) of Proposition 24.27 the derivatives of g and h coincide on $\mathbb{R}^m_{\lambda = c}$. It follows that a derivation on the space of germs of smooth functions at p can be defined in exactly the same way as before, and thus the arguments from Chapters 2 and 3 go through without change to show that the tangent space T_pM at p is again an m-dimensional vector space.
- On the other hand, the tangent space to ∂M at $p \in \partial M$ can be identified with an $(m-1)$-dimensional subspace of T_pM. Indeed, if we let $\iota \colon \partial M \hookrightarrow M$ denote the inclusion then with the notation as above, $x \circ \iota|_{U \cap \partial M}$ is a chart on ∂M and thus

$$D\iota(p)(T_p\partial M) = Dx(p)^{-1}\left(T_z\mathbb{R}^m_{p=c}\right) \tag{24.3}$$

(Exercise: Prove that the right-hand side of (24.3) does not depend on the choice of half-space chart x.)
We usually suppress the $D\iota(p)$ map and thus think of $T_p\partial M$ as an actual subspace of T_pM.

- If N is a smooth manifold (with or without boundary) and $M \subset N$ is a subset endowed with a topology and a smooth structure making it into a smooth manifold with boundary such that the inclusion $M \hookrightarrow N$ is an embedding then M is said to be an **embedded submanifold with boundary**. Immersed submanifolds with boundary are defined similarly. If M is a smooth manifold with boundary then ∂M is an embedded submanifold of M—this follows immediately from the definition.[8]
- Both the Whitney Embedding Theorem 7.1 and the Whitney Approximation Theorem 7.12 still work for manifolds with boundary.
- A vector field X on a smooth manifold with boundary M is said to be **tangent to** ∂M if $X(p) \in T_p\partial M$ for each $p \in \partial M$. For vector fields that are tangent to M, Theorem 9.10 still works.
- The notion of a fibre bundle still makes sense if the base space is allowed to have boundary. In particular, vector bundles over manifolds with boundary are defined entirely analogously. Things go wrong however if the fibre is allowed to have boundary.[9]
- Tensors and differential forms are defined in exactly the same way.

We will however go through one aspect in detail, since this will be important in our treatment of the global Stokes' Theorem in Chapter 27. Suppose M is a manifold with boundary and $\pi\colon E \to M$ is a vector bundle over M. An orientation $\mathfrak{o}$ of E is, as before, determined by a non-vanishing section $\mu \in \Gamma(\det E^*)$. Given such a section μ, we can restrict it to obtain a section $\mu|_{\partial M}$ of the bundle $\det E|_{\partial M} \to \partial M$, where $E|_{\partial M} = \pi^{-1}(\partial M)$. This is a subbundle of E since ∂M is an embedded submanifold of M, cf. Definition 17.18.

For the special case $E = TM$, this gives us an orientation of the bundle $TM|_{\partial M} \to \partial M$. This however is *not* the same thing as an orientation of ∂M as a manifold—this would be an orientation of the bundle $T\partial M \to \partial M$.

Definition 24.35. Let M be a smooth manifold with boundary of dimension m, and let $p \in \partial M$. A tangent vector $\xi \in T_pM$ is said to be **outward pointing** if for some half-space chart $x\colon U \to Q$ about p, with $Q \subset \mathbb{R}^m_{\lambda \geq c}$ an open set and $z := x(p) \in \mathbb{R}^m_{\lambda = c}$, one has

$$\lambda\big(\mathcal{J}_z^{-1}(Dx(p)\xi)\big) < 0.$$

To unwrap this: $Dx(p)$ is a linear map $T_pM \to T_z\mathbb{R}^m_{\lambda\geq c} = T_z\mathbb{R}^m$. Applying the dash-to-dot map $\mathcal{J}_z^{-1}$ we obtain a vector $\mathcal{J}_z^{-1}(Dx(p)\xi) \in \mathbb{R}^m$, which λ can then eat to produce a real number. It follows from part (iii) of Proposition 24.27 that the property of being outward pointing is independent of the choice of half-space chart x.

[8] Exercise: Investigate how the Implicit Function Theorem 6.10 behaves with respect to manifolds with boundary. What is the correct notion of a slice chart in this setting?

[9] Exercise: Why?

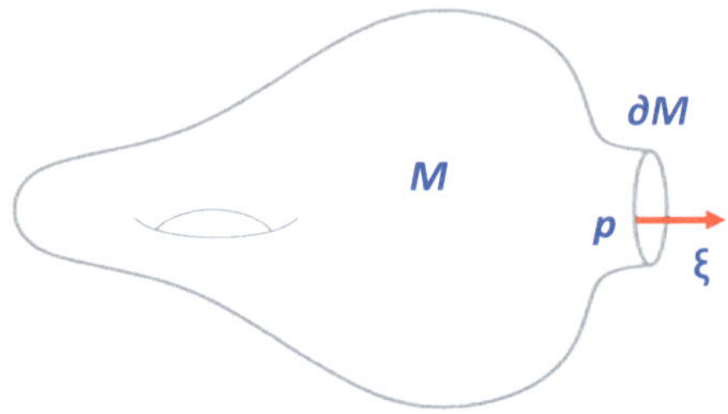

Figure 24.1: An outward pointing vector ξ

The definition is rather clearer if we take a standard half-space chart. Then the condition that $\xi \in T_p M$ is outward pointing is simply that

$$dx_p^1(\xi) > 0,$$

as illustrated in Figure 24.1.

The sign change is because $\mathbb{R}^m_- = \mathbb{R}^m_{-u^1 \geq 0}$.

Similarly an **inward pointing** vector is one for which $dx_p^1(\xi) < 0$.

The notion of outward and inward pointing vectors allow us to decompose $T_p M$ as:

$$T_p M = \{\text{outward pointing}\} \cup \{\text{inward pointing}\} \cup T_p \partial M,$$

since in such a chart x, one has

$$T_p \partial M = \left\{ \xi \in T_p M \mid dx_p^1(\xi) = 0 \right\}$$

by (24.3). Similarly a section X of $TM|_{\partial M}$ is said to be **outward pointing** if $X(p)$ is outward pointing for every p.

Example 24.36. Let M be a manifold with boundary and let $p \in \partial M$. Let (U, x) denote a standard half-space chart about p. Then $\frac{\partial}{\partial x^1}$ is an outward pointing section of $TM|_{\partial M}$ over $U \cap \partial M$.

In fact, via a standard partition of unity argument, one can produce outward pointing sections defined on the entire boundary.

Lemma 24.37. *Let M be a smooth manifold with boundary. Then there exists a section $X \in \Gamma(TM|_{\partial M})$ of the bundle $TM|_{\partial M} \to \partial M$ which is outward pointing at every $p \in \partial M$.*

Proof. We may assume M has a standard smooth atlas $\{x_a \colon U_a \to Q_a \mid a \in A\}$. Let $\{\kappa_a \mid a \in A\}$ denote a partition of unity subordinate to $\{U_a \mid a \in A\}$. Set

$$X := \sum_a \kappa_a \frac{\partial}{\partial x_a^1}.$$

This is outward pointing by Example 24.36. ■

Let us now use this to define the induced orientation.

Definition 24.38. Let $(M, \mathfrak{o})$ be an oriented smooth manifold with boundary. Let $\mu \in \Omega^m(M)$ be a volume form representing $\mathfrak{o}$. Let X be an outward pointing section. Then we can view $i_X \mu$ as an element of $\Omega^{m-1}(\partial M)$. Since X is outward pointing, $i_X \mu$ is nowhere vanishing on ∂M. Thus $i_X \mu$ determines an orientation of ∂M, which we rather suggestively write as $\partial \mathfrak{o}$. We call $\partial \mathfrak{o}$ the **induced orientation**.

On Problem Sheet J you are asked to show that the induced orientation is well-defined, i.e. independent of the choice of X and μ. Note that if $(X_1, \ldots, X_n)$ is a positively oriented frame[10] for $TM|_U$ such that $U \cap \partial M$ is non-empty and such that $X_1|_{U \cap \partial M}$ is outward pointing. Then $(X_2, \ldots, X_n)$ is a positively oriented local frame for $TM|_{\partial M}$ over $U \cap \partial M$ with respect to the induced orientation

Remark 24.39. In the case $m = 1$, the boundary ∂M is a discrete set of points. We only defined orientations for vector spaces of positive dimension, but this can still be made sense of if we simply think of a boundary point p being positively oriented if the function $\mu(X)$ is positive at p and negatively oriented otherwise.

Here is an extension of Corollary 24.16 for manifolds with boundary. This is where it is important to make the distinction between a standard atlas and a normal one.

Proposition 24.40. *Let M be an oriented smooth manifold with boundary of dimension m. Then M admits a positively oriented smooth atlas* (*that is, one such that* (24.2) *holds*). *If $m \geq 2$ then M admits a positively oriented standard smooth atlas.*

Proof. If x is a chart with local coordinates $(x^1, \ldots, x^m)$ that is not positively oriented then we replace it with a new chart $(x^1, -x^2, \ldots, x^m)$. If $m \geq 2$ and x is a standard half-space chart then the new chart is also a standard half-space chart. This goes wrong for $m = 1$ however, since in this case it changes a $\mathbb{R}^1_-$ half-space chart into a $\mathbb{R}^1_+$ half-space chart. ■

For the rest of this book, **all** manifolds (topological or smooth) should be assumed **not** to have boundary, unless it is explicitly said that they do.

[10] I.e. $\mu(X_1, \ldots, X_n) > 0$ over U.

Problem Sheet J

A (♣) indicates the problem is non-examinable. This does not necessarily imply it is difficult.

PROBLEM J.1 Let E and F be two vector bundles over M and let $\{U_a \mid a \in A\}$ be an arbitrary open covering of M. Suppose we are given a collection

$$\{\zeta_a \colon \Gamma(U_a, E) \to \Gamma(U_a, F) \mid a \in A\}$$

of local operators such that

$$\zeta_a^{U_a \cap U_b} \equiv \zeta_b^{U_a \cap U_b}, \qquad \text{if } U_a \cap U_b \neq \emptyset.$$

Prove there exists a unique local operator $\zeta \colon \Gamma(E) \to \Gamma(F)$ such that

$$\zeta^{U_a} = \zeta_a, \qquad \forall\, a \in A.$$

PROBLEM J.2 Let $\varphi \colon M \to N$ denote a smooth map. Let $A \in \mathcal{T}^{0,k}(N)$. Using the Tensor Criterion Theorem 21.5, regard A as a $C^\infty(N)$-multilinear function

$$\underbrace{\mathfrak{X}(N) \times \cdots \times \mathfrak{X}(N)}_{k \text{ copies}} \to C^\infty(N).$$

and similarly regard $\varphi^*(A)$ as a $C^\infty(M)$-multilinear function

$$\underbrace{\mathfrak{X}(M) \times \cdots \times \mathfrak{X}(M)}_{k \text{ copies}} \to C^\infty(M).$$

Suppose $X_i \in \mathfrak{X}(M)$ is φ-related to $Y_i \in \mathfrak{X}(N)$ for $i = 1, \ldots, s$. Prove that

$$(\varphi^* A)(X_1, \ldots, X_k) = A(Y_1, \ldots, Y_k) \circ \varphi$$

as functions $M \to N$.

PROBLEM J.3 Let V be a vector space and suppose $\omega \in \bigwedge^h V^*$ and $\theta \in \bigwedge^k V^*$. Let $v_i \in V$ for $i = 1, \ldots, h+k$. Prove that if we identify ω with an element of $\operatorname{Alt}_h(V)$, θ with an element of $\operatorname{Alt}_k(V)$, and $\omega \wedge \theta$ with an element of $\operatorname{Alt}_{h+k}(V)$, one has:

$$(\omega \wedge \theta)(v_1, \ldots, v_{h+k}) = \frac{1}{h!k!} \sum_{\varrho \in \mathfrak{S}_{h+k}} \operatorname{sgn}(\varrho)\omega\big(v_{\varrho(1)}, \ldots, v_{\varrho(h)}\big)\, \theta\big(v_{\varrho(h+1)}, \ldots, v_{\varrho(h+k)}\big)$$

or equivalently

$$(\omega \wedge \theta)(v_1, \ldots, v_{h+k}) = \sum_{\varrho \in \mathrm{Shuffle}(h,k)} \mathrm{sgn}(\varrho)\omega\big(v_{\varrho(1)}, \ldots, v_{\varrho(h)}\big)\,\theta\big(v_{\varrho(h+1)}, \ldots, v_{\varrho(h+k)}\big).$$

Problem J.4 Let $\omega \in \bigwedge^h V^*$ and $\theta \in \bigwedge^k V^*$. Prove that

$$i_v(\omega \wedge \theta) = i_v\omega \wedge \theta + (-1)^h \omega \wedge i_v\theta.$$

Problem J.5 Let $(M, \mathfrak{o})$ be an oriented smooth manifold with boundary. Let $\mu \in \Omega^m(M)$ be a volume form representing $\mathfrak{o}$. Let X be an outward pointing section.

(i) Prove that $i_X\mu$ restricts to define a volume form on ∂M.
(ii) Let $\partial\mathfrak{o}$ denote the orientation of ∂M determined by $i_X\mu$. Prove that (as the notation suggests) $\partial\mathfrak{o}$ only depends on $\mathfrak{o}$, and not on the particular choice of μ and X.

Problem J.6

(i) Prove that S^m is orientable.
(ii) Prove that any Lie group is orientable.
(iii) Prove that $\mathbb{R}P^m$ is orientable if and only if n is odd. *Hint*: Consider the antipodal map $x \mapsto -x$ on S^m.

Problem J.7 Let

$$\mathbb{R}^m_- := \mathbb{R}^m_{u^1 \leq 0}, \qquad \mathbb{H}^m := \mathbb{R}^m_{u^m \geq 0}.$$

We can identify both $\partial\mathbb{R}^m_-$ and $\partial\mathbb{H}^m$ with $\mathbb{R}^{m-1}$. Endow both $\mathbb{R}^m_-$ and $\mathbb{H}^m$ with the standard orientation that they inherit from $\mathbb{R}^m$. Show that the induced orientation on $\partial\mathbb{R}^m_-$ is equal to the standard orientation on $\mathbb{R}^{m-1}$ for all m, but that the induced orientation on $\partial\mathbb{H}^m$ agrees with the standard orientation of $\mathbb{R}^{m-1}$ only when m is even.[11]

Problem J.8

(i) Let V be a vector space of dimension n. A **symplectic form** on V is an element $\omega \in \mathrm{Alt}_2(V) \cong \bigwedge^2 V^*$ which is **non-degenerate** in the sense that $i_v(\omega) \equiv 0$ if and only if $v = 0$. Prove that if a symplectic form exists then $n = 2k$ is necessarily an even number.
(ii) A **symplectic manifold** is a smooth manifold M equipped with a *closed* differential 2-form ω such that ω_p is a symplectic form on T_pM for every $p \in M$. Prove that any symplectic manifold is orientable.
(iii) Let M be a smooth manifold. Define a 1-form $\Theta \in \Omega^1(T^*M)$ on the cotangent bundle via the formula:

$$\Theta_{p,\lambda}(\zeta) = \lambda\big(D\pi(p,\lambda)\zeta\big),$$

for $p \in M$, $\lambda \in T^*_pM$, and $\zeta \in T_{(p,\lambda)}T^*M$. Prove that $\omega := d\Theta$ is a symplectic form on T^*M. Thus every cotangent bundle is a symplectic manifold.

[11] *Remark*: This is the main reason we take our "standard" half-space to be $\mathbb{R}^m_-$, not $\mathbb{H}^m$, cf. Remark 24.20.

(♣)Problem J.9 Let M and N be smooth manifolds. Prove that if M has boundary and N does not, then $M \times N$ is a smooth manifold with boundary. Prove that if both M and N have non-empty boundary, then $M \times N$ is not a smooth manifold with boundary,

Bonus Problem(s) for Sheet J

These problem(s) are hard, and are included for enthusiasts only. Solutions will not be provided.

(♣)Problem J.10 After making appropriate modifications, reprove all results in the book for manifolds with boundary.

Chapter 25
Smooth Singular Cubes

In this chapter we begin our discussion of *integration* on manifolds. We introduce the notion of a *singular cube* in a manifold and explain how a differential form can be integrated over such a cube. We then mirror the construction of singular homology[1] and show how to create the (smooth) cubical chain complex of M. We will use this formalism next chapter en route to proving the manifold version of Stokes' Theorem.

DEFINITION 25.1. Let us abbreviate by C^k the closed cube[2] $[0, 1]^k$, thought of as sitting inside $\mathbb{R}^k$. For $k = 0$, $C^0 = \{0\}$ is a point. Let M be a smooth manifold with boundary.[3] A **smooth singular k-cube** (often shortened to: a "singular k-cube" or just a "k-cube") in M is a smooth map $c\colon C^k \to M$. Thus a singular 0-cube is simply a point $c(0)$ in M.

A map $c\colon C^k \to M$ is **smooth** if for each point $x \in C^k$ there is a chart $\varphi\colon V \to \mathbb{R}^m_+$ of M with $c(x) \in V$ and an open neighbourhood U of x in $\mathbb{R}^k$ such that there is a smooth extension $\tilde{c}\colon U \to \mathbb{R}^m$ of $\varphi \circ c|_{U \cap C^k} : U \cap C^k \to \mathbb{R}^m$.[4]

For $k = 0$, we declare that any map $c\colon C^0 \to M$ is smooth.

REMARK 25.2. The adjective "singular" is meant to draw your attention to the fact that c need not be injective nor an immersion. Indeed, a constant map is a valid smooth singular k-cube! Moreover if $k > \dim M$ then no singular k-cube can be an immersion.

The next example is more important than you would first guess.

---

[1] Don't worry if you have not seen homology before. We will not use any algebraic topology in this book. Exception: the bonus section to Chapter 27.

[2] In the past we used $\mathbb{I}^k$ for the open cube $(-1, 1)^k$; here it is more convenient to work on $[0, 1]$ itself, so we choose different notation.

[3] Recall that any smooth manifold can be viewed as a smooth manifold with boundary whose boundary is empty; so any definition or theorem relating to smooth manifolds with boundary also apply to smooth manifolds.

[4] In the case where M is a smooth manifold (without boundary) this definition of smooth is equivalent to that given in Definition 7.14.

W. J. Merry, *Lectures on Differential Geometry I*,
https://doi.org/10.1007/978-3-032-03733-6_25

EXAMPLE 25.3. We let $\mathsf{i}_k : C^k \hookrightarrow \mathbb{R}^k$ denote the inclusion and call i_k the **standard smooth singular k-cube**.

REMARK 25.4. We will often regard i_k as taking values in C^k, not $\mathbb{R}^k$, and hence identify i_k with the identity map $C^k \to C^k$. Strictly speaking however in this case i_k is not a smooth singular cube, since the range space C^k is not a smooth manifold with boundary (for $k > 1$). For the most part we shall ignore this pedantry.

REMARK 25.5. You might hope that the machinery of smooth manifolds with boundary would allow us to forego the tedious extension business. This works fine for $k = 1$: $C^1 = [0, 1]$ is a smooth manifold with boundary, and a singular 1-cube is simply a smooth map $C^1 \to M$ between manifolds. However for $k = 2$ it goes wrong: C^2 is not a manifold with boundary (see Problem J.9). It is however a **smooth manifold with corners**, which is defined as you might expect: instead of a half space atlas one works with a *quarter space atlas*. If M is a smooth manifold with corners then its boundary ∂M is a smooth manifold with boundary,[5] and the boundary of the boundary is then a smooth manifold without boundary.

Sadly however this *still* isn't enough, since for $k \geq 3$ the space C^k is not a smooth manifold with corners either. The correct notion is that of a **stratified manifold**, which, roughly speaking is a manifold which is allowed to have "boundary-like" pieces of arbitrarily high codimension. A manifold with boundary is a stratified manifold with only codimension one strata, and a manifold with corners is a stratified manifold with only codimension one and two strata. In general, C^k is a stratified manifold with k different strata.

That said, developing the entire theory of stratified manifolds just to dispense with the need to talk about extensions is somewhat inefficient, even by our standards, so we will stick with the extensions. This will therefore be a minor annoyance throughout the chapter.

DEFINITION 25.6. Let $k > 0$, and let $\omega \in \Omega^k(C^k)$ denote a k-form on C^k i.e. a smooth[6] map $C^k \to \Lambda^k(\mathbb{R}^k)^*$ (this amounts to a section of the trivial bundle $\Lambda^k T^*\mathbb{R}^k \to \mathbb{R}^k$ restricted to C^k). We can write $\omega = f\, du^1 \wedge \cdots \wedge du^k$ for some $f \in C^\infty(C^k)$. We define the **integral of** ω to be the Riemann integral of f:

$$\int_{C^k} \omega := \int_{C^k} f$$

We emphasise the right-hand side is the *normal Riemann integral* of the function f.

We now transfer this to manifolds. We first note that if M is a smooth manifold with boundary, c is a smooth singular k-cube in M where $k > 0$, and $\omega \in \Omega^k(M)$ is a k-form, then as in Definition 22.20 we can form the pullback form $c^*\omega$. To see that this is a k-form on C^k note that by definition ω has locally in the chart a smooth extension to an open subset, and this local extension can be pulled back by some $\tilde{c}$ as in Definition 25.1 to yield the local extension of $c^*\omega$.

[5] For C^2, one has ∂C^2 equal to the union of the edges, and $\partial(\partial C^2)$ equal to the four vertices.

[6] As in Definition 7.14.

DEFINITION 25.7. Let M be a smooth manifold with boundary. Let $k > 0$, let c be a smooth singular k-cube in M and let $\omega \in \Omega^k(M)$ denote a k-form. We define the **integral of ω over c** to be the real number

$$\int_c \omega := \int_{C^k} c^*\omega.$$

It would be sufficient if ω was only defined on some neighbourhood of the image of c for this to make sense. For $k = 0$, the definition is simpler; in this case ω is just a function f, and

$$\int_c f := f(c(0)).$$

REMARK 25.8. If we write $c^*\omega = f\, du^1 \wedge \cdots \wedge du^k$ then the function f is given explicitly by

$$f = c^*\omega\left(\frac{\partial}{\partial u^1}, \dots, \frac{\partial}{\partial u^k}\right).$$

This expression is well-defined since $c^*\omega$ is a section of a bundle which has an extension to a smooth section on an open neighbourhood.

Thus an alternative formula is

$$\int_c \omega = \int_{C^k} c^*\omega\left(\frac{\partial}{\partial u^1}, \dots, \frac{\partial}{\partial u^k}\right),$$

where again the right-hand side is just a normal Riemann integral.

REMARK 25.9. Since for any singular k-cube c we have $c = c \circ \mathrm{i}_k$, we have using (22.6) that

$$\begin{aligned}\int_c \omega &= \int_{c\circ \mathrm{i}_k} \omega \\ &= \int_{C^k} (c \circ \mathrm{i}_k)^*(\omega) \\ &= \int_{C^k} \mathrm{i}_k^*(c^*\omega) \\ &= \int_{\mathrm{i}_k} c^*\omega.\end{aligned}$$

DEFINITION 25.10. Let M be a manifold with boundary.

A singular k-cube $c\colon C^k \to M$ is said to be **degenerate** if there exists $1 \le i \le k$ such that c does not depend on u^i.

A singular k-cube $c\colon C^k \to M$ is said to be **non-degenerate** if it is not degenerate.

Thus a 0-cube is never degenerate, and a 1-cube is degenerate if and only if it is a constant map.

On Problem Sheet K you will prove.

LEMMA 25.11. *If M is a manifold with boundary and $c\colon C^k \to M$ is a degenerate singular k-cube then $\int_c \omega = 0$ for any $\omega \in \Omega^k(M)$.*

The next result is also on Problem Sheet K. You should think of it as a version of the usual change of variables formula from multivariable calculus:

Proposition 25.12 (Change of Variables). *Let M be a manifold with boundary and let $c\colon C^k \to M$ be a smooth singular k-cube in M and let $\varphi\colon C^k \to C^k$ be an orientation preserving diffeomorphism.*[7] *Let $\tilde{c} := c \circ \varphi$. Then*

$$\int_c \omega = \int_{\tilde{c}} \omega.$$

Let us now consider formal sums of singular cubes.

Definition 25.13. Let M be a manifold with boundary and let $\mathsf{Q}_k(M)$ denote the (infinite-dimensional) free vector space generated by the collection of all the smooth singular k-cubes in M. Thus an element of $\mathsf{Q}_k(M)$ is a formal finite sum $\mathsf{q} = \sum_i a_i\, c_i$ where $a_i \in \mathbb{R}$ and the c_i are smooth singular k-cubes.[8] We call an element $\mathsf{q} \in \mathsf{Q}_k(M)$ a **smooth singular k-chain**, or (sometimes just a k-chain).

Example 25.14. Since a 0-cube in M is just a point in M, the space $\mathsf{Q}_0(M)$ can be thought as the infinite-dimensional vector space with basis the points of M. In particular, if $p, q \in M$ then the expression $p - q$ makes sense in $\mathsf{Q}_0(M)$, even though it does not in M.

Warning 25.15. The space $\mathsf{Q}_0(M)$ is a vector space with basis the points in M. Thus (by definition) there are no relations between different elements. This can be confusing, particularly if the manifold M happens to be a submanifold of Euclidean space where it *does* make sense to add points together. As an example, let us take $M = \mathbb{R}^m$. Let $v, w \in \mathbb{R}^m$ be two vectors. Then in $\mathbb{R}^m$, we can add v and w together to get a new vector $v+w$. However in $\mathsf{Q}_0(\mathbb{R}^m)$, the three elements v, w and $v+w$ are *linearly independent* and thus it is *not* true that $v + w = (v + w)$! A similar issue occurs with scalar multiplication. If this confuses you, consider writing the addition and multiplication operations in $\mathsf{Q}_0(\mathbb{R}^m)$ with a different colour or underlining them.

Thus if $v, w \in \mathbb{R}^m$ and $a \in \mathbb{R}$ then

$$v \underline{+} w \neq v + w, \qquad \underline{a}v \neq \underline{1}(av).$$

Luckily most of the time this shouldn't be confusing, since typically on manifolds one cannot add points together, and thus the notation is unambiguous.

Definition 25.16. Let M be a manifold with boundary. We define the integral of a k-form over a k-chain in M by linearity: if $\mathsf{q} = \sum_i a_i\, c_i$ then

$$\int_{\mathsf{q}} \omega := \sum_i a_i \int_{c_i} \omega.$$

We will also need the concept of the *boundary* of a chain.

---

[7] As usual, think of this as meaning that φ is the restriction to C^k of an orientation preserving diffeomorphism of some neighbourhood.

[8] To explain the notation: "cube" sounds like it begins with a "Q".

DEFINITION 25.17. Let $1 \leq i \leq k$. Define the *i*th **front face map**

$$\mathsf{f}_{i,k} : C^{k-1} \to C^k, \qquad (u^1, \ldots, u^{k-1}) \mapsto (u^1, \ldots, u^{i-1}, 0, u^i, \ldots u^{k-1})$$

and the *i*th **back face map**

$$\mathsf{b}_{i,k} : C^{k-1} \to C^k, \qquad (u^1, \ldots, u^{k-1}) \mapsto (u^1, \ldots, u^{i-1}, 1, u^i, \ldots u^{k-1})$$

When k is clear from the context, we write simply f_i and b_i instead.

DEFINITION 25.18. Let M be a manifold with boundary. Fix $1 \leq i \leq k$ and let $c : C^k \to M$ denote a singular k-cube. The ith **front face** of c is the smooth singular $(k-1)$-cube $c \circ \mathsf{f}_i : C^{k-1} \to M$. Similarly the *i*th **back face** of c is the smooth singular $(k-1)$-cube $c \circ \mathsf{b}_i : C^{k-1} \to M$. We abbreviate

$$\mathsf{f}_i c := c \circ \mathsf{f}_i, \qquad \mathsf{b}_i c := c \circ \mathsf{b}_i.$$

Writing out the definitions yields the following statement.

LEMMA 25.19. *Let M be a manifold with boundary and let $c : C^k \to M$ be a smooth singular k-cube. Let $1 \leq i < j \leq k$. Then:*

$$\begin{aligned} \mathsf{f}_i(\mathsf{f}_j c) &= \mathsf{f}_{j-1}(\mathsf{f}_i c), \\ \mathsf{b}_i(\mathsf{b}_j c) &= \mathsf{b}_{j-1}(\mathsf{b}_i c), \\ \mathsf{f}_i(\mathsf{b}_j c) &= \mathsf{b}_{j-1}(\mathsf{f}_i c), \\ \mathsf{b}_i(\mathsf{f}_j c) &= \mathsf{f}_{j-1}(\mathsf{b}_i c). \end{aligned} \tag{25.1}$$

Thus by definition

$$\mathsf{f}_{i,k} = \mathsf{f}_i \mathsf{i}_k, \qquad \mathsf{b}_{i,k} = \mathsf{b}_i \mathsf{i}_k.$$

DEFINITION 25.20. Let M be a manifold with boundary and let $c : C^k \to M$ be a smooth singular k-cube for $k > 0$. We define the **boundary** of c, written ∂c, to be the element of $\mathsf{Q}_{k-1}(M)$ given by

$$\partial c := \sum_{i=1}^{k} (-1)^i (\mathsf{f}_i c - \mathsf{b}_i c).$$

This works for $k = 0$ too if we define $\mathsf{Q}_{-1}(M) := 0$.

We define the boundary of a 0-cube to be the zero of the vector space $\mathsf{Q}_{-1}(M)$. Note that if a cube c is degenerate then ∂c is a linear combination of degenerate cubes. We then extend ∂ to arbitrary k-chains by linearity. Thus we may think of ∂ as a linear map $\mathsf{Q}_k(M) \to \mathsf{Q}_{k-1}(M)$ for all $k \geq 1$.

REMARK 25.21. Thus this is yet another meaning of the symbol ∂. This one is not as confusing as the topological boundary and the manifold boundary (cf. Remark 24.17), since c is a function, and thus there can be no ambiguity about what is meant.

EXAMPLE 25.22. Let $c : [0, 1] \to M$ be a 1-cube. Then $\mathsf{f}_1 c$ is the 0-cube $c(0)$ and $\mathsf{b}_1 c$ is the 0-cube $c(1)$. Thus $\partial c = c(1) - c(0)$. Remember the subtraction is taking place in $\mathsf{Q}_0(M)$, not M itself!

PROPOSITION 25.23. *The boundary operator squares to zero*: $\partial^2 = 0$.

Proof. Since ∂ is linear, it suffices to show that $\partial(\partial c) = 0$ for any cube c. For this we compute:

$$\begin{aligned}\partial(\partial c) &= \partial\left(\sum_{i=1}^{k}(-1)^i(\mathsf{f}_i c - \mathsf{b}_i c)\right)\\ &= \sum_{i=1}^{k}\sum_{j=1}^{k-1}(-1)^{i+j}\Big(\mathsf{f}_j(\mathsf{f}_i c) - \mathsf{f}_j(\mathsf{b}_i c) - \mathsf{b}_j(\mathsf{f}_i c) + \mathsf{b}_j(\mathsf{b}_i c)\Big)\end{aligned}$$

Using the face relations (25.1), we see that the first and fourth terms cancel in pairs, and the second and third terms cancel each other. ■

Proposition 25.23 allows us to play a similar game to the definition (Definition 23.2) of the de Rham cohomology groups.[9]

DEFINITION 25.24. Let M be a manifold with boundary and let $Q_k(M)$ be the vector space generated by the collection of all the smooth singular k-cubes in M as in Definition 25.13. Let $D_k(M)$ denote the subspace of $Q_k(M)$ generated by the degenerate cubes, and let

$$\bar{Q}_k(M) := Q_k(M)\Big/D_k(M)$$

As we have previously noted, if a k-cube c is degenerate then ∂c is a linear combination of degenerate cubes; thus ∂ maps $D_k(M)$ to $D_{k-1}(M)$ and we can also regard ∂ as a linear map $\bar{Q}_k(M) \to \bar{Q}_{k-1}(M)$.[10]

Now let $\bar{\mathsf{q}}$ be the equivalence class in $\bar{Q}_k(M)$ of a smooth singular k-chain, q, in M. We say $\bar{\mathsf{q}}$ is **closed** if $\partial\bar{\mathsf{q}} = 0$ and that $\bar{\mathsf{q}}$ is **exact** if $\bar{\mathsf{q}} = \partial\bar{\mathsf{p}}$ for some $\bar{\mathsf{p}}$ in $\bar{Q}_{k+1}(M)$. Then every exact chain is also closed (as $\partial^2 = 0$), and thus we can form the quotient vector space:

$$H_k^{\text{cube}}(M;\mathbb{R}) := \frac{\{\bar{\mathsf{q}} \in \bar{Q}_k(M) : \bar{\mathsf{q}} \text{ is closed}\}}{\{\bar{\mathsf{q}} \in \bar{Q}_k(M) : \bar{\mathsf{q}} \text{ is exact}\}}.$$

We call $H_k^{\text{cube}}(M;\mathbb{R})$ the **kth cubical singular homology group** of M.[11]

If $\bar{\mathsf{q}}$ is closed, we denote by $[\bar{\mathsf{q}}]$ its equivalence class in $H_k^{\text{cube}}(M;\mathbb{R})$. The reason for modding out degenerate cubes will be explained in the bonus section to Chapter 27. Unlike the groups $H^k_{\text{dR}}(M)$, which are certainly zero for $k > \dim M$, a priori the groups

[9] Using the terminology from algebraic topology, Proposition 25.23 tells us that $(Q_\bullet(M), \partial)$ is a **chain complex**. We can thus take its homology.

[10] Strictly speaking, the boundary of a degenerate 1-cube is not degenerate, since there are no degenerate 0-cubes; but since the boundary is zero it is a member of the zero vector space $D_0(M)$.

[11] The word "homology" is used because ∂ decreases the degree: $\partial\colon \mathsf{Q}_k \to \mathsf{Q}_{k-1}$. In the de Rham construction the exterior differential d increases the degree: $d\colon \Omega^k \to \Omega^{k+1}$. Thus we use the word "cohomology".

$H_k^{\mathrm{cube}}(M; \mathbb{R})$ could be non-zero for arbitrarily high k. However this is not the case. In fact, as we will explain in the bonus section to Chapter 27, there is an isomorphism

$$H_k^{\mathrm{cube}}(M; \mathbb{R}) \cong H_{\mathrm{dR}}^{n-k}(M), \qquad \forall k \geq 0. \tag{25.2}$$

The isomorphism (25.2) is one way of stating **Poincaré Duality**. This is one of the cornerstones of modern algebraic topology.

Chapter 26
Stokes' Theorem

In this chapter we state and prove Stokes' Theorem, which is one of the cornerstones of modern differential geometry. (Fun Fact: The only connection Stokes has with "Stokes' Theorem" is that he decided to set it as a problem in an exam.) In fact, we will prove two versions of Stokes' Theorem: a local version (Theorem 26.2) using the language of smooth singular cubes from the last chapter, and a global version (Theorem 26.16) which concerns integration over the entire manifold.

As a warm up, we translate the Fundamental Theorem of Calculus into the language of chains. This can be thought of as the one-dimensional version of the Local Stokes' Theorem.

Theorem 26.1 (The Fundamental Theorem of Calculus for Singular 1-Chains) *Let $f \colon \mathbb{R} \to \mathbb{R}$ be differentiable and let q be a singular 1-chain in $\mathbb{R}$. Then*

$$\int_{\mathsf{q}} df = \int_{\partial \mathsf{q}} f$$

Proof. By linearity of the integral, we may assume that q is a single singular 1-cube c. Then we compute

$$\begin{aligned} \int_c df &= \int_0^1 c^*(df)\left(\frac{\partial}{\partial t}\right) dt \\ &= \int_0^1 (f \circ c)'(t) dt \\ &= f(c(1)) - f(c(0)) \\ &= \int_{c(1)} f - \int_{c(0)} f \\ &= \int_{\partial c} f, \end{aligned}$$

W. J. Merry, *Lectures on Differential Geometry I*,
https://doi.org/10.1007/978-3-032-03733-6_26

where the third equality used the usual Fundamental Theorem of Calculus that you have known since kindergarten. ■

Here is the general Local Stokes' Theorem.

Theorem 26.2 (The Local Stokes' Theorem) *Let M be a smooth manifold with boundary. Let $q \in \mathsf{Q}_k(M)$ and $\omega \in \Omega^{k-1}(M)$. Then*

$$\int_{\mathsf{q}} d\omega = \int_{\partial \mathsf{q}} \omega.$$

Note Theorem 26.1 is the special case $M = \mathbb{R}$ and $k = 1$. The proof is not particularly difficult, but it is somewhat fiddly, and hence it is deferred to the bonus section below.

Definition 26.3. Let M be a smooth manifold. Then for $0 \leq k \leq m$ we can think of integration as defining a bilinear map

$$\int : \mathsf{Q}_k(M) \times \Omega^k(M) \to \mathbb{R}, \qquad (\mathsf{q}, \omega) \mapsto \int_{\mathsf{q}} \omega.$$

Corollary 26.4. *The bilinear form $\int$ is also well-defined on the (co)homology level, that is, the map*

$$\int : H_k^{\text{cube}}(M; \mathbb{R}) \times H_{\text{dR}}^k(M) \to \mathbb{R}, \qquad ([q], [\omega]) \mapsto \int_{\mathsf{q}} \omega$$

is well-defined.

Proof. We already know that $\int_{\mathsf{q}} \omega$ vanishes whenever q is degenerate (Lemma 25.11). Thus we need only show that if q is a closed non-degenerate k-chain and ω is a closed k-form, then for any non-degenerate $(k+1)$-chain p and any $(k-1)$-form θ, one has

$$\int_{\mathsf{q}+\partial \mathsf{p}} (\omega + d\theta) = \int_{\mathsf{q}} \omega.$$

To see this we compute

$$\begin{aligned} \int_{\mathsf{q}+\partial \mathsf{p}} (\omega + d\theta) &= \int_{\mathsf{q}} \omega + \int_{\partial \mathsf{p}} \omega + \int_{\mathsf{q}} d\theta + \int_{\partial \mathsf{p}} d\theta \\ &= \int_{\mathsf{q}} \omega + \int_{\mathsf{p}} d\omega + \int_{\partial \mathsf{q}} \theta + \int_{\mathsf{p}} (d^2\theta) \\ &= \int_{\mathsf{q}} \omega + 0, \end{aligned}$$

where the second equality used the local Stokes' Theorem and the last used the assumption that q and ω are closed. ■

We move onto the global version of Stokes' Theorem, starting with an explanation of how to make sense of the expression $\int_{M,\mathfrak{o}} \omega$. Unlike the local version, this will only work when M is oriented and ω is a compactly supported differential form of top degree $m = \dim M$.

DEFINITION 26.5. Let $(M, \mathfrak{o})$ be an oriented manifold. A singular cube $c\colon C^m \to M$ is said to be **orientation preserving** if there exists a neighbourhood U of C^m in $\mathbb{R}^m$ (or $\mathbb{R}^m_+$ when M has boundary) and an orientation preserving[1] embedding $\tilde{c}\colon U \to M$ such that $\tilde{c}|_{C^m} = c$. Note that $\tilde{c}$ is thus a diffeomorphism onto its image.

REMARK 26.6. If $(M, \mathfrak{o})$ is an oriented manifold, we can always find an open cover of M such that each open set U in that cover is contained in the interior of the image of an orientation preserving singular cube $c\colon C^m \to M$. Indeed suppose $x\colon V \to O$ is a positively oriented chart (cf. Definition 24.15) such that $C^m \subset O$. Set $c := x^{-1}|_{C^m}$ and let $U := x^{-1}\big((\frac{1}{3}, \frac{2}{3})^m\big)$. The collection of all open sets U as (V, x) ranges over all such charts forms an open cover with the desired properties.

DEFINITION 26.7. Let M be a smooth manifold, and let $\omega \in \Omega(M)$. The **support** of ω is defined in the same way as normal:

$$\operatorname{supp}(\omega) := \overline{\{p \in M \mid \omega_p \neq 0\}}.$$

A differential form ω is said to have **compact support** if $\operatorname{supp}(\omega)$ is compact. We denote by $\Omega_c(M) \subset \Omega(M)$ the subset of differential forms with compact support, and $\Omega^k_c(M)$ the differential k-forms with compact support. Note that by definition of the exterior differential, we have

$$\operatorname{supp}(d\omega) \subseteq \operatorname{supp}(\omega).$$

The next lemma is the reason why global integration works on oriented manifolds.

LEMMA 26.8. *Let $(M, \mathfrak{o})$ be an oriented manifold and $\omega \in \Omega^m(M)$. Let $c_1, c_2\colon C^m \to M$ be two orientation preserving singular cubes, and assume that*

$$\operatorname{supp}(\omega) \subset \operatorname{int}(\operatorname{im} c_1) \cap \operatorname{int}(\operatorname{im} c_2)$$

Then

$$\int_{c_1} \omega = \int_{c_2} \omega.$$

Proof. This almost follows from Proposition 25.12, since $c_2^{-1} \circ c_1$ is almost an orientation preserving diffeomorphism of C^m. The only issue is that $c_2^{-1} \circ c_1$ may not be defined on all of C^m. However, since $\operatorname{supp}(\omega) \subset \operatorname{int}(\operatorname{im} c_1) \cap \operatorname{int}(\operatorname{im} c_2)$, the proof of Proposition 25.12 goes through without change to give

$$\int_{c_2} \omega = \int_{c_2 \circ c_2^{-1} \circ c_1} \omega = \int_{c_1} \omega$$

as required. ∎

[1] Here we always implicitly assume $C^m \subset \mathbb{R}^m$ carries its standard orientation.

Thus we can unambiguously make the following definition.

DEFINITION 26.9. Let $(M, \mathfrak{o})$ be an oriented manifold and $\omega \in \Omega^m(M)$. Assume that ω has support in the interior of the image of some orientation preserving singular cube c. We define

$$\int_{M,\mathfrak{o}} \omega := \int_c \omega.$$

The following lemma is immediate from the definitions.

LEMMA 26.10. *If c is an orientation reversing singular cube and ω has support in* $\operatorname{im} c$ *then*

$$\int_{M,\mathfrak{o}} \omega = -\int_c \omega.$$

Thus

$$\int_{M,\mathfrak{o}} \omega = -\int_{M,-\mathfrak{o}} \omega.$$

We can use a partition of unity to extend this to an arbitrary $\omega \in \Omega_c^m(M)$. We first give the definition, and then prove it is well defined.

DEFINITION 26.11. Let $(M, \mathfrak{o})$ be an oriented manifold and let $\omega \in \Omega_c^m(M)$. Let $\{U_a \mid a \in A\}$ be an open cover with the property that each U_a is contained in the interior of the image of some orientation preserving singular cube (cf. Remark 26.6). Let $\{\kappa_a \mid a \in A\}$ be a partition of unity subordinate to this cover. We define

$$\int_{M,\mathfrak{o}} \omega := \sum_{a\in A} \int_{M,\mathfrak{o}} \kappa_a\, \omega. \tag{26.1}$$

Note this is a finite sum since ω has compact support and $\operatorname{supp}(\kappa_a)$ is locally finite.

LEMMA 26.12. *The sum* (26.1) *is well defined*: *if $\{V_b \mid b \in B\}$ is another open cover with the property that each V_b is contained in the interior of the image of some orientation preserving singular cube and $\{\nu_b \mid b \in B\}$ is a partition of unity subordinate to that cover then for any $\omega \in \Omega_c^m(M)$ one has*:

$$\sum_{a\in A} \int_{M,\mathfrak{o}} \kappa_a\, \omega = \sum_{b\in B} \int_{M,\mathfrak{o}} \nu_b\, \omega$$

Proof. Since

$$\sum_{a\in A} \kappa_a(p) = \sum_{b\in B} \nu_b(p) = 1, \qquad \forall\, p \in M,$$

we have using linearity of the integral that

$$\begin{aligned}\sum_{a\in A}\int_{M,\mathfrak{o}}\kappa_a\,\omega &= \sum_{a\in A}\int_{M,\mathfrak{o}}\left(\sum_{b\in B}\nu_b\right)\kappa_a\,\omega\\ &= \sum_{a\in A}\sum_{b\in B}\int_{M,\mathfrak{o}}\nu_b\,\kappa_a\,\omega\\ &= \sum_{b\in B}\int_{M,\mathfrak{o}}\left(\sum_{a\in A}\kappa_a\right)\nu_b\,\omega\\ &= \sum_{b\in B}\int_{M,\mathfrak{o}}\nu_b\,\omega\end{aligned}$$

where the rearrangement of the sums is justified as everything is a finite sum. ■

We now know how to integrate a top-dimensional differential form with compact support on an oriented manifold. Let us extend this to oriented manifolds with boundary. For this we use the following trick:

DEFINITION 26.13. Let $(M, \mathfrak{o})$ be an oriented smooth manifold with boundary. An orientation preserving singular cube $c\colon C^m \to M$ is said to be **adapted to the boundary** if either $\operatorname{im} c \subset \operatorname{int}(M)$ or

$$\partial M \cap \operatorname{im} c = \operatorname{im}(\mathfrak{f}_1 c),$$

where as usual $\mathfrak{f}_1 c\colon C^{m-1} \to M$ is the first front face.

LEMMA 26.14. *Let $(M, \mathfrak{o})$ be an oriented smooth manifold with boundary, and let $c\colon C^m \to M$ be a singular cube which is adapted to the boundary such that* $\operatorname{im} c \cap \partial M \neq \emptyset$. *Then $\mathfrak{f}_1 c$ is an orientation reversing singular cube for $(\partial M, \partial\mathfrak{o})$.*[2]

This is *not* a typo—we really do want $\mathfrak{f}_1 c$ to reverse orientation! As we shall see, the minus sign will eventually cancel, since the coefficient of $\mathfrak{f}_1 c$ in ∂c is -1.

Proof. We need only check that $\mathfrak{f}_1 c$ is orientation reversing with respect to the induced orientation. Let (u^i) denote the standard coordinates on C^m. Since c is a diffeomorphism, we may take c^{-1} as a $\mathbb{R}^m_+$ half-space chart on M (remember c is really defined on an open neighbourhood of C^m). Let $x^i := u^i \circ c^{-1}$ denote the local coordinates of this chart. Since c is orientation preserving, $\left\{\frac{\partial}{\partial x^1}, \dots, \frac{\partial}{\partial x^m}\right\}$ is a positive oriented local frame of TM. Note that $\frac{\partial}{\partial x^1}$ is an *inward* pointing section,[3] and thus the frame $\left\{\frac{\partial}{\partial x^2}, \dots, \frac{\partial}{\partial x^m}\right\}$ is a negatively oriented frame for $T(\partial M)$. This shows that $\mathfrak{f}_1 c$ is orientation reversing. ■

To define integration on a oriented smooth manifold with boundary we need to extend some of our earlier definitions and results. We first note that in a similar way to Remark 26.6, if M is a smooth manifold with boundary then we can always find an

[2] Recall $\partial\mathfrak{o}$ denotes the induced orientation of ∂M.

[3] The reason why $\frac{\partial}{\partial x^1}$ is inward pointing (in contrast to Example 24.36) is that x is an $\mathbb{R}^m_+$ chart not a $\mathbb{R}^m_-$ chart.

open cover of M with the property that each open set is contained in the interior of the image of a orientation preserving singular cube which is adapted to the boundary.[4] We now note that Definition 26.7, Lemma 26.8, Definition 26.9 and Lemma 26.10[5] can all be extended to manifolds with boundaries.[6]

We can now extend the definition of integration to manifolds with boundary.

DEFINITION 26.15. Let M be an oriented smooth manifold with boundary, and let $\omega \in \Omega_c^m(M)$. Let $\{U_a \mid a \in A\}$ be an open cover with the property that each U_a is contained in the interior of the image of some orientation preserving singular cube which is adapted to the boundary. Let $\{\kappa_a \mid a \in A\}$ be a partition of unity subordinate to this cover. We define

$$\int_{M,\mathfrak{o}} \omega := \sum_{a \in A} \int_{M,\mathfrak{o}} \kappa_a \, \omega.$$

The same proof as Lemma 26.12 shows this is well-defined.

We are finally ready to state and prove the Global Stokes' Theorem.

Theorem 26.16 (The Global Stokes' Theorem) *Let $(M, \mathfrak{o})$ be an oriented smooth manifold with boundary, and let $\omega \in \Omega_c^{m-1}(M)$. Then*

$$\int_{M,\mathfrak{o}} d\omega = \int_{\partial M, \partial\mathfrak{o}} \omega.$$

Proof. We prove the result in two steps. Let $\{U_a \mid a \in A\}$ be an open cover with the property that each U_a is contained in the interior of the image of some orientation preserving singular cube which is adapted to the boundary.

1. First assume that $\operatorname{supp}(\omega)$ is contained in one of the sets U_a, which itself is contained in the interior of the image of some orientation preserving singular cube c which is adapted to the boundary. If $\operatorname{im} c \cap \partial M = \emptyset$ then the result is immediate from the Local Stokes' Theorem 26.2, since

$$\int_{M,\mathfrak{o}} d\omega = \int_c d\omega = \int_{\partial c} \omega = 0,$$

since $\operatorname{supp}(\omega)$ does not intersect the image of ∂c. But also clearly $\int_{\partial M, \partial\mathfrak{o}} \omega = 0$ since $\operatorname{supp}(\omega)$ does not intersect ∂M.

Now assume that $\operatorname{im} c \cap \partial M \neq \emptyset$. Then we have:

[4] Note that this does not mean that every point in M is in the image of an interior point of some cube C. Instead it means that each point is in the image of $\operatorname{int}(C) \cup \operatorname{int}(\mathsf{f}_1(C))$ for some cube C which is adapted to the boundary.

[5] Lemma 26.10 is true for manifolds with boundaries, but we don't actually use this in what follows.

[6] Theorem 26.2 is already applicable to manifolds with boundaries.

$$
\begin{aligned}
\int_{M,\mathfrak{o}} d\omega &= \int_{c} d\omega \\
&= \int_{\partial c} \omega \quad \text{by Theorem 26.2} \\
&= \sum_{i=1}^{m} (-1)^i \left(\int_{\mathsf{f}_i c} \omega - \int_{\mathsf{b}_i c} \omega \right) \\
&= - \int_{\mathsf{f}_1 c} \omega,
\end{aligned}
$$

since $\operatorname{supp}(\omega)$ misses all faces apart from $\mathsf{f}_1 c$ as c is adapted to the boundary. Thus by Lemmas 26.10 and 26.14 we have

$$
\begin{aligned}
\int_{M,\mathfrak{o}} d\omega &= - \int_{\mathsf{f}_1 c} \omega \\
&= (-1)^2 \int_{\partial M, \partial\mathfrak{o}} \omega \\
&= \int_{\partial M, \partial\mathfrak{o}} \omega.
\end{aligned}
$$

2. Now we prove the general case. Let $\{\kappa_a \mid a \in A\}$ be a partition of unity subordinate to the open cover $\{U_a \mid a \in A\}$. Then by definition,

$$
\begin{aligned}
\int_{\partial M, \partial\mathfrak{o}} \omega &= \sum_{a \in A} \int_{\partial M, \partial\mathfrak{o}} \kappa_a\, \omega \\
&= \sum_{a \in A} \int_{M} d(\kappa_a\, \omega) \\
&= \sum_{a \in A} \int_{M} d\kappa_a \wedge \omega + \kappa_a\, d\omega \\
&= \int_{M,\mathfrak{o}} d\omega + \sum_{a \in A} \int_{M} d\kappa_a \wedge \omega \\
&= \int_{M,\mathfrak{o}} d\omega + \int_{M,\mathfrak{o}} d\left(\sum_{a \in A} \kappa_a \right) \wedge \omega \\
&= \int_{M,\mathfrak{o}} d\omega + 0,
\end{aligned}
$$

where the second equation used Step 1 and the fourth and last equalities used the fact that $\sum_{a \in A} \kappa_a \equiv 1$, and the interchange of summation and integral is always justified as these are always finite sums as $\operatorname{supp}(\omega)$ is compact. This completes the proof. ■

Bonus Material for Chapter 26

In this bonus section we prove the Local Stokes' Theorem, which for convenience we restate here.

Theorem 26.17 (The Local Stokes' Theorem) *Let M be a smooth manifold with boundary. Let $q \in \mathsf{Q}_k(M)$ and $\omega \in \Omega^{k-1}(M)$. Then*

$$\int_{\mathsf{q}} d\omega = \int_{\partial \mathsf{q}} \omega.$$

Proof. We prove the result in three steps.

1. Let us first consider the case where $M = \mathbb{R}^k$ and $c = \mathsf{i}_k$ is the standard cube from Example 25.3.[7] This actually represents most of the work. By linearity we may assume that ω is of the form

$$\omega = g\, du^1 \wedge \cdots \wedge \widehat{du^j} \wedge \cdots \wedge du^k,$$

where the caret indicates we skip the term du^j. In this first step, we come up with a nice formula for the right-hand side $\int_{\partial \mathsf{i}_k} \omega$.

We have by definition that

$$\int_{\partial \mathsf{i}_k} \omega = \sum_{i=1}^{k} (-1)^i \left(\int_{\mathsf{f}_i} \omega - \int_{\mathsf{b}_i} \omega \right). \tag{26.2}$$

We now claim that:

$$\int_{\mathsf{f}_i} \omega = \begin{cases} \int_{C^{k-1}} g \circ \mathsf{f}_i, & i = j, \\ 0, & i \neq j. \end{cases} \tag{26.3}$$

The proof of (26.3) is a little fiddly. One way to argue this is as follows: from Remark 25.8 and Corollary 19.26 we have that

$$\int_{\mathsf{f}_i} \omega = \int_{C^{k-1}} (g \circ \mathsf{f}_i) \cdot \det A,$$

where $A = (A^h_l)$ is the $(k-1) \times (k-1)$ matrix whose entries are given by

$$A^h_l = D_l(u^h \circ \mathsf{f}_i), \qquad \text{for } 1 \leq l \leq k-1 \text{ and } 1 \leq h \leq k,\ h \neq j.$$

The function $u^h \circ \mathsf{f}_i$ is given by

[7] The Local Stokes' Theorem in the case $M = \mathbb{R}^k$ and $c = \mathsf{i}_k$ is in fact just a fancy way of expressing Fubini's Theorem (and the Fundamental Theorem of Calculus).

$$u^h \circ \mathsf{f}_i \mathsf{i}_k(u^1, \ldots, u^{k-1}) = u^h(u^1, \ldots, u^{i-1}, 0, u^i, \ldots u^{k-1}).$$

Thus if $i = j$ then $A^h_l = \delta^h_l$ and thus $\det A = 1$. However if $i \neq j$ then the entire ith row (A^i_l) is zero (since $u^i \circ \mathsf{f}_i$ is a constant function), and thus $\det A = 0$. This proves (26.3). Together with a similar formula for the back face, we see that (26.2) reduces to

$$\int_{\partial \mathsf{i}_k} \omega = (-1)^j \int_{C^{k-1}} g \circ \mathsf{f}_j - g \circ \mathsf{b}_j. \tag{26.4}$$

By Fubini's Theorem and the Fundamental Theorem of Calculus:

$$\begin{aligned}
&\int_{C^{k-1}} g \circ \mathsf{b}_j - g \circ \mathsf{f}_j \\
&= \int_0^1 \cdots \int_0^1 \Big(g(u^1, \ldots, 1, \ldots, u^k) - g(u^1, \ldots, 0, \ldots, u^k) \Big) du^1 \cdots \widehat{du^j} \cdots du^k \\
&= \int_{C^k} \frac{\partial g}{\partial u^j}.
\end{aligned}$$

Thus we conclude from (26.4) that

$$\int_{\partial \mathsf{i}_k} \omega = (-1)^{j-1} \int_{C^k} \frac{\partial g}{\partial u^j}.$$

2. We now consider the term $\int_{\mathsf{i}_k} d\omega$. Since $dg = \frac{\partial g}{\partial u^j} du^j$ we have (writing the summation signs for clarity)

$$\begin{aligned}
\int_{\mathsf{i}_k} d\omega &= \int_{\mathsf{i}_k} dg \wedge du^1 \wedge \cdots \wedge \widehat{du^j} \wedge \cdots \wedge du^k \\
&= \int_{\mathsf{i}_k} \sum_{i=1}^k \frac{\partial g}{\partial u^i} du^i \wedge du^1 \wedge \cdots \wedge \widehat{du^j} \wedge \cdots \wedge du^k) \\
&= (-1)^{j-1} \int_{\mathsf{i}_k} \frac{\partial g}{\partial u^j} du^1 \wedge \cdots \wedge du^k \\
&= (-1)^{j-1} \int_{C^k} \frac{\partial g}{\partial u^j}
\end{aligned}$$

This completes the proof for $M = \mathbb{R}^k$ and $c = \mathsf{i}_k$.

3. In the general case, again by linearity we may assume $\mathsf{q} = c$ is a singular k-cube. Then

$$
\begin{aligned}
\int_{\partial c} \omega &= \sum_{i=1}^{k} (-1)^i \left(\int_{\mathsf{f}_i c} \omega - \int_{\mathsf{b}_i c} \omega \right) \\
&= \sum_{i=1}^{k} (-1)^i \left(\int_{c \circ \mathsf{f}_i} \omega - \int_{c \circ \mathsf{b}_i} \omega \right) \\
&= \sum_{i=1}^{k} (-1)^i \left(\int_{C^{k-1}} (c \circ \mathsf{f}_i)^* \omega - \int_{C^{k-1}} (c \circ \mathsf{b}_i)^* \omega \right) \\
&= \sum_{i=1}^{k} (-1)^i \left(\int_{C^{k-1}} \mathsf{f}_i^* (c^* \omega) - \int_{C^{k-1}} \mathsf{b}_i^* (c^* \omega) \right) \\
&= \sum_{i=1}^{k} (-1)^i \left(\int_{\mathsf{f}_i} c^* \omega - \int_{\mathsf{b}_i} c^* \omega \right) \\
&= \int_{\partial \mathsf{i}_k} c^* \omega \\
&= \int_{\mathsf{i}_k} d(c^* \omega)
\end{aligned}
$$

by the previous step. But since $c^* \circ d = d \circ c^*$ by Lemma 23.4, we have

$$
\int_{\mathsf{i}_k} d(c^* \omega) = \int_{\mathsf{i}_k} c^* (d\omega) = \int_c d\omega,
$$

where we used Remark 25.9 at the end. This completes the proof. ■

Note that the general case follows from the special case simply by unravelling the formalism. Thus the Local Stokes' Theorem really is just Fubini's Theorem in disguise.

Chapter 27
The Poincaré Lemma and the de Rham Theorem

In this final chapter of Volume I we return to the de Rham cohomology of a smooth manifold. We show that de Rham cohomology is a *homotopy invariant*, and use this to prove the Poincaré Lemma, which states that any closed form is locally exact. In the bonus section we prove the de Rham Theorem, which can be thought of as a massive generalisation of the homotopy invariance property.

We begin with the following application of the global version of Stokes' Theorem.

PROPOSITION 27.1. *Let* $(M, \mathfrak{o})$ *be a oriented manifold,*[1] *and let* $\omega \in \Omega_c^{m-1}(M)$. *Then* $\int_{M,\mathfrak{o}} d\omega = 0$.

Proof. M is also a smooth manifold with boundary whose boundary is the empty set. The claim is now thus immediate from the Global Stokes' Theorem 26.16. ∎

COROLLARY 27.2. *Let* M *be an oriented connected compact smooth manifold. Then* $H^m_{\mathrm{dR}}(M) \neq 0$.

Proof. Let μ be a volume form. Then for any orientation preserving cube, we have $\int_c \mu > 0$. Thus $\int_{M,\mathfrak{o}} \mu > 0$. The form μ is closed (as $d\mu = 0$ for dimension reasons). If μ was exact then $\int_{M,\mathfrak{o}} \mu = 0$ by Proposition 27.1. Thus μ is a closed non-exact form, and hence defines a non-zero element in $H^m_{\mathrm{dR}}(M)$. ∎

REMARK 27.3. In the bonus section below we will improve Corollary 27.2 and show that for an oriented connected compact smooth manifold M the class $[\mu]$ actually generates $H^m_{\mathrm{dR}}(M)$, and hence $H^m_{\mathrm{dR}}(M) \cong \mathbb{R}$. Together with Lemma 23.3, this implies that

$$H^0_{\mathrm{dR}}(M) \cong H^m_{\mathrm{dR}}(M).$$

[1] Recall by the convention at the end of Chapter 24, all manifolds are assumed not to have boundary unless explicitly stated to the contrary.

W. J. Merry, *Lectures on Differential Geometry I*,
https://doi.org/10.1007/978-3-032-03733-6_27

This is not an accident. It is a special case of **Poincaré Duality** (together with Universal Coefficients Theorem for Cohomology). We briefly discuss this at the end of the bonus section.

The main step in proving the homotopy invariance property of de Rham cohomology is the following innocuous looking statement.

PROPOSITION 27.4. *Let M be a smooth manifold. Define for $t \in [0, 1]$ a smooth map*[2]

$$e_t \colon M \to M \times [0, 1], \qquad e_t(x) := (x, t).$$

Then the induced maps on de Rham cohomology

$$e_0^*,\ e_1^* \colon H^k_{\mathrm{dR}}(M \times [0, 1]) \to H^k_{\mathrm{dR}}(M)$$

coincide for all $0 \le k \le m + 1$.

Proof. We prove the result in two steps.

1. Fix $1 \le k \le m + 1$. In this step we construct a map

$$\mathsf{h} \colon \Omega^k(M \times [0, 1]) \to \Omega^{k-1}(M)$$

such that for every differential k-form $\omega \in \Omega^k(M \times [0, 1])$, one has

$$\mathsf{h}(d\omega) + d(\mathsf{h}\omega) = e_1^*\omega - e_0^*\omega$$

as elements of $\Omega^k(M)$. Let Y denote the vector field on $M \times [0, 1]$ whose value at (p, t) is

$$Y(p, t) = \left(0, \frac{\partial}{\partial t}\Big|_t\right).$$

(Compare (9.3) in Chapter 9—we are using slightly different notation to simplify the formulae to come). The desired map h is then given by

$$\mathsf{h}(\omega) := \int_0^1 e_t^* i_Y \omega \, dt, \tag{27.1}$$

for $\omega \in \Omega^k(M \times [0, 1])$. That is, for any $p \in M$, $\mathsf{h}(\omega)_p \in \bigwedge^{k-1} T_p^* M$ given by

$$\mathsf{h}(\omega)_p = \int_0^1 e_t^*(i_Y \omega)_{(p,t)} \, dt,$$

where the integrand is thought of as a function of t on the vector space $\bigwedge^{k-1} T_p^* M$. Note that this is just a normal Riemann integral on a vector space, not an integral on a manifold! By choosing local coordinates, we see that the integral defines a smooth $(k - 1)$-form on M. To compute $d(\mathsf{h}\omega)$ it suffices to work locally. In local coordinates (x^i) we can express $\mathsf{h}(\omega)$ as a sum

---

[2] Here we view $M \times [0, 1]$ as a smooth manifold with boundary.

$$\mathsf{h}(\omega) = \sum_I \left(\int_0^1 f_I(p,t)\, dt \right) dx^I, \tag{27.2}$$

using the notation introduced in the proof of Theorem 23.1. Applying d to such a term and differentiating under the integral sign[3] gives

$$\begin{aligned} d(\mathsf{h}\omega) &= \sum_I \sum_j \frac{\partial}{\partial x^j} \left(\int_0^1 f_I(p,t)\, dt \right) dx^j \wedge dx^I \\ &= \sum_I \left(\int_0^1 \sum_j \frac{\partial f_I}{\partial x^j}(p,t)\, dt \right) dx^j \wedge dx^I. \end{aligned}$$

By comparing (27.1) and (27.2), we see from this last expression that

$$d(\mathsf{h}\omega) = \int_0^1 d\big(e_t^* i_Y \omega\big) dt.$$

Thus using Lemma 23.4 and Cartan's Magic Formula (Theorem 23.12) we see that

$$\begin{aligned} \mathsf{h}(d\omega) + d(\mathsf{h}\omega) &= \int_0^1 \Big(e_t^* i_Y d\omega + d\big(e_t^* i_Y \omega\big) \Big) dt \\ &= \int_0^1 e_t^* \big(i_Y d\omega + d(i_Y \omega) \big) dt \\ &= \int_0^1 e_t^* \mathcal{L}_Y \omega\, dt. \end{aligned}$$

Let Φ_t denote the flow of Y. Then $\Phi_t(x,s) = (x, t+s)$, and thus $e_t = \Phi_t \circ e_0$ and we can compute the Lie derivative as

$$\begin{aligned} e_t^* \mathcal{L}_Y \omega &= e_0^* \Phi_t^* (\mathcal{L}_Y \omega) \\ &= e_0^* \left(\frac{d}{dt} \Phi_t^* \omega \right) \\ &= \frac{d}{dt} e_0^* \Phi_t^* \omega \\ &= \frac{d}{dt} e_t^* \omega. \end{aligned}$$

where the second equation used Problem K.5. Thus by the (normal) Fundamental Theorem of Calculus we obtain

$$\begin{aligned} \mathsf{h}(d\omega) + d(\mathsf{h}\omega) &= \int_0^1 \frac{d}{dt} e_t^*(\omega)\, dt \\ &= e_1^* \omega - e_0^* \omega. \end{aligned}$$

---

[3] This is sometimes referred to as the Leibniz integral rule [9]. In this expression we include the summation signs for clarity.

2. We now complete the proof. Suppose $[\omega] \in H^k_{\mathrm{dR}}(M \times [0,1])$. Fix a representative ω of $[\omega]$. Then

$$e_1^*[\omega] - e_0^*[\omega] = [\mathsf{h}(d\omega)] + [d(\mathsf{h}\omega)].$$

Since ω is closed, the first term is zero. The second term is exact, and hence zero in $H^k_{\mathrm{dR}}(M)$. This completes the proof. ■

Why is Proposition 27.4 useful? The next statement makes this abundantly clear.

Theorem 27.5 *Let M and N be two smooth manifolds and suppose φ and ψ are two homotopic smooth maps from M to N. Then the induced maps φ^* and ψ^* on the de Rham cohomology groups are the same.*

Proof. To say that φ and ψ are homotopic means there is a *continuous* map $H: M \times [0,1] \to N$ such that $H(\cdot,0) = \varphi$ and $H(\cdot,1) = \psi$. In fact, by the Homotopy Whitney Approximation Theorem 7.16, we may assume H is a *smooth* map. Then the induced maps on de Rham cohomology satisfy

$$\begin{aligned}\varphi^* &= (H \circ e_0)^* \\ &= e_0^* \circ H^* \\ &= e_1^* \circ H^* \\ &= (H \circ e_1)^* \\ &= \psi^*,\end{aligned}$$

where the third equality used Proposition 27.4. This completes the proof. ■

Two topological spaces X and Y are said to be **homotopy equivalent** if there exist continuous maps $f: X \to Y$ and $g: Y \to X$ such that both $f \circ g$ and $g \circ f$ are homotopic to the respective identity maps.

Corollary 27.6 (Homotopy invariance of de Rham cohomology). *Let M and N be smooth manifolds that are homotopy equivalent. Then M and N have isomorphic de Rham cohomology groups.*

Proof. Let $f: M \to N$ and $g: N \to M$ be continuous maps such that $f \circ g$ and $g \circ f$ are homotopic to the identity maps. By the Whitney Approximation Theorem 7.12 we can find smooth maps $\varphi: M \to N$ and $\psi: N \to M$ such that φ is homotopic to f and ψ is homotopic to g. Then $\varphi \circ \psi$ and $\psi \circ \varphi$ are homotopic to the identity maps. By Theorem 27.5, $(\varphi \circ \psi)^*$ and $(\psi \circ \varphi)^*$ coincide with the maps induced by the identity. Since id^* is clearly the identity, we see that φ^* is an inverse to ψ^*. The claim follows. ■

Remark 27.7. A particular case of Corollary 27.6 tells us that the de Rham cohomology cannot see the smooth structure on a topological manifold M. This is surprising, since the data used to define it (the differential forms) very much depend on the choice of smooth structure.

A topological space is **contractible** if it is homotopy equivalent to a point.

COROLLARY 27.8. *Let M be contractible. Then $H^k_{\mathrm{dR}}(M) = 0$ for all $k \geq 1$.*

Proof. It is clear this is true for M equal to a point. Now apply Corollary 27.6. ∎

REMARK 27.9. This shows that de Rham cohomology cannot distinguish Euclidean spaces: $H^k_{\mathrm{dR}}(\mathbb{R}^m)$ is independent of m (since all Euclidean spaces are contractible). Thus a lot of information is lost when passing to de Rham cohomology.

Perhaps the most useful corollary of this is the following statement, which is classically called the **Poincaré Lemma**.

COROLLARY 27.10 (THE POINCARÉ LEMMA). *Let M be a smooth manifold and let $\omega \in \Omega^k(M)$ be a closed differential form of degree $k > 0$. For any point $p \in M$ there exists a neighbourhood U of p such that $\omega|_U$ is an exact form in $\Omega^k(U)$.*

Proof. Every point in a manifold admits a contractible neighbourhood. ∎

Bonus Material for Chapter 27

In this bonus section we prove that the de Rham cohomology agrees with singular cohomology—this is usually referred to as the **de Rham Theorem**. There are many different ways to prove this result. Perhaps the neatest is via *sheaf cohomology*, but this is a little bit too far afield.

> This bonus section assumes you are familiar with singular (co)homology and some basic homological algebra.

CONVENTION. All our homology and cohomology groups should be understood to have coefficients in $\mathbb{R}$ for the remainder of this section. We will not comment on this further.

Let X be a topological space. You are hopefully familiar with the singular chain complex of X. This is normally defined by looking at singular *simplices* (i.e. continuous maps $\Delta^k \to X$, where Δ^k is the kth standard simplex). However one can equally well carry out the construction using singular *cubes* instead. The resulting algebraic invariant is the same (more on this later). Let us recall the definitions in the continuous category.

DEFINITION 27.11. Let X be a topological space. A **singular k-cube** in X is a *continuous* map $c\colon C^k \to X$. We let $Q_k(X)$ denote the (infinite-dimensional) vector space with basis all the singular k-cubes in X.

REMARK 27.12. Note in the continuous category there is no need to require c to extend to a map on an open neighbourhood.

DEFINITION 27.13. A singular k-cube $c\colon C^k \to X$ is said to be **degenerate** if there exists $1 \leq i \leq k$ such that c does not depend on u^i. We let $D_k(X)$ denote the subspace of $Q_k(X)$ generated by the degenerate cubes, and we let

$$\bar{Q}_k(X) := Q_k(X) \big/ D_k(X)$$

denote the quotient space.

Thus for instance

$$c\colon C^3 \to \mathbb{R}, \qquad c(u^1, u^2, u^3) := u^1 + u^3$$

is a degenerate singular 3-cube in $\mathbb{R}$. The front and back faces of a cube are defined in the same way as Definition 25.18, and this allows us to define the boundary operator as before:

DEFINITION 27.14. Let $c\colon C^k \to X$ be a singular k-cube for $k > 0$. We define the **boundary** of c, written ∂c, to be the element of $Q_{k-1}(X)$ given by

$$\partial c := \sum_{i=1}^{k} (-1)^i (\mathsf{f}_i c - \mathsf{b}_i c).$$

We define the boundary of a 0-cube to be the real number 0. We then extend ∂ to arbitrary k-chains by linearity. As before, this works for $k = -1$ too if we define $Q_{-1}(X) := 0$. Thus we may think of ∂ as a linear map $Q_k(X) \to Q_{k-1}(X)$ for all $k \geq 0$.

Since ∂ maps $D_k(X)$ to $D_{k-1}(X)$, we can also regard it as a linear map

$$\partial\colon \bar{Q}_k(X) \to \bar{Q}_{k-1}(X), \qquad k \geq 0.$$

The same argument from Proposition 25.23 gives:

PROPOSITION 27.15. *The boundary operator squares to zero*: $\partial^2 = 0$. *Thus* $(\bar{Q}_\bullet(X), \partial)$ *is a chain complex of vector spaces.*

DEFINITION 27.16. The **cubical singular homology groups** $H_k^{\text{cube}}(X; \mathbb{R})$ are defined to be the homology of this chain complex.

REMARK 27.17. Why bother with quotienting out by the degenerate cubes? After all, $(Q_\bullet(X), \partial)$ is also a chain complex, so we could just take its homology instead. To see why this quotienting out by the degenerate cubes is superior, consider the case where X is a one point space $\{*\}$. It is easy to see that $H_k^{\text{cube}}(\{*\}) = 0$ for $k > 0$ and $H_0^{\text{cube}}(\{*\}) = \mathbb{R}$, as one would hope—this is a necessary requirement in order for $H_\bullet^{\text{cube}}$ to be "a homology theory" in the sense of Eilenberg-Steenrod. However if one does not quotient out by the degenerate cubes, this ceases to be the case.[4]

[4] Exercise: Why?

All the properties of the singular chain complex (built with singular simplices) continue to hold without change (homotopy invariance, long exact sequence, excision. Mayer-Vietoris, etc.).

REMARK 27.18. In fact, sometimes the proof gets *easier* for singular cubes. For instance, one of the key steps in establishing excision for singular simplices is the concept of **barycentric subdivision**, which allows one to chop up a singular simplex into smaller ones whose diameter can be made arbitrarily small. This is quite involved, and rather unpleasant. On the other hand, it is not very hard to work out how to chop up a singular cube into smaller ones!

REMARK 27.19. For us the main reason we preferred cubes over simplices in Chapter 25 is that it is much easier to define an integral over a cube (it is just a nested sequence of integrals $\int_0^1 \cdots \int_0^1$), whereas this is messier for simplices.

REMARK 27.20. At a much more advanced level, there are compelling reasons both to prefer using simplices and to prefer using cubes. This concerns **cubical sets** and **simplicial sets** in homotopy theory. However this all goes way beyond this book, so we won't discuss it.

It is not completely obvious why the resulting cubical singular homology groups agree with the normal singular homology groups. It can be proved directly using the *Acyclic Models Theorem*, or it can deduced from the uniqueness result for Eilenberg-Steenrod homology theories.

Let us now return to manifolds. If M is a smooth manifold then the vector spaces $Q_k(M)$ defined in this chapter do *not* coincide with the vector spaces $Q_k(M)$ defined in Chapter 25. This is because in Chapter 25 we insisted on *smooth* maps. Let us temporarily write $Q_k^\infty(M)$ for the smooth singular k-cubes, and $H_k^{\text{cube},\infty}(M)$ for the homology of the chain complex $(\bar{Q}_k^\infty(M), \partial)$. It is not obvious that the two groups coincide, but luckily they do:

Theorem 27.21 *Let M be a smooth manifold. Then the inclusion $Q_k^\infty(M) \to Q_k(M)$ induces an isomorphism*

$$H_k^{\text{cube}}(M;\mathbb{R}) \cong H_k^{\text{cube},\infty}(M;\mathbb{R}), \qquad \forall\, k \geq 0.$$

Proof (Sketch). The proof is a standard induction-type argument on the complexity of M. We proceed in six steps.

1. Suppose M is a single point. This is trivial.

2. Suppose M is an open contractible subspace of $\mathbb{R}^m$. This follows from Step 1 and the Whitney Approximation Theorem 7.12, which allows us to assume the contraction of M is smooth.

3. Suppose $M = U \cup V$, where U and V are open in M and the theorem is assumed to be true for U, V and $U \cap V$. We apply naturality of the Mayer-Vietoris sequence to see that the following diagram commutes, where we omit the coefficient group $\mathbb{R}$ and rotate the diagram so that it fits on the page:

$$\begin{array}{ccc}
H_k^{\text{cube}}(U \cap V) & \longrightarrow & H_k^{\text{cube},\infty}(U \cap V) \\
\downarrow & & \downarrow \\
H_k^{\text{cube}}(U) \oplus H_k^{\text{cube}}(V) & \longrightarrow & H_k^{\text{cube},\infty}(U) \oplus H_k^{\text{cube},\infty}(V) \\
\downarrow & & \downarrow \\
H_k^{\text{cube}}(M) & \longrightarrow & H_k^{\text{cube},\infty}(M) \\
\downarrow & & \downarrow \\
H_{k-1}^{\text{cube}}(U \cap V) & \longrightarrow & H_{k-1}^{\text{cube},\infty}(U \cap V)
\end{array}$$

The Five Lemma then completes the proof.

4. Now assume $M = \bigcup_i U_i$, where $U_i \subset U_{i+1}$ is an open set, and the theorem is true for each U_i. Then the theorem follows for M via an abstract argument using filtered colimits:

$$\begin{aligned}
H_k^{\text{cube}}(M;\mathbb{R}) &\cong \varinjlim H_k^{\text{cube}}(U_i;\mathbb{R}) \\
&= \varinjlim H_k^{\text{cube},\infty}(U_i;\mathbb{R}) \\
&\cong H_k^{\text{cube},\infty}(M;\mathbb{R}).
\end{aligned}$$

5. Now assume M is an arbitrary open subset of $\mathbb{R}^m$. Then we can write M as a countable union of convex open subsets. For any finite union, the theorem holds by applying Step 3 and induction, and then Step 4 gives the result for M itself.

6. The general case: since we can cover M by charts, it follows from Step 5 and Zorn's Lemma that there exists a maximal open subset $U \subset M$ for which the theorem is true. If $U \neq M$, then we an find a chart domain V such that V is not contained in U. Then by Step 3 and Step 5, the theorem is true for $U \cup V$. This contradicts maximality of U. ■

With this out of the way, we shall drop the ∞ from the notation and just write $\bar{Q}_k(M)$ for the groups defined in Chapter 25. Let us now recall how one constructs the *cohomology groups* from the homology groups.

DEFINITION 27.22. Let X be a topological space. Set

$$Q^k(X) := \operatorname{Hom}(\bar{Q}_k(X);\mathbb{R})$$

and define $d\colon Q^k(X) \to Q^{k+1}(X)$ by

$$d(\mathsf{a})(\mathsf{q}) := \mathsf{a}(\partial \mathsf{q}), \qquad \alpha \in Q^k(X),\ \mathsf{q} \in \bar{Q}_{k+1}(X).$$

Then $d^2 = 0$, and hence $(Q^\bullet(X), d)$ is a cochain complex. Its homology is denoted by $H^k_{\text{cube}}(X;\mathbb{R})$ and referred to as the **cubical singular cohomology** of X.

The next lemma is just a restatement of Corollary 26.4.

LEMMA 27.23. *Let M be a smooth manifold. Then integration induces a cochain map*

$$\Phi\colon \Omega^\bullet(M) \to Q^\bullet(M), \qquad \Phi[\omega][q] := \int_q \omega.$$

We can now state the main result of this chapter.

Theorem 27.24 (The de Rham Theorem) *The integration cochain map Φ induces a natural isomorphism $H^k_{\mathrm{dR}}(M) \to H^k_{\mathrm{cube}}(M;\mathbb{R})$.*

Singular cohomology is a topological invariant. Thus Corollary 27.6 is an immediate consequence of the de Rham Theorem. Moreover the de Rham Theorem implies that when M is compact and oriented, $H^m_{\mathrm{dR}}(M) \cong \mathbb{R}$, since the same is true of singular cohomology (cf. Remark 27.3).

Proof of the de Rham Theorem (Sketch). The proof proceeds in the same fashion as Theorem 27.21.

1. Suppose M is an open convex subset of $\mathbb{R}^m$. Then the theorem follows from Corollary 27.8 and standard properties of singular cohomology.

2. Suppose $M = U \cup V$, where U and V are open in M and the theorem is assumed to be true for U, V and $U \cap V$. The theorem will gain follow for M via a standard argument using the Mayer-Vietoris sequences and the Five Lemma, apart from the fact that we have not constructed the Mayer-Vietoris sequence in de Rham cohomology. Let us rectify this. We denote by

$$\iota_U : U \cap V \hookrightarrow U, \qquad \iota_V : U \cap V \hookrightarrow V$$

and

$$e_U : U \hookrightarrow M, \qquad e_V : V \hookrightarrow M$$

the inclusions. One then defines

$$\alpha\colon \Omega^\bullet(M) \to \Omega^\bullet(U) \oplus \Omega^\bullet(V), \qquad \alpha(\omega) := (e_U^*(\omega), e_V^*(\omega))$$

and

$$\beta\colon \Omega^\bullet(U) \oplus \Omega^\bullet(V) \to \Omega^\bullet(U \cap V), \qquad \beta(\omega_1, \omega_2) := \iota_U^*(\omega_1) - \iota_V^*(\omega_2).$$

Then we claim that

$$0 \to \Omega^\bullet(M) \xrightarrow{\alpha} \Omega^\bullet(U) \oplus \Omega^\bullet(V) \xrightarrow{\beta} \Omega^\bullet(U \cap V) \to 0$$

is exact. The only claim that isn't clear is why β should be surjective. To see this, let $\{\kappa_U, \kappa_V\}$ be a partition of unity subordinate to the open cover $\{U, V\}$. Then if ω is a k-form on $U \cap V$, we can think of $\kappa_U\omega$ and $\kappa_V\omega$ as k-forms on U and V respectively, and $\beta(\kappa_U\, \omega, -\kappa_V\, \omega) = \omega$.

3. Now assume $M = \bigcup_i U_i$, where $U_i \subset U_{i+1}$ is an open set such that $\bar{U}_i$ is compact for each i, and assume the theorem is true for each U_i. Then the theorem is also true for M. The proof of this is considerably harder than the proof of Step 4 of Theorem 27.21, since now we are working with cohomology, and thus instead of filtered colimits we have filtered *limits*. It is a sad fact of life that limits are less well behaved than colimits, and are not exact functors from diagrams of vector spaces to diagrams of vector spaces. Consequently we need to worry about the first right *derived* functors $R^1 \varprojlim$. But this is not too bad: since $\bar{U}_i$ is compact one has $R^1 \varprojlim \Omega^k(U_i) = 0$ for all k and i, and thus we have a natural short exact sequence

$$0 \to R^1 \varprojlim H^{k-1}_{\mathrm{dR}}(U_i) \to H^k_{\mathrm{dR}}(M) \to \varprojlim H^k_{\mathrm{dR}}(U_i) \to 0.$$

A similar sequence holds for $H^k_{\mathrm{cube}}(M)$, and naturality of the two sequences allow us to conclude this step.

4. Now assume M is an arbitrary open subset of $\mathbb{R}^m$. Then we can write M as a countable union of convex open subsets. For any finite union, the theorem holds by applying Step 2 and induction, and then Step 3 gives the result for M itself.

5. The general case: this follows from the previous step, since M has a countable basis of open sets diffeomorphic to open sets of Euclidean space. The proof is complete. ■

From a purely topological point of view, the thing that makes manifolds "special" is the following duality between homology and cohomology.

Theorem 27.25 (Poincaré Duality) *Let M be a compact connected orientable manifold M. Then*

$$H^k_{\mathrm{cube}}(M;\mathbb{R}) \cong H^{\mathrm{cube}}_{m-k}(M;\mathbb{R}),$$

and hence there is a non-degenerate pairing

$$H^k_{\mathrm{cube}}(M;\mathbb{R}) \times H^{m-k}_{\mathrm{cube}}(M;\mathbb{R}) \to \mathbb{R}.$$

Poincaré Duality is false if we move outside the category of manifolds (e.g. to finite cell complexes). For de Rham cohomology, this pairing is particularly easy to understand: it is induced from the pairing

$$\Omega^k(M) \times \Omega^{m-k}(M) \to \mathbb{R}, \qquad (\omega, \theta) \mapsto \int_M \omega \wedge \theta.$$

We leave it to you to investigate how to prove this.

Problem Sheet K

A (♣) indicates the problem is non-examinable. This does not necessarily imply it is difficult.

PROBLEM K.1. A singular k-cube $c\colon C^k \to M$ is said to be **degenerate** if there exists $1 \le i \le k$ such that c does not depend on u^i. Prove that if $c\colon C^k \to M$ is a degenerate singular k-cube then $\int_c \omega = 0$ for any $\omega \in \Omega^k(M)$.

PROBLEM K.2. Let $c\colon C^k \to M$ be a smooth singular k-cube in M and let $\varphi\colon C^k \to C^k$ be an orientation preserving diffeomorphism.[5] Let $\tilde{c} := c \circ \varphi$. Prove that for any $\omega \in \Omega^k(M)$, one has

$$\int_c \omega = \int_{\tilde{c}} \omega.$$

PROBLEM K.3. Prove that there does not exist a compact symplectic[6] manifold (M, ω) without boundary with the property that ω is exact.

PROBLEM K.4. Find a closed $(m-1)$-form on $\mathbb{R}^m \setminus \{0\}$ that is not exact.

PROBLEM K.5. Let M be a smooth manifold, let $X \in \mathfrak{X}(M)$, and let A be a tensor field. Let Φ_t denote the flow of X. Prove that

$$\frac{d}{dt}\bigg|_{t=t_0} \Phi_t^* A = \Phi_{t_0}^*(\mathcal{L}_X A).$$

PROBLEM K.6. Let $\varphi\colon M \to N$ be a diffeomorphism of connected oriented manifolds and let $\omega \in \Omega_c^m(N)$. Prove that

$$\int_M \varphi^*(\omega) = \pm \int_N \omega,$$

where the + signs occurs if and only if φ is orientation preserving.

PROBLEM K.7. Let G be a compact connected Lie group.[7]

(i) Prove there exists a unique **normalised left-invariant** volume form μ on G, i.e. a volume form μ such that $\int_G \mu = 1$ and $l_g^* \mu = \mu$ for all $g \in G$.

(ii) This allows us to define the integral of a *function* on G via:

$$\int_G f := \int_G f\mu, \qquad f \in C^\infty(G).$$

---

[5] As usual, think of this as meaning that φ is the restriction to C^k of an orientation preserving diffeomorphism of some neighbourhood.

[6] See Problem J.8 if you forgot the definition of a symplectic manifold.

[7] Recall G is orientable by part (ii) of Problem J.6.

Prove that for all $f \in C^\infty(G)$ and $g \in G$, one has

$$\int_G f = \int_G (f \circ l_g) = \int_G (f \circ r_g), \qquad \forall\, f \in C^\infty(G),\ g \in G.$$

(♣)PROBLEM K.8. For this problem you may assume that for any compact connected orientable smooth manifold M^m, one has $H^m_{\mathrm{dR}}(M) \cong \mathbb{R}$, and that an explicit isomorphism is given by

$$\int : H^m_{\mathrm{dR}}(M) \to \mathbb{R}, \qquad [\omega] \mapsto \int_M \omega.$$

(This is a special case of the de Rham Theorem 27.24.)

Let $\varphi \colon M \to N$ be a smooth map between compact connected orientable smooth manifolds of dimension m. Then $\varphi^* \colon H^m_{\mathrm{dR}}(N) \to H^m_{\mathrm{dR}}(M)$ is a linear map between one-dimensional vector spaces, and hence is multiplication by a number. We call this number the **degree** of φ. Explicitly,

$$\int_M \varphi^* \omega = \deg(\varphi) \int_N \omega, \qquad \omega \in \Omega^m(N).$$

(i) Let $q \in N$ denote a regular value of φ. Given $p \in \varphi^{-1}(q)$, let

$$\mathrm{sgn}_p(\varphi) := \begin{cases} +1, & \text{if } D\varphi(p) \text{ is orientation preserving,} \\ -1, & \text{if } D\varphi(p) \text{ is not orientation preserving.} \end{cases}$$

Prove that

$$\deg(\varphi) = \sum_{p \in \varphi^{-1}(q)} \mathrm{sgn}_p(\varphi).$$

Thus $\deg(\varphi)$ is an integer.

(ii) Prove the **Hairy Ball Theorem**: if m is even then any vector field on S^m has at least one zero.

Appendix A
Solutions to Exercises

Problem Sheet A

Problem A.1. Let $\varphi\colon M \to N$ be a continuous map between two smooth manifolds. Prove that the following two statements are equivalent:

(i) For every point $p \in M$, if (U, x) is **any** chart on M with $p \in U$ and (V, y) is **any** chart on N with $\varphi(U) \subseteq V$, the composition

$$y \circ \varphi \circ x^{-1}\colon x(U) \to y(V)$$

is of class C^k.

(ii) For every point $p \in M$, **there exists** a chart (U, x) on M with $p \in U$ and a chart (V, y) on N with $\varphi(U) \subseteq V$ such that the composition

$$y \circ \varphi \circ x^{-1}\colon x(U) \to y(V)$$

is of class C^k.

Solution. The nontrivial direction is from (ii) to (i).

For this purpose we pick any charts (U, x) and (V, y) as in (i) and let (U_q, x_q), $(V_{\varphi(q)}, y_{\varphi(q)})$ be a pair of charts as in (ii) for every $q \in U$. We need to check that

$$y \circ \varphi \circ x^{-1}\colon x(U) \to y(V)$$

is of class C^k (in the Euclidean space). The latter is a local property, thus it suffices to prove it in a neighbourhood of each $z \in x(U)$.

For every $z \in x(U)$ however, there exists a $q \in U$ such that $z \in x(U \cap U_q)$. Indeed, q belongs to both U and U_q, and $\bigcup_{q \in U} U_q$ covers U.

W. J. Merry, *Lectures on Differential Geometry I*,
https://doi.org/10.1007/978-3-032-03733-6

It is easy to check that $y \circ \varphi \circ x^{-1}$ sends $x(U \cap U_q)$ to $y(V \cap V_{\varphi(q)})$. But then we get, when restricted to $x(U \cap U_q)$:

$$y \circ \varphi \circ x^{-1} = (y \circ y_{\varphi(q)}^{-1}) \circ (y_q \circ \varphi \circ x_q^{-1}) \circ (x_q \circ x^{-1})\colon x(U \cap U_q) \to y(V \cap V_{\varphi(q)}),$$

with

$$y \circ y_{\varphi(q)}^{-1}\colon y_{\varphi(q)}(V \cap V_{\varphi(q)}) \to y(V \cap V_{\varphi(q)})$$

and

$$x_q \circ x^{-1}\colon x(U \cap U_q) \to x_q(U \cap U_q).$$

By definition of a smooth manifold, the two maps above are smooth (i.e. of class C^∞). Moreover, $y_{\varphi(q)} \circ \varphi \circ x_q^{-1}$ is of class C^k by (ii). Thus it follows that around z the map $y \circ \varphi \circ x^{-1}$ is of class C^k. Since z was chosen arbitrarily, this proves that $y \circ \varphi \circ x^{-1}$ is of class C^k and hence finishes the proof.

PROBLEM A.2. Prove that the set $\mathrm{GL}(m)$ of invertible $m \times m$ matrices is a smooth manifold of dimension m^2.

SOLUTION. The space of $m \times m$ matrices is homeomorphic to $\mathbb{R}^{m^2}$, and hence is a smooth manifold of dimension m^2. The determinant function

$$\det\colon \{m \times m \text{ matrices}\} \to \mathbb{R}$$

is continuous, and $\mathrm{GL}(m) = \det^{-1}(\mathbb{R} \setminus \{0\})$ is thus an open subset. The claim follows from Example 1.6.

PROBLEM A.3. Let M and N be two smooth manifolds of dimension m and n respectively. Prove that $M \times N$ is a smooth manifold of dimension $m + n$. Deduce that the m-dimensional torus:

$$T^m := \underbrace{S^1 \times \cdots \times S^1}_{m \text{ times}} \subset \mathbb{R}^{2m}.$$

is a compact smooth manifold of dimension m.

SOLUTION. If $\mathcal{X} = \{x_a\colon U_a \to O_a \mid a \in A\}$ and $\mathcal{Y} = \{y_b\colon V_b \to O_b \mid b \in B\}$ are smooth atlases on M and N respectively then

$$\mathcal{X} \times \mathcal{Y} := \{(x_a, y_b)\colon U_a \times V_b \to O_a \times O_b \mid (a, b) \in A \times B\}$$

is a smooth atlas on $M \times N$. It follows from point set topology that $M \times N$ is again separable and metrisable, thus proving that $M \times N$ is a smooth manifold of the required dimension.

Therefore, the m-dimensional torus T^m is a smooth manifold as S^1 is a smooth manifold by Proposition 1.16.

PROBLEM A.4. Let $\mathbb{R}P^m$ denote m-dimensional real projective space, that is, the space of lines through the origin in $\mathbb{R}^{m+1}$. Prove that $\mathbb{R}P^m$ is a compact smooth manifold of dimension m.

SOLUTION. For convenience we use the definition of $\mathbb{R}P^n$ via equivalence classes, i.e. we define

$$\mathbb{R}P^m = \{[x] \,|\, x \in \mathbb{R}^{m+1} \setminus \{0\}\},$$

where for x, y two elements in $\mathbb{R}^{m+1} \setminus \{0\}$ we define

$$x \sim y : \iff \exists \lambda \in \mathbb{R} \setminus \{0\} \text{ such that } x = \lambda y.$$

We will also adopt the notation $[x] = [x_1 : \ldots : x_{m+1}]$ for the equivalence class of $x = (x_1, \ldots, x_{m+1})$.

We equip $\mathbb{R}P^m$ with the final topology of the projection

$$\pi : \mathbb{R}^{m+1} \setminus \{0\} \to \mathbb{R}P^m, \; \pi(x) = [x]$$

and define an open cover $\{U_i\}_{i=1,\ldots,m+1}$ by setting

$$U_i = \{[x_1 : \ldots : x_{n+1}] \mid x_i \neq 0\},$$

for all $i = 1, \ldots, m+1$. Furthermore, for each i we define functions

$$\sigma_i : U_i \to \mathbb{R}^m, \; [x_1 : \ldots : x_{m+1}] \mapsto \frac{1}{x_i}(x_1, \ldots, x_{i-1}, x_{i+1}, \ldots, x_{m+1})$$

and claim that

$$\mathcal{X} := \{\sigma_i : U_i \to \mathbb{R}^m \,|\, i = 1, \ldots, m+1\}$$

defines a smooth atlas on $\mathbb{R}P^m$. Here it is more convenient to write the charts as σ_i instead of x_i. It is straightforward to show that each σ_i is well defined and defines a bijection. Continuity of σ_i can be shown using the universal property of the final topology on $\mathbb{R}P^m$. Thus, in order to conclude that the σ_i's are all homeomorphisms it suffices to find a continuous inverse τ_i. Such a τ_i can be defined by

$$\tau_i : \mathbb{R}^m \to U_i, \; \tau_i(x_1, \ldots, x_m) = [x_1 : \ldots : x_{i-1} : 1 : x_i : \ldots : x_m].$$

Note that all the τ_i's are indeed continuous, for they can all be written as a composition of continuous maps. Therefore the σ_i's are all homeomorphisms and we are only left to show the compatibility condition: Let σ_i and σ_j be two charts defined as above and assume without loss of generality that $i < j$. We need to show that the transition map

$$\sigma_i \circ \sigma_j^{-1} \colon \sigma_j(U_i \cap U_j) \to \sigma_i(U_i \cap U_j)$$

is a diffeomorphism in the usual calculus sense, but plugging in the definitions shows that

$$\begin{aligned}\sigma_i \circ \sigma_j^{-1}(x_1, \ldots, x_m) &= \sigma_i[x_1 : \ldots : x_{j-1} : 1 : x_j : \ldots : x_m] \\ &= \frac{1}{x_i}(x_1, \ldots, x_{i-1}, x_{i+1}, \ldots, x_{j-1}, 1, x_j, \ldots, x_n),\end{aligned}$$

which is clearly a diffeomorphism on $\sigma_j(U_i \cap U_j)$. This proves the claim that $\mathcal{X}$ is an atlas. The quickest way to see why $\mathbb{R}P^m$ is compact, metrisable and separable is by viewing $\mathbb{R}P^m$ as the sphere S^n with the antipodal identification

$$\mathbb{R}P^m = S^m/(-x \sim x),$$

which readily finishes the proof.[1]

Problem A.5. Let $G(k, m)$ denote the set of k-dimensional linear subspaces of $\mathbb{R}^m$. We call $G(k, m)$ a Grassmannian manifold. Prove that $G(k, m)$ is a compact smooth manifold and compute its dimension.

Solution. We first determine a model for $G(k, m)$. We choose the obvious approach: Every element of $G(k, m)$ is given by the image $A(\mathbb{R}^k)$ of some injective linear map $\mathbb{R}^k \to \mathbb{R}^m$, represented by a $m \times k$ matrix A of rank k (i.e. maximal rank). We will denote the space of such matrices A by $\mathcal{M}(k, m)$. Of course, for any invertible $k \times k$ matrix B we will have $A(\mathbb{R}^k) = AB(\mathbb{R}^k)$, so in fact we have a map

$$\begin{aligned}\mathcal{M}(k, m)/\!\sim \longrightarrow G(k, m) \\ [A] \mapsto A(\mathbb{R}^k),\end{aligned} \tag{A.1}$$

where $A \sim \widetilde{A}$ if and only if there exists an invertible $k \times k$ matrix B with $A = \widetilde{A}B$.

We noted above that every element of $G(k, m)$ is in the image of (A.1) and by linear algebra (A.1) is also injective (think about why this is true). Hence, our model of $G(k, m)$ will be the space $\mathcal{M}(k, m)/\!\sim$.

[1] More generally, we have the two following general facts: if X is a separable topological space, then the quotient space $X/\!\sim$ is again separable (take the projection of the countable dense subset), and if X is metrisable, then so is its quotient $X/\!\sim$ *if* the later is Hausdorff—this a doable exercise in point-set topology, also see [11].

Now, for integers $1 \leq i_1 < \cdots < i_k \leq m$ we denote by $U_{i_1,\ldots,i_k} \subset \mathcal{M}(k,m)/\sim$ the set of classes $[A]$ such that the i_1'th, i_2'th up to i_k'th row form a set of linearly independent vectors in $\mathbb{R}^k$. Note that $U_{i_1,\ldots,i_k}$ is well-defined because one element of a class in $\mathcal{M}(k,m)/\sim$ satisfies the mentioned condition if and only if all elements of that class do.

Note that if we equip $\mathcal{M}(k,m)$ with the obvious topology it inherits from $\mathbb{R}^{k\cdot m}$, and $\mathcal{M}(k,m)/\sim$ with the quotient topology, then the $U_{i_1,\ldots,i_k}$s are open subsets (check this). Note also that $\{U_{i_1,\ldots,i_k}\}_{1\leq i_1<\ldots<i_k\leq m}$ is an open cover of $\mathcal{M}(k,m)/\sim$. By Gaussian elimination every class $\alpha \in U_{1,\ldots,k} \subset \mathcal{M}(k,m)/\sim$ contains a unique element of the form

$$\left(\frac{\mathrm{id}_{k\times k}}{Z_\alpha}\right)$$

for some (unique) $(m-k)\times k$ matrix Z_α. Hence, we can define a bijective map

$$\begin{aligned}\sigma_{1,\ldots,k}\colon U_{1,\ldots,k} &\to \mathbb{R}^{k(m-k)}\\ \alpha &\mapsto Z_\alpha,\end{aligned}$$

which is clearly a homeomorphism. We have a bijection

$$\varphi_{i_1,\ldots,i_k}\colon U_{i_1,\ldots,i_k} \to U_{1,\ldots,k}$$

which is given by permuting the rows of every element of $U_{i_1,\ldots,i_k}$ with the unique permutation which takes $(i_1,\ldots,i_k)$ to $(1,\ldots,k)$ and preserves the relative order of $\{1,\ldots,n\}\backslash\{i_1,\ldots,i_k\}$. We can then define

$$\sigma_{i_1,\ldots,i_k}\colon U_{i_1,\ldots,i_k} \to \mathbb{R}^{k(m-k)}$$

as the composition $\sigma_{i_1,\ldots,i_k} := \sigma_{1,\ldots,k}\circ\varphi_{i_1,\ldots,i_k}$, which is again a homeomorphism. The transition maps

$$\sigma_{i_1,\ldots,i_k}\circ\sigma_{j_1,\ldots,j_k}^{-1}\colon \sigma_{j_1,\ldots,j_k}(U_{j_1,\ldots,j_k}\cap U_{i_1,\ldots,i_k}) \to \sigma_{i_1,\ldots,i_k}(U_{j_1,\ldots,j_k}\cap U_{i_1,\ldots,i_k})$$

are given by

$$\sigma_{1,\ldots,k}\circ\varphi_{i_1,\ldots,i_k}\circ\varphi_{j_1,\ldots,j_k}^{-1}\circ\sigma_{1,\ldots,k}^{-1}.$$

These maps are smooth simply because they are compositions of the following smooth maps: $\sigma_{1,\ldots,k}^{-1}$ takes a $(m-k)\times k$ matrix Z to the $m\times k$ matrix

$$\left(\frac{\mathrm{id}_{k\times k}}{Z}\right).$$

The map $\varphi_{i_1,\ldots,i_k}\circ\varphi_{j_1,\ldots,j_k}^{-1}$ simply permutes rows of this matrix to get a matrix

$$\left(\frac{C}{D}\right)$$

for an invertible $k \times k$ matrix C and a $(m-k) \times k$ matrix D. The map $\sigma_{1,\ldots,k}^{-1}$ sends this matrix to the matrix

$$DC^{-1} \in \mathbb{R}^{(m-k)k},$$

so it is clearly also smooth.

We now argue that $G(k, m)$ is in fact compact. We do this by induction on k. Note first that $G(1, m) = \mathbb{R}P^{m-1}$, which was seen to be compact in the previous exercise. Hence, it suffices to prove the induction step, i.e. that $G(k+1, m)$ is compact if $G(k, m)$ is. For this purpose, define the set[2]

$$\mathcal{S} := \bigcup_{V \in G(k,m)} (S^{m-1} \cap V^{\perp}) \times \{V\} \subset S^{m-1} \times G(k, m).$$

$\mathcal{S}$ is easily seen to be a compact subset of $S^{m-1} \times G(k, m)$. In particular, if $G(k, m)$ is compact then so is $\mathcal{S}$. The map

$$\begin{aligned} \mathcal{S} &\to G(k+1, m) \\ (v, V) &\mapsto \mathbb{R}v + V \end{aligned} \tag{A.2}$$

is both surjective (use the Gram-Schmidt orthogonalization procedure) and continuous. To see the latter, define S to be the preimage of $\mathcal{S}$ under the quotient map $\mathcal{M}(k, m) \to \mathcal{M}(k, m)/\sim$. Then the map (A.2) is the map on the quotient induced by the (obviously continuous) composition of maps

$$\begin{aligned} S \mapsto \mathcal{M}(k+1, m) &\to G(k+1, m) \\ (v, A) \mapsto \left(v \,\middle|\, A\right) &\mapsto \left[\left(v \,\middle|\, A\right)\right] \end{aligned}$$

In particular $G(k+1, m)$ is compact if S is. Since (A.2) is both continuous and surjective, and $\mathcal{S}$ is compact it follows that $G(k+1, m)$ is compact. Separability and metrizability follows from the footnote in the exercise on the projective space.

PROBLEM A.6. Let X denote the union of the x-axis and the y-axis in $\mathbb{R}^2$. Prove that X is *not* a topological manifold.

SOLUTION. Since X has the subspace topology inherited from $\mathbb{R}^2$, it is clear that any *connected* neighbourhood $U \subset X$ of $(0, 0)$ intersects both axes and that $U \setminus \{(0, 0)\}$ has four connected components. Suppose that X is a topological manifold. Then X must be 1-dimensional, since every point outside of the origin is contained in a neighbourhood homeomorphic to an open subset of $\mathbb{R}$. We may thus choose U such that there exists a homeomorphism $\sigma \colon U \xrightarrow{\sim} O \subset \mathbb{R}^1$, where O is a connected open subset of $\mathbb{R}$, i.e.,

---

[2] Here $V^{\perp} \subset \mathbb{R}^m$ denotes the set of vectors which are perpendicular to V with respect to the Euclidean inner product on $\mathbb{R}^{m+1}$.

a bounded open interval. But then σ induces a homeomorphism from $U \setminus \{(0,0)\}$ to $O \setminus \{\sigma(0,0)\}$. Since the former has four connected components and the latter has only two, this yields a contradiction.

PROBLEM A.7. Let Y denote the "pinched torus". Prove that Y is *not* a topological manifold.

SOLUTION. Let $y \in Y$ denote the "pinched point" on Y. Suppose Y is a topological manifold. Clearly $Y \setminus \{y\}$ is two-dimensional, so Y must be as well. There is therefore an open neighbourhood U of y which is homeomorphic to an open disc B in $\mathbb{R}^2$. But then $U \setminus \{y\}$ is homeomorphic to B minus a point. Since the former is disconnected and the latter is connected, this is a contradiction.

PROBLEM A.8. Show that the smooth atlas on $\mathbb{R}$ consisting of the single chart $\sigma\colon \mathbb{R} \to \mathbb{R}$ given by $\sigma(x) = x^3$ defines a smooth structure that is *different* to the "standard" smooth structure (the latter is the smooth structure containing the identity map as a chart). Prove however that both the smooth structures belong to the same diffeomorphism class.

SOLUTION. To prove that the maps

$$\sigma\colon \mathbb{R} \to \mathbb{R} \quad x \mapsto x^3 \quad \text{and} \quad \iota\colon \mathbb{R} \to \mathbb{R} \quad x \mapsto x$$

define charts on $\mathbb{R}$ belonging to different smooth structures, it is sufficient to check that either $\sigma \circ \iota^{-1}$ or $\iota \circ \sigma^{-1}$ are not smooth in the ordinary sense. Since

$$(\iota \circ \sigma^{-1})(x) = x^{1/3}$$

is not smooth (because its derivative $x^{-2/3}/3$ is singular at 0), the conclusion is reached.

To prove that the differentiable structures on $\mathbb{R}$ induced by σ and ι, which we denote by $\mathcal{X}_\sigma$ and $\mathcal{X}_\iota$ respectively, belong to the same diffeomorphism class, we need to find a diffeomorphism $\varphi\colon (\mathbb{R}, \mathcal{X}_\sigma) \to (\mathbb{R}, \mathcal{X}_\iota)$. Since σ and ι provide global charts for $\mathbb{R}$, and it is not important which chart is chosen to check differentiability, it is sufficient to find a bijective map $\varphi\colon \mathbb{R} \to \mathbb{R}$ so that $\iota \circ \varphi \circ \sigma^{-1}$ and $\sigma \circ \varphi^{-1} \circ \iota^{-1}$ are smooth in the ordinary sense. By choosing $\varphi(x) = x^3$, we see that

$$(\iota \circ \varphi \circ \sigma^{-1})(x) = (x^{1/3})^3 = x \quad \text{and} \quad (\sigma \circ \varphi^{-1} \circ \iota)(x) = (x^{1/3})^3 = x$$

are smooth as required, hence the conclusion is reached.

Problem Sheet B

PROBLEM B.1. Let $\{V_a \mid a \in A\}$ be a family of vector spaces indexed by a set A, and let W be a fixed set. Suppose that for each $a \in A$ we are given a bijection $\ell_a \colon V_a \to W$ such that for any $a, b \in A$, the composition $\ell_b^{-1} \circ \ell_a \colon V_a \to V_b$ is a linear isomorphism. Prove that there is a unique vector space structure on W such that each ℓ_a is a linear isomorphism.

SOLUTION. We must define a zero element, addition and scalar multiplication on W. Let $a \in A$. Since we want l_a to be a linear isomorphism, we must do the following:

- Define $0 \in W$ via $0 := l_a(0)$.[3]
- For each $v, w \in W$, define $v + w := l_a\big(l_a^{-1}(v) + l_a^{-1}(w)\big)$.
- For each $\lambda \in \mathbb{R}$ and $v \in W$, define $\lambda v := l_a\big(\lambda l_a^{-1}(v)\big)$.

This satisfies the axioms of a vector space since V_a is itself a vector space. Moreover, the definitions are forced if we want l_a to be linear, whence the uniqueness of the resulting vector space structure on W.

It remains to show that the vector space structure on W is independent of the choice of $a \in A$. Let $b \in A$. The independence of "+" on W follows from the following computation:

$$\begin{aligned} l_a\big(l_a^{-1}(v) + l_a^{-1}(w)\big) &= l_a\big(l_a^{-1} \circ l_b \circ l_b^{-1}(v) + l_a^{-1} \circ l_b \circ l_b^{-1}(w)\big) \\ &= l_a \circ l_a^{-1} \circ l_b\big(l_b^{-1}(v) + l_b^{-1}(w)\big) \\ &= l_b\big(l_b^{-1}(v) + l_b^{-1}(w)\big). \end{aligned}$$

The proof that the definition of scalar multiplication on W is also independent from the choice of $a \in A$ is similar.

(♣) PROBLEM B.2. Let M be a smooth manifold of dimension m with maximal smooth atlas $\mathcal{X}$. Given a point $p \in M$, let $\mathcal{X}_p \subset \mathcal{X}$ denote the set of charts (U, x) such that $p \in U$. Define an equivalence relation on $\mathbb{R}^m \times \mathcal{X}_p$ by saying

$$(\xi, x) \sim (\zeta, y) \qquad \Leftrightarrow \qquad D(y \circ x^{-1})(x(p))\xi = \zeta.$$

(i) Prove that this is indeed a well-defined equivalence relation.

[3] Note that the zero element of W is already uniquely determined by the second bullet point since $v + w = v$ if and only if $l_a^{-1}(v) + l_a^{-1}(w) = l_a^{-1}(v)$. We include its definition for clarity.

(ii) Write $[\xi, x]$ to denote the equivalence class of (ξ, x), and let $\mathcal{T}_pM$ denote the set of equivalence classes. Prove that for every $x \in \mathcal{X}_p$ the map $\ell_x : \mathbb{R}^m \to \mathcal{T}_pM$ given by

$$\ell_x \xi := [\xi, x]$$

is a bijection. Deduce that $\mathcal{T}_pM$ admits a unique vector space structure such that each ℓ_x is a linear isomorphism.

(iii) Let $(U, x) \in \mathcal{X}_p$, and let $\tilde{\ell}_x : \mathbb{R}^m \to T_pM$ denote the linear isomorphism defined by

$$\tilde{\ell}_x e_i = \left.\frac{\partial}{\partial x^i}\right|_p .$$

Prove that there exists a linear isomorphism $\kappa_p : \mathcal{T}_pM \to T_pM$ which in addition satisfies

$$\kappa_p \circ \ell_x = \tilde{\ell}_x,$$

for every $(U, x) \in \mathcal{X}_p$. Deduce that $\mathcal{T}_pM$ is another equivalent way to define the tangent space of a manifold.

Solution. We start by showing that $\sim$ is an equivalence relation. Reflexivity is quickly verified:

$$D(x \circ x^{-1})(x(p))\xi = D(\mathrm{id})(x(p))\xi = \xi.$$

Assuming that $(\xi, x) \sim (\zeta, y)$, we get symmetry by plugging in $\zeta = D(y \circ x^{-1})(x(p))\xi$ and applying the chain rule

$$\begin{aligned} D(x \circ y^{-1})(y(p))\zeta &= D(x \circ y^{-1})(y(p)) \circ D(y \circ x^{-1})(x(p))\xi \\ &= D(\underbrace{x \circ y^{-1} \circ y \circ x^{-1}}_{=\mathrm{id}})(x(p))\xi = \xi. \end{aligned}$$

Transitivity follows from a similar chain rule argument. This concludes the proof of the first part.

For the second part it is convenient to invoke the result of the previous problem with $A = \mathcal{X}_p$ and $W = \mathcal{T}_pM$. Therefore we are only left to show that each $\ell_x : \mathbb{R}^m \to \mathcal{T}_pM$ defines a bijection and that $\ell_y^{-1} \circ \ell_x$ is a linear isomorphism for every $x, y \in \mathcal{X}_p$. Injectivity readily follows from the observation

$$(\xi, x) \sim (\zeta, x) \iff \xi = \zeta.$$

For surjectivity let us pick an arbitrary $(\zeta, y) \in \mathbb{R}^m \times \mathcal{X}_p$.

Simply setting $\xi := D(x \circ y^{-1})(y(p))\zeta$ grants $(\xi, x) \sim (\zeta, y)$ and hence

$$l_x(\xi) = [(\xi, x)] = [(\zeta, y)].$$

A similar argument as in the surjectivity proof above shows that

$$l_y^{-1} \circ \ell_x(\xi) = \underbrace{D(y \circ x^{-1})(x(p))}_{\text{linear isomorphism}}\,\xi.$$

This proves the second part.

For last part we make an educated guess for κ_p, namely

$$\kappa_p \colon \mathcal{T}_p M \to T_p M, \qquad [(\xi, x)] \mapsto \sum_{i=1}^{n} \xi^i \left.\frac{\partial}{\partial x^i}\right|_p,$$

where $\xi = (\xi^1, \ldots, \xi^m)$ and the $x^i = u^i \circ x$ denote the local coordinates of x (see Definition 2.5). The fact that κ_p is well defined follows can be seen as follows: pick $(\zeta, y) \sim (\xi, x)$ and let y^i denote the local coordinates of y. We then have

$$\left.\frac{\partial}{\partial y^j}\right|_p (x^k) = D_j(x^k \circ y^{-1})(x) = \langle e_k, D(x \circ y^{-1})(x) e_j \rangle,$$

where $\langle \cdot, \cdot \rangle$ denotes the usual Euclidean inner product. Using bilinearity of the inner product one obtains the identity

$$\sum_j \zeta^j \left.\frac{\partial}{\partial y^j}\right|_p (x^k) = \langle e_k, D(x \circ y^{-1})(x)\zeta \rangle = \xi^k = \sum_i \xi^i \left.\frac{\partial}{\partial x^i}\right|_p (x^k),$$

which proves well-definedness of κ_p.

The relation $\kappa_p \circ l_x = \tilde{l}_x$ is easily verified. The map κ_p is a linear isomorphism, for both ℓ_x and $\tilde{l}_x$ are linear isomorphisms. This concludes the proof.

Problem B.3. Let E and F be vector spaces and assume that $\ell \colon E \to F$ is a linear map. Prove that for any $p \in E$ the following diagram commutes:

$$\begin{array}{ccc} E & \xrightarrow{\ \ell\ } & F \\ {\scriptstyle \mathcal{J}_p}\downarrow & & \downarrow{\scriptstyle \mathcal{J}_{\ell(p)}} \\ T_p E & \xrightarrow[D\ell(p)]{} & T_{\ell(p)} F \end{array}$$

Solution. Fix $p, \xi \in E$. With $\gamma(t) = p + t\xi \in E$ we compute:

$$\begin{aligned} \mathcal{J}_{l(p)} \circ l(\xi) &= \dot{\widetilde{\gamma}}(0), \\ Dl(p) \circ \mathcal{J}_p(\xi) &= Dl(p) \circ \dot{\gamma}(0) \\ &= Dl(p) \circ D\gamma(0) \left[\frac{\partial}{\partial t}\bigg|_t \right] \\ &= D(l \circ \gamma)(0) \left[\frac{\partial}{\partial t}\bigg|_t \right] \\ &= \dot{\widehat{\gamma}}(0), \end{aligned}$$

with $\widetilde{\gamma}(t) = l(p) + tl(\xi)$ and $\widehat{\gamma}(t) = (l \circ \gamma)(t) = l(p + t\xi)$. Note that in the last row of these equations we make use of the manifold version of the chain rule. Since l is linear we have $\widetilde{\gamma}(t) = l(p) + tl(\xi) = l(p + t\xi) = \widehat{\gamma}(t)$, so it follows that $\mathcal{J}_{l(p)} \circ l(\xi) = Dl(p) \circ \mathcal{J}_p(\xi)$ for all $p, \xi \in E$, which exactly says that the diagram commutes.

Problem B.4. Let M be a smooth manifold of dimension m. Prove that the cotangent bundle T^*M is naturally a smooth manifold of dimension $2m$.

Solution. To simplify the presentation of this proof we will use the Einstein Summation Convention.[4] Let $\{e^i\}$ denote the basis of the dual space $(\mathbb{R}^m)^*$ which is dual to the standard basis $\{e_i\}$, i.e.

$$e^i(e_j) = \delta^i_j.$$

Of course, $(\mathbb{R}^m)^* \cong \mathbb{R}^m$ (an isomorphism is provided by $e^i \mapsto e_i$), but in order to make the formalism work we need to use the dual space.[5] If $x \colon U \to O$ is a chart for M, we build a chart for T^*M as follows: let $\pi \colon T^*M \to M$ be the projection and define $\tilde{x} \colon \pi^{-1}(U) \to O \times (\mathbb{R}^m)^*$ to be defined as

$$\tilde{x}(p, \xi^*) = \left(x(p), \xi^* \left(\frac{\partial}{\partial x^i}\bigg|_p \right) e^i \right),$$

This map is a bijection and its image is an open subset of $\mathbb{R}^m \times (\mathbb{R}^m)^* \cong \mathbb{R}^{2m}$. Suppose now that $\mathcal{X} := \{x_a \colon U_a \to O_a \mid a \in A\}$ is an atlas on M. If $x = \sigma_a \colon U_a \to O_a$ and $y = \sigma_b \colon U_b \to O_b$ are charts for M whose domains have non-empty intersection,

---

[4] The Einstein Summation Convention is a way of reducing the number of summation signs in a statement, and is explained at the beginning of Chapter 8. We use it here simply because it makes the proof easier to follow.

[5] An easy way to check whether the formalism is correct is to see whether the Einstein Summation Convention works. If the indices are not in the correct position (upper versus lower) then something is wrong.

denoting by x^i and y^j the corresponding local coordinates, and $(z, q) = (z, q_i\, e^i) \in x(U_a \cap U_b) \times (\mathbb{R}^m)^*$, then if $p := x^{-1}(z) \in U_a \cap U_b$, we obtain

$$\tilde{x}^{-1}(z, q) = \left(p, q_i\, dx^i\big|_p\right),$$

and hence

$$\tilde{y} \circ \tilde{x}^{-1}(z, q) = \left(\underbrace{y(x^{-1}(z))}_{=y(p)}, q_i\, dx^i\Big|_p \left(\frac{\partial}{\partial y^j}\Big|_p\right) e^j\right) \tag{B.1}$$

Since

$$dx^i\Big|_p \left(\frac{\partial}{\partial y^j}\Big|_p\right) = \frac{\partial x^i}{\partial y^j}\Big|_p$$

we have

$$\tilde{y} \circ \tilde{x}^{-1}(z, q) = \left(y(p), q_i\, \frac{\partial x^i}{\partial y^j}\Big|_p e^j\right)$$

Now $\frac{\partial x^i}{\partial y^j}\big|_p$ is the (i, j)-th entry of the matrix $D(x \circ y^{-1})(y(p)) = D(x \circ y^{-1})(y(x^{-1}(z))$ (see Remark 4.5) and hence (B.1) shows that

$$\tilde{y} \circ \tilde{x}^{-1}(z, q) = \left(y \circ x^{-1}(z), D(x \circ y^{-1})(y(p))^T [q]\right),$$

where we denote by A^T the *transpose* of the matrix A (pay attention to the ordering of x and y on the right-hand side!).

Thus the charts $\{\tilde{\sigma}_a : \pi^{-1}(U_a) \to O_a \times (\mathbb{R}^m)^* \mid a \in A\}$ then induce, thanks to Proposition 1.17, a smooth manifold structure on T^*M.

Problem Sheet C

Problem C.1. Let M and N be smooth manifolds. Prove that for all $(p, q) \in M \times N$ there is a canonical isomorphism

$$T_{(p,q)}(M \times N) = T_p M \times T_q N.$$

Solution. Fix $(p, q) \in M \times N$ and let $\iota_M : M \to M \times N$ (resp. ι_N) denote the obvious inclusion and $\pi_M : M \times N \to M$ (resp. π_N) the obvious projection. Define

$$T : T_p M \times T_q N \to T_{(p,q)}(M \times N),\ T(v, w) = D\iota_M(p)[v] + D\iota_N(q)[w].$$

To conclude that T is a canonical isomorphism we observe that the composition with $(D\pi_M(p, q), D\pi_N(p, q))$ gives the identity:

$$\begin{aligned}
(D\pi_M(p,q), &D\pi_N(p,q)) \circ T(v,w) \\
&= (D\pi_M(p,q), D\pi_N(p,q))\,(D\iota_M(p)[v] + D\iota_N(q)[w]) \\
&= \left(D(\underbrace{\pi_M \circ \iota_M}_{=\mathrm{id}})(p)[v], D(\underbrace{\pi_N \circ \iota_N}_{=\mathrm{id}})(q)[w] \right) \\
&= (v,w).
\end{aligned}$$

Since $\dim(T_{(p,q)}(M \times N)) = \dim(M \times N) = \dim(T_pM \times T_qN)$ this forces T to be a linear isomorphism and also shows that T is canonical.

Problem C.2. Let $\varphi\colon M \to N$ be a smooth map. Prove that $D\varphi\colon TM \to TN$ is also smooth. Prove that if $\varphi\colon M \to N$ is an embedding then so is $D\varphi\colon TM \to TN$.

Solution. To prove that the derivative of φ is a smooth map, let p_0 be a point in M, let (U, x) be a chart for M around p_0 and let (V, y) be a chart around $\varphi(p_0)$. Pick $(\pi_M^{-1}(U), \tilde{x})$ and $(\pi_N^{-1}(v), \tilde{y})$ the associated charts for TM and TN respectively (see proof of Theorem 5.6), and recall that

$$\tilde{x}\colon TU \to \mathbb{R}^m \times \mathbb{R}^m,\ (p,\xi) \mapsto (x(p), (dx^i)_p(\xi)e_i),$$

where here we are using Einstein's Summation Convention.[6] Pick $(v_1, v_2) \in \mathbb{R}^m \times \mathbb{R}^m$, define $(p,\xi) := \tilde{x}^{-1}(v_1, v_2)$. Then we have

$$(\tilde{y} \circ D\varphi \circ \tilde{x}^{-1})(v_1, v_2) = \Big((y \circ \varphi \circ x^{-1}(v_1), D(y \circ \varphi \circ x^{-1})(v_1)[v_2]\Big)\Big).$$

This expression is smooth, since $y \circ \varphi \circ x^{-1}$ is smooth by smoothness of φ and the fact that $D(y \circ \varphi \circ x^{-1})(v_1)$ is the Jacobian matrix of $y \circ \varphi \circ x^{-1}$ at v_1, which has entries that depend smoothly on v_1. This proves that $D\varphi\colon TM \to TN$ is a smooth map.

Assume now that $\dim M = m$ and $\dim N = n$. If φ is an embedding, $\varphi(M)$ is an m-dimensional embedded submanifold of N.

In order to simplify the notations, we will show that if Σ is any embedded submanifold of N, then $T\Sigma$ is an embedded submanifold of TN. This will imply the result once we have shown that $D\varphi$ defines a homeomorphism of TM onto $T(\varphi(M))$, where the latter is endowed with the subspace topology in TN.

We start by showing the latter: since φ is an embedding, we know that the differential $D\varphi(p)$ at any point $p \in M$ is of full rank, hence that φ is a local diffeomorphism onto its image by the Inverse Function Theorem 5.10.

---

[6] See the beginning of Chapter 8.

But $\varphi: M \to \varphi(M)$ is a bijection, which with the above implies that φ is a global diffeomorphism onto its image. Applying the first part of the proof above to $\varphi: M \to \varphi(M)$ shows that $D\varphi: TM \to T(\varphi(M))$ is a smooth map. A smooth inverse to this $D\varphi$ is given by $D(\varphi^{-1}): T(\varphi(M)) \to TM$ (use the chain rule!).

We go back to proving that $T\Sigma$ is embedded in TN. To say that a subset $\Sigma \subset N$ is an k-dimensional embedded submanifold is equivalent to saying that, for every $q \in \Sigma$, there exists a slice chart adapted to Σ, that is, a chart (V, y) of N so that

$$y(V \cap \Sigma) = y(V) \cap (\mathbb{R}^k \times \{0, \ldots, 0\}), \tag{C.1}$$

where the number of zeroes is $n - k$.[7]

Any such slice chart induces a respective chart $(\pi_N^{-1}(V), \tilde{y})$ for TN, and we claim that this is also a slice chart adapted to $T\Sigma$. Indeed, we can write for any $\zeta \in T_qN \subset \pi_N^{-1}(V)$,

$$\tilde{y}(q, \zeta) = \left(y(q), d(y^i)_q(\zeta)\, e_i \right),$$

but then condition (C.1) implies that the vector ζ is tangent to Σ if and only if

$$0 = d(y^{k+1})_q(\zeta) = d(y^{k+2})_q(\zeta) = \ldots = d(y^n)_q(\zeta).$$

We obtain in other words that

$$\tilde{y}(\pi_N^{-1}(V) \cap T\Sigma) = \big(y(V) \cap (\mathbb{R}^k \times \{0, \ldots, 0\})\big) \times \mathbb{R}^k \times \{0, \ldots, 0\},$$

where each of the two chunks of zeroes contains $n - k$ elements. Up to a shift of the coordinates, this means precisely that we have obtained a slice chart adapted to $T\Sigma$. Since such chart can be produced around any $(q, \zeta) \in T\Sigma$, we conclude that $T\Sigma$ is an embedded submanifold of TN by Proposition 6.7.

PROBLEM C.3. Let $\varphi: M \to N$ be an injective immersion with M compact. Prove that φ is an embedding. Give an example to show that this need not be true if M is not compact.

SOLUTION. In order to prove that an injective immersion $\varphi: M \to N$ is an embedding, if M is compact, we need to show that φ is a homeomorphism onto its image $\varphi(M)$, where the latter is equipped with the subspace topology from N. Clearly $\varphi: M \to \varphi(M)$ is bijective, so it suffices to show that it and its inverse are continuous. Given an open subset $O \subset \varphi(M)$ there exists an open subset $V \subset N$ such that

$$O = V \cap \varphi(M).$$

[7] This is just the chart applied to the equation in Proposition 6.7.

Here we simply use the definition of the subspace topology. Hence, $\varphi^{-1}(O) = \varphi^{-1}(V) \subset M$ is open because φ is assumed to be smooth (and in particular continuous) when viewed as a map $M \to N$. This shows that $\varphi\colon M \to \varphi(M)$ is continuous.

To see that $\varphi^{-1}\colon \varphi(M) \to M$ is continuous, it suffices to show that $(\varphi^{-1})^{-1}(C) = \varphi(C) \subset \varphi(M)$ is closed for every closed subset $C \subset M$. Since M is compact, a closed subset $C \subset M$ is also compact (with the subspace topology). Since $\varphi\colon M \to \varphi(M)$ is continuous, it maps compact subsets to compact subsets. Hence $(\varphi^{-1})^{-1}(C) = \varphi(C) \subset \varphi(M)$ is closed for every closed subset $C \subset M$. We conclude that $\varphi\colon M \to \varphi(M)$ is a homeomorphism which was exactly what we needed to show.[8]

The example we consider is that of $\mathbb{R}$ immersed (but not embedded!) into $\mathbb{R}^2$. The immersion $\varphi\colon \mathbb{R} \to \mathbb{R}^2$ takes the negative (infinite) end of $\mathbb{R}$ to the line $\{(0, y) \mid -1.5 < y < 1.5\}$, then "swipes around" and continuously into the curve $\{(t, \sin(\frac{1}{t}) \mid 0 < t < 1\}$ (see Figure C.1). It is not difficult to imagine that this can be done in such a way that φ is an immersion. Now note that $\varphi\colon \mathbb{R} \to \varphi(\mathbb{R})$ will not be a homeomorphism if the latter space is equipped with the subspace topology from $\mathbb{R}^2$. To see this, note that (in this topology) the point $(0, 0) \in \varphi(\mathbb{R})$ has a basis of neighbourhoods all of which contain infinitely many connected components. This is not the case for any element in $\mathbb{R}$...

Problem C.4. Let $\mathcal{O}$ be an open subset in $\mathbb{R}^m$ and suppose $f\colon \mathcal{O} \to \mathbb{R}$ is smooth. Define $g\colon \mathcal{O} \to \mathbb{R}^{m+1}$ by

$$g(x) = (x, f(x)).$$

Prove that g is a smooth embedding, and hence that $g(\mathcal{O})$ is a smooth embedded m-dimensional submanifold of $\mathbb{R}^{m+1}$. (We call $g(\mathcal{O})$ the **graph** of f.)

Solution. Define a map $\phi\colon \mathcal{O} \times \mathbb{R} \to \mathcal{O} \times \mathbb{R}$ by

$$\phi(x, y) = (x, y - f(x))$$

Since f is smooth, it follows that ϕ is as well. Since the inverse

$$\phi^{-1}(u, v) = (u, v + f(u))$$

is also smooth, we deduce that ϕ is a diffeomorphism. We have

$$g = \phi^{-1} \circ \iota\colon \mathcal{O} \to \mathcal{O} \times \mathbb{R} \subseteq \mathbb{R}^m \times \mathbb{R},$$

where $\iota\colon \mathcal{O} \hookrightarrow \mathcal{O} \times \mathbb{R}$ is the inclusion $u \mapsto (u, 0)$. Since ι and ϕ are both smooth embeddings, so is g. Furthermore we have

[8] Note that in this proof we never used the manifold structure. Indeed, it's a point-set topological fact that any continuous bijective map from a compact topological space into a Hausdorff space is automatically a homeomorphism.

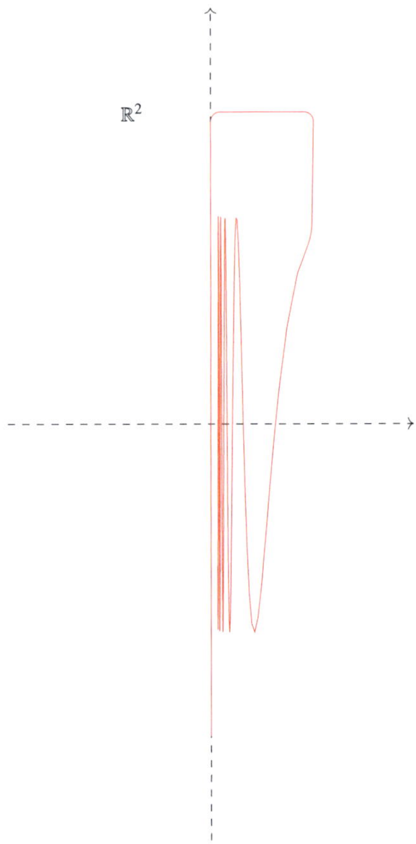

Fig. C.1: A copy of $\mathbb{R}$ immersed (but not embedded) in $\mathbb{R}^2$

$$\phi\big(g(\mathcal{O})\big) \cap (\mathcal{O} \times \mathbb{R}) = \{(u, v) \in \mathcal{O} \times \mathbb{R} \mid v = 0\},$$

therefore the map ϕ defines an explicit (global) slice chart for $g(\mathcal{O})$ in $\mathcal{O} \times \mathbb{R}$. By Proposition 6.7 this concludes the proof.

PROBLEM C.5. Let $i\colon S^m \hookrightarrow \mathbb{R}^{m+1}$ denote the inclusion. Prove that

$$Di(p)[T_pS^m] = \mathcal{J}_p(p^\perp),$$

where $\mathcal{J}_p\colon \mathbb{R}^{m+1} \to T_p\mathbb{R}^{m+1}$ is the dash-to-dot map and

$$p^\perp := \left\{q \in \mathbb{R}^{m+1} \mid \langle p, q\rangle = 0\right\},$$

for $\langle\cdot,\cdot\rangle$ the standard Euclidean dot product.

SOLUTION. Previously we defined the isomorphism

$$\mathcal{J}_p : \mathbb{R}^{m+1} \to T_p\mathbb{R}^{m+1}$$

by setting $\mathcal{J}_p(v) = \dot{\gamma}(0)$ with $\gamma(t) = p + tv$, where $\dot{\gamma}(0)$ is viewed as a lincar derivation on the space of germs at x. What this is really saying is that the inverse

$$\mathcal{J}_p^{-1} : T_p\mathbb{R}^{m+1} \to \mathbb{R}^{m+1}$$

has the following simple description: given any $v \in T_p\mathbb{R}^{m+1}$ there exists a smooth curve $\widetilde{\gamma} : (-\epsilon, \epsilon) \to \mathbb{R}^{m+1}$ such that $\widetilde{\gamma}(0) = p$ and $\dot{\widetilde{\gamma}}(0) = v$. Then

$$\mathcal{J}_p^{-1}(v) = \widetilde{\gamma}'(0),$$

where we view $\widetilde{\gamma}'(0)$ as an element of $\mathbb{R}^{m+1}$ in the sense of elementary calculus—this is essentially the content of the "dash to dot" result (Lemma 4.12).

Now fix $q \in T_pS^m$ and choose a smooth curve $\gamma : (-\epsilon, \epsilon) \to S^m$ with $\gamma(0) = p$ and $\gamma'(0) = q$. Since $S^m = \{y \in \mathbb{R}^{m+1} \mid |y| = 1\}$ we have $|i(\gamma(t))|^2 = 1$ for all $t \in (-\epsilon, \epsilon)$. From elementary calculus it now follows that

$$\begin{aligned}
0 &= \frac{d}{dt}\Big|_{t=0} |i(\gamma(t))|^2 \\
&= \frac{d}{dt}\Big|_{t=0} \langle i(\gamma(t)), i(\gamma(t))\rangle \\
&= \langle (i \circ \gamma)'(0), i(\gamma(0))\rangle + \langle i(\gamma(0)), (i \circ \gamma)'(0)\rangle \\
&= 2\langle (i \circ \gamma)'(0), p\rangle,
\end{aligned}$$

where $(i \circ \gamma)'(0)$ is understood in the sense of elementary calculus. By the manifold version of the chain rule and the above remark we have $(i \circ \gamma)'(0) = \mathcal{J}_p^{-1} \circ Di(p)[q]$. So, since $q \in T_pS^m$ was arbitrary, the above computation implies

$$Di(p)[T_pS^m] \subset \mathcal{J}_p(p^\perp).$$

Moreover, since both $Di(p)[T_pS^m]$ and $\mathcal{J}_p(p^\perp)$ are linear subspaces of $T_p\mathbb{R}^{m+1}$ of dimension m, it follows from elementary linear algebra that in fact

$$Di(p)[T_pS^m] = \mathcal{J}_p(p^\perp).$$

(♣) PROBLEM C.6. Prove that $\operatorname{Nor} M$ is an embedded n-dimensional submanifold of $T\mathbb{R}^n = \mathbb{R}^{2n}$, where $\operatorname{Nor} M$ is as in Definition 7.7

SOLUTION. Endow $\operatorname{Nor} M \subset \mathbb{R}^{2n}$ with the subspace topology. Let $p \in M$. Since M is an embedded submanifold of $\mathbb{R}^n$, there exists a slice chart $x : U \subset \mathbb{R}^n \to \mathcal{O}$ around p

with

$$M \cap U = \{q \in U \mid x^{m+1}(q) = \cdots = x^k(q) = 0\}.$$

For each $q \in U$, the vectors $e_j := \partial/\partial x^j|_q$ form a basis of $T_q\mathbb{R}^n$, with $(e_j)_{j=1,\ldots,m}$ forming a basis of the subspace $T_qM \subset T_q\mathbb{R}^n$. Let q^i denote the standard coordinates on $\mathbb{R}^n$. Then

$$e_j = \sum_{i=1}^{n} \frac{\partial}{\partial x^j}\bigg|_q (q^i) \cdot \frac{\partial}{\partial q^i}\bigg|_q.$$

Let $E(q) = \left(\partial/\partial x^j|_y(q^i)\right)_{ij}$ denote the corresponding change of basis matrix.[9]

Define a function $\phi\colon U \times \mathbb{R}^n \to x(U) \times \mathbb{R}^n$ by

$$\phi(q, y) = \left(x^1(q), \ldots, x^n(q), y \cdot e_1, \ldots, v \cdot e_n\right),$$

where $y \cdot e_j$ is the Euclidean dot product after identifying $T_q\mathbb{R}^n$ with $\mathbb{R}^n$ via the basis $(\partial/\partial q^i|_q)$. This means that if $y = \sum_{i=1}^n v^i e^i$, then

$$y \cdot e_j = \sum_{i=1}^{n} v^i \frac{\partial}{\partial x^j}\bigg|_y (q^i).$$

Since $q \mapsto \partial/\partial x^j|_y(q^i)$ defines a smooth function on U, it follows that ϕ is smooth - note that the e_j also implicitly depend on q, but smoothly. The derivative of ϕ at a point (q, y) is

$$D\phi(q, y) = \begin{pmatrix} \left(\frac{\partial x^i}{\partial q^j}(q)\right)_{ij} & 0 \\ * & E(y) \end{pmatrix}.$$

Since this is invertible, it follows from the Inverse Function Theorem 5.10 that ϕ is a local diffeomorphism. If $\phi(q, y) = \phi(q', y')$, then $y = y'$ because $x\colon U \to \mathcal{O}$ is injective. By definition of ϕ, we have $y \cdot e_j = y' \cdot e_i$ for each i, which implies that $y - y'$ is orthogonal to the span of $(e_1, \ldots, e_n)$.

Since the e_i form a basis of $T_q\mathbb{R}^n \cong \mathbb{R}^n$, it follows that $y - y' = 0$. Thus ϕ is injective and hence defines a smooth chart on $U \times \mathbb{R}^n$.

By construction, we have $(q, y) \in \operatorname{Nor} M$ if and only if $\phi(q, y)$ is contained in

$$\{(z, w) \in \mathbb{R}^n \times \mathbb{R}^n \mid z^{m+1} = \cdots = z^n = 0, w^1 = \cdots = w^m = 0\}.$$

It follows that ϕ is a slice chart for $\operatorname{Nor} M$. Proposition 6.7 implies that $\operatorname{Nor} M$ is an embedded n-dimensional submanifold of $\mathbb{R}^{2n}$, as desired.

---

[9] Between

$$\left\{\frac{\partial}{\partial q^i}\bigg|_q\right\}_i \text{ and } \left\{\frac{\partial}{\partial x^j}\bigg|_q\right\}_j.$$

PROBLEM C.7. Let $\varphi\colon M \to N$ be a smooth map. Assume that the rank of φ is constant on all of M.

(i) Assume that φ is injective. Prove that φ is an immersion.
(ii) Assume that φ is surjective. Prove that φ is a submersion.
(iii) Assume that φ is bijective. Prove that φ is a diffeomorphism.

SOLUTION. We look at each part in turn.

(i) If φ has rank $k < m$ (i.e. φ is *not* an immersion) on a neighbourhood of p then it is clear that φ is not injective near p, since with charts x and y as in Corollary 6.16, the map $\kappa = y \circ \varphi \circ x^{-1}$ is not injective, since $(0, \dots, 0, \varepsilon)$ is mapped to $(0, \dots, 0)$ for all ε small.

(ii) If φ has rank $k < n$ (i.e. φ is *not* a submersion) then by Sard's Theorem 6.14, $\varphi(M)$ is nowhere dense in N, and thus certainly φ is not surjective.

(iii) Finally, if φ is bijective, then φ must have rank $k = m = n$, and thus by the Inverse Function Theorem 5.10 φ is locally a diffeomorphism in a neighbourhood of every point, and hence also a global diffeomorphism.

Problem Sheet D

PROBLEM D.1. Let $\mathcal{O} \subset \mathbb{R}^m$ be an open set.

(i) Prove that the dash-to-dot maps induce a diffeomorphism between $T\mathcal{O}$ and $\mathcal{O} \times \mathbb{R}^m$.
(ii) Prove that there is a bijective correspondence between vector fields on $\mathcal{O}$ and smooth function $\mathcal{O} \to \mathbb{R}^m$. Namely, given a vector field X associate the function f defined by
$$f(p) := \mathcal{J}_p^{-1}(X(p)), \qquad \forall\, p \in \mathcal{O}.$$
(iii) Let X and f be associated as above, and let γ be a smooth curve in $\mathcal{O}$. Prove that γ is an integral curve of f in the sense of (9.2), i.e. $\gamma' = f(\gamma)$, if and only if γ is an integral curve of X in the sense of (9.3), i.e. $\dot{\gamma} = X(\gamma)$.

SOLUTION. For part (i) we define
$$\mathcal{J}\colon \mathcal{O} \times \mathbb{R}^m \to T\mathcal{O}, \ \mathcal{J}(p, \xi) := \mathcal{J}_p\, \xi.$$

From Lemma 4.14 we already know that $\xi \mapsto \mathcal{J}_p\, \xi$ is smooth. Moreover, fixing $\xi \in \mathbb{R}^m$ and taking a closer look at the description in local coordinates in Lemma 4.14 reveals that $p \mapsto \mathcal{J}_p\, \xi$ is also smooth. This proves that $\mathcal{J}$ is smooth.

Now we exhibit a smooth inverse to $\mathcal{J}$. For that purpose we first observe that $T_p\mathcal{O} = T_p\mathbb{R}^m$. Indeed, an open subset always defines a smooth submanifold of the same dimension as its ambient manifold.[10] With this we can define an inverse

[10] Prove this! It is not hard.

$$\mathcal{J}^{-1}\colon T\mathcal{O} = \cup_{p\in\mathcal{O}} T_p\mathbb{R}^m \to \mathcal{O}\times\mathbb{R}^m,\ \mathcal{J}^{-1}(p,\zeta) = (p, \mathcal{J}_p^{-1}\,\zeta).$$

Since $\mathcal{J}_p$ is a linear isomorphism, we do know that $\mathcal{J}_p^{-1}$ is a smooth map as well. This shows that $\mathcal{J}^{-1}$ is a smooth inverse to $\mathcal{J}$ and finishes the first part.

Part (ii) follows from the first result: given a function $f\colon \mathcal{O}\to\mathbb{R}^m$, we define a vector field $X\colon \mathcal{O}\to T\mathcal{O}$ by

$$X(p) = \mathcal{J}(p, f(p)) = (p, \mathcal{J}_p\, f(p)).$$

Note that X is indeed a vector field by smoothness of $\mathcal{J}$ and $\pi(p, \mathcal{J}_p\,\xi)) = p$. On the other hand, every vector field Y defines a smooth function g as described above, namely

$$g(p) = \mathcal{J}_p^{-1}(Y(p)) = \mathcal{J}^{-1}(p, Y(p)).$$

The one to one correspondence readily follows from $\mathcal{J}$ being a diffeomorphism.

Part (iii) is an immediate consequence of the "dash to dot" Lemma (Lemma 4.12): start with an integral curve γ of X and observe that

$$\gamma'(t) = \mathcal{J}_{\gamma(t)}^{-1}\,\dot\gamma(t) = \mathcal{J}_{\gamma(t)}^{-1}\,X(\gamma(t)) = f(\gamma(t)).$$

In the last step we use the definition of f (see part (ii)). The other direction follows similarly by applying $\mathcal{J}$ instead of its inverse.

PROBLEM D.2. Let M be a smooth manifold, let $p\in M$, and let $\xi\in T_pM$. Let U be any open set containing p. Prove that there exists a vector field $X\in\mathfrak{X}(U)$ such that $X(p)=\xi$.

SOLUTION. First consider an arbitrary subset $A\subset M$ endowed with the subspace topology. A **smooth vector field along** A is a continuous map $X\colon A\to TM$ such that

1. the section property $\pi(X(p)) = p$ holds for all $x\in A$, and
2. for each $x\in A$, there exists a neighbourhood V of p in M and a smooth vector field $\tilde X$ on V that agrees with X on $V\cap A$.

The statement in the exercise follows from the following more general statement:

Lemma. Let $A\subset M$ be a closed subset, and let X be a smooth vector field along A. Given any open subset $U\subset M$ containing A, there exists a smooth global vector field $\tilde X$ on M such that $\tilde X|_A = X$ and $\operatorname{supp}\tilde X\subset U$.

Proof. By assumption, for each $p\in A$, we may choose a neighbourhood $V_p\subset U$ of p and a smooth vector field $\tilde X_p$ of V_p that agrees with X on $V_p\cap A$. Then the family of sets $\{V_p \mid p\in A\}\cup\{M\setminus A\}$ is an open cover of M. Let $\{\lambda_p \mid p\in A\}\cup\{\lambda_0\}$ be a partition of unity subordinate to this open cover, with $\operatorname{supp}\lambda_x\subset V_p$ and $\operatorname{supp}\lambda_0\subset M\setminus A$.

For each $p \in A$, the product $\lambda_p \tilde{X}_p$ is a smooth vector field on V_p. We consider it as a smooth vector field on M by extending by 0 on $M \setminus \operatorname{supp} \lambda_p$. Consider the function $\tilde{X}\colon M \to TM$ defined by

$$\tilde{X}(p) = \sum_{q \in A} \lambda_q(p) \tilde{X}_q(p).$$

Each term in the above sum is smooth. Since $\{\operatorname{supp} \lambda_q \mid q \in A\}$ is locally finite, the sum has only finitely many terms in a neighbourhood of any point of M and thus defines a smooth function. If $x \in A$, then $\lambda_0(x) = 0$ and $\lambda_q(p)\tilde{X}_q(p) = X(p)$ for each q such that $\lambda_q(p) \neq 0$. Thus

$$\tilde{X}(p) = \sum_{q \in A} \lambda_q(p) X(p) = \Big(\lambda_0(p) + \sum_{q \in A} \lambda_q(p)\Big) X(p) = X(p);$$

hence $\tilde{X}$ is a global smooth vector field extending X. Finally, consider the following equalities:

$$\operatorname{supp} \tilde{X} = \overline{\bigcup_{x \in A} \operatorname{supp} \lambda_x} = \bigcup_{x \in A} \operatorname{supp} \lambda_x \subset U.$$

Thus $\operatorname{supp} \tilde{X} \subset U$. This concludes the proof. ■

The function $X\colon \{p\} \to TM,\ x \mapsto \xi$ is a smooth vector field along $\{p\}$ since it can be extended, for example, to a constant-coefficient vector field in a chart of p. We thus obtain the desired result by applying the above proposition with $A = \{p\}$.[11]

Problem D.3. Let M be a smooth manifold and let $W \subset M$ be a non-empty open set. Prove that the Lie bracket $[\cdot, \cdot]$ on $\mathfrak{X}(W)$ satisfies the Jacobi identity.

Solution. Let $X, Y, Z \in \mathfrak{X}(W)$. We compute:

$$\begin{aligned}
&[X,[Y,Z]] + [Y,[Z,X]] + [Z,[X,Y]] \\
&\qquad = X[Y,Z] - [Y,Z]X + Y[Z,X] - [Z,X]Y + Z[X,Y] - [X,Y]Z \\
&\qquad = XYZ - XZY - YZX + ZYX + YZX \\
&\qquad\qquad - ZXY + XZY + ZXY - ZYX - XYZ + YXZ.
\end{aligned}$$

Note that the product of vector fields in the above equations is given by composition, i.e., $XY := X \circ Y$, where X and Y are viewed as functions from $C^\infty(W)$ to itself. In the

[11] Note that in order to prove the statement in the exercise, one can simply extend to a constant coefficient vector field in a coordinate neighbourhood U of x with value v at x, multiply by a bump function on U, and then extend the resulting vector field on U by zero to a vector field on M. We include the general statement above as a nice illustration of the usefulness of partitions of unity, and one can see that its proof generalises the aforementioned simpler argument.

last expression above, all of the terms cancel in pairs. Thus $[X,[Y,Z]]+[Y,[Z,X]]+[Z,[X,Y]]=0$ and $\mathfrak{X}(W)$ satisfies the Jacobi identity.

Problem D.4. Let M be a smooth manifold and let (U,x) be a chart on M with local coordinates (x^i). Fix $X,Y\in\mathfrak{X}(U)$, and write $X=X^i\frac{\partial}{\partial x^i}$ and $Y=Y^i\frac{\partial}{\partial x^i}$. Prove that

$$[X,Y]=\left(X^i\frac{\partial Y^j}{\partial x^i}-Y^i\frac{\partial X^j}{\partial x^i}\right)\frac{\partial}{\partial x^j},$$

where $\frac{\partial Y^j}{\partial x^i}$ and $\frac{\partial X^j}{\partial x^i}$ are the functions from Definition 8.4.

Solution. We recall that the Lie bracket $[X,Y]$ of two vector fields $X,Y\in\mathfrak{X}(U)$ is a derivation defined by

$$[X,Y]\colon C^\infty(U)\to C^\infty(U),$$
$$[X,Y](f)=X(Y(f))-Y(X(f)).$$

Let us write $X=X^i\frac{\partial}{\partial x^i}$ and $Y=Y^j\frac{\partial}{\partial x^j}$ and view them both as derivations. We pick some arbitrary smooth function $f\in C^\infty(U)$ and feed it to Y to obtain, for each $p\in U$:

$$Y(f)(p)=\left(Y^j\frac{\partial}{\partial x^j}\right)(f)(p)=Y^j(p)\,\frac{\partial f}{\partial x^j}(p).$$

For notation reasons it is convenient to set $g_j=\frac{\partial f}{\partial x^j}\in C^\infty(U)$. Since we are using the Einstein Summation Convention we are forced to use a lower subscript for g. Invoking the derivation property grants us

$$\begin{aligned}X(Y(f))(p)&=X^i(p)\frac{\partial}{\partial x^i}\left(Y^j\,g_j\right)(p)\\&=X^i(p)\left(\frac{\partial Y^j}{\partial x^i}(p)\,g_j(p)+\frac{\partial g_j}{\partial x^i}(p)\,Y^j(p)\right),\end{aligned}\tag{D.1}$$

and by symmetry, after swapping the i's and j's indices, we also get

$$\begin{aligned}Y(X(f))(p)&=Y^i(p)\frac{\partial}{\partial x^i}\left(X^j\,g_j\right)(p)\\&=Y^i(p)\left(\frac{\partial X^j}{\partial x^i}(p)\,g_j(p)+\frac{\partial g_j}{\partial x^i}(p)\,X^j(p)\right).\end{aligned}\tag{D.2}$$

Now we compute the difference (D.1)–(D.2):

$$X(Y(f)) - Y(X(f)) = \left(X^i \frac{\partial Y^j}{\partial x^i}\frac{\partial f}{\partial x^j} - Y^i \frac{\partial X^j}{\partial x^i}\frac{\partial f}{\partial x^j}\right) + \left(\underbrace{X^i \frac{\partial g_j}{\partial x^i} Y^j - Y^i \frac{\partial g_j}{\partial x^i} X^j}_{=0}\right).$$

This concludes the proof.

PROBLEM D.5. Let M be a smooth manifold and let $W \subset M$ be a non-empty open set. Let $X, Y \in \mathfrak{X}(W)$, and let $f, g \in C^\infty(W)$. Prove that

$$[fX, gY] = fg[X,Y] + fX(g)Y - gY(f)X.$$

SOLUTION. In order to make the next computation more readable we will use dots "$\cdot$" to emphasise that we are taking the product of two smooth functions on W. This has the advantage that it becomes very clear where the derivation property (i.e. the Leibniz-rule) is used.

We pick some $h \in C^\infty(W)$ and finish the proof with the following computation:

$$\begin{aligned}
[f\,X, g\,Y](h) &= f \cdot X\left((g\,Y)(h)\right) - g \cdot Y\left((f\,X)(h)\right) \\
&= f \cdot X\left(g \cdot Y(h)\right) - g \cdot Y\left(f \cdot X(h)\right) \\
&= f \cdot \{X(g) \cdot Y(h) + g \cdot X(Y(h))\} - g \cdot \{Y(f) \cdot X(h) + f \cdot Y(X(h))\} \\
&= (f \cdot g)\left([X,Y]\,(h)\right) + f \cdot X(g) \cdot Y(h) - g \cdot Y(f) \cdot X(h).
\end{aligned}$$

PROBLEM D.6. Let $\varphi\colon M \to N$ be a smooth map. Let $X \in \mathfrak{X}(M)$ and $Y \in \mathfrak{X}(N)$. We say that X and Y are **φ-related** if

$$D\varphi(p)X(p) = Y(\varphi(p)), \qquad \forall\, p \in M.$$

Note that if φ is a diffeomorphism then any $X \in \mathfrak{X}(M)$ is φ-related to $\varphi_* X$.

(i) Prove that $X \in \mathfrak{X}(M)$ and $Y \in \mathfrak{X}(N)$ are φ-related if and only if for every open set $V \subset N$ and every smooth function $f \in C^\infty(V)$, one has

$$X(f \circ \varphi) = Y(f) \circ \varphi.$$

(ii) Let $X_i \in \mathfrak{X}(M)$ and $Y_i \in \mathfrak{X}(N)$ for $i = 1, 2$ be vector fields. Assume X_i is φ-related to Y_i for each $i = 1, 2$. Prove that $[X_1, X_2]$ is φ-related to $[Y_1, Y_2]$.

SOLUTION.

(i) Recall that the derivative of φ satisfies, for every p in M and every neighbourhood $V \subset N$ of $\varphi(p)$,

$$X(f \circ \varphi)(p) = D\varphi(p)[X(p)](f) \quad \text{for every } f \in C^\infty(V).$$

Since two tangent vectors $W_1, W_2 \in T_{\varphi(p)}N$ coincide if and only if

$$W_1(f) = W_2(f) \quad \text{for every } f \in C^\infty(V),$$

where $V \subset N$ is any (equiv. some) neighbourhood of $\varphi(p)$, we deduce that the vector field Y is φ-related to X if and only if $Y(f) \circ \varphi(p) = X(f \circ \varphi)(p)$ for every $p \in M$ and $f \in C^\infty(V)$, where V is any (equiv. some) neighbourhood of $\varphi(p)$ in N.

(ii) By (i) it suffices to prove that, for every open set $V \subset N$ (so that $V \cap \varphi(M) \neq \emptyset$, of course) and every $f \in C^\infty(V)$,

$$[X_1, X_2](f \circ \varphi) = [Y_1, Y_2](f) \circ \varphi.$$

Since Y_i is φ-related to X_i we see that

$$X_1(X_2(f \circ \varphi)) = X_1(Y_2(f) \circ \varphi) = Y_1(Y_2(f)) \circ \varphi,$$

and similarly we see that $X_2(X_1(f \circ \varphi)) = Y_2(Y_1(f)) \circ \varphi$. Consequently we deduce that

$$\begin{aligned}[X_1, X_2](f \circ \varphi) &= X_1(X_2(f \circ \varphi)) - X_2(X_1(f \circ \varphi)) \\ &= Y_1(Y_2(f)) \circ \varphi - Y_2(Y_1(f)) \circ \varphi \\ &= [Y_1, Y_2](f) \circ \varphi,\end{aligned}$$

as desired.

(♣) PROBLEM D.7. Let $M \subset N$ be an (immersed or embedded) submanifold and let $p \in M$. We say that a vector field $Y \in \mathfrak{X}(N)$ is **tangent to** M **at** p if $Y(p) \in T_pM \subset T_pM$. We say Y is **tangent to** M if it is tangent to M at every point $p \in M$.

(i) Assume $M \subset N$ is an embedded submanifold. Prove that $Y \in \mathfrak{X}(N)$ is tangent to M if and only if $Y(f)|_M \equiv 0$ for every function $f \in C^\infty(N)$ such that $f|_M \equiv 0$.
Now assume $M \subset N$ is merely an immersed submanifold. Let $\iota\colon M \hookrightarrow N$ denote the inclusion. Assume that $Y \in \mathfrak{X}(N)$ is tangent to M. Prove there exists a unique $X \in \mathfrak{X}(M)$ such that X is ι-related to Y.
Continue to assume that $M \subset N$ is an immersed submanifold. Suppose $Y_1, Y_2 \in \mathfrak{X}(N)$ are tangent to M. Prove that $[Y_1, Y_2]$ is tangent to M.

SOLUTION. We apply Proposition 6.7, according to which there exists a chart $x\colon U \to \mathcal{O}$ on N for every $p \in M$, with $p \in U \overset{\text{open}}{\subset} N$ and $0 = x(p) \in \mathcal{O} \overset{\text{open}}{\subset} \mathbb{R}^n$ such that

$$x(M \cap U) = \mathcal{O} \cap (\mathbb{R}^m \times \{0\}^{n-m}).$$

We will sometimes use the notation $z = (z_1, z_2) \in \mathbb{R}^n \times \mathbb{R}^{m-n} = \mathbb{R}^m$. Now let $Y^x \in \mathfrak{X}(\mathcal{O})$ be the pushforward of Y:

$$Y^x(z) = Dx(q)[Y(q)],$$

where $z = x(q)$. We can expand Y^x in the basis $\frac{\partial}{\partial x^1}, \frac{\partial}{\partial x^2}, \ldots, \frac{\partial}{\partial x^n}$:

$$Y^x(z) = Z^i(z) \left.\frac{\partial}{\partial x^i}\right|_y \quad \forall\, z \in \mathcal{O}.$$

Here we use the Einstein summation convention and $Z^i \in C^\infty(\mathcal{O})$ for every $i = 1, \ldots, n$. Clearly Y being tangent to M is equivalent to $Z^i(z_1, 0) = 0$ for all $(z_1, 0) \in \mathcal{O}$ and every $i = m+1, \ldots, n$.

Assume first that Y is tangent to M, i.e.

$$Z^i(z_1, 0) = 0 \quad \forall\, (z_1, 0) \in \mathcal{O},\; i = m+1, \ldots, n. \tag{D.3}$$

If $f \in C^\infty(N)$ satisfies $f|_M \equiv 0$ then

$$f \circ x^{-1}(z_1, 0) = 0 \quad \forall\, (z_1, 0) \in \mathcal{O}. \tag{D.4}$$

In particular we can compute that for $q \in M$ and $x(q) = z = (z_1, 0)$:

$$\begin{aligned} Y(f)|_q &= Y^\varphi(f \circ x^{-1})|_{(z_1,0)} \\ &= \sum_{i=1}^m Z^i(z_1, 0) \left.\frac{\partial(f \circ x^{-1})}{\partial x^i}\right|_{(z_1,0)} + \sum_{i=m+1}^n Z^i(z_1, 0) \left.\frac{\partial(f \circ x^{-1})}{\partial x^i}\right|_{(z_1,0)} \\ &= 0 \end{aligned}$$

The first sum vanishes because of condition (D.4) and the second vanishes because of condition (D.3).

Assume now that $Y(f)|_M \equiv 0$ for every $f \in C^\infty(N)$ with $f|_M \equiv 0$. We need to show that Y is tangent to M. We do so by contradiction. Hence, assume for contradiction that Y were not tangent to M. Then there exists some chart x as above about p and some $l \in \{m+1, \ldots, n\}$ such that

$$Z^l(x(p)) \neq 0.$$

Now define $\widetilde{f}(z_1, \ldots, z_n) = z_l$. By Lemma 3.2 there exists a compactly supported function $\eta \in C^\infty(\mathcal{O})$ such that $\mathrm{supp}(\eta) \subset \mathcal{O}$ and $\eta \equiv 1$ on $\overline{B} \subset \mathcal{O}$, where B is a small open ball in $\mathcal{O}$ containing $0 = x(p) \in \mathcal{O}$. The function

$$f(z) := \eta(z)\widetilde{f}(z)$$

is a compactly supported function on $\mathcal{O}$, so by extending $f \circ x|_U$ by 0 outside of U we can view $f \circ x \in C^\infty(N)$. Note that $f \circ x|_M \equiv 0$ because $\widetilde{f}|_{\mathbb{R}^m \times \{0\}^{n-m}} \equiv 0$. Now we can compute

$$Y(f \circ x)(p) = Y^x(f)(x(p)) = Z^i(x(p)) \left.\frac{\partial f}{\partial x^i}\right|_{x(p)}$$
$$= Z^i(x(p)) \left.\frac{\partial x^l}{\partial x^i}\right|_{x(p)} = Z^l(x(p)) \neq 0.$$

In this computation we use that $f = x_l$ on a neighbourhood of p. This contradicts our assumption that $Y(f)|_M \equiv 0$ for all $f \in C^\infty(N)$ with $f|_M \equiv 0$ and hence finishes the proof of the first part.

For the second part, let $X \in \mathfrak{X}(M)$ and $Y \in \mathfrak{X}(N)$ be ι-related. i.e.

$$D\iota(p)[X(p)] = Y(\iota(p)) \quad \forall\, p \in M. \tag{D.5}$$

Since $\iota\colon M \to N$ is an immersion, $D\iota(p)\colon T_pM \to T_{\iota(p)}N$ is injective for every $p \in M$. The assumption that Y is tangent to M means that $Y(\iota(p)) \in \text{Image}(D\iota(p))$ for every $p \in M$. Hence (D.5) defines X pointwise uniquely. That X is smooth follows from $X(p) = (D\iota(p))^{-1}Y(\iota(p))$ and the fact that ι is a local embedding (cf. Proposition 6.3).

The $[Y_1, Y_2]$ being tangent to M is clearly a local property: We need to show that for a given $p \in M \subset N$ we have $[Y_1, Y_2](p) \in T_pM$. Since any immersion is a local embedding, there exists a neighbourhood $U \subset M$ of p such that $U \subset N$ is an *embedded* submanifold. Since $T_pU = T_pM$ (as explained in the first exercise) it suffices to show that $[Y_1, Y_2](p) \in T_pU$. Given $f \in C^\infty(N)$ with $f|_U \equiv 0$ we compute using (i) and the fact that Y_1 and Y_2 are tangent to U:

$$[Y_1, Y_2](f)(p) = Y_1(Y_2(f))(p) - Y_2(Y_1(f))(p)$$
$$= Y_1(0)(p) - Y_2(0)(p) = 0$$

This shows that $[Y_1, Y_2](x) \in T_pU$ and thus finishes the exercise.

Problem Sheet E

Problem E.1. Let $J_0 \in \text{Mat}(2n)$ denote the matrix

$$J_0 := \begin{pmatrix} 0 & -I \\ I & 0 \end{pmatrix},$$

where I is the $n \times n$ identity matrix. The **symplectic linear group** $\text{Sp}(2n)$ consists of the matrices A such that $A^T J_0 A = J_0$. Prove that $\text{Sp}(2n)$ is a Lie group. Compute its dimension, and compute its Lie algebra $\mathfrak{sp}(2n)$.

Solution. We follow the same approach as in Proposition 10.18.

First of all we observe that the matrix $A^T J_0 A$ is antisymmetric, i.e. its transpose is equal to minus itself. In other words, the expression $A^T J_0 A$ lives inside the vector space

$$\text{Antsym}(2n) = \left\{ B \in \text{Mat}(2n) \middle|\, B^T = -B \right\}.$$

It is straightforward to see that the (real) vector space dimension is given by

$$\dim_{\mathbb{R}} \operatorname{Antsym}(2n) = \frac{2n(2n+1)}{2} - 2n = \frac{2n(2n-1)}{2}.$$

We now define the smooth function

$$\varphi\colon \operatorname{Mat}(2n) \longrightarrow \operatorname{Antsym}(2n),\ A \mapsto A^T J_0 A.$$

In view of $\varphi^{-1}(J_0) = \operatorname{Sp}(2n)$, we want to show that J_0 is a regular value of φ so that $\operatorname{Sp}(2n)$ inherits a smooth manifold structure by means of the Implicit Function Theorem, more precisely Theorem 6.10.

Let $A_0 \in \varphi^{-1}(J_0)$, $B \in \operatorname{Mat}(2n)$ and compute

$$\begin{aligned} D\varphi(A_0)\left(\mathcal{J}_{A_0}B\right) &= \frac{d}{dt}\Big|_{t=0}(\varphi\left(A_0 + t \cdot B\right))(t) \\ &= \frac{d}{dt}\Big|_{t=0}\left(tA_0^T J_0 B + tB^T J_0 A_0 + A_0^T J_0 A_0 + t^2 B^T J_0 B\right)(t) \\ &= B^T J_0 A_0 + A_0^T J_0 B. \end{aligned} \tag{E.1}$$

For any antisymmetric matrix C we can now define $B = -\frac{1}{2}A_0 J_0 C$ and plug it into the differential above. Using the fact that $A_0 = A_0^T J_0 A_0$ one then obtains

$$D\varphi(A_0)[\mathcal{J}_{A_0}B] = C,$$

which shows that $D\varphi(A_0)$ is surjective for any element A_0 in the preimage $\varphi^{-1}(J_0)$ and equivalently that J_0 is a regular value of φ. Thus the Implicit Function Theorem tells us that $\operatorname{Sp}(2n)$ is a smooth manifold of dimension

$$\dim(\operatorname{Sp}(2n)) = 4n^2 - \dim(\operatorname{Antsym}(2n)) = 4n^2 - \frac{4n^2 - 2n}{2} = 2n^2 + n.$$

Proposition 10.15 tells us that $\operatorname{Sp}(2n)$ is automatically a Lie group and therefore we are only left to compute the Lie algebra $\mathfrak{sp}(2n)$:

$$\begin{aligned} \mathfrak{sp}(2n) &= T_I \operatorname{Sp}(2n) && \text{by Definition 11.1} \\ &\cong \ker D\varphi(I) && \text{by Proposition 6.15} \\ &= \left\{B \in \operatorname{Mat}(2n) \middle| B^T J_0 + J_0 B = 0\right\} && \text{by (E.1)} \end{aligned}$$

Problem E.2. Let G be a Lie group with Lie algebra $\mathfrak{g}$. Suppose $\gamma\colon \mathbb{R} \to G$ is a smooth curve with $\gamma(0) = e$. Set $\xi := \gamma'(0)$. Prove that γ is a one-parameter subgroup if and only if

$$\Phi_t^\xi = r_{\gamma(t)}$$

(this is an equality of diffeomorphisms of G).

SOLUTION. We will prove that the following three are equivalent

(i) γ is a one-parameter subgroup,
(ii) $\gamma(t) = \gamma^{\xi}(t)$,[12]
(iii) $\Phi_t^{\xi} = r_{\gamma(t)}$.

In particular, the equivalence between (i) and (iii) proves the statement of the exercise.

We already know that condition (i) is equivalent to condition (ii) by Proposition 11.14. To see that (iii) implies (ii) observe that if (iii) holds then

$$\gamma^{\xi}(t) = \Phi_t^{\xi}(e) = r_{\gamma(t)}e = \gamma(t).$$

so that condition (ii) holds. Now assume that (ii) holds, and fix $g \in G$. Then since X_{ξ} is left-invariant, one can easily check that $g\gamma^{\xi} = l_g \circ \gamma^{\xi}$ is another integral curve of X_{ξ} with initial condition g, and thus by uniqueness of integral curves one last time we have $r_{\gamma(t)}(g) = g\gamma^{\xi}(t) = \Phi_t^{\xi}(g)$. Since g was arbitrary, (iii) holds. This concludes the proof.

PROBLEM E.3. Prove that the Lie bracket on $\mathfrak{gl}(n)$ is given by matrix commutation, i.e.

$$[A, B] = AB - BA, \qquad \forall\, A, B \in \mathfrak{gl}(n) = \mathrm{Mat}(n).$$

SOLUTION. Fix $C, D \in \mathrm{GL}(n)$. Using the canonical "dash to dot" map, we want to prove

$$[\mathcal{J}_I(C), \mathcal{J}_I(D)] = \mathcal{J}_I(CD) - \mathcal{J}_I(DC) = \mathcal{J}_I(CD - DC),$$

where the left-hand side $[\cdot,\cdot]$ is the Lie bracket in $\mathfrak{gl}(n)$. This is sufficient since every $A, B \in \mathfrak{gl}(n) = T_I\,\mathrm{GL}(n)$ can be written as $A = \mathcal{J}_I(C)$ and $B = \mathcal{J}_I(D)$.

Let X_C and X_D denote the unique left-invariant vector fields on $\mathrm{GL}(n)$ determined by $X_C(I) = \mathcal{J}_I(C)$ and $X_D(I) = \mathcal{J}_I(D)$. By definition of the Lie bracket on $\mathfrak{gl}(n)$ one has

$$[\mathcal{J}_I(C), \mathcal{J}_I(D)] = [X_C, X_D](I).$$

Fix an arbitrary matrix $F \in \mathrm{GL}(n)$. Since $l_F\colon \mathrm{GL}(n) \longrightarrow \mathrm{GL}(n)$ is itself a linear map, one has, by Problem B.3, that

$$\mathcal{J}_F \circ l_F = Dl_F \circ \mathcal{J}_I = \mathcal{J}_I$$

thus

$$\mathcal{J}_F(FC) = Dl_F\,\mathcal{J}_I(C) = Dl_F\,X_C(I) = X_C(F), \quad \forall C \in \mathrm{GL}(n),$$

by left-invariance of X_C. Now let $u^i_j\colon \mathrm{GL}(n) \to \mathbb{R}$ denote the function that assigns to a matrix F its (i, j)-th entry. Then

[12] See notation after Proposition 11.13.

$$X_D(u_j^i)(F) = X_D(F)(u_j^i) = \mathcal{J}_F(FD)(u_j^i),$$

which is equal to the (i, j)-th entry of FD. Thus as functions we have

$$X_D(u_j^i) = u_j^i \circ r_D,$$

where r_D is right-translation by D. Let $\gamma(t) = I + tC$, so that $\mathcal{J}_I(C) = \dot{\gamma}(0)$. Then

$$X_C(I)(X_D(u_j^i)) = \dot{\gamma}(0)(u_j^i \circ r_D) = \frac{d}{dt}\Big|_{t=0} u_j^i(D + tCD),$$

which is the (i, j)-th entry of CD. Thus

$$[X_C, X_D](I)(u_j^i) = X_C(I)(X_D(u_j^i)) - X_D(I)(X_C(u_j^i))$$

is the (i, j)th entry of $CD - DC$. This shows that

$$[\mathcal{J}_I(C), \mathcal{J}_I(D)](u_j^i) = \mathcal{J}_I(CD - DC)(u_j^i).$$

Since for any F, one has[13]

$$\mathcal{J}_I(F) = \mathcal{J}_I(F)(u_j^i) \frac{\partial}{\partial u_j^i},$$

the claim follows.

PROBLEM E.4. Let $\varphi\colon M \to N$ be a smooth map between two smooth manifolds. Let $X \in \mathfrak{X}(M)$ and $Y \in \mathfrak{X}(N)$, and assume X and Y are φ-related in the sense of Problem D.6. Let Φ_t and Ψ_t denote the respective flows, with domains $M_t \subset M$ and $N_t \subset N$ respectively. Prove that $\varphi(M_t) \subset N_t$ and that

$$\Psi_t \circ \varphi = \varphi \circ \Phi_t, \qquad \text{on } M_t.$$

Deduce that if φ is a diffeomorphism then for any vector field X with flow Φ_t, the flow of $\varphi_* X$ is given by $\varphi \circ \Phi_t \circ \varphi^{-1}$.

SOLUTION. Recall that, by definition, a point x belongs to M_t when the solution of the Cauchy problem

$$\begin{cases} \dot{\gamma}_p(s) = X(\gamma_p(s)), \\ \gamma_p(0) = p, \end{cases}$$

[13] In this formula, in order to make the Einstein Summation Convention work, in the term $\frac{\partial}{\partial u_j^i}$, the i is a lower index and the j is an upper index…

is defined for $s = t$. If we want to prove that $\varphi(p)$ belongs to N_t we need to show that the solution of the Cauchy problem

$$\begin{cases} \dot{\eta}_p(s) = Y(\eta_p(s)), \\ \eta_p(0) = \varphi(p), \end{cases} \tag{E.2}$$

is defined for $s = t$. We claim that the solution to such a problem is the curve

$$\eta_p(s) = \varphi(\gamma_p(s)).$$

Indeed, there holds $\eta_p(0) = \varphi(\gamma_p(0)) = \varphi(p)$, and since X and Y are φ-related, we can compute:

$$\dot{\eta}_p(s) = D\varphi(\gamma_p(s))[X(\gamma_p(s))] = Y(\varphi(\gamma_p(s))) = Y(\eta_p(s)).$$

Hence η_p is the (unique) solution of (E.2), and, since γ_p is defined at $s = t$, so is ζ_x. Note moreover, by definition of the flows Φ_t and Ψ_t, we simply have

$$\Phi_t(p) = \gamma_p(t) \quad \text{and} \quad \Psi_t(\varphi(p)) = \eta_p(t),$$

hence, as $\eta_p(t) = \varphi(\gamma_p(t))$, we also showed that

$$(\Psi_t \circ \varphi)(p) = (\varphi \circ \Phi_t)(p) \quad \text{for } p \in M_t.$$

Finally note that if φ is a diffeomorphism, we can compose both sides of the above relation with φ^{-1} and notice that Y is φ-related to X precisely if $Y = \varphi_\star(p)$.

(♣) Problem E.5. Let X and Y be vector fields on a smooth manifold M with flows Φ_t and Ψ_t respectively. Prove that $[X, Y] \equiv 0$ if and only if the two flows commute, i.e.

$$\Phi_t \circ \Psi_s(p) = \Psi_s \circ \Phi_t(p)$$

for all s, t small and for all p in the domains of both $\Phi_t \circ \Psi_s$ and $\Psi_s \circ \Phi_t$.

Solution. Recall first of all the fundamental identity

$$\begin{aligned} [X, Y](p) &= \mathcal{L}_X Y(p) \\ &= \frac{d}{d\varepsilon}\left(D(\Phi_{-\varepsilon})(\Phi_\varepsilon(p))[Y(\Phi_\varepsilon(p))] \right)\Big|_{\varepsilon=0} \qquad \text{for } x \in M. \end{aligned} \tag{E.3}$$

If the flows of X and Y commute, we deduce that Y *is* X*-invariant*, that is, for every $p \in M$ and every sufficiently small values s, t so that the expressions below are defined, we have

$$D(\Phi_s)(p)[Y(p)] = \frac{d}{dt}(\Phi_s \circ \Psi_t)(p)\bigg|_{t=0} = \frac{d}{dt}(\Psi_t \circ \Phi_s)(p)\bigg|_{t=0} = Y(\Phi_s(p)).$$

In particular, for any $p \in M$, if ε is sufficiently small we may choose $p = \Phi_{-\varepsilon}(q)$ and thus deduce that

$$Y(p) = D(\Phi_{-\varepsilon})(\Phi_\varepsilon(p))[Y(\Phi_\varepsilon(p))].$$

But $Y(p)$ does not depend on ε, so plugging this in (E.3) shows $[X, Y](p) = 0$ and thus $[X, Y] = 0$ as p was arbitrary.

Conversely, let us assume that the bracket of X and Y vanishes. Let us prove first that Y is X-invariant, that is, for every $p \in M$ there holds

$$D(\Phi_\varepsilon)(p)[Y(p)] = Y(\Phi_\varepsilon(p)),$$

for every sufficiently small value of ε so that this expression is defined. This is equivalent to proving that the vector field

$$D(\Phi_{-\varepsilon})(\Phi_\varepsilon(p))[Y(\Phi_\varepsilon(p))]$$

is constant with respect to ε.[14] In the computation that follows, we use the so-called semigroup property of the flow:

$$\Phi_{s+t}(p) = \Phi_s(\Phi_t(p)),$$

which implies, when differentiated and applied to Y, that

$$D(\Phi_{s+t})(p)[Y(p)] = D(\Phi_s)(\Phi_s(p))[D\Phi_s(p)[Y(p)]],$$

for every s, t so that the above expressions are defined. We may then compute:

$$\begin{aligned}
&\frac{d}{d\varepsilon}\left(D(\Phi_{-\varepsilon})(\Phi_\varepsilon(p))[Y(\Phi_\varepsilon(p))]\right)\bigg|_{\varepsilon=\varepsilon_0} \\
&= \frac{d}{d\varepsilon}\left(D(\Phi_{-\varepsilon-\varepsilon_0})(\Phi_{\varepsilon+\varepsilon_0}(p))[Y(\Phi_{\varepsilon+\varepsilon_0}(p))]\right)\bigg|_{\varepsilon=0} \\
&= \frac{d}{d\varepsilon}\left(D(\Phi_{-\varepsilon_0})(\Phi_{\varepsilon_0}(p))\big[D(\Phi_{-\varepsilon})(\Phi_{\varepsilon+\varepsilon_0}(p))[Y(\Phi_{\varepsilon+\varepsilon_0}(p))]\big]\right)\bigg|_{\varepsilon=0} \\
&= \frac{d}{d\varepsilon}\left(D(\Phi_{-\varepsilon_0})(\Phi_{\varepsilon_0}(p))\left[D(\Phi_{-\varepsilon})(\Phi_\varepsilon(\Phi_{\varepsilon_0}(p))[Y(\Phi_\varepsilon(\Phi_{\varepsilon_0}(p))]\right]\right)\bigg|_{\varepsilon=0} \\
&= D(\Phi_{-\varepsilon_0}(\Phi_{\varepsilon_0}(p))\frac{d}{d\varepsilon}\left[D(\Phi_{-\varepsilon})(\Phi_\varepsilon(\Phi_{\varepsilon_0}(p))[Y(\Phi_\varepsilon(\Phi_{\varepsilon_0}(p))]\right]\bigg|_{\varepsilon=0} \\
&= D(\Phi_{\varepsilon_0}(\Phi_\varepsilon(p))\big[[X, Y](\Phi_{\varepsilon_0}(p))\big]\bigg|_{\varepsilon=0} = 0,
\end{aligned}$$

[14] Check this! Note: you might want to use that Φ_0 is the identity.

and hence Y is X-invariant. As a consequence, if for fixed t and p we consider the curve $\eta(s) = (\Phi_t \circ \Psi_s)(p)$, differentiating σ it with respect to s yields

$$\begin{aligned}\dot{\eta}(s) &= D(\Phi_t)(\Psi_s(p))[Y(\Psi_s(p))]\\ &= Y(\Phi_t(\Psi_s(p)))\\ &= Y(\eta(s)),\end{aligned}$$

and since $\eta(0) = \Phi_t(p)$, we deduce that η is the integral curve for Y at the point $\Phi_t(p)$, that is $\eta(s) = \Psi_s(\Phi_t(p))$, and from the definition of η, this means precisely that the flows of X and Y commute.

PROBLEM E.6. Prove that if a Lie group is abelian then its Lie algebra is abelian.

SOLUTION. Let G be an abelian Lie group with Lie algebra $\mathfrak{g}$. We first show that the inversion map $i\colon G \to G$ is a Lie group isomorphism. Observe that

$$i(gh) = (gh)^{-1} = h^{-1}g^{-1} = g^{-1}h^{-1} = i(g)i(h),$$

so i is a group homomorphism, hence a Lie group homomorphism since it is smooth by assumption. But i is clearly bijective, thus it follows from Corollary 10.13 that i is a Lie group isomorphism, as desired. We now apply Proposition 11.7 to deduce that $Di(e)\colon \mathfrak{g} \to \mathfrak{g}$ is a Lie algebra isomorphism.

Recall that we can identify $T_{(e,e)}(G \times G)$ with $\mathfrak{g} \oplus \mathfrak{g}$ as in Problem C.1. We need the following lemma:

LEMMA Let G be a Lie group with multiplication map $m\colon G \times G \to G$ and inversion map $i\colon G \to G$. Then

(a) $Dm(e,e)\colon \mathfrak{g} \oplus \mathfrak{g} \to \mathfrak{g}$ is the map $(X,Y) \mapsto X + Y$.
(b) $Di(e)\colon \mathfrak{g} \to \mathfrak{g}$ is the map $X \mapsto -X$.

Proof. Using the fact that $Dm(e,e)$ is linear as well as the chain rule, we obtain

$$\begin{aligned}Dm(e,e)(X,Y) &= Dm(e,e)(X,0) + Dm(e,e)(Y,0)\\ &= Dm_1(e)(X) + Dm_2(e)(Y),\end{aligned}$$

where m_1 and m_2 are the smooth maps from G to itself defined respectively by $x \mapsto m(a,e)$ and $b \mapsto m(e,b)$. Since $m_1 = m_2 = \mathrm{Id}_G$, this proves (a).

Now consider the maps $p\colon G\times G \to G\times G$ given by $p(g,h) = (g,h^{-1})$ and $\triangle\colon G \to G \times G$ given by $\triangle(g) = (g,g)$. Then the composite $n := m \circ p \circ \triangle$ is the constant map $g \mapsto e$. We thus have

$$
\begin{aligned}
0 &= Dn(e)(X) \\
&= Dm(e,e)\Big(Dp(e,e)\big(D\Delta(e)(X)\big)\Big) \\
&= Dm(e,e)\big(Dp(e,e)(X,X)\big) \\
&= Dm(e,e)\big(X, Di_e(X)\big) \\
&= X + Di(e)(X),
\end{aligned}
$$

where we have used (a) and the chain rule for manifolds. This proves (b). ■

Using part (b) of the lemma and the fact that $Di(e)$ is an isomorphism of Lie algebras, we obtain

$$[X,Y] = [-X,-Y] = [Di(e)(X), Di(e)(Y)] = Di(e)[X,Y] = -[X,Y],$$

and hence $[X,Y] = 0$ for all $X, Y \in \mathfrak{g}$. Thus $\mathfrak{g}$ is abelian.

(♣) PROBLEM E.7. Let G be a Lie group and suppose H is a subgroup of G which is also an embedded submanifold. Prove that H is closed in G (as a subspace).

SOLUTION. Choose a sequence $(h_k)_{k\in\mathbb{N}}$ such that

$$h_k \to g \in G$$

for $k \to \infty$. We need to show that $g \in H$. Since the identity element $e \in H \subset G$ we can choose a slice chart $\varphi\colon U \to \mathcal{O}$ with $e \in U \subset G$ and $0 = \varphi(e) \in \mathcal{O} \subset \mathbb{R}^n$. Recall that this means that $\varphi(H \cap U) = \mathcal{O} \cap (\mathbb{R}^k \times \{0\}^{n-k})$ for k equal to the dimension of H (as a submanifold). Choose an open subset $e \in W \subset U$ such that $\overline{W} \subset U$. Note that this implies (by the properties of a slice chart) that

$$\overline{W} \cap H \text{ is a closed subset of } G. \tag{E.4}$$

Now, since the map $f(g_1, g_2) = g_1^{-1} g_2$ is continuous, there exists an open non-empty set $e \in V \subset G$ such that

$$V \times V \subset f^{-1}(W).$$

By continuity of f again we have

$$g^{-1} h_k = f(g, h_k) \overset{k\to\infty}{\longrightarrow} e.$$

In particular, there exists $K \in \mathbb{N}$ such that $g^{-1} h_k \in V$ for all $k \geq K$. For $j \geq K$ we therefore have

$$h_K^{-1} h_j = (g^{-1} h_K)^{-1} (g^{-1} h_j) = f(g^{-1} h_K, g^{-1} h_j) \in W,$$

which implies

$$h_K^{-1} g = \lim_{j \to \infty} h_K^{-1} h_j \in \overline{W}.$$

Since each $h_j^{-1} h_K \in H$ (for $j \geq K$) it follows from (E.4) that $h_K^{-1} g \in H$. In particular, $g = (h_K)(h_K^{-1} g) \in H$ which finishes the proof.

Problem Sheet F

Problem F.1.

(i) Let M be a smooth manifold. Assume there exist vector fields $X_1, \ldots, X_m \in \mathfrak{X}(M)$ such that $\{X_i(p)\}$ is a basis of $T_p M$ for every $p \in M$. Prove that the tangent bundle TM is diffeomorphic to $M \times \mathbb{R}^m$.
(ii) Let G be a Lie group with Lie algebra $\mathfrak{g}$. Prove that TG is diffeomorphic to $G \times \mathfrak{g}$.

Solution. For each point in the tangent bundle $(p, \xi) \in TM$ we write

$$\xi = \xi^i(p)\, X_i(p) \in T_p M,$$

where $\xi^i(p) \in \mathbb{R}$ are the unique coordinates of ξ with respect to the basis $\{X_i(p)\}$ of $T_p M$. This allows us to define

$$\varphi \colon TM \to M \times \mathbb{R}^m,\ (p, \xi) \mapsto (p, \xi^i(x)\, e_i).$$

The smoothness of the map φ is a direct consequence of the smoothness of the vector fields X_i. Alternatively, one can check directly that this map is smooth, which boils down to the proof of Theorem 5.6.

In order to see that φ defines a diffeomorphism we consider the function

$$\psi \colon M \times \mathbb{R}^m \to TM,\ (q, \zeta) = (q, \zeta^j\, e_j) \mapsto (q, \zeta^j\, X_j(q)).$$

The map ψ is also smooth (same reason as before) and it is an inverse to φ:

$$(\psi \circ \varphi)(p, \xi) = \psi(p, \xi^j(p)\, e_j) = (p, \xi^j(p)\, X_j(p)) = (p, \xi).$$

For part (ii): in view of part (i) it suffices to find vector fields $X_1, \ldots X_m \colon G \to TG$ that form a basis of $T_g G$ at every point $g \in G$. We fix a basis

$$(v_1, \ldots, v_n) \subset \mathfrak{g} = T_e G$$

and set

$$X_i(g) = Dl_g(e)[v_i],$$

for all $i = 1, \ldots, m$. Recall that $l_g \colon G \to G,\ b \mapsto ab$ is a diffeomorphism, in particular $Dl_g(e) \colon T_e G \to T_a G$ is a vector space isomorphism and therefore our vector fields do indeed satisfy the assumptions in the first part of the exercise. We can also write down the diffeomorphism φ explicitly:

$$\varphi\colon TG \to G \times \mathfrak{g} = G \times T_eG,\ (g, u) = (g, u^i\, X_i(g)) \mapsto (g, u^i\, v_i).$$

Problem F.2. Let $A \in \mathfrak{gl}(m)$. Prove that the matrix exponential

$$e^A := \sum_{k=0}^{\infty} \frac{1}{k!} A^k$$

converges and defines an element of $\mathrm{GL}(m)$. Prove that $A \mapsto e^A$ is the exponential map of $\mathrm{GL}(m)$.

Solution. We first check that the series converges. The norm on $\mathfrak{gl}(m)$, obtained via identification with $\mathbb{R}^{n^2}$, is given by $|A| = \sqrt{\sum_{i,j} a_{ij}^2}$. This is called the *Frobenius norm* on $\mathfrak{gl}(m)$. The norm is submultiplicative, i.e. $|AB| \le |A||B|$, and hence by induction $|A^k| \le |A|^k$. Using this, we apply the Weierstrass M-test to deduce that the matrix exponential $\exp(A)$ converges uniformly on any bounded subset of $\mathfrak{gl}(m)$, by comparison with the series $\sum_k (1/k!)c^k = e^c$ for suitable $c \in \mathbb{R}$.

The exponential map, which we denote by

$$\widetilde{\exp}\colon \mathfrak{gl}(m) \to \mathrm{GL}(m),$$

in order to distinguish it from the matrix exponential, is defined via $\widetilde{\exp}(A) := \gamma(1)$, where γ is the unique integral curve of X_A at the identity I. The integral curve γ is determined by the initial value problem:

$$\dot{\gamma}(t) = X_A(\gamma(t)),\ \ \gamma(0) = I. \tag{F.1}$$

One can show that $\mathcal{J}_B^{-1}(X_A(B)) = B\mathcal{J}_I^{-1}(A)$ for each $B \in \mathrm{GL}(m)$. Indeed, define $\eta(t) = I + t \cdot \mathcal{J}_I^{-1}A$. Then by definition of the $\mathcal{J}$-map we have $\dot{\eta}(0) = A$, thus using the chain rule we get

$$\begin{aligned}
X_A(B) &= Dl_B(I)(X_A(I)) \\
&= Dl_B(I)(A) \\
&= Dl_B(I)(\dot{\eta}(0)) \\
&= Dl_B(I) \circ D\eta(0)\frac{\partial}{\partial t}\Big|_{t=0} \\
&= D(l_B \circ \eta)(0)\,\frac{\partial}{\partial t}\Big|_{t=0} \\
&= \dot{\beta}(0),
\end{aligned}$$

where $\beta(t) = l_B \circ \eta(t) = B + tB\mathcal{J}_I^{-1}(A)$. Applying $\mathcal{J}_B^{-1}$ to the above then proves $\mathcal{J}_B^{-1}(X_A(B)) = B\mathcal{J}_I^{-1}(A)$. Therefore (F.1) becomes the matrix equation

$$\gamma'(t) = \mathcal{J}_{\gamma(t)}^{-1}\dot{\gamma}(t) = \gamma(t)\mathcal{J}_I^{-1}A. \tag{F.2}$$

Let us abbreviate $\underline{A} = \mathcal{J}_I^{-1}A$. We claim that $\gamma(t) := \exp(t\underline{A})$ satisfies (F.2). To prove the claim, we differentiate the series $\exp(t\underline{A})$ formally term-by-term to obtain

$$\gamma'(t) = \sum_{k=1}^{\infty} \frac{k}{k!}\underline{A}^k = \Big(\sum_{k=1}^{\infty} \frac{1}{(k-1)!}t^{k-1}\underline{A}^{k-1}\Big)\underline{A} = \gamma(t)\underline{A}.$$

The differentiated series also converges uniformly on bounded sets since it only differs from the series for $\gamma(t)$ by a factor of $\underline{A}$. The term-by-term differentiation is thus justified, and we see that γ satisfies (F.2), as desired. All in all this means that

$$\widetilde{\exp} \circ \mathcal{J}_I = \exp\,.$$

It only remains to show that $\gamma(t)$ is invertible for all t so that, in particular, $\exp(\underline{A}) \in \mathrm{GL}(m)$. Define $\sigma(t) := \gamma(t)\gamma(-t)$. Then σ is a smooth curve in $\mathrm{GL}(m)$ and satisfies

$$\sigma'(t) = \big(\gamma(t)\underline{A}\big)\gamma(-t) - \gamma(t)\big(\underline{A}\gamma(-t)\big) = 0.$$

It follows that σ is the constant curve $\sigma(t) = I$. It follows that $\gamma(t)$ is invertible with inverse $\gamma(-t)$. This finishes the proof.

Problem F.3. Let G be a Lie group with Lie algebra $\mathfrak{g}$. Prove that for $\xi, \zeta \in \mathfrak{g}$ one has $\mathrm{ad}_\xi(\zeta) = [\xi, \zeta]$.

Solution. First note that by definition

$$\begin{aligned} \mathrm{ad}_\xi(\zeta) &= D(\mathrm{Ad})(e)[\xi][\zeta] \\ &= \left(\frac{d}{dt}\Big|_{t=0} \mathrm{Ad}(\exp(t\xi))\right)[\zeta] \\ &= \frac{d}{dt}\Big|_{t=0} \mathrm{Ad}_{\exp(t\xi)}(\zeta) \\ &= \frac{d}{dt}\Big|_{t=0} D(c_{\exp(t\xi)})(e)[\zeta], \end{aligned}$$

where c_g is the conjugation by $g \in G$. Now writing $c_g = r_{g^{-1}} \circ l_g$ and letting X_ξ and Φ_t^ξ be the left-invariant vector field associated to ξ and the flow thereof, respectively, we have:

$$
\begin{aligned}
\left.\frac{d}{dt}\right|_{t=0} D(c_{\exp(t\xi)})(e)[\zeta] &= \left.\frac{d}{dt}\right|_{t=0} Dr_{\exp(-t\xi)} \circ Dl_{\exp(t\xi)}(e)[X_\zeta(e)] \\
&= \left.\frac{d}{dt}\right|_{t=0} Dr_{\exp(-t\xi)}(\exp(t\xi))[X_\zeta(\exp(t\xi))] \\
&= \left.\frac{d}{dt}\right|_{t=0} D\Phi^\xi_{-t}(\Phi^\xi_t(e))[X_\zeta(\Phi^\xi_t(e)] \\
&= \mathcal{L}_{X_\xi} X_\zeta(e).
\end{aligned}
$$

where the second equality used the left-invariance of X_ζ, the third equality used Proposition 12.2 and the last equality is the definition of the Lie derivative (Definition 10.3). Then finally by Proposition 10.4 and Theorem 11.5 we have

$$
\mathcal{L}_{X_\xi} X_\zeta(e) = [X_\xi, X_\zeta](e) = [\xi, \zeta].
$$

Problem F.4. Let σ be a smooth action of a Lie group G on a smooth manifold M.

(i) Prove that σ is proper if and only if the following condition holds: if (p_k) is a sequence in M and (g_k) is a sequence in G such that both (p_k) and $(\sigma_{g_k}(p_k))$ converge, then a subsequence of (g_k) converges.

(ii) Deduce that if G is compact then every smooth action is proper.

Solution. First we prove the " $\implies$ " direction: Let $p, q \in M$ be arbitrary and let (p_k) and (g_k) satisfy the conditions above. Define the set

$$
K := \{(\sigma_{g_k}(p_k), p_k)\}_{k \in \mathbb{N}} \subset M \times M
$$

and observe that it is compact. Since σ is proper the set $(\sigma, \mathrm{id})^{-1}K$ is compact as well. But the sequence (g_k, p_k) is contained in the compact set $(\sigma, \mathrm{id})^{-1}K$, thus it admits a convergent subsequence (g_{k_n}, p_{k_n}). In particular, g_{k_n} converges to some $g \in G$

Conversely, let $K \subset M \times M$ be any compact set and let (g_k, p_k) be an arbitrary sequence in $(\sigma, \mathrm{id})^{-1}K$. But then $(\sigma_{g_k}(p_k), p_k)$ belongs to K and thus admits a convergent subsequence. It follows, up to passing to a subsequence, that $p_k \to p$ and $g_k \to g$ as well by the hypothesis. This proves that $(\sigma, \mathrm{id})^{-1}K$ is compact and hence that σ is proper.

For the second part observe that if G is compact, then any sequence (g_k) in G admits a convergent subsequence. In particular, using the equivalence in part (i), we deduce that any action $\sigma \colon G \times M \to M$ must be proper.

Problem F.5. Let G be a Lie group and H be a closed subgroup (possibly equal to G). Let H act on G via left (or right) translations. Prove that this action is proper.

SOLUTION. We only prove the statement for left translations—the proof for right translations is analogous.

In order to show that $l\colon H \times G \to G$ is a proper action we will use the equivalent description from part (i) of Problem F.4:

We pick $g_k \to g$ in G, and another sequence (h_k) in H such that $l_{h_k}(g_k)$ converges. This means that $h_k \cdot g_k \to a$ for some $a \in G$, in particular

$$h_k = ag_k^{-1} \to ag^{-1},$$

by smoothness of r^{-1} and $g_k \to g$. But $H \subset G$ is closed, hence the limit ag^{-1} belongs to H and thus the equivalent description of proper actions in Problem F.4 is satisfied.[15]

PROBLEM F.6. Let σ be a proper smooth action of a Lie group G on a smooth manifold M. Prove that the orbits $\mathsf{orb}_\sigma(p)$ are closed subsets of M.

SOLUTION. Let $\sigma_{g_k}(p) \to x$. We need to show that x belongs to $\mathsf{orb}_\sigma(p)$, i.e. that there exists some $g \in G$ such that $\sigma_g(p) = x$. But using the equivalent description of properness in part (i) of Problem F.4 we deduce that g_k admits a convergent subsequence g_{k_n} with limit, say $g \in G$. But then one necessarily has $\sigma_{g_{k_n}}(p) \to \sigma_g(p) = x$ by uniqueness of limits. This proves that $\mathsf{orb}_\sigma(p)$ is closed and since p was arbitrary this finishes the proof.

PROBLEM F.7. Let Δ be an integrable distribution on a smooth manifold M, and let L be a connected integral manifold of Δ. Assume that L is closed in M. Prove that L is a leaf of foliation induced by Δ.

SOLUTION. By the Global Frobenius Theorem (Theorem 15.4) we know that the distribution Δ is induced by some foliation $\mathcal{F}$. Let L be as above and denote by L' a leaf of $\mathcal{F}$ such that $L \subseteq L'$.

To prove $L = L'$ we will show that L is open and closed with respect to the subspace topology of L'—this suffices since L' is connected and L is non-empty.

It is clear that $L \subseteq L'$ is closed since $L \subseteq M$ is already closed. To prove that $L \subseteq L'$ is open, we pick an arbitrary $p \in L$ and a shifted slice chart (U, x) around p (with slice $L(c)$) as in the statement of the Local Frobenius Theorem (Theorem 14.13).

We claim that the set

---

[15] Note that it's crucial that the sequence converges within H as G takes the role of the manifold M here!

$$L \cap U$$

is an open neighbourhood in L'. The claim asserts that every point $p \in L$ is an interior point, hence that $L \subseteq L'$ is open. Therefore we are only left to prove the claim.

First of all observe that

$$L' \cap U = L(c),$$

by the Local Frobenius Theorem and the maximal property of the leaf L'. This means that $L(c)$ is open in the subspace topology of L'. At the same time, $L \cap U \subseteq L(c)$ which with the above inclusions proves

$$L \cap U = L \cap L(c).$$

Note that L is immersed in M by assumption, and so is $L \cap U = L \cap L(c)$ then. In particular, $L \cap U = L \cap L(c) \hookrightarrow L(c)$ is an injective immersion, which defines an embedding by dimension reasons. In particular $L \cap U$ is open in $L(c)$ and since the latter is open in L' (as seen above) we conclude that $L \cap U \subseteq L'$ is open and the claim is proven.

Problem F.8. Let $\varphi \colon M \to N$ be a surjective submersion. Prove that the connected components of the preimages $\varphi^{-1}(p)$ as p ranges over N defines an $(m-n)$-dimensional foliation of M.

Solution. Since φ is a submersion, as a consequence of the Implicit Function Theorem, for every $p \in N$, each connected component of $\varphi^{-1}(p)$ is an embedded $(m-n)$-dimensional submanifold of M and, if $q \in \varphi^{-1}(p)$, there holds

$$T_q(\varphi^{-1}(p)) = \ker(D\varphi(q)),$$

where we are identifying $T_q(\varphi^{-1}(p))$ as a subspace of T_qM. As a consequence, if we set

$$\Delta_q = T_q(\varphi^{-1}(p)),$$

we have a mapping that to each $q \in M$ assigns a $(m-n)$-dimensional subspace of T_qM. To verify that this mapping is smooth, we proceed as follows: given $q_0 \in M$, by the Implicit Function Theorem we may choose a local chart (U, x) about q_0 in M and a local chart (V, y) around $\varphi(q_0)$ in N so that the local representative of φ with respect to these charts is given by the projection onto the first n-coordinates, namely (recall that $m \geq n$)

$$(y \circ \varphi \circ x^{-1})(u^1, \dots, u^m) = (u^1, \dots, u^n), \quad \text{for } v \in \sigma(U).$$

Hence, Δ may be written as

$$\Delta_q = \ker(D\varphi(q)) = \operatorname{span}\left\{ \left.\frac{\partial}{\partial x^{n+1}}\right|_q, \ldots, \left.\frac{\partial}{\partial x^m}\right|_q \right\} \quad \text{for } q \in U,$$

which implies that $q \mapsto \Delta_q$ is smooth about q_0, and hence that Δ is a $(m-n)$-dimensional distribution on M.

Finally, let L be any connected integral manifold of Δ and denote by $\iota\colon L \to M$ the inclusion map. For fixed $q_0 = \iota(z_0) \in M \cap \iota(L)$, choosing a local chart (U, x) about q_0 as above, we have that, for every z in a sufficiently small and connected neighbourhood V of z_0, so that $\iota(V) \subset U$, there holds

$$\begin{aligned} \xi \in T_z L &\iff D\iota(z)\xi \in \Delta_{\iota(z)} \\ &\iff D\iota(z)\xi(u^i) = 0 \quad \text{for } i = 1, \ldots, n, \end{aligned}$$

for every $z \in V$ and $\xi \in T_z L$. Consequently

$$z \mapsto x^i(\iota(z)) \quad \text{is constant for every } i = 1, \ldots, n \text{ on } V,$$

in particular $x^i(q_0) = x^i(\iota(z_0)) = x^i(\iota(z))$ for all $z \in V$, $i = 1, \ldots, n$. Since q_0 was arbitrary, this proves that every sufficiently small open set V in L is mapped by ι into a connected component of some level set $\varphi^{-1}(p)$. However, since $\iota(L)$ is connected (as L is) we obtain that the whole $\iota(L)$ belongs to a single connected component of a fixed level set $\varphi^{-1}(q)$.

This proves that the family $\{\varphi^{-1}(p)\}_{p\in N}$ defines a $(m-n)$-dimensional foliation on M.

Problem Sheet G

Problem G.1. Show that the real projective space $\mathbb{R}P^{n-1}$ can be seen as the homogeneous space $\mathrm{SO}(n)/\mathrm{O}(n-1)$.

Solution. The Lie group action of $\mathrm{SO}(n)$ on $\mathbb{R}P^{n-1}$ is defined as follows: for $A \in \mathrm{SO}(n)$ and $[x] = [x^1, \ldots, x^n] \in \mathbb{R}P^{n-1}$, we set

$$\mathrm{SO}(n) \times \mathbb{R}P^{n-1} \to \mathbb{R}P^{n-1} \quad (A, [x]) \mapsto A[x] := [Ax]$$

Since A is a linear map, this map is well defined and it is a Lie group action.

It is a transitive action: if $[x], [y]$ are elements of $\mathbb{R}P^{n-1}$, let x_0 be one of the points in the equivalence class of $[x]$ so that $x_0 \in S^{n-1}$ (recall that there are precisely two of them, and in fact $\mathbb{R}P^{n-1} \simeq S^{n-1}/\{p \mapsto -p\}$) and likewise let y_0 be one of the points in the equivalence class of $[y]$ so that $y_0 \in S^{n-1}$. The map $A \in \mathrm{SO}(n)$ so that $Ax_0 = y_0$ will then map $[x_0]$ to $[y_0]$ via the above action.

We deduce that, if H is the isotropy group of a fixed element $[x]$ for the above action, then $\mathbb{R}P^{n-1} \simeq \mathrm{SO}(n)/H$, so let us compute the isotropy group of the n-th canonical vector $[0, \dots, 1] = [e_n]$. We have that $A[e_n] = [e_n]$ precisely if $Ae_n = e_n$ or $Ae_n = -e_n$. Since $A \in \mathrm{SO}(n)$, it must then have the form

$$A = \begin{pmatrix} \begin{pmatrix} & & \\ & B & \\ & & \end{pmatrix} & \begin{matrix} 0 \\ \vdots \\ 0 \end{matrix} \\ 0 \cdots 0 & 1 \end{pmatrix} \quad \text{or} \quad A = \begin{pmatrix} \begin{pmatrix} & & \\ & B' & \\ & & \end{pmatrix} & \begin{matrix} 0 \\ \vdots \\ 0 \end{matrix} \\ 0 \cdots 0 & -1 \end{pmatrix}$$

where $B, B' \in O(n-1)$ with $\det(B) = 1$ and $\det(B') = -1$. But then we may identify H with $O(n-1)$ via the map

$$A \mapsto A|_{\mathbb{R}^{n-1}},$$

where $A|_{\mathbb{R}^{n-1}}$ is the restriction of A to the first $n-1$ components. Such identification is compatible with the group action we are considering, and consequently we conclude that $\mathbb{R}P^{n-1} \simeq \mathrm{SO}(n)/O(n-1)$.

Problem G.2. Let σ be a smooth free action of G on M. Assume that the quotient space M/G admits the structure of a smooth manifold such that the quotient map $\rho\colon M \to M/G$ is a smooth submersion. Prove that σ is a proper action.

Solution. The idea is to use Problem F.4 to prove that the action σ is proper.

First of all observe that since ρ is a submersion we find an open neighbourhood $U \subseteq M$ for every point $p \in U$ and another neighbourhood $V \subseteq M/G$ of $\rho(p)$ with a local section

$$\psi\colon V \subset M/G \to U \subseteq M,$$

see Proposition 6.13. Intuitively, ψ is a preferred choice of representatives in each orbit $[p]$, thus we expect $V \times G$ to be diffeomorphic to U "moved around by the action". More precisely, we claim that

$$\alpha\colon V \times G \to W = \bigcup_{g \in G} \sigma_g(U), \quad ([p], g) \mapsto \sigma_g \circ \psi[p]$$

is a diffeomorphism.

For injectivity, assume $\sigma_g \circ \psi[p] = \sigma_h \circ \psi[q]$. If $[p] = [q]$, then freeness and the previous equation imply $g = h$. If $[p] \neq [q]$, then $\psi[p] \neq \psi[q]$ and thus, since we consider the quotient by G, $\sigma_g \circ \psi[p] \neq \sigma_h \circ \psi[q]$ which is a contradiction.

Surjectivity is clear (after possible shrinking U to $\psi(V)$). Smoothness follows from that of σ and ψ. To conclude that α defines a diffeomorphism is suffices to show that the differential is of full rank at every point (see Inverse Function Theorem 5.10).

Now we show that the action is proper. Let $(p_k) \to p$ in M and $(g_k) \subseteq G$ such that $\sigma_{g_k}(p_k)$ converges. Since the p_k will eventually all lie inside U, we can compute

$$\alpha^{-1}(\sigma_{g_k}(p_k)) = (\rho(p_k), g_k).$$

The left hand side and $\rho(p_k)$ converge, which forces g_k to converge as well. By Problem F.4 this proves that σ is a proper action.

PROBLEM G.3. Let $\pi_1 : P \longrightarrow M$ and $\pi_2 : Q \longrightarrow N$ be two principal G-bundles. Suppose (φ, Φ) is a principal bundle morphism from P to Q such that φ is a diffeomorphism. Prove that Φ is also a diffeomorphism.

SOLUTION. We begin by showing that Φ is surjective. Let $v \in Q$ be any element lying over, say, $q \in N$. Now let $u \in P$ be some element in the fibre over $\varphi^{-1}(q)$. Then $\Phi(u) \in Q_q$ and since the fibres are precisely the G-orbits there exists some $g \in G$ such that

$$\Phi(u) \cdot g = v.$$

Using the G-equivariance of principle bundle morphisms we deduce

$$\Phi(u \cdot g) = v,$$

which proves that Φ is surjective.

For the injectivity we pick two distinct elements $u_1, u_2 \in P$ and let us assume without loss of generality that they lie in the same fibre over p—otherwise their images also lie in distinct fibres, since φ is injective, and then there is nothing to prove. Therefore assume

$$\Phi(u_1), \Phi(u_2) \in Q_{\varphi(p)}.$$

Since G acts freely and transitively on P_p there exists a unique $g \in G \setminus \{e\}$ such that $u_1 = u_2 \cdot g$ and consequently

$$\Phi(u_1) = \Phi(u_2 \cdot g) = \Phi(u_2) \cdot g.$$

But G also acts freely on Q and since $g \neq e$ we conclude

$$\Phi(u_1) \neq \Phi(u_2).$$

In order to conclude the proof we will show that at any point $u \in P$ the differential $D\Phi(u)$ is an isomorphism—once we have shown this, by the Implicit Function Theorem it follows that Φ is a local diffeomorphism around any u and since Φ itself is already bijective one obtains that Φ is a (global) diffeomorphism.

So we need to prove that $D\Phi(u)$ is a linear isomorphism.

We fix any point $u \in P$ and pick a bundle chart

$$\epsilon \colon \pi_1^{-1}(U) \rightarrow G,$$

where U is an open neighbourhood of $p := \pi_1(u)$. Denote $g = \epsilon(u)$ and observe that $D\Phi(u)$ can be viewed as

$$D(\Phi \circ (\pi_1, \epsilon)^{-1})(p, g) = \left(D\varphi(p), D\left(\Phi\big|_{P_p} \circ (\epsilon_p)^{-1}\right)(g)\right)$$

under the identification $T_pM \times T_gG \cong T_uP$. We already know that $D\varphi(p)\colon T_pM \to T_{\varphi(p)}N$ is a linear isomorphism as $\varphi\colon M \to N$ is a diffeomorphism by assumption, thus it suffices to show that

$$\Phi\big|_{P_p} \circ (\epsilon)_p^{-1} : G \to Q_{\varphi(p)}$$

is a diffeomorphism. Set

$$v := \Phi\big|_{P_p} \circ (\epsilon)_p^{-1}(e) \in Q_{\varphi(p)}$$

and define the unique G-equivariant smooth map

$$\Psi_c : Q_{\varphi(p)} \to G \text{ with } \Psi_c(c) = e.$$

Since the bundle chart ϵ is G-equivariant by definition, we have that the whole composition

$$\Psi_c \circ \Phi\big|_{P_p} \circ (\epsilon)_p^{-1} : G \to G$$

is G-equivariant, satisfying

$$\Psi_c \circ \Phi\big|_{P_p} \circ (\epsilon)_p^{-1}(e) = e.$$

Therefore

$$\Psi_c \circ \Phi\big|_{P_p} \circ (\epsilon)_p^{-1} = \mathrm{id}_G$$

by uniqueness of such a G-equivariant map, which then concludes the proof.

Problem G.4. Let σ be an effective action of a Lie group G on a smooth manifold L. Assume we are given two fibre bundles

$$L \to E \xrightarrow{\pi_1} M, \qquad \text{and} \qquad L \to F \xrightarrow{\pi_2} M$$

Let $\{U_a \mid a \in A\}$ be an open cover of M such that both E and F admit G-bundle atlases over the U_a. Let

$$g^1_{ab} : U_a \cap U_b \to G, \qquad \text{and} \qquad g^2_{ab} : U_a \cap U_b \to G$$

denote the transition functions of E and F with respect to these bundle atlases. Prove that E and F are isomorphic as G-fibre bundles if and only if there exists a family $f_a\colon U_a \to G$ of smooth functions such that

$$f_a(p) \circ g^1_{ab}(p) = g^2_{ab}(p) \circ f_b(p), \qquad \forall\, p \in U_a \cap U_b,\ \forall\, a, b \in A.$$

Solution. Suppose first that we are given a G-fibre bundle isomorphism $\Phi\colon E_1 = E \to E_2 = F$. For each $i = 1, 2$ we denote the corresponding bundle chart $\epsilon_a^i\colon \pi_i^{-1}(U_a) \to L$. Let $p \in U_a$ and define

$$\nu_a(p) := (\epsilon_a^2)_p \circ \Phi \circ (\epsilon_a^1)_p^{-1} \in \mathrm{Diff}(L).$$

Recall that Φ covers the identity on M, hence $\nu_a\colon U_a \to \mathrm{Diff}(L)$ coincides with the map f_a^b of Eq. (17.1). In particular, by Definition 17.3 we have functions[16] $f_a\colon U_a \to G$ such that

$$\nu_a(p) = \sigma_{f_a(p)} \in \mathrm{Diff}(L).$$

Now by $\epsilon_{ab}^i\colon U_a \cap U_b \to \mathrm{Diff}(L)$ the transition functions. We compute

$$\begin{aligned}
\nu_a(p) \circ \epsilon_{ab}^1(p) &= \left((\epsilon_a^2)_p \circ \Phi \circ (\epsilon_a^1)_p^{-1}\right) \circ \left((\epsilon_a^1)_p \circ (\epsilon_b^1)_p^{-1}\right)\\
&= (\epsilon_a^2)_p \circ \Phi \circ (\epsilon_b^1)_p^{-1}\\
&= (\epsilon_a^2)_p \circ (\epsilon_b^2)_p^{-1} \circ (\epsilon_b^2)_p \circ \Phi \circ (\epsilon_b^1)_p^{-1}\\
&= \epsilon_{ab}^2(p) \circ \nu_b(p).
\end{aligned}$$

Since $\epsilon_{ab}^i(p) = \sigma_{g_{ab}^i(p)}$ the above equation reads as

$$\sigma_{f_a(p)} \circ \sigma_{g_{ab}^1(p)} = \sigma_{g_{ab}^2(p)} \circ \sigma_{f_b(p)}.$$

Since σ is effective this shows that the family f_a has the desired property.

Conversely, we first define a candidate Φ locally, i.e. for each a pick a $f_a\colon U_a \to G$ as in the statement and define

$$\tilde{\Phi}_a\colon U_a \times L \to U_a \times L,\ \tilde{\Phi}_a(p, u) := (p, \sigma_{f_a(p)} u).$$

With this we set

$$\Phi\colon E_1 \to E_2,\ \Phi(u) := (\hat{\epsilon}_a^2)^{-1} \circ \tilde{\Phi}_a \circ \hat{\epsilon}_a^1(u)$$

whenever $u \in \pi_1^{-1}(U_a)$. To see that Φ is well defined we pick $p \in U_a \cap U_b$ and show that Φ (defined via a) composed with the inverse Φ (defined via b) is the identity. Indeed, for u over p we have

$$\begin{aligned}
&(\hat{\epsilon}_a^2)_p^{-1} \circ (\tilde{\Phi}_a)_p \circ (\hat{\epsilon}_a^1)_p \circ (\hat{\epsilon}_b^1)_p^{-1} \circ (\tilde{\Phi}_b)_p^{-1} \circ (\hat{\epsilon}_b^2)_p\\
&\qquad = (\epsilon_a^2)_p^{-1} \circ \sigma_{f_a(p)} \circ \epsilon_{ab}^1(p) \circ \sigma_{f_b(p)^{-1}} \circ (\epsilon_b^2)_p\\
&\qquad = (\epsilon_a^2)_p^{-1} \circ \sigma_{f_a(p)} \circ \sigma_{g_{ab}(p)} \circ \sigma_{f_b(p)^{-1}} \circ (\epsilon_b^2)_p,
\end{aligned}$$

but $\sigma_{f_a(p)} \circ \sigma_{g_{ab}^1(p)} \circ \sigma_{f_b(p)^{-1}} = \sigma_{g_{ab}^2(p)}$ by assumption. On the other hand the latter is also equal to $\epsilon_{ab}^2(p)$, thus plugging that in proves that Φ is well defined. Moreover, Φ has been constructed locally to satisfy all the properties of a G-fibre bundle isomorphism. This finishes the proof.

---

[16] There is an unfortunate notation clash between f's in the statement here and the f's in Chapter 17.

Problem G.5. Let $V \to E \xrightarrow{\pi} M$ be a vector bundle. Prove that it is possible to reduce the structure group from $\mathrm{GL}(V)$ to $\mathrm{O}(V)$. Find an example where it is not possible to reduce the structure group from $\mathrm{GL}(V)$ to $\mathrm{GL}^+(V)$.

Solution. Fix an inner product $\langle\cdot,\cdot\rangle$ on V and a smooth family of inner products $\langle\cdot,\cdot\rangle_p$ on each fibre E_p. (One can construct such a smooth family by working locally and gluing together everything via a partition of unity. Buzz word: Riemannian metric.)

We need to modify each transition function

$$\epsilon_{ab}\colon U_a \cap U_b \to \mathrm{GL}(V).$$

Pick $\{v_i\}$ an orthonormal basis of V. Then

$$X_i^a\colon U_a \to \pi^{-1}(U_a),\ X_i^a(p) := (\epsilon_a)^{-1}\, v_i$$

is a (local) vector field and the $\{X_i^a(p)\}$ define a basis of E_p for all $p \in U_a$—such $\{X_i^a\}$ is often referred to as a (local) frame. For each $p \in U_a$ we apply the Gram-Schmidt procedure to the basis $\{X_i^a(p)\}$ to obtain an orthonormal basis with respect to $\langle\cdot,\cdot\rangle_p$. This change of basis can be described by smooth family of matrices

$$S_p^a\colon E_p \to E_p.$$

The smoothness follows from smoothness of the X_i and smoothness of the Gram-Schmidt algorithm. In particular we get a smooth map

$$S^a\colon \pi^{-1}(U_a) \to \pi^{-1}(U_a),\ S^a(p,u) = S_p^a(u).$$

We redefine

$$\bar{\epsilon}_a\colon \epsilon_a \circ (S^a)^{-1}\colon \pi^{-1}(U_a) \to V.$$

But then $(\bar{\epsilon}_a)_p$ maps the orthonormal basis $S_p^a(X_i^a(p))$ to v_i, which is a orthornomal basis of V. In particular, the transition function $\bar{\epsilon}_{ab}(p) := (\bar{\epsilon}_a)_p \circ (\bar{\epsilon}_b)_p^{-1}$ maps orthonormal bases to orthonormal bases, i.e. $\bar{\epsilon}_{ab}(p) \in \mathrm{O}(p)$. This concludes the proof.

An example of a vector bundle whose structure group $\mathrm{GL}(V)$ does not reduce to $\mathrm{GL}^+(V)$ is given by the Möbius bundle, see Example 17.6. Indeed, $\mathrm{GL}(\mathbb{R}) = \mathbb{R} \setminus \{0\}$, and $\mathrm{GL}^+(\mathbb{R}) = \mathbb{R}_{>0}$, but the transition function g_{21} takes values ± 1 on U_1 and U_2, respectively. We will see in Problem G.6 (i.e. use Problem G.4) that changing only one of the two signs changes the isomorphism class. Therefore the structure group cannot be reduced to $\mathbb{R}_{>0}$.

(♣) Problem G.6. Prove that there are exactly two vector bundles of rank 1 over S^1 (up to vector bundle isomorphism).

Solution. We claim that up to vector bundle isomorphisms the only rank 1 vector bundles over S^1 are the Möbius bundle

$$\mathbb{R} \longrightarrow E^M \longrightarrow S^1$$

(see Example 17.6 and exercise above) and the trivial bundle

$$\mathbb{R} \longrightarrow E^T = S^1 \times \mathbb{R} \longrightarrow S^1.$$

Let $\mathbb{R} \longrightarrow E \longrightarrow S^1$ be an arbitrary rank 1 vector bundle over S^1. We proceed as follows: first we try to prove that E is isomorphic to the trivial bundle E^T using Problem G.4. Whenever the former fails, we argue why $E \cong E^M$ has to hold.

Start by picking bundle charts $\epsilon_a : \pi^{-1}(U_a) \longrightarrow \mathbb{R}$. Due to compactness of S^1 (and up to intersecting) we can assume that $\{U_a\}$ is a finite collection of open intervals $\{U_{a_i}\}_{i=0,\ldots,k}$ on S^1 so that only U_{a_i} and $U_{a_{i+1}}$ intersect nontrivially (and also $U_{a_k} \cap U_{a_0} \neq \emptyset$). Moreover, after potentially intersecting with the U_1 and U_2 from Example 17.6, we can assume

$$\bigcup_{i=0}^{l} U_{a_l} \subseteq U_1, \quad \bigcup_{j=l+1}^{k} U_{a_j} \subseteq U_2.$$

By Problem G.4 we know that $E \cong E^T$ is equivalent to the existence of a finite collection of functions $f_{a_i} : U_{a_i} \longrightarrow \mathrm{GL}(\mathbb{R}) \cong \mathbb{R} \setminus \{0\}$ such that

$$f_{a_i} \circ g_{a_i,a_{i+1}}(p) = g^T_{a_i,a_{i+1}}(p) \circ f_{a_{i+1}}, \ \forall i = 0, \ldots, k-1 \text{ and } p \in U_{a_i} \cap U_{a_{i+1}}$$

and for $i = k$:

$$f_{a_k} \circ g_{a_k,a_0}(p) = g^T_{a_k,a_0}(p) \circ f_{a_0}, \ \forall p \in U_{a_k} \cap U_{a_0}.$$

We try to construct such a family f_{a_i}, but first we make some observations: under the isomorphism $\mathrm{GL}(\mathbb{R}) \cong \mathbb{R} \setminus \{0\}$ the above composition becomes multiplication. By assumption, the domains of our transition functions are open intervals, hence connected subsets of S^1, thus $g_{a_i,a_{i+1}}$ is either positive or negative. After rescaling we can assume $g_{a_i,a_{i+1}}(p) \in \{\pm 1\}$. Also observe that $g^T_{a_i,a_{i+1}}(p) = 1$.

Now we make an inductive argument for the construction of the f_{a_i}'s .

Pick f_{a_0} equal to 1 and f_{a_1} so that the equation above is satisfied—there is a unique choice and it is either 1 or -1.

Pick f_{a_2} the same way, i.e. via the equation above. We do this inductively until we reach $f_{a_{k-1}}$. The latter determines f_{a_k}

However f_{a_k} also has to satisfy

$$f_{a_k} \cdot g_{a_k,a_0}(p) = g^T_{a_k,a_0}(p) \cdot f_{a_0} = f_{a_0}.$$

If this works, then E is isomorphic to E^T. If not, we need to replace $g^T_{a_k,a_0}(p) = 1$ by $g^M_{a_k,a_0}(p) = -1$.

So we have a well defined choice of f_{a_i}'s such that the "trivial" equations holds for $i = 0, \ldots, k-1$ and the "Möbius" equation holds for $i = k$. Since all the U_{a_j} for

$j = l, \dots k$ are contained in U_2, we can flip the signs of the corresponding f_{a_j}'s and replace $g^T_{a_j,a_{j+1}}$ by $g^M_{a_j,a_{j+1}}$ (for $j = l, \dots k-1$). On U_1 the transition functions of the E^M and E^T agree. All in all this means that we have found a choice of f_{a_i}'s such that

$$f_{a_i} \circ g_{a_i,a_{i+1}}(p) = g^M_{a_i,a_{i+1}}(p) \circ f_{a_{i+1}}, \ \forall i = 0, \dots, k-1 \text{ and } p \in U_{a_i} \cap U_{a_{i+1}}$$

and for $i = k$:

$$f_{a_k} \circ g_{a_k,a_0}(p) = g^M_{a_k,a_0}(p) \circ f_{a_0}, \ \forall p \in U_{a_k} \cap U_{a_0}.$$

This means that whenever $E \ncong E^T$ we have $E \cong E^M$. This proves that E has precisely two isomorphism classes and finishes the proof.

(♣) Problem G.7. Prove that the Klein bottle is a fibre bundle over S^1 with fibre S^1. Prove however that the Klein bottle is not a principal S^1-bundle over S^1.

Solution. We define the Klein bottle

$$K = [0,1]^2/\sim$$

by making the following identifications on the boundary of $[0,1]^2$

- $(x,0) \sim (x,1)$
- $(0,y) \sim (1,1-y)$

and denote equivalence classes in K with square brackets, i.e. $[x,y] \in K$. Checking that K is a smooth manifold is not hard and is left to the reader. Now we define the projection

$$\pi\colon K \to S^1 = \mathbb{R}/\mathbb{Z}, \ [x,y] \mapsto x$$

and claim that this is a fibre bundle with fibre S^1. (Sanity check: $\pi^{-1}(x) = \{[x,y] \mid y \in [0,1]\} \cong S^1$ as $[x,1] = [x,0]$.)

Indeed, $\pi\colon K \to S^1$ is a smooth surjection and we can construct two bundle charts ϵ_1 and ϵ_2 as follows: Let $U_1 := (0,1)$ be the open arc viewed inside S^1 and $U_2 = \left[0, \frac{1}{2}\right) \cup \left(\frac{1}{2}, 1\right]$ also viewed inside S^1. The map

$$\epsilon_1 : \pi^{-1}(U_1) \to S^1, \ [x,y] \mapsto y$$

is obviously well defined as $[x,1] = [x,0]$ and $0 = 1$ in $S^1 = \mathbb{R}/\mathbb{Z}$, and also smooth. It readily follows that

$$(\pi, \epsilon_1)\colon \pi^{-1}(U_1) \to U_1 \times S^1$$

defines a diffeomorphism, thus proving that ϵ_1 is a bundle chart.

For $\epsilon_2 \colon \pi^{-1}(U_2) \to S^1$ one has to take the identification $(0, y) \sim (1, 1-y)$ into account, which is the reason why the same choice as for ϵ_1 does not work. However, we claim that the following choice works:

$$\epsilon_2([x, y]) = \begin{cases} y, & \text{if } x < \frac{1}{2} \\ -y, & \text{if } x > \frac{1}{2}. \end{cases}$$

The only classes $[x, y] \in \pi^{-1}(U_2)$ that have several representatives and matter for the definition of α are those of the form

$$[0, y] = \{(0, y), (1, 1-y)\}.$$

We check

$$\epsilon_2([0, y]) = y \text{ and } \epsilon_2([1, 1-y]) = y - 1,$$

but $y - 1$ equals y if viewed in S^1 and therefore ϵ_2 is well defined (note that it is crucial to exclude $\frac{1}{2}$ to make this definition work!). Just as for ϵ_1, smoothness of ϵ_2 is clear. It is also not hard to see that

$$(\pi, \epsilon_2) \colon \pi^{-1}(U_2) \to U_2 \times S^1$$

defines a diffeomorphism by taking a closer look at what ϵ_2 does on the two preimages $V_+ = \pi^{-1}\left[0, \frac{1}{2}\right)$ and $V_- = \pi^{-1}\left(\frac{1}{2}, 1\right]$, namely:

$$(\pi, \epsilon_2)\Big|_{V_\pm} = \pm\, \mathrm{id}_{V_\pm}.$$

In order to show that the K is *not* a principal S^1-bundle we will make use of the following nice fact:

Claim A principal G-bundle $G \to P \to M$ is trivial if and only if admits a global section.

Indeed, if P is a trivial principal bundle, then there exists a fibre preserving diffeomorphism $\Psi \colon M \times G \to P$. But then $M \to P$, $s(p) := \Psi(p, e)$ defines a global section. Conversely, let $s \colon M \to P$ be a global section. We claim that

$$\Psi \colon M \times G \to P, \ \Psi(p, g) := \tau_g(s(p))$$

is a principal bundle isomorphism, where τ denotes the fibre preserving right action on P (see Proposition 16.18). We start by showing that Ψ is equivariant:[17]

[17] The corresponding action τ' on the trivial principal bundle $M \times G$ is right multiplication, i.e. $\tau'_g = r_g$.

$$\begin{aligned}(\Psi \circ \tau'_g)(p,h) &= \Psi(p, r_g(h)) \\ &= \Psi(p, hg) \\ &= \tau_{hg}(s(p)) \\ &= \tau_g \circ \tau_h(s(p)) \\ &= (\tau_g \circ \Psi)(p,h).\end{aligned}$$

Moreover, Ψ is fibre-preserving as τ is a fibre-preserving action. Invoking Problem G.3 and observing that Ψ covers the identity then automatically implies that Ψ is a diffeomorphism, and hence with the above, that Ψ defines a principal G-bundle isomorphism. This proves the claim.

With the claim in hand it suffices to establish a global section $s: S^1 \to K$ in order to conclude that K is not a principal S^1-bundle. Indeed, if K admits a global section *and* is a principal S^1-bundle, then the claim above implies that K isomorphic to the torus $S^1 \times S^1$ as a principal bundle. In particular, the total spaces K and $S^1 \times S^1$ are diffeomorphic as manifolds, but K and $S^1 \times S^1$ are not even homeomorphic[18]—contradiction!

To construct a global section, we first define $f(x) = x$ on $[0, 1]$. Then

$$s: S^1 \to K,\ s(x) := (x, f(x))$$

is well defined since $f(0) = 0$, $f(1) = 1$ and $(0, 1)$ is equivalent to $(1, 0)$ inside K. This s is readily checked to be a section, which concludes the proof.

(♣) Problem G.8. Find a non-trivial principal S^1-bundle over $\mathbb{R}P^2$.

Solution. We give a description of a non-trivial principal S^1-bundle over $\mathbb{R}P^2$: Define the *involution*

$$\iota: S^2 \times S^1 \to S^2 \times S^1,\ \iota(p,q) = (-p,-q).$$

Define P to be the quotient $S^2 \times S^1/\sim$ where $(p', g') \sim (p, q) :\iff \iota(p', q') = (p, q)$. This is a well defined equivalence relation since ι is a bijective. Define

$$\pi: P \to \mathbb{R}P^2,\ [p,q] \mapsto [p].$$

This is a well defined map and the preimage $\pi^{-1}[p]$ is diffeomorphic to $\mathbb{R}P^1 \cong S^1$ via

$$\pi^{-1}[p] \to \mathbb{R}P^1,\ [p,q] \mapsto [q].$$

A somewhat tedious check shows that the bundle at hand is not trivial.

18 Exercise: why?

Problem G.9. Suppose σ is a smooth transitive action of a Lie group G on M, so that M is the homogeneous space G/H for an appropriate subgroup H of G. Prove that the subgroup of G acting trivially on M is the largest normal subgroup $N(H)$ of G contained in H. Let $\overline{G}$ and $\overline{H}$ denote the quotient groups $G/N(H)$ and $H/N(H)$ respectively. Prove that $\overline{G}$ acts effectively and transitively on M, and M is the homogeneous space $\overline{G}/\overline{H}$.

Solution. Denote by $N \subset G$ the subset of elements which act trivially on M. Suppose $g' \in N$ and fix any $g \in G$. Then for all $p \in M$ we have

$$\sigma_{gg'g^{-1}}(p) = \sigma_g(\sigma_{g'}(\sigma_{g^{-1}}(p)) = \sigma_g(\sigma_{g^{-1}}(p)) = \sigma_e(p) = p.$$

This shows that N is a normal subgroup. Clearly $N \subset H$, since $H = \{g \in G \mid \sigma_g(x_0) = x_0\}$ for some fixed $x_0 \in M$. Suppose $N_0 \leq G$ is a normal subgroup satisfying $N_0 \subset H$. We need to show that this implies $N_0 \subset N$. Choose $g_0 \in N_0$ and $y \in M$. Then, since G acts transitively on M there is a $g \in G$ such that $\sigma_g x_0 = y$. Since N_0 is normal in G we have $gg_0g^{-1} \in N_0 \subset H$, which implies $\sigma_{g^{-1}g_0g} x_0 = x_0$, or $\sigma_{g_0} y = y$. This shows that every element of N_0 fixes every element of M. Hence, $N_0 \subset N$ which proves the first claim.

Since $N = N(H)$ is a normal subgroup of G it is well-known from algebra that $\bar{G} := G/N$ carries a canonical group structure. It is obvious that $\bar{G}$ acts on M in a canonical way. This action is said to be effective if $\sigma_{\bar{g}} x = x$ for all $x \in M$ implies $\bar{g} = \mathrm{id}$. Suppose $\bar{g} \in \bar{G}$ satisfies $\sigma_{\bar{g}} x = x$ for all $x \in M$. Then the representative $g \in G$ of the class $\bar{g} \in G$ fixes every element of M, so that $g \in N$. Hence, $\bar{g}$ is the multiplicative unit in $\bar{G}$, which proves that the action $\bar{G} \times M \to M$ is effective. $\bar{H}$ is the isotropy group of the (transitive) action, so $M = \bar{G}/\bar{H}$.

(♣) Problem G.10. Let $\pi \colon P \to M$ be a principal G-bundle with corresponding right action τ. Let $H \subset G$ be a Lie subgroup, and let $Q \subset P$ be a subset such that:

(i) The restriction $\pi|_Q \colon Q \to M$ is surjective.
(ii) If $q \in Q$ and $h \in H$ then $\tau_h(q) \in Q$.
(iii) For all $p \in M$, the action of H on $Q \cap P_p$ is transitive.
(iv) For all $p \in M$, there exists a neighbourhood U of p and a smooth local section $\psi \colon U \to P$ of π such that $\psi(x) \in Q$ for all $x \in U$.

Prove that $\pi|_Q \colon Q \to M$ is a principal H-bundle, and that moreover Q is a principal H-subbundle of P.

Solution. Fix $p \in M$, U and $\psi \colon U \to P$ as in the last point above. For every point $v \in Q \cap \pi^{-1}(U)$, there exists a unique $h \in H$ such that $\psi(\pi(v)) \cdot h = v$.[19] Define $\epsilon \colon Q \cap \pi^{-1}(U) \to H$ by

[19] Here the right action τ is written by "multiplying" from the right.

$$\epsilon(v) = \epsilon(\psi(\pi(v)) \cdot h) := h$$

Then $\hat{\epsilon} = (\pi|_Q, \epsilon)$ defines a bijection between $Q \cap \pi^{-1}(U)$ and $U \times H$. We define a topology and a smooth structure on Q by declaring the $\hat{\epsilon}$ to be diffeomorphisms.

Now we apply Proposition 16.18 to deduce that Q is a principal H-bundle. Indeed, the bundle charts are equivariant: for $v \in Q_q$ and $h' = \epsilon(v)$ we have:

$$\begin{aligned}
(\epsilon)_q \circ \tau_h(v) &= (\epsilon)_q(v \cdot h) \\
&= (\epsilon)_q(\psi(\pi(v)) \cdot h' \cdot h) \\
&= (\epsilon)_q(\psi(\pi(v)) \cdot r_h(h')) \\
&= r_h(h') \\
&= r_h \circ (\epsilon)_q(v).
\end{aligned}$$

The right action of τ restricted is of course still fibre preserving and free when viewed on Q. The sections ψ are smooth as maps into Q, since Q is weakly embedded in P. Therefore Proposition 16.18 is applicable, which proves that $\pi|_Q : Q \to M$ is a principal H-bundle.

The H-subbundle part (see Definition 17.22) follows readily, since $U \times H \subset U \times G$ is a weakly embedded, because so is $H \subset G$ as a Lie subgroup, by Remark 15.6.

Problem Sheet H

Problem H.1. Let V, W and U be vector spaces. Prove there are natural isomorphisms $V \otimes W \cong W \otimes V$ and $(U \otimes V) \otimes W \cong U \otimes (V \otimes W)$.

Solution. To prove that $V \otimes W \cong W \otimes V$, consider the bilinear map $F : V \times W \to W \otimes V$ given by $F(v, w) = w \otimes v$. By the universal property of the tensor product there is an associated linear map $T_F : V \otimes W \to W \otimes V$ defined by $T_F(v \otimes w) = w \otimes v$ and then extended by linearity. Such a map is invertible, the inverse being defined by $T_F^{-1}(w \otimes v) = v \otimes w$ and then extended by linearity. This map is then the required isomorphism.

To prove that $(U \otimes V) \otimes W \cong U \otimes (V \otimes W)$, consider the map $G : (U \otimes V) \times W \to U \otimes (V \otimes W)$ (which we regard as a function of two variables, the first on $U \otimes V$ and the second on W) defined by $G(u \otimes v, w) = u \otimes (v \otimes w)$, and extended by linearity with respect to its first argument. This map is bilinear and so by the universal property of the tensor product again, there is an associated linear map $T_G : (U \otimes V) \otimes W \to U \otimes (V \otimes W)$ defined by $T_G((u \otimes v) \otimes w) = u \otimes (v \otimes w)$ and extended by linearity. Since such a map is invertible, (the inverse being defined by $T_G^{-1}(u \otimes (v \otimes w)) = (u \otimes v) \otimes w$ and then extended by linearity), and so it provides the required isomorphism.

Problem H.2. Let V and W be vector spaces. Prove that for any $A \in \mathrm{Alt}_h(V, W)$ there is a unique linear map $a\colon \bigwedge^h V \to W$ such that the following diagram commutes:

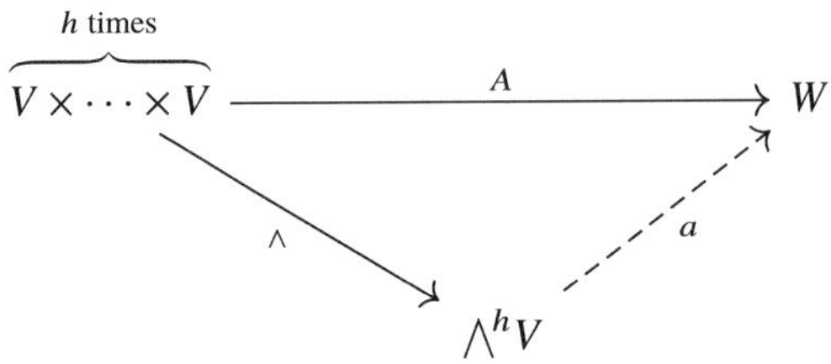

Prove moreover that $\bigwedge^h V$ is uniquely characterised by this property.

Solution. Any alternating r-linear is by definition multilinear which allows us to define a map

$$\tilde{a}\colon T^{h,0}(V) = V \otimes \cdots \otimes V \longrightarrow W,\ \tilde{a}(v_1 \otimes \cdots \otimes v_r) := A(v_1, \ldots, v_r).$$

This map is well defined and linear by the very definition of the tensor product and multilinearity of A.

Now we claim that $\tilde{a}\colon T^{h,0}(V) \longrightarrow W$ factors through $\bigwedge^h(V) \cong T^{h,0}(V)\big/ I^h(V)$. Indeed, any element of the form $v_1 \otimes \cdots \otimes v_i \otimes \cdots \otimes v_i \otimes \cdots \otimes v_r$ gets mapped to 0 by $\tilde{a}$ simply because the v_i appears twice and A is alternating, i.e.

$$A(v_1, \ldots, v_i, \ldots, v_i, \ldots, v_r) = 0.$$

Therefore

$$I^h(V) \subset \ker \tilde{a}$$

and we end up with a linear map $a\colon \bigwedge^h(V) \longrightarrow W$ such that the following diagram commutes

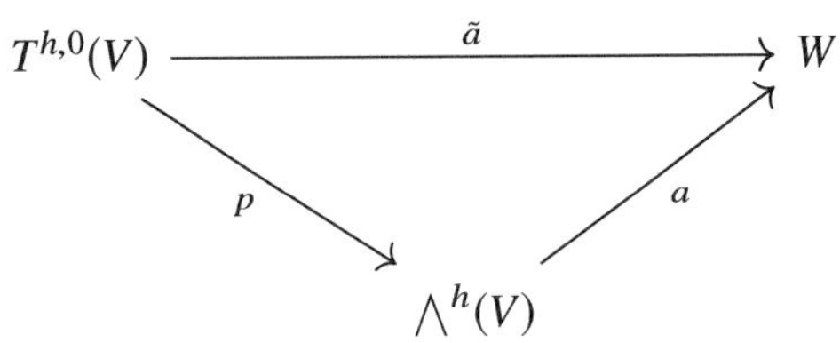

where p the obvious projection.

Moreover, the linear map a fits into the desired commutative diagram ($a \circ \wedge = T \circ (p \circ \otimes) = \tilde{a} \circ \otimes = A$) and we are only left to show that such a a is unique, but that's straightforward: The decomposable elements $v_1 \wedge \cdots \wedge v_r$ generate $\bigwedge^h(V)$ and as $a(v_1 \wedge \cdots \wedge v_r)$ is already determined to be equal to $A(v_1, \ldots, v_r)$ the uniqueness of a follows.

The proof of the second statement about the unique characterisation of $\bigwedge^h(V)$ is identical to the second part in the proof of Lemma 19.2.

Problem H.3. Let V be a vector space of dimension n with basis $\{e_1, \ldots, e_n\}$. Prove that

$$\{e_{i_1} \wedge \cdots \wedge e_{i_h} \mid 1 \leq i_1 < \cdots < i_h \leq n\}$$

is a basis of $\bigwedge^h V$ and $\bigwedge^h V = 0$ for $h > n$. Deduce that $\dim \bigwedge^h V - \binom{n}{h}$ and that $\dim \bigwedge V = 2^n$.

Solution. We start by noticing that an element $A \in \mathrm{Alt}_h(V)$ has components (with respect to the chosen basis on V) given by

$$A_{i_1 \cdots i_h} = A(e_{i_1}, \ldots, e_{i_h}) \quad \text{for every } 1 \leq i_1 \leq \cdots \leq i_r \leq k.$$

Since A is alternating, we deduce that if σ is a permutation of any set of indices $\{i_1, \ldots, i_h\}$, then $A_{\sigma(i_1) \cdots \sigma(i_h)} = \mathrm{sign}(\sigma) A_{i_1 \cdots i_h}$ ($\mathrm{sign}(\sigma)$ is the signature of σ). In particular A is determined completely by the values $A_{i_1 \cdots i_h}$ for $1 \leq i_1 < i_2 < \ldots i_h \leq n$ and consequently a basis for $\mathrm{Alt}_h(V)$ is given, for this choice of indices, by the maps

$$E^{i_1, \ldots, i_h}(e_{j_1}, \ldots, e_{j_h}) = \begin{cases} \mathrm{sign}(\eta) & \text{if } \{i_1, \ldots, i_r\} = \{j_1, \ldots, j_r\}, \\ 0 & \text{else,} \end{cases}$$

for every $1 \leq j_1 \leq \ldots \leq j_h \leq n$ where $\mathrm{sign}(\eta)$ is the signature of the permutation sending $(i_1, \ldots, i_h)$ to $(j_1, \ldots j_h)$. With the canonical vector-space identification between $\mathrm{Alt}_r(V)$ and $\bigwedge^h(V^*)$ given by Proposition 19.8, $E^{i_1, \ldots, i_h}$ corresponds to the element $e^{i_1} \wedge \cdots \wedge e^{i_h}$. Consequently, $\{e^{i_1} \wedge \cdots \wedge e^{i_h}\}_{1 \leq i_1 < \cdots < i_h \leq n}$ must then be a basis for $\bigwedge^h(V^*)$.

The same reasoning can be applied to $\mathrm{Alt}_h(V^*)$ and $\bigwedge^h(V)$ (or also simply recalling the canonical identification $V \cong V^{**}$), and so the conclusion is reached.

Finally, it is clear that $\bigwedge^h(V) = 0$ for $h > n$ since eg. $e_1 \wedge \cdots \wedge e_n \wedge e_i = 0$ for any i as the e_i term appears twice.

Problem H.4. Let $\varphi\colon M \to N$ be a smooth map, and suppose $L \to E \xrightarrow{\pi} N$ is a fibre bundle. Set

$$\varphi^* E := \{(p, u) \in M \times E \mid \varphi(p) = \pi(u)\},$$

with projection $\mathrm{pr}_1 \colon \varphi^* E \to M$.

(i) Prove that $\varphi^* E$ is a fibre bundle over M with fibre L such that (φ, pr_2) is a fibre bundle morphism:

$$\begin{array}{ccc} \varphi^* E & \xrightarrow{\mathrm{pr}_2} & E \\ {\scriptstyle \mathrm{pr}_1}\downarrow & & \downarrow{\scriptstyle \pi} \\ M & \xrightarrow[\varphi]{} & N \end{array}$$

(ii) Prove that

$$T_{(p,u)}(\varphi^*E) = \{(\xi,\zeta) \in T_pM \times T_uE \mid D\varphi(p)\xi = D\pi(p)\zeta\}.$$

(iii) Suppose $\psi\colon K \to M$ is another smooth map. Prove that $\psi^*(\varphi^*E)$ and $(\varphi \circ \psi)^*E$ are isomorphic fibre bundles over K.

(iv) Assume now that E has structure group G. Prove that φ^*E has structure group G. Deduce that the pullback of a vector bundle is a vector bundle, and the pullback of a principal bundle is a principal bundle.

(v) Prove that the isomorphism from part (iii) can be taken to be an isomorphism of G-fibre bundles (or a subgroup thereof).

Solution. (i) We prove the assertion in two steps. First we show that φ^*E carries the structure of a smooth manifold and then we show that for any point there is an open neighbourhood over which φ^*E is trivial.

Define the smooth map

$$\phi : M \times E \to N \times N \text{ by } \phi(p,u) = (\varphi(p), \pi(u))$$

and denote the diagonal of $N \times N$ by $\Delta = \{(q,q) \in N \times N\}$. Then we may write $\varphi^*E = \phi^{-1}(\Delta)$. By problem C.8 it suffices to show that ϕ is transverse to Δ in order to conclude that φ^*E is an embedded submanifold of $M \times E$.
We say that the maps $\varphi : M \to N$ and $\pi : E \to N$ are transverse to each other if

$$T_{\varphi(p)}N = d\varphi(p)[T_pM] + d\pi(u)[T_uE] \text{ for any } (p,u) \in \varphi^*E.$$

One may directly verify that ϕ is transverse to Δ if and only if φ and π are transverse to each other. The latter of these two conditions is satisfied since π is a submersion. We conclude that $\phi^{-1}(\Delta) = \varphi^*E$ carries a smooth manifold structure. Moreover, problem C.8 also asserts that

$$\dim \varphi^*E = \dim M \times E - \dim N \times N + \dim \Delta = \dim M + \dim E - \dim N.$$

Now suppose $\Phi = (\pi, \alpha) : \pi^{-1}(U) \to U \times L$ locally trivializes the bundle $\pi : E \to N$. Then consider the set $V = \varphi^{-1}(U)$ and the map

$$\Psi : \mathrm{pr}_1^{-1}(V) \to V \times L, \Psi(p,u) := (p, \mathrm{pr}_2 \circ \Phi(u)).$$

This is a diffeomorphism since

$$\Psi(p,u) = (p, \mathrm{pr}_2 \circ \Phi(u)) = (p, \alpha(u)),$$

where α is assumed to be a diffeomorphism.
Therefore, $\mathrm{pr}_1 : \varphi^*E \to M$ is a fibre bundle with fibre L.

(ii) We first show that

$$T_{(p,u)}(\varphi^*E) \subset \{(\xi,\zeta) \in T_pM \times T_uU \mid D\varphi(p)\xi = D\pi(u)\zeta\}.$$

For this let $(\xi, \zeta) \in T_{(p,u)}(\varphi^* E)$. Choose a curve $\gamma : (-\epsilon, \epsilon) \to \varphi^* E$ with

$$\gamma(0) = (p, u) \text{ and } \dot{\gamma}(0) = (\xi, \zeta).$$

We may write $\gamma(t) = (p(t), u(t))$. Then since $\gamma(t) \in \varphi^* E$ for every t one has that

$$\varphi(p(t)) = \pi(u(t)).$$

Differentiating both sides at $t = 0$ yields

$$D\varphi(p)\xi = D\pi(u)\zeta,$$

which shows the desired inclusion.

Now we proceed with a dimension argument. For this consider the linear map

$$A : T_p M \times T_u E \to T_{\varphi(p)} N, A(\xi, \zeta) := D\varphi(p)\xi - D\pi(u)\zeta.$$

Note that this is well-defined since $D\varphi(p)\xi$, $D\pi(u)\zeta \in T_{\varphi(p)} N$. Also observe that $\ker A$ coincides with the set on the RHS. Now A is surjective as π is a submersion. Therefore, one has

$$\dim \ker A = \dim(T_p M \times T_u E) - \dim T_{\varphi(p)} N = \dim M + \dim E - \dim N.$$

As was shown in (i), this is the same as $\dim \varphi^* E = \dim T_{(p,u)} \varphi^* E$. This completes our proof.

(iii) Our goal is to find a diffeomorphism $\sigma : (\varphi \circ \psi)^* E \to \psi^*(\varphi^* E)$ along the identity $\mathrm{id} : K \to K$ which restricts to a linear isomorphism on the fibres.

By definition of the pullback bundle we have

$$(\varphi \circ \psi)^* E = \{(q, u) \in K \times E \mid (\varphi \circ \psi)(p) = \pi(u)\}$$

and

$$\begin{aligned} \psi^*(\varphi^* E) &= \{(q, (p, u)) \in K \times \varphi^* E \mid \psi(q) = \mathrm{pr}_1(p, u) = p\} \\ &= \{(q, p, u) \in K \times M \times E \mid \psi(q) = p,\ \varphi(p) = \pi(u)\} \end{aligned}$$

We define the following smooth maps

$$\Phi : (\varphi \circ \psi)^* E \to \psi^*(\varphi^* E), \quad \Phi(q, u) = (q, \psi(q), u)$$

and

$$\Psi : \psi^*(\varphi^* E) \to (\varphi \circ \psi)^* E, \quad \Psi(q, p, u) = (q, u)$$

One directly sees that these are well-defined and are inverse to each other. Moreover, since the footpoint maps of both bundles are given by projections onto first factor, both Ψ and Φ can be seen to cover the identity map on K.

Next we have a look at the fibres of both bundles over some point $q \in K$.

$$((\varphi \circ \psi)^* E)_q = \{q\} \times E_{(\varphi \circ \psi)(a)} \text{ and } (\psi^*(\varphi^* E))_q = \{q\} \times \{\psi(q)\} \times E_{(\varphi \circ \psi)(q)}.$$

Both maps Ψ and Φ map fibres isomorphically to fibres.

(iv) Let $\pi : E \to N$ be a (G, σ)-fibre bundle. Suppose we are given two bundle charts

$$\alpha' : \mathrm{pr}_1^{-1}(V_\alpha) \to L$$

$$\beta' : \mathrm{pr}_1^{-1}(V_\beta) \to L$$

with $V_\alpha \cap V_\beta \neq \emptyset$. Recall that

$$V_\alpha = \varphi^{-1}(U_\alpha), V_\beta = \varphi^{-1}(U_\beta)$$

$$\alpha' = \alpha \circ \mathrm{pr}_2, \beta' = \beta \circ \mathrm{pr}_2,$$

where $\alpha : \pi^{-1}(U_\alpha) \to L$ and $\beta : \pi^{-1}(U_\beta) \to L$ are two bundle charts for $E \to N$. Denote by $\epsilon_{\alpha'\beta'} : V_\alpha \cap V_\beta \to \mathrm{Diff}(L)$ and $\epsilon_{\alpha\beta} : U_\alpha \cap U_\beta \to G$ the corresponding transition function.

For $p \in V_\alpha \cap V_\beta$ and $z \in L$ one then has

$$\begin{aligned}
\epsilon_{\alpha'\beta'}(p)(z) &= \alpha'|_{(\varphi^* E)_p} \circ \beta'|^{-1}_{(\varphi^* E)_p}(z) \\
&= (\alpha \circ \mathrm{pr}_2 |_{(\varphi^* E)_p}) \circ (\beta \circ \mathrm{pr}_2 |_{(\varphi^* E)_p})^{-1}(z) \\
&= (\alpha|_{E_{\varphi(p)}} \circ \beta|_{E_{\varphi(p)}})(z) \\
&= \epsilon_{\alpha\beta}(\varphi(p))(z).
\end{aligned}$$

Therefore, one has that $\epsilon_{\alpha'\beta'}$ maps into G and moreover,

$$\epsilon_{\alpha'\beta'}(p)(z) = \sigma(\epsilon_{\alpha'\beta'}(p), z) = \sigma(\epsilon_{\alpha\beta}(\varphi(p), z).$$

This shows that also the pullback bundle $\varphi^* E$ is a (G, σ)-fibre bundle.

A vector bundle is just a $\mathrm{GL}(L)$-fibre bundle with the fibre L being a vector space. But then the pullback bundle is also a $\mathrm{GL}(L)$-fibre bundle with the same fibre, thus it is a vector bundle.

Now let a principal G-bundle $P \to N$ be given. For $u \in P$ and $g \in G$ denote by $u \cdot g$ the free right action of G on P. Then define a right G-action on the pullback $\varphi^* P$ by $(p, u) \cdot g = (p, u \cdot g)$. Then the freeness of this action follows readily from the freeness of the G-action on P. We are only left to show the G-equivariance of the bundle charts of $\mathrm{pr}_1 : \varphi^* P \to M$. Let $\alpha : \pi^{-1}(U) \to G$ be a bundle chart on P and consider again the bundle chart $\alpha' : \varphi^{-1}(U) \to G$, defined by $\alpha'(p, u) = \alpha(u)$. Then the following computation finishes the proof:

$$\begin{aligned}
\alpha'((p, u) \cdot g) &= \alpha'(p, u \cdot g) \\
&= \alpha(u \cdot g) \\
&= \alpha(u)g \\
&= \alpha'(p, u)g.
\end{aligned}$$

(v) Let α and β be two bundle charts on $E \to N$. Let

$$\alpha' : U_\alpha \to L, \alpha' := \alpha \circ \mathrm{pr}_2,$$

$$\beta'' : V_\beta \to L, \beta'' := \beta \circ \mathrm{pr}_3$$

be any charts of $(\varphi \circ \psi)^* E$ and $\psi^*(\phi^* E)$, respectively, where we view the latter as a submanifold of $K \times M \times E$. Let

$$\mathrm{pr}_3 : K \times M \times E$$

be the projection onto the third factor. Let

$$\Psi : (\varphi \circ \psi)^* E \to \psi^*(\varphi^* E)$$

be given as above. Then consider the transition function which assigns to every $q \in K$ the map

$$f_\alpha^\beta(q) = \beta''|_{(\psi^*(\varphi^* E))_q} \circ \Psi \circ \alpha'|^{-1}_{((\varphi\circ\psi)^* E)_q}.$$

A direct computation shows that

$$f_\alpha^\beta(p)(z) = \beta|_{E_{(\varphi\circ\psi)(q)}} \circ \alpha|^{-1}_{E_{\varphi\circ\psi}(q)}(z) = \epsilon_{\alpha\beta}(\varphi \circ \psi)(q)(z).$$

Thus, the map Ψ serves as an isomorphism of the given G-bundles.

PROBLEM H.5. Let $L \to E \xrightarrow{\pi_1} M$ and $K \to F \xrightarrow{\pi_2} N$ be fibre bundles with structure groups G and H respectively. Prove that $(\pi_1, \pi_2)\colon E \times F \to M \times N$ is another fibre bundle with fibre $L \times K$ and structure group $G \times H$. We call this the **external product** of E and F.

SOLUTION. Let σ, τ be the actions of G, H on L, K, respectively. Then $G \times H$ acts on $L \times K$ by $\sigma \times \tau$, i.e. the map

$$(\sigma \times \tau)_{(g,h)}(u, v) = (\sigma_g(u), \tau_h(v))$$

is a smooth action. This action is clearly faithful if both σ and τ are. We have a (G, σ) bundle atlas $\{(U_\alpha, \epsilon_\alpha)\}_{\alpha\in A}$ and an (H, τ) bundle atlas $\{(V_\beta, \eta_\beta)\}_{\beta\in B}$ for E and F, which means that the transition functions can be written as

$$\epsilon_{\alpha\alpha'}(p) = \sigma_{g_{\alpha\alpha'}(p)} \in \mathrm{Diff}(L)$$

$$\eta_{\beta\beta'}(q) = \tau_{h_{\beta\beta'}(q)} \in \mathrm{Diff}(K)$$

for some smooth maps $g_{\alpha\alpha'} : U_\alpha \cap U_{\alpha'} \to G$ and $h_{\beta\beta'} : V_\beta \cap V_{\beta'} \to H$.

We consider the map $\pi_1 \times \pi_2 : E \times F \to M \times N$. Clearly, $\{U_\alpha \times V_\beta\}_{(\alpha,\beta)\in A\times B}$ is an open cover of $M \times N$ such that

$$(\pi_1 \times \pi_2)^{-1}(U_\alpha \times V_\beta) = \pi_1^{-1}(U_\alpha) \times \pi_2^{-1}(V_\beta)$$

is diffeomorphic via $\hat{\epsilon}_\alpha \times \hat{\eta}_\beta$ to

$$(U_\alpha \times L) \times (V_\beta \times K) \cong (U_\alpha \times V_\beta) \times (L \times K).$$

Thus $E \times F$ has the structure of an $L \times K$ fibre bundle with the bundle atlas

$$\{(U_\alpha \times V_\beta, \epsilon_\alpha \times \eta_\beta)\}_{(\alpha,\beta)\in A\times B}.$$

Finally, given a non-empty intersection

$$U_\alpha \times V_\beta \cap U_{\alpha'} \times V_{\beta'} = (U_\alpha \cap U_{\alpha'}) \times (V_\beta \cap V_{\beta'}),$$

the transition function at a point (p, q) in this open set is

$$\begin{aligned}(\epsilon_{\alpha\alpha'} \times \eta_{\beta\beta'})(p, q) &= \epsilon_{\alpha\alpha'}(p) \times \eta_{\beta\beta'}(q) \\ &= \sigma_{g_{\alpha\alpha'}(p)} \times \tau_{h_{\beta\beta'}(q)} \\ &= (\sigma \times \tau)_{(g_{\alpha\alpha'}\times h_{\beta\beta'})(p,q)},\end{aligned}$$

where

$$g_{\alpha\alpha'} \times h_{\beta\beta'} : (U_\alpha \cap U_{\alpha'}) \times (V_\beta \cap V_{\beta'}) \to G \times H$$

is a smooth map. This shows that $G \times H$ is naturally a structure group for $E \times F$.

(♣) PROBLEM H.6. Let $\iota\colon M \to M \times M$ denote the diagonal map $p \mapsto (p, p)$.

(i) Let E and F be two vector bundles over M. Prove that

$$E \oplus F = \iota^*(E \times F).$$

(ii) Let P and Q be two principal bundles over M. Prove that

$$P \star Q = \iota^*(P \times Q).$$

SOLUTION. We start with (i). Let $V \to E \xrightarrow{\pi_1} M$ and $W \to F \xrightarrow{\pi_2} M$ be two vector bundles on the same manifold M. Let $\{U_a, \epsilon_a\}$ be a vector bundle atlas for E and $\{U_a, \gamma_a\}$ be a vector bundle atlas for F. We can suppose that both atlases have the same open cover $\{U_a\}$ by taking mutual intersections. The direct sum

$$V \oplus W \to E \oplus F \xrightarrow{\pi_1 \oplus \pi_2} M,$$

can be constructed explicitly by setting

$$E \oplus F := \bigsqcup_{p \in M} E_p \oplus F_p,$$

and $(\pi_1 \oplus \pi_2)(v_p, w_p) := p$ for every $(v_p, w_p) \in E_p \oplus F_p$. The associated vector bundle atlas takes the form

$$\{U_a, \epsilon_a \oplus \gamma_a\}, \text{ where } (\epsilon_a \oplus \gamma_a)(u_1, u_2) := (\epsilon_a(u_1), \gamma_a(u_2)) \in V \oplus W,$$

for every $(u_1, u_2) \in (\pi_1 \oplus \pi_2)^{-1}(U_a)$. On the other side, the external product

$$V \times W \to E \times F \xrightarrow{\pi_1 \times \pi_2} M \times M,$$

can be constructed by taking

$$E \times F := \bigsqcup_{(p,q) \in M \times M} E_p \times F_q,$$

and $(\pi_1 \times \pi_2)(v_p, w_q) := (p, q)$ for every $(v_p, w_q) \in E_p \times F_q$. We can endow $E \times F$ with the vector bundle atlas

$$\{U_a \times U_b, \epsilon_a \times \gamma_b\}, \text{ where } (\epsilon_a \times \gamma_b)(u_1, u_2) := (\epsilon_a(u_1), \gamma_b(u_2)) \in V \times W,$$

for every $(u_1, u_2) \in (\pi_1 \times \pi_2)^{-1}(U_a \times U_b)$. By definition

$$\begin{aligned} \iota^*(E \times F) &= \Big\{(p, (u_1, u_2)) \in M \times E \times F : (\pi_1 \times \pi_2)(u_1, u_2) = \iota(p) = (p, p)\Big\} \\ &= \bigsqcup_{p \in M} \{p\} \times E_p \times F_p. \end{aligned}$$

The induced vector bundle atlas on $\iota^*(E \times F)$ can be taken of the form

$$\{\iota^{-1}(U_a \times U_b), (\epsilon_a \times \gamma_b) \circ \mathrm{pr}_2\},$$

and since $\iota^{-1}(U_a \times U_b) = U_a \cap U_b$ and $\{U_a\}$ is a open cover of M, we can reduce it to the smaller atlas

$$\{U_a, (\epsilon_a \times \gamma_a) \circ \mathrm{pr}_2\}.$$

Then, for every $(p, (v_p, w_p)) \in \iota^*(E \times F)$ since

$$((\epsilon_a \times \gamma_a) \circ \mathrm{pr}_2)(p, (v_p, w_p)) = (\epsilon_a \times \gamma_a)(v_p, w_p) = (\epsilon_a \oplus \gamma)(v_p, w_p),$$

it follows that $\mathrm{pr}_2 : \iota^*(E \times M) \to E \oplus F$ is a smooth vector bundle isomorphism, finishing the proof of point (i).

We now attack point (ii). Let $G \to P \xrightarrow{\pi_1} M$ and $H \to Q \xrightarrow{\pi_2} M$ be two principal bundles on the same manifold M, with associated actions on the total space τ_1, respectively τ_2. Recall that the fibre bundle

$$G \times H \to P \star Q \xrightarrow{\pi_1 \times \pi_2} M,$$

is a principal bundle constructed by setting

$$P \star Q := \bigsqcup_{p \in M} P_p \times Q_p,$$

with associated action on the total space $(\tau_1 \star \tau_2)$ naturally defined as

$$(\tau_1 \star \tau_2)((g, h), (u_1, u_2)) := (\tau_1(g, u_1), \tau_2(h, u_2)),$$

for every $((g, h), (u_1, u_2)) \in (G \times H) \times (P \star Q)$. We can endow the external product

$$G \times H \to P \times Q \xrightarrow{\pi_1 \times \pi_2} M \times M,$$

with a structure of principal bundle by defining the action

$$(\tau_1 \times \tau_2)((g, h), (u_1, u_2)) := (\tau_1(g, u_1), \tau_2(h, u_2)),$$

for all $((g, h), (u_1, u_2)) \in (G \times H) \times (P \times Q)$. As in part (i), it is easy to see that

$$\iota^*(P \times Q) = \bigsqcup_{p \in M} \{p\} \times P_p \times Q_p.$$

Moreover, it follows from its construction that $\iota^*(P \times Q)$ is a principal bundle with action

$$\tilde{\tau}((g, h), (p, u_1, u_2)) = (p, \tau_1(g, u_1), \tau_2(h, u_2)).$$

Then, the fibre bundle morphism

$$\mathrm{pr}_2 : \iota^*(P \times Q) \to (P \star Q)$$

$\mathrm{pr}_2(p, (u_1, u_2)) := (u_1, u_2)$ is $(\tilde{\tau}, \tau_1 \star \tau_2)$-equivariant (smoothness can be proved exactly as in point (i)). In fact

$$\begin{aligned}
(\mathrm{pr}_2 \circ \tilde{\tau}((g, h), \cdot))(p, u_1, u_2) &= \mathrm{pr}_2(\tau_1(g, u_1), (h, u_2)) \\
&= (\tau_1(g, u_1), \tau_2(h, u_2)) \\
((\tau_1 \star \tau_2)((g, h), \cdot) \circ \mathrm{pr}_2)(p, u_1, u_2) &= (\tau_1 \star \tau_2)((g, h), (u_1, u_2)) \\
&= (\tau_1(g, u_1), \tau_2(h, u_2)),
\end{aligned}$$

for every $((g, h), (u_1, u_2)) \in (G \times H) \times (P \star Q)$. By problem G.3 we have immediately that pr_2 is a principal bundle isomorphism, as wished.

PROBLEM H.7. Let E be a vector bundle over M and F a vector bundle over N. Suppose $\Psi\colon E \to F$ is any smooth map that maps each fibre E_p for $p \in M$ linearly onto some fibre F_q for $q \in N$. Prove that $\Psi = \Phi_1 \circ \Phi_2$ where Φ_1 is a vector bundle homomorphism and Φ_2 is a vector bundle morphism along a map $\varphi\colon M \to N$.

Solution. Denote by $\pi_1 : E \to M$ and $\pi_2 : F \to N$ the projections and let $s : M \to E$ be the zero section (see proof of Proposition 20.4). The map

$$\varphi := \pi_2 \circ \Psi \circ s$$

is a smooth map $M \to N$ which is covered by Ψ. Now note that there is a natural induced map $\Phi_1 : E \to \varphi^\star F$ which covers the identity $M \to M$, namely $\Phi_1(p, \xi) = (p, \Psi(p, \xi))$. With this definition we clearly have

$$\begin{array}{ccc} E_1 & \xrightarrow{\quad\Phi_1\quad} & \varphi^\star F \\ & \searrow^{\pi_1} \quad {}^{\pi}\swarrow & \\ & M & \end{array}$$

where $\pi : \varphi^\star F \to M$ is the induced projection map. But there is another map $\Phi_2 : \varphi^\star F \to F$ covering φ, namely $\Phi_2(p, \zeta) = (\varphi(p), \zeta)$. Now everything fits into a commutative diagram as wanted:

$$\begin{array}{ccccc} E & \xrightarrow{\Phi_1} & \varphi^\star F & \xrightarrow{\Phi_2} & F \\ & {}_{\pi_1}\searrow & \downarrow{\scriptstyle \pi} & & \downarrow{\scriptstyle \pi_2} \\ & & M & \xrightarrow{\quad\varphi\quad} & N. \end{array}$$

Problem H.8. Let $V \to E \xrightarrow{\pi_1} M$ and $W \to F \xrightarrow{\pi_2} M$ be two vector bundles over M, and let $\Phi : E \to F$ be a vector bundle homomorphism.

(i) Assume Φ is injective on each fibre. Consider the quotient vector space

$$\overline{E}_p := F_p / \Phi_p(E_p).$$

Prove that $\overline{E} := \bigsqcup_{p \in M} \overline{E}_p$ is a vector bundle over M with fibre W/V.

(ii) Assume that Φ is surjective on each fibre. Let

$$Z_p := \ker \Phi_p \subset E_p.$$

Prove that $Z := \bigsqcup_{p \in M} Z_p$ is a vector bundle over M. What is the fibre?

Solution. Choose a covering $\mathcal{U} = \{U\}$ of open subsets of M such that E_2 is trivial over each U, i.e. we have diffeomorphisms

$$\begin{array}{ccc} \pi_2^{-1}(U) & \xrightarrow{\qquad\cong\qquad} & U \times W \\ & {}_{\pi_2}\searrow \quad \swarrow_{\text{projection}} & \\ & U & \end{array}$$

which restrict to linear isomorphisms on fibres. Since the restriction of Φ to fibres is assumed to be an injective map, $\Phi(E_p) \subset F_p$ is a k_1-dimensional linear subspace, where $k_1 = \dim V$. Over a subset U we may therefore view $\Phi(E_p)$ as a k_1-dimensional linear subspace of W. Hence, we may for each point $p \in U$ choose a basis $w_1(p), \ldots, w_{k_1}(p)$ of W. By perhaps restricting to an open subset $U' \subset U$, this basis can in fact be chosen in such a way that they depend smoothly on $p \in U'$. By the Gram-Schmidt orthogonalisation procedure[20] this set of vectors $w_1(p), \ldots, w_{k_1}(p)$ can be extended to a basis $w_1(p), \ldots, w_{k_1}(p), w_{k_1+1}(p), \ldots w_{k_2}(p)$ for W and this basis depends smoothly on $p \in U'$. Therefore we have shown that there exist bundle charts

$$\begin{array}{ccc} \pi_2^{-1}(U) & \xrightarrow{\cong} & U \times W/V \\ & \searrow^{\pi_2} \quad \swarrow_{\text{projection}} & \\ & U & \end{array}$$

given by sending p to $w_{k_1+1}(p), \ldots w_{k_2}(p)$. Since we can cover M by such subsets U', we conclude that $\overline{E}$ is a vector bundle over M with fibre W/V.

Fix a point $p_0 \in M$ and choose an open neighbourhood U of p_0 such that both E and F are trivial over U. For all $p \in U$ we can then view $\Phi_p : E_p \to F_p$ as a surjective linear map $V \to W$, and hence it is represented by a matrix A_p. In particular, $A_{p_0} : V \to W$ is surjective, so there exists $v_1, \ldots v_{k_2} \in V$ such that $A_{p_0}(v_1), \ldots A_{p_0}(v_{k_2}) \in W$ is a basis. By continuity it follows that also

$$A_p(v_1), \ldots A_p(v_{k_2}) \in W$$

is a basis for all $x \in U$, after possibly shrinking U. Set $V' := \mathrm{Span}(v_1, \ldots v_{k_2}) \subseteq V$. The concatenation (over U)

$$Z \hookrightarrow U \times V \to U \times V/V'$$

of the inclusion and the quotient map is a fibrewise linear isomorphism by linear algebra reasons. Since $p_0 \in M$ was arbitrary we have shown that Z is locally trivial and admits a smooth structure, so it is a vector bundle with fibre V/V'.

(♣) Problem H.9. Let $\varphi\colon M \to N$ be a smooth map and suppose $\pi\colon E \to N$ is a vector bundle, which we illustrate pictorially as:

$$\begin{array}{ccc} & & E \\ & & \downarrow{\scriptstyle \pi} \\ M & \xrightarrow[\varphi]{} & N \end{array} \tag{H.1}$$

[20] Similar to Problem G.5.

A **solution** of the diagram (H.1) is a vector bundle $\pi_1 : F \to M$ together with a vector bundle morphism $\phi : F \to E$ along φ:

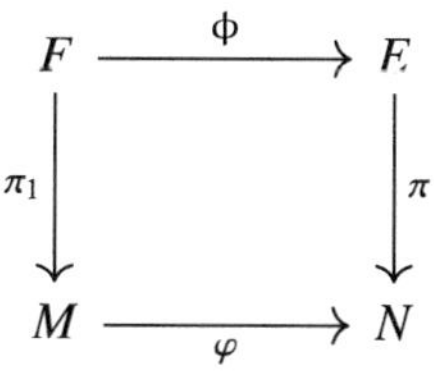

As we have seen, one possible solution is the pullback bundle $\varphi^* E$:

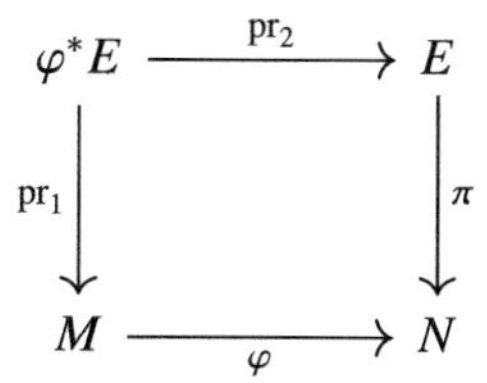

The aim of this problem is to prove that the pullback bundle can be characterised as a solution to a *universal mapping problem*: namely, that $\varphi^* E$ is the "most efficient" solution in the following sense:

Suppose $\pi_1 : F \to M$ and ϕ is *any* solution to (H.1). Prove there exists a *unique* vector bundle homomorphism $\psi : F \to \varphi^* E$ such that the following diagram commutes:

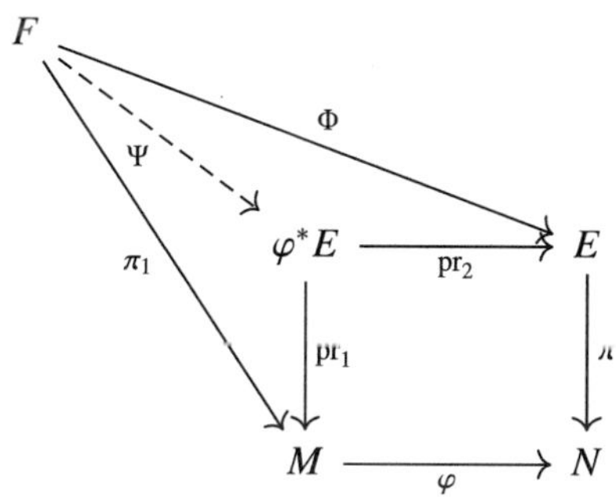

Prove moreover that $\varphi^* E$ is uniquely determined by this property. Explicitly this means that if $\widetilde{\pi} : \widetilde{E} \to M$ and $\widetilde{\phi}$ is another solution to the diagram (H.1) with the property that for *any* solution $\pi_1 : F \to M$ and ϕ of (H.1) there exists a *unique* vector bundle homomorphism $\widetilde{\psi} : F \to \widetilde{E}$ such that the following commutes:

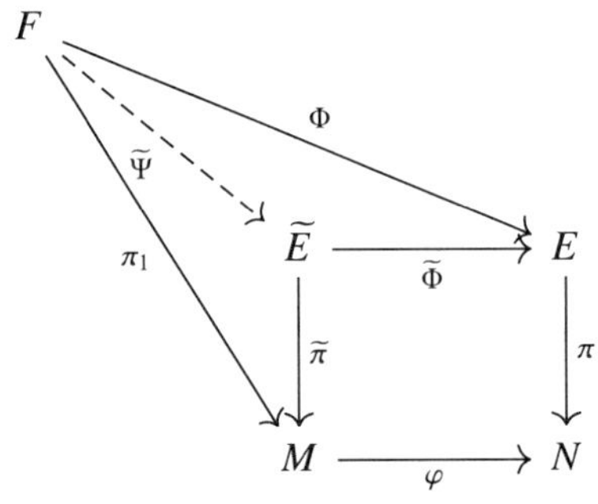

then in fact $\widetilde{E}$ is isomorphic as a vector bundle over M to φ^*E.

Solution. In the language of category theory we are looking for a certain bundle morphism (Ψ, id_M) between the two objects $\pi_1 \colon E_1 \longrightarrow M$ and $\mathrm{pr}_1 \colon \varphi^*E \longrightarrow M$ in the category of vector bundles VectBundles. Suppose first that there exists a bundle morphism (Ψ, id_M) fitting into the diagram from the statement (modulo the uniqueness part), i.e.

$$\Psi \colon F \to \varphi^*E,$$

is smooth and satisfies

$$\begin{cases} \mathrm{pr}_2 \circ \Psi &= \Phi \\ \mathrm{pr}_1 \circ \Psi &= \pi_1 . \end{cases}$$

The definition of the pullback bundle φ^*E and the fact that Ψ takes values in φ^*E imply

$$\forall \zeta \in F \colon \ \Psi(\zeta) = \big(\mathrm{pr}_1(\Psi(\zeta)), \mathrm{pr}_2(\Psi(\zeta))\big) .$$

With the commutative relations above this becomes

$$\Psi(\zeta) = (\pi_1(\zeta), \Phi(\zeta)),$$

which proves uniqueness. Existence follows by reading the last few lines backwards and finishes the first part of the exercise.

For the second part of the exercise we pick the vector bundle $\tilde{\pi} \colon \tilde{E} \longrightarrow M$ together with the bundle morphism $(\tilde{\Phi}, \varphi)$ just as in the statement. In particular, the vector bundle $\tilde{\pi} \colon \tilde{E} \longrightarrow M$ is a solution to (H.1) and so the first part of the exercise grants us the following commutative diagram

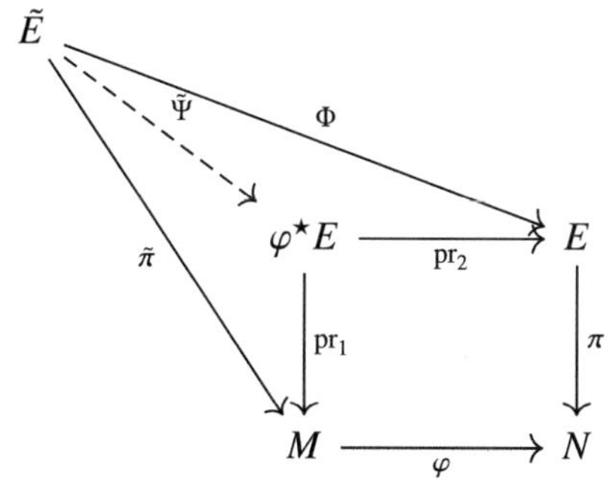

(H.2)

Due to the assumptions on $\tilde{\pi} : \tilde{E} \longrightarrow M$ however, we can "stick" yet another commutative diagram on top of (H.2), namely:

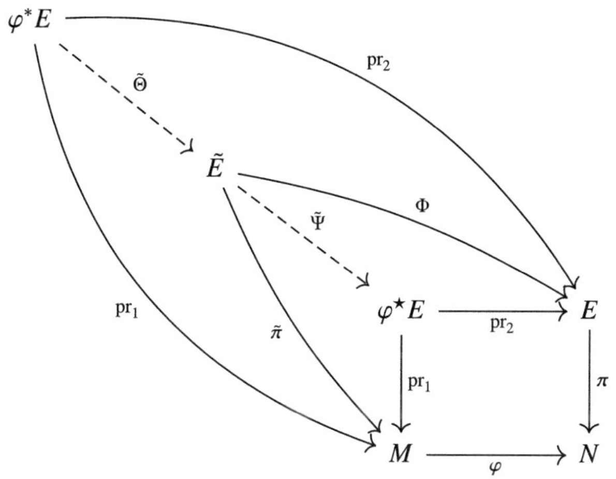

This means that $\mathrm{pr}_1 : \varphi^* E \longrightarrow M$ and pr_2 is the *unique* solution of (H.1) by the first part of the proof which shows

$$\tilde{\Psi} \circ \tilde{\Theta} = \mathrm{id}_{\varphi^* E},$$

simply because both $\mathrm{pr}_1 : \varphi^* E \longrightarrow M$ and $\mathrm{id}_{\varphi^* E}$ solve (H.1).

Similarly, reversing the roles[21] of $\varphi^* E$ and $\tilde{E}$ then proves

$$\tilde{\Theta} \circ \tilde{\Psi} = \mathrm{id}_{\tilde{E}} \, .$$

This concludes the proof.

[21] Note that reversing the roles also uses the first part of the proof. Why?

Problem Sheet I

(♣) PROBLEM H.10. Let $\varphi\colon M \to N$ be a smooth map.

(i) Let $A \in \mathcal{T}^{h,k}(M)$ denote a tensor of type (h, k). Let (U, x) and (V, y) denote two charts on M with $U \cap V \neq \emptyset$. Then one can write

$$A = f^{i_1\cdots i_h}_{j_1\cdots j_k} \frac{\partial}{\partial x^{i_1}} \otimes \cdots \otimes \frac{\partial}{\partial x^{i_h}} \otimes dx^{j_1} \otimes \cdots \otimes dx^{j_k}$$

and

$$A = g^{i_1\cdots i_h}_{j_1\cdots j_k} \frac{\partial}{\partial y^{i_1}} \otimes \cdots \otimes \frac{\partial}{\partial y^{i_h}} \otimes dy^{j_1} \otimes \cdots \otimes dy^{j_k}$$

for smooth functions $f^{i_1\cdots i_h}_{j_1\cdots j_k} \in C^\infty(U)$ and $g^{i_1\cdots i_h}_{j_1\cdots j_k} \in C^\infty(V)$. Investigate the relationship between

$$f^{i_1\cdots i_h}_{j_1\cdots j_k}|_{U\cap V} \qquad \text{and} \qquad g^{i_1\cdots i_h}_{j_1\cdots j_k}|_{U\cap V}.$$

(ii) Let $\omega \in \Omega^k(M)$ denote a differential k-form. Let (U, x) and (V, y) denote two charts on M with $U \cap V \neq \emptyset$. Then one can write

$$\omega = f_{i_1\cdots i_k} dx^{i_1} \wedge \cdots \wedge dx^{i_k}$$

and

$$\omega = g_{i_1\cdots i_k} dy^{i_1} \wedge \cdots \wedge dy^{i_k}$$

for smooth functions $f_{i_1\cdots i_k} \in C^\infty(U)$ and $g_{i_1\cdots i_k} \in C^\infty(V)$. Investigate the relationship between

$$f_{i_1\cdots i_k}|_{U\cap V} \qquad \text{and} \qquad g_{i_1\cdots i_k}|_{U\cap V}.$$

(iii) Assume φ is a diffeomorphism, and let $A \in \mathcal{T}^{h,k}(N)$. Let (U, x) be a chart on M and (V, y) a chart on N with $\varphi(U) \subset V$. Write

$$\varphi^* A = f^{i_1\cdots i_h}_{j_1\cdots j_k} \frac{\partial}{\partial x^{i_1}} \otimes \cdots \otimes \frac{\partial}{\partial x^{i_h}} \otimes dx^{j_1} \otimes \cdots \otimes dx^{j_k}$$

and

$$A = g^{i_1\cdots i_h}_{j_1\cdots j_k} \frac{\partial}{\partial y^{i_1}} \otimes \cdots \otimes \frac{\partial}{\partial y^{i_h}} \otimes dy^{j_1} \otimes \cdots \otimes dy^{j_k}$$

for smooth functions $f^{i_1\cdots i_h}_{j_1\cdots j_k} \in C^\infty(U \cap \varphi^{-1}(V))$ and $g^{i_1\cdots i_h}_{j_1\cdots j_k} \in C^\infty(V)$. Investigate the relationship between

$$f^{i_1\cdots i_h}_{j_1\cdots j_k} \qquad \text{and} \qquad g^{i_1\cdots i_h}_{j_1\cdots j_k}.$$

(iv) Let $\omega \in \Omega^k(N)$. Let (U, x) be a chart on M and (V, y) a chart on N such that $\varphi(U) \subset V$. Then one can write

$$\varphi^*\omega = f_{i_1\cdots i_k} dx^{i_1} \wedge \cdots \wedge dx^{i_k}$$

and

$$\omega = g_{i_1 \cdots i_k} dy^{i_1} \wedge \cdots \wedge dy^{i_k}$$

for smooth functions $f_{i_1 \cdots i_k} \in C^\infty(U \cap \varphi^{-1}(V))$ and $g_{i_1 \cdots i_k} \in C^\infty(V)$. Investigate the relationship between

$$f_{i_1 \cdots i_k} \qquad \text{and} \qquad g_{i_1 \cdots i_k}.$$

(v) Conclude that local coordinates are horrible.

Solution. We start with (i). First of all observe that each $\frac{\partial}{\partial x^i} \in T(U \cap V)$ can be written in terms of the local frame field $\frac{\partial}{\partial y^1}, \dots \frac{\partial}{\partial y^n}$, i.e.

$$\frac{\partial}{\partial x^r} = dy^s \left(\frac{\partial}{\partial x^r} \right) \frac{\partial}{\partial y^s} = \frac{\partial y^s}{\partial x^r} \frac{\partial}{\partial y^s},$$

and similarly

$$dx^r = dx^r \left(\frac{\partial}{\partial y^s} \right) dy^s = \frac{\partial x^r}{\partial y^s} dy^s.$$

Thus on $U \cap V$ we get

$$\begin{aligned} A &= f^{i_1 \cdots i_h}_{j_1 \cdots j_k} \Big(\frac{\partial}{\partial x^{i_1}} \otimes \cdots \otimes \frac{\partial}{\partial x^{i_h}} \otimes dx^{j_1} \otimes \cdots \otimes dx^{j_k} \\ &= f^{i_1 \cdots i_h}_{j_1 \cdots j_k} \frac{\partial y^{s_1}}{\partial x^{i_1}} \frac{\partial}{\partial y^{s_1}} \otimes \cdots \otimes \Big(\frac{\partial y^{s_h}}{\partial x^{i_h}} \frac{\partial}{\partial y^{s_h}} \Big) \\ &\qquad \otimes \Big(\frac{\partial x^{j_1}}{\partial y^{r_1}} dy^{r_1} \Big) \otimes \cdots \otimes \Big(\frac{\partial x^{j_k}}{\partial y^{r_k}} dy^{r_k} \\ &= f^{i_1 \cdots i_h}_{j_1 \cdots j_k} \left(\frac{\partial y^{s_1}}{\partial x^{i_1}} \cdots \frac{\partial y^{s_h}}{\partial x_{i_h}} \right) \cdot \left(\frac{\partial x^{j_1}}{\partial y^{r_1}} \cdots \frac{\partial x^{j_k}}{\partial y^{r_k}} \right) \frac{\partial}{\partial y^{s_1}} \otimes \cdots \otimes \frac{\partial}{\partial y^{s_h}} \\ &\qquad \otimes dy^{r_1} \otimes \cdots \otimes dy^{r_k}. \end{aligned}$$

This then proves

$$g^{s_1 \cdots s_h}_{r_1 \cdots r_k} = f^{i_1 \cdots i_h}_{j_1 \cdots j_k} \left(\frac{\partial y^{s_1}}{\partial x^{i_1}} \cdots \frac{\partial y^{s_h}}{\partial x^{i_h}} \right) \cdot \left(\frac{\partial x^{j_1}}{\partial y^{r_1}} \cdots \frac{\partial x^{j_k}}{\partial y^{r_k}} \right), \qquad \text{on } U \cap V.$$

The relation for (ii) follows in a similar fashion, namely

$$\begin{aligned} \omega &= f_{i_1 \cdots i_k} dx^{i_1} \wedge \cdots \wedge dx^{i_k} \\ &= f_{i_1 \cdots i_k} \frac{\partial x^{i_1}}{\partial y^{j_1}} \cdots \frac{\partial x^{i_k}}{\partial y^{j_k}} dy^{j_1} \wedge \cdots \wedge dy^{j_k}, \end{aligned}$$

and hence

$$g_{j_1 \cdots j_k} = f_{i_1 \cdots i_k} \frac{\partial x^{i_1}}{\partial y^{j_1}} \cdots \frac{\partial x^{i_k}}{\partial y^{j_k}}, \qquad \text{on } U \cap V.$$

For (iii) it will be convenient to view A and its pullback $\varphi^* A$ as multilinear maps at every footpoint (cf. Proposition 19.8). We first observe that $f_{j_1\cdots j_k}^{i_1\cdots i_h}$ is given by

$$\begin{aligned} f_{j_1\cdots j_k}^{i_1\cdots i_h} &= \varphi^*(A)\left(dx^{i_1}, \cdots, dx^{i_h}, \frac{\partial}{\partial x^{j_1}}, \cdots, \frac{\partial}{\partial x^{j_k}}\right) \\ &= A_\varphi\left(dx^{i_1} \circ D\varphi^{-1}, \cdots, dx^{i_h} \circ D\varphi^{-1}, D\varphi\left[\frac{\partial}{\partial x^{j_1}}\right], \cdots, D\varphi\left[\frac{\partial}{\partial x^{j_k}}\right]\right), \end{aligned}$$

where the second equality is just definition. Now let us denote by $(D\varphi^{-1})^i_j(y)$ the (i, j)-th entry of the matrix representative $D\varphi^{-1}(y)\colon T_y N \longrightarrow T_{\varphi^{-1}(y)} M$ with respect to the two bases $\frac{\partial}{\partial y^r}\big|_y$ and $\frac{\partial}{\partial x^s}\big|_{\varphi^{-1}(y)}$. Adopting the analogue notation for $D\varphi^i_j(x)$ and omitting footpoints again we get

$$f_{j_1\cdots j_k}^{i_1\cdots i_h} = A_\varphi\left((D\varphi^{-1})^{i_1}_{r_1}\, dy^{r_1}, \cdots, (D\varphi^{-1})^{i_h}_{r_h}\, dy^{r_h}, D\varphi^{s_1}_{j_1}\, \frac{\partial}{\partial y^{s_1}}, \cdots, D\varphi^{s_k}_{j_k}\, \frac{\partial}{\partial y^{s_k}}\right).$$

Now by multilinearity of A we conclude

$$f_{j_1\cdots j_k}^{i_1\cdots i_h} = (D\varphi^{-1})^{i_1}_{r_1} \cdots (D\varphi^{-1})^{i_h}_{r_h} \cdot D\varphi^{s_1}_{j_1} \cdots D\varphi^{s_k}_{j_k}\, g^{s_1,\cdots s_k}_{r_1\cdots r_h}.$$

Part (iv) follows from a very similar argument to the one in (iii) [22]

$$\begin{aligned} f_{i_1\cdots i_k} &= \varphi^*\omega\left(\frac{\partial}{\partial x^{i_1}}, \cdots \frac{\partial}{\partial x^{i_k}}\right) \\ &= \omega\left(D\varphi\left[\frac{\partial}{\partial x_{i_1}}\right], \cdots D\varphi\left[\frac{\partial}{\partial x_{i_k}}\right]\right) \\ &= D\varphi^{j_1}_{i_1} \cdots D\varphi^{j_h}_{i_h}\, g_{j_1,\cdots j_h}. \end{aligned}$$

Part (v) is left as an exercise.

Problem I.2. Let $\pi\colon E \longrightarrow M$ be a vector bundle. An operator $\zeta\colon \Gamma(E) \longrightarrow \Gamma(E)$ is said to satisfy the **Leibniz rule** if there exists a vector field X on M such that for any $f \in C^\infty(M)$ and $s \in \Gamma(E)$ one has

$$\zeta(fs) = (Xf)s + f\zeta(s).$$

Prove that an operator satisfying the Leibniz rule is a local operator but—provided $X \neq 0$—not a point operator.

[22] Actually (ii) (resp. (iv)) are special cases of (i) (resp. (iii)).

SOLUTION. To prove that ζ is a local operator, let $U \subset M$ be an open set and let $s \in \Gamma(E)$ be a section of E vanishing on U. Let $p_0 \in U$ and pick a corresponding $f \in C^\infty(M)$ so that

$$f(x) = \begin{cases} 0 & \text{at } p = p_0, \\ 1 & \text{in } M \setminus U. \end{cases}$$

Such function can be constructed because $\{p_0\}$ is a compact subset of U, and since s vanishes on U, we have

$$f\,s \equiv s \quad \text{in } M.$$

Consequently,

$$\zeta(s)(p_0) = \zeta(f\,s)(p_0) = \underbrace{X(f)(p_0)\,s(p_0)}_{=0} + \underbrace{f(p_0)}_{=0}\,\zeta(s)s(p_0) = 0,$$

and since $p_0 \in U$ was chosen arbitrarily, we conclude that $\zeta(s)$ vanishes on U.

To prove that, unless $X \equiv 0$, ζ is never a point operator, let p_0 be any point in M with $X(p_0) \neq 0$ and let e be a section of E so that $e(p_0) \neq 0$.

Since X is not zero at p_0, we can always construct a function $f \in C^\infty(M)$ so that $f(p_0) = 0$ and $X(f)(p_0) \neq 0$, as follows:

let (U, x) be local chart at p_0 so that $x(p_0) = 0$ and in whose local coordinates $x^1, \ldots, x^m$ we have $X(x) = \frac{\partial}{\partial x^1}\big|_x$ (cf. Corollary 14.2). The required function f is then defined in this coordinate patch as $f(x^1, \ldots, x^m) = x^1$, and then extended arbitrarily to a smooth function on all of M using a bump function. Now consider the section $s \in \Gamma(E)$ given by $s = f\,e$: we see that

$$s(p_0) = f(p_0)\,e(p_0) = 0, \quad \text{but} \quad \zeta(s)(p_0) = X(f)(p_0)\,e_0(p_0) = e(p_0) \neq 0,$$

so ζ is not a point operator.

PROBLEM I.3. Let M be a smooth manifold and let $E_1, \ldots, E_k$ and E be vector bundles over M. Let $\zeta\colon \Gamma(E_1) \times \cdots \times \Gamma(E_k) \to \Gamma(E)$ be a $C^\infty(M)$-multilinear operator. Prove that for each $p \in M$ there is a unique $\mathbb{R}$-multilinear map

$$\Phi_p\colon (E_1)_p \times \cdots \times (E_k)_p \to E_p$$

such that for all $s_i \in \Gamma(E_i)$ one has

$$\Phi_p(s_1(p), \ldots s_k(p)) = \zeta(s_1, \ldots, s_k)(p).$$

Solution. Fix $p \in M$ and let $v_i \in (E_i)_p$ for all $i = 1, \dots, k$. Extend each v_i to a global section $s_i \in \Gamma(E_i)$ (here we use Lemma 20.11). Define

$$\Phi_p(v_1, \dots, v_k) := \zeta(s_1, \dots s_k)(p). \tag{I.1}$$

We claim that this definition is independent of the extensions s_i of v_i. To see this, let $\hat{s}_i$ be another extension of v_i. Then

$$\chi(s_1, \dots, s_i - \hat{s}_i, \dots s_k)(p) = 0$$

because ζ is a point operator and $s_i(p) - \hat{s}_i(p) = v_i - v_i = 0$. Now multilinearity of ζ implies

$$\zeta(s_1, \dots, s_i, \dots s_k)(x) = \zeta(s_1, \dots, \hat{s}_i, \dots s_k)(x).$$

This shows that (I.1) defines Φ_p independently of the extensions s_i. Clearly, $\mathbb{R}$-multilinearity of Φ_p follows from $C^\infty(M)$-multilinearity of ζ.

Problem I.4. Let M be a smooth manifold and let $U \subset M$ be a non-empty open set. Prove that there is a canonical identification between $\mathcal{T}^{h,k}(U)$ and $C^\infty(U)$-multilinear functions

$$\underbrace{\Omega^1(U) \times \cdots \times \Omega^1(U)}_{h \text{ copies}} \times \overbrace{\mathfrak{X}(U) \times \cdots \times \mathfrak{X}(U)}^{k \text{ copies}} \to C^\infty(U). \tag{I.2}$$

Solution. The idea is to continue the proof of Theorem 21.5 with Problem I.3 taking the role of Proposition 20.24 (which is the key result to prove the Hom-Gamma Theorem). We start with ζ a $C^\infty(U)$-multilinear function as in (I.2). Recalling $\Gamma(TU) = \mathfrak{X}(U)$, $\Gamma(T^*U) = \Omega^1(U)$ and $\Gamma(U \times \mathbb{R}) = C^\infty(U)$ we can invoke Problem I.3 to get a *unique* $\mathbb{R}$-multilinear map

$$\Phi_p \colon \underbrace{T_p^*U \times \cdots T_p^*U}_{h\text{-times}} \times \overbrace{T_pU \times \cdots \times T_pU}^{k\text{-times}} \to \mathbb{R},$$

such that for every tuple of 1-forms $(\omega_1, \dots, \omega_h)$ and vector fields $(X_1, \dots, X_k)$ on U one has

$$\Phi_p\left(\omega_1(p), \dots, \omega_h(p), X_1(p), \dots, X_k(p)\right) = \zeta\left(\omega_1, \dots, \omega_h, X_1, \dots, X_k\right)(p).$$

For each $p \in U$ we have $\Phi_p \in \mathrm{Mult}_{h,k}(T_p^*U) \cong \mathrm{Mult}_{k,h}(T_pU) \cong T^{h,k}(T_pU)$ (cf. Proposition 19.8). The map

$$\Phi \colon U \to T^{h,k}(TU),\ p \mapsto \Phi_p$$

satisfies the section property. Using a similar argument as the one in the proof of Proposition 20.24 we see that Φ is smooth, thus proving that Φ is a tensor field on U of type (h, k), i.e.

$$\Phi \in \mathcal{T}^{h,k}(U).$$

This concludes the proof.

(♣) PROBLEM I.5. This problem introduces the **vertical bundle** of a fibre bundle.

(i) Let $\pi\colon E \to M$ be a fibre bundle with fibre L. Let

$$VE := \bigsqcup_{u \in E} \left\{\ker D\pi(u)\colon T_u E \to T_{\pi(u)} M\right\}$$

with projection map $\pi_V\colon VE \to E$. Prove that VE is a vector bundle over E of rank $l = \dim L$.

(ii) Assume now that $\pi\colon E \to M$ is a vector bundle. Prove that the vertical bundle VE is isomorphic as a vector bundle to the pullback bundle $\pi^* E \to E$.

(iii) Continue to assume that $\pi\colon E \to M$ is a vector bundle. Prove that composition $\pi \circ \pi_V\colon VE \to M$ is another vector bundle over M which is isomorphic to the direct sum bundle $E \oplus E$.

(iv) Continue to assume that $\pi\colon E \to M$ is a vector bundle. View M as an embedded submanifold of TM via the zero section. Consider the composite bundle $\pi \circ \pi_V\colon VE \to M$. Prove that there is a vector bundle morphism $VE \to TE$ over the inclusion of M into TM.

SOLUTION. For (i), we define bundle charts on VE as follows: let $\epsilon_a\colon \pi^{-1}(U_a) \to L$ be a bundle chart for E, and let $\sigma\colon \tilde{U} \to O \subset \mathbb{R}^l$ be a coordinate chart on F. We identify the tangent bundle TO with $O \times \mathbb{R}^l$, define $W = \pi^{-1}(U_a) \cap (\epsilon_a)^{-1}(\tilde{U})$ and set

$$\left(\epsilon_{a,V} := \mathrm{pr}_2 \circ D\sigma \circ D\epsilon_a\right)\colon \pi_V^{-1}(W) \to \mathbb{R}^l. \tag{I.3}$$

This choice induces a smooth structure on VE. To check that the resulting smooth manifold is a vector bundle over E, we must verify that the transition functions obtained from the ϵ_V are linear isomorphisms. Consider another bundle chart $\epsilon_b\colon \pi^{-1}(U_b) \to L$ and coordinate chart $\tau\colon \tilde{V} \to O'$ such that

$$\left(\pi^{-1}(U_a) \cap (\epsilon_a)^{-1}(\tilde{U})\right) \cap \left(\pi^{-1}(U_b) \cap (\epsilon_\beta)^{-1}(\tilde{V})\right) \neq \emptyset, \tag{I.4}$$

and define $\epsilon_{b,V}$ in analogy. Consider an element u in the left hand side of (I.4). Let $p := \pi(u)$. By Proposition 6.15, the inclusion $\iota_p\colon E_p \to E$ yields an isomorphism $D\iota_p(u)\colon T_u E_p \xrightarrow{\sim} \ker D\pi(u) \subset T_u E$. We have the following commutative diagram

$$\begin{array}{ccc} T_u E_p & \xrightarrow[\sim]{D\iota_p(u)} & \ker D\pi(u) \\ & {\scriptstyle D((\epsilon_a)_p)(u)} \searrow \quad \swarrow {\scriptstyle D(\epsilon_a)(u)} & \\ & T_{(\epsilon_a)_p(u)}L. & \end{array}$$

Since $(\epsilon_a)_p : E_p \to L$ is a diffeomorphism, its derivative is an isomorphism and thus it follows that $D(\epsilon_a)(u)\big|_{\ker D\pi(u)}$ is also an isomorphism.

$$\epsilon_{a,V}|_{(VE)_u} : VE_u = \ker D\pi(u) \to \mathbb{R}^l$$

is simply the composite $D\sigma\big(\varepsilon_a(u)\big) \circ D(\varepsilon_a)(u)|_{\ker D\pi(u)}$. So we conclude that $\epsilon_{a,V}|_{(VE)_u}$ is a linear isomorphism. The map $\epsilon_{b,V}|_{(VE)_u}$ has a similar description, and thus

$$\epsilon_{a,b}(p) = \alpha_V|_{(VE)_p} \circ \left(\beta_V|_{(VE)_p}\right)^{-1}$$

is a linear isomorphism. We conclude that VE is a rank l vector bundle over E, as desired.

For (ii), we first note that by definition

$$\pi^*E = \{(u, v) \in E \times E \mid \pi(u) = \pi(v)\}.$$

In other words, elements of the pullback bundle π^*E consist of pairs $(u, v) \in E \times E$ such that $u, v \in E_p$ for some $p \in M$. Since each E_p is a vector space, for any $(u, v) \in \pi^*E$ the map $t \mapsto \pi(u+tv)$ from $\mathbb{R}$ to M is constant in t. It follows that $\frac{d}{dt}|_{t=0}(u+tv) \in \ker D\pi(u)$. We thus have a map

$$\begin{aligned} \mathcal{J} : \pi^*E &\to VE \\ (u, v) &\mapsto \frac{d}{dt}\bigg|_{t=0}(u + tv). \end{aligned}$$

To see that $\mathcal{J}$ is an isomorphism of vector bundles, it suffices to show that it restricts to a linear isomorphism $(\pi^*E)_u \xrightarrow{\sim} (VE)_u$ for every $u \in E$. This is easy to see since the map $(u, v) \mapsto v$ identifies $(\pi^*E)_u$ with $E_{\pi(u)}$ and, again by Proposition 6.15, we may identify $(VE)_u = \ker D\pi(u)$ with $T_u E_{\pi(u)} \cong E_{\pi(u)}$. All of these identifications are linear, and after making these identifications, the map $\mathcal{J}|_{(\pi^*E)_u}$ becomes the identity map on $E_{\pi(u)}$. It follows that $\mathcal{J}|_{(\pi^*E)_u}$ is a linear isomorphism; hence $\mathcal{J}$ is an isomorphism of vector bundles.

For (iii), we first observe that $\pi^*E = E \oplus E$ as vector bundles over M. Indeed, as stated in the proof of (ii), for each element $(u, v) \in \pi^*E$, we have $u, v \in E_p$, where $p := \pi(u) = \pi(v)$. Hence we can view (u, v) as an element of $E_p \oplus E_p = (E \oplus E)_p$. Conversely, any $(u, v) \in E_p \oplus E_p$ can be viewed as an element of π^*E. We may thus identify π^*E and E as sets. To see that they are equal as vector bundles, it suffices to observe that they have the same bundle charts. Indeed, given a bundle chart $\epsilon : \pi^{-1}(U) \to L$ on E, the corresponding chart on $\mathrm{pr}_1 : \pi^*E \to E$ is given by

$$\epsilon \circ \mathrm{pr}_2 : \mathrm{pr}_1^{-1}\big(\pi^{-1}(U)\big) \to L,$$

which in turn corresponds to the bundle chart

$$(\epsilon \circ \mathrm{pr}_1, \epsilon \circ \mathrm{pr}_2) \colon \mathrm{pr}_1^{-1}\big(\pi^{-1}(U)\big) \to L \times L \tag{I.5}$$

on the composite bundle $\pi \circ \mathrm{pr}_1 : \pi^* E \to M$. Since

$$\mathrm{pr}_1^{-1}\big(\pi^{-1}(U)\big) = \pi^*\pi^{-1}(U) = \pi^{-1}(U) \oplus \pi^{-1}(U),$$

we see that (I.5) is precisely the bundle chart on $E \oplus E$ given by

$$(\epsilon, \epsilon) \colon \pi^{-1}(U) \oplus \pi^{-1}(U) \to L \times L.$$

Hence $\pi^* E = E \oplus E$ as vector bundles over M, as claimed.

We may thus view the map $\mathcal{J}$ from part (ii) as morphism $\mathcal{J} : E \oplus E \to VE$ of vector bundles over M. Since $\mathcal{J}$ is smooth, it only remains to show that $\mathcal{J}$ restricts to a linear isomorphism $E_p \oplus E_p \to (VE)_u$ for each $p \in M$. We first describe the vector space structure on $(VE)_u$ as a subspace of $(TE)_u$. The latter has vector space structure given by

$$\zeta + \xi := Da(u, v)(\zeta, \xi) \in T_{u+v}E, \qquad u, v \in E_p, \quad \zeta \in T_u E, \quad \xi \in T_v E, \tag{I.6}$$

where $a \colon E \oplus E \to E$ is the bundle map given on fibres by $(u, v) \mapsto u + v$. For $(u_1, v_1), (u_2, v_2) \in E_p \oplus E_p$ and $\lambda \in \mathbb{R}$, we need to show that

$$\mathcal{J}(u_1 + \lambda u_2, v_1 + \lambda v_2) = \mathcal{J}(u_1, v_1) + \lambda \mathcal{J}(u_2, v_2).$$

This amounts to showing the equality

$$\frac{d}{dt}\Big|_{t=0}(u_1 + \lambda u_2 + t(v_1 + \lambda v_2)) = \frac{d}{dt}\Big|_{t=0}(u_1 + tv_1) + \lambda \frac{d}{dt}\Big|_{t=0}(u_2 + tv_2), \tag{I.7}$$

where the addition on the right-hand side is as defined in (I.6). We denote the path $t \mapsto u + tv$ by $\gamma^{u,v} : \mathbb{R} \to E$. Then by definition

$$\frac{d}{dt}\Big|_{t=0}(u + tv) = D\gamma^{u,v}(0)\Big(\frac{d}{dt}\Big|_{t=0}\Big).$$

Using the chain rule, we compute

$$\begin{aligned}
\frac{d}{dt}\Big|_{t=0}&(u_1 + tv_1) + \lambda \frac{d}{dt}\Big|_{t=0}(u_2 + tv_2) \\
&= Da(u_1, \lambda u_2)\Big[D\gamma^{u_1,v_1}(0)\Big(\frac{d}{dt}\Big|_{t=0}\Big), D\gamma^{\lambda u_2,\lambda v_2}(0)\Big(\frac{d}{dt}\Big|_{t=0}\Big)\Big] \\
&= D\big(a \circ (\gamma^{u_1,v_1}, \gamma^{\lambda u_2,\lambda v_2})\big)(0)\Big(\frac{d}{dt}\Big|_{t=0}\Big) \\
&= D(\gamma^{u_1+\lambda u_2, v_1+\lambda v_2})(0)\Big(\frac{d}{dt}\Big|_{t=0}\Big) \\
&= \frac{d}{dt}\Big|_{t=0}(u_1 + \lambda u_2 + t(v_1 + \lambda v_2)).
\end{aligned}$$

This yields (I.7). Thus $\mathcal{J}|_{E_p \oplus E_p}$ is linear. Since it is also bijective onto $(VE)_p$, it must be a linear isomorphism. Hence $\mathcal{J}\colon E \oplus E \to VE$ is a vector bundle isomorphism over M, as desired.

For (iv), first recall that we can view M as an embedded submanifold of TM via the 0-section

$$s_0\colon M \to TM, \quad x \mapsto (x, 0).$$

Let $\epsilon\colon \pi^{-1}(U) \to \mathbb{R}^k$ be a bundle chart on U. Note that in the case of vector bundles, the associated chart as defined in (I.3) has the simpler description

$$\epsilon_V := \mathrm{pr}_2 \circ D\epsilon\colon \pi_V^{-1}\left(\pi^{-1}(U)\right) \to \mathbb{R}^k.$$

The associated bundle chart of $\pi \circ \pi_V$ is defined to be $(\epsilon \circ \pi_V, \mathrm{pr}_2 \circ D\epsilon)$. Since the π_V is just the restriction of the natural map $TE \to E$, it follows that $\epsilon \circ \pi_V = \mathrm{pr}_1 \circ D\epsilon$. Thus the bundle chart $(\epsilon \circ \pi_V, \mathrm{pr}_2 \circ D\epsilon)$ on the composite is simply

$$D\epsilon\colon \pi_V^{-1}\left(\pi^{-1}(U)\right) \to T\mathbb{R}^k = \mathbb{R}^k \times \mathbb{R}^k.$$

Since $D\epsilon$ is also a bundle chart for $D\pi\colon TE \to TM$, this shows that the natural inclusion $VE \to TM$ is a vector bundle morphism along $s_0\colon M \to TM$.

PROBLEM I.6. Let M be a smooth manifold.

(i) Suppose $A \in \mathfrak{T}^{1,1}(M) \cong \Gamma(\mathrm{End}(TM))$. Prove there exists a unique tensor derivation ζ_A on M with the property that $\zeta_A(Y)(p) = A_p(Y(p))$ for any vector field Y and satisfies $\zeta_A(f) = 0$ for any function f.
(ii) Let ξ be an arbitrary tensor derivation. Prove that there exists a vector field X on M and $A \in \mathfrak{T}^{1,1}(M)$ such that $\xi = \mathcal{L}_X + \zeta_A$. Deduce that the space of tensor derivations on M can be identified with $\mathfrak{X}(M) \times \Gamma(\mathrm{End}(TM))$.

SOLUTION. For part (i) let us suppose first that ζ_A exists and let us see how it must behave on functions, that is to say how $\zeta_A(f)$ behaves for a fixed $f \in C^\infty(M)$. On the one hand, since A is a tensor, its C^∞-multilinearity implies that for every $X \in \mathfrak{Y}(M)$ it must satisfy

$$\zeta_A(fY) = A(fY) = fA(Y).$$

On the other hand, since ζ_A is a derivation it must also hold that

$$\zeta_A(fY) = \zeta_A(f)Y + f\zeta_A(Y) = \zeta_A(f)Y + fA(Y).$$

For both condition to hold it is then necessary that

$$\zeta_A(f)Y = 0 \quad \text{for every } X \in \mathfrak{X}(M),$$

and consequently that $\zeta_A(f) = 0$. In other words, ζ_A has to vanish identically on functions. To prove that ζ_A exists we now set, by definition,

$$\begin{aligned} \zeta_A(f) &= 0 \quad \text{for every } f \in C^\infty(M), \\ \zeta_A(Y) &= A(Y) \quad \text{for every } Y \in \mathfrak{X}(M), \end{aligned}$$

and observe that ζ_A extends (uniquely) to a tensor derivation by Proposition 22.11.

For part (ii): if ξ is any derivation, it is in particular a derivation on functions and consequently, there exists some vector field $X \in \mathfrak{X}(M)$ so that

$$\xi(f) = \mathcal{L}_X(f) = X(f) \quad \text{for every } f \in C^\infty(M).$$

It is sufficient to check that the difference $\zeta' = \xi - \mathcal{L}_X$ restricted to vector fields defines a $(1, 1)$-tensor, for then the result follows from part (i) above: if $\zeta'|_{\mathfrak{X}(M)}$ is a $(1, 1)$-tensor, i.e. an element in $\mathcal{T}^{1,1}(M) = \Gamma(\mathrm{End}(TM))$, we can set $A(Y) = \zeta'(Y)$, $A \in \mathcal{T}^{1,1}(M)$ and invoke part (i) to obtain a unique tensor derivation ζ_A that vanishes on functions and satisfies $\zeta_A(Y) = A(Y)$. Since ζ' satisfies both these conditions, we get by uniqunes $\zeta_A = \zeta' = \xi - \mathcal{L}_X$, thus

$$\xi = \zeta_A + \mathcal{L}_X.$$

So we need to prove that ζ' defines a $(1, 1)$-tensor. Using the Hom-Gamma Theorem 20.25 we can identify

$$\mathcal{T}^{1,1}(M) = \Gamma(\mathrm{End}(TM)) = \mathrm{Hom}(\mathfrak{X}(M), \mathfrak{X}(M)).$$

Thus it suffices to show that $\zeta'|_{\mathfrak{X}(M)}$ is $C^\infty(M)$-linear, but that is an easy check:

$$\begin{aligned} \zeta'(fY) &= \xi(fY) - \mathcal{L}_X(fY) \\ &= \xi(f)Y + f\xi(Y) - \underbrace{\mathcal{L}_X(f)}_{=\xi(f)}Y - f\mathcal{L}_X(Y) \\ &= f(\xi(Y) - \mathcal{L}_X(Y)) \\ &= f\zeta'(Y). \end{aligned}$$

(♣) Problem I.7. Let (h_0, k_0), (h_1, k_1) and (h_2, k_2) be three pairs of non-negative integers. Suppose we are given a $C^\infty(M)$-bilinear operator

$$\mathcal{A} \colon \mathcal{T}^{h_0,k_0}(M) \times \mathcal{T}^{h_1,k_1}(M) \to \mathcal{T}^{h_2,k_2}(M).$$

Assume in addition that $\mathcal{A}$ has the property that if $\varphi \colon U \to V$ is a local diffeomorphism between open sets of M then the corresponding local operators

$$\varphi^*\left(\mathcal{A}^V(A, B)\right) = \mathcal{A}^U(\varphi^* A, \varphi^*(B)).$$

Prove that for every vector field X on M, one has

$$\tilde{\mathcal{L}}_X\big(\mathcal{A}(A, B)\big) = \mathcal{A}\big(\tilde{\mathcal{L}}_X A, B\big) + \mathcal{A}\big(A, \tilde{\mathcal{L}}_X B\big).$$

Solution. Denote by $\varphi_t \colon M \to M$ the flow generated by the vector field X. For notational reasons we will omit the basepoints in the following computation:

$$\begin{aligned}
\tilde{\mathcal{L}}_X(\mathcal{A}(A, B)) &= \lim_{t\to 0} \frac{\varphi_t^*\mathcal{A}(A, B) - \mathcal{A}(A, B)}{t} \\
&= \lim_{t\to 0} \frac{\varphi_t^*\mathcal{A}^V(A, B) - \mathcal{A}^U(A, B)}{t} \\
&\overset{(1)}{=} \lim_{t\to 0} \frac{\mathcal{A}^U(\varphi_t^* A, \varphi_t^* B) - \mathcal{A}^U(A, B)}{t} \\
&\overset{(2)}{=} \lim_{t\to 0} \frac{\mathcal{A}(\varphi_t^* A - A, \varphi_t^* B) + \mathcal{A}(A, \varphi_t^* B) - \mathcal{A}(A, B)}{t} \\
&\overset{(3)}{=} \lim_{t\to 0} \mathcal{A}\left(\frac{\varphi_t^* A - A}{t}, \varphi_t^* B\right) + \mathcal{A}\left(A, \frac{\varphi_t^* B - B}{t}\right) \\
&\overset{(4)}{=} \mathcal{A}(\tilde{\mathcal{L}}_X(A), B) + \mathcal{A}(A, \tilde{\mathcal{L}}_X(B)).
\end{aligned}$$

Equality (1) is justified by our assumptions, (2) and (3) are just bilinearity and (4) uses the C^∞_M-bilinear assumption on $\mathcal{A}$, or more precisely the fact that $\mathcal{A}$ is smooth in both entries so that we can take the limit inside.

Problem Sheet J

Problem J.1. Let E and F be two vector bundles over M and let $\{U_a \mid a \in A\}$ be an arbitrary open covering of M. Suppose we are given a collection

$$\{\zeta_a \colon \Gamma(U_a, E) \to \Gamma(U_a, F) \mid a \in A\}$$

of local operators such that

$$\zeta_a^{U_a \cap U_b} \equiv \zeta_b^{U_a \cap U_b}, \qquad \text{if } U_a \cap U_b \neq \emptyset.$$

Prove there exists a unique local operator $\zeta \colon \Gamma(E) \to \Gamma(F)$ such that

$$\zeta^{U_a} = \zeta_a, \qquad \forall\, a \in A.$$

Solution. We define ζ by $\zeta(s)(p) = \zeta_a(s|_{U_a})(p)$ if $p \in U_a$. The first equation shows that the choice of U_a does not matter, hence ζ is well defined and satisfies the last condition. In particular ζ is uniquely determined. We are only left to show that ζ is a local operator. For that purpose let s be a section on E vanishing on an open set U. If $p \in U \cap U_a$, then

$$\zeta(s)(p) = \zeta_a(s|_{U_a})(p) = 0,$$

since ζ_a is a local operator and $s|_{U_a} \equiv 0$.Therefore $\zeta(s)$ vanishes for all $p \in U$, which proves that ζ is a local operator.

PROBLEM J.2. Let $\varphi\colon M \to N$ denote a smooth map. Let $A \in \mathcal{T}^{0,k}(N)$. Using the Tensor Criterion Theorem 21.5, regard A as a $C^\infty(N)$-multilinear function

$$\underbrace{\mathfrak{X}(N) \times \cdots \times \mathfrak{X}(N)}_{k \text{ copies}} \to C^\infty(N).$$

and similarly regard $\varphi^*(A)$ as a $C^\infty(M)$-multilinear function

$$\underbrace{\mathfrak{X}(M) \times \cdots \times \mathfrak{X}(M)}_{k \text{ copies}} \to C^\infty(M).$$

Suppose $X_i \in \mathfrak{X}(M)$ is φ-related to $Y_i \in \mathfrak{X}(N)$ for $i = 1, \ldots, s$. Prove that

$$(\varphi^* A)(X_1, \ldots, X_k) = A(Y_1, \ldots, Y_k) \circ \varphi$$

as functions $M \to N$.

SOLUTION. We compute: for a point $p \in M$ we set $q = \varphi(p)$ and see

$$\begin{aligned}
\varphi^\star(A)(X_1, \ldots, X_k)(p) &= A_{\varphi(p)}(D\varphi(p)X_1(p), \ldots, D\varphi(p)X_k(p)) \\
&= A_q(D\varphi(\varphi^{-1}(q))X_1(\varphi^{-1}(q)), \ldots, D\varphi(\varphi^{-1}(q))X_k(\varphi^{-1}(q))) \\
&= A_q(Y_1(q), \ldots, Y_k(q)) \\
&= A(Y_1, \ldots, Y_k) \circ \varphi(q)
\end{aligned}$$

This solves the problem.

PROBLEM J.3. Let V be a vector space and suppose $\omega \in \bigwedge^h V^*$ and $\theta \in \bigwedge^k V^*$. Let $v_i \in V$ for $i = 1, \ldots, h+k$. Prove that if we identify ω with an element of $\mathrm{Alt}_h(V)$, θ with an element of $\mathrm{Alt}_k(V)$, and $\omega \wedge \theta$ with an element of $\mathrm{Alt}_{h+k}(V)$, one has:

$$\begin{aligned}
&(\omega \wedge \theta)(v_1, \ldots, v_{h+k}) \\
&\quad = \frac{1}{h!k!} \sum_{\varrho \in \mathfrak{S}_{h+k}} \mathrm{sgn}(\varrho)\omega\big(v_{\varrho(1)}, \ldots, v_{\varrho(h)}\big)\, \theta\big(v_{\varrho(h+1)}, \ldots, v_{\varrho(h+k)}\big)
\end{aligned}$$

or equivalently

$$
\begin{aligned}
&(\omega \wedge \theta)(v_1, \ldots, v_{h+k}) \\
&\quad = \sum_{\varrho \in \mathrm{Shuffle}(h,k)} \operatorname{sgn}(\varrho)\omega\big(v_{\varrho(1)}, \ldots, v_{\varrho(h)}\big)\, \theta\big(v_{\varrho(h+1)}, \ldots, v_{\varrho(h+k)}\big).
\end{aligned}
$$

Solution. Recall Proposition 19.24, which gives a canonical isomorphism

$$\alpha\colon \bigwedge^h V^* \xrightarrow{\sim} \mathrm{Alt}_h(V).$$

Explicitly, this is given on decomposable elements by

$$\alpha(p^1 \wedge \cdots \wedge p^h)(v_1, \ldots, v_h) = \det\big(p^i(v_j)\big).$$

Using the Leibniz formula for the determinant, we obtain

$$\alpha(p^1 \wedge \cdots \wedge p^h)(v_1, \ldots, v_h) = \sum_{\varrho \in \mathfrak{S}_h} \operatorname{sgn}(\varrho) p^1(v_{\varrho(1)}) \cdots p^r(v_{\varrho(h)}).$$

Making the identifications specified in the problem, this yields[23]

$$
\begin{aligned}
&(\omega \wedge \theta)(v_1, \ldots v_{h+k}) \\
&\quad = \frac{1}{h!k!} \sum_{\varrho \in \mathfrak{S}_{h+k}} \operatorname{sgn}(\varrho)\omega(v_{\varrho(1)}, \ldots, v_{\varrho(h)})\theta(v_{\varrho(r+1)}, \ldots, v_{\varrho(h+k)}).
\end{aligned}
\tag{J.1}
$$

Now we prove that the latter is equal to the last equation in the statement. Observe that given any permutation $\pi \in \mathfrak{S}_{h+k}$, there exists a unique permutation σ of $\mathfrak{S}_h$ and a unique permutation τ on the letters $h+1, \ldots, h+k$ such that $\varrho := \pi \circ \sigma \circ \tau$ is an (h, k)-shuffle. Let us write $\mathfrak{S}'_k$ for the permutations on the letters $h+1, \ldots, h+k$. Both sides are linear in ω and θ, and hence we may assume that both ω and ϑ are decomposable, say:

$$\omega = p^1 \wedge \cdots \wedge p^h, \qquad \theta = q^1 \wedge \cdots \wedge q^k.$$

Then if $v_1, \ldots, v_{h+k} \in V$, we have (using the Leibniz formula for the determinant) that

$$
\begin{aligned}
&\big(p^1 \wedge \cdots \wedge p^h \wedge q^1 \wedge \cdots \wedge q^k\big)(v_1, \ldots, v_{h+k}) \\
&\quad = \sum_{\pi \in \mathfrak{S}_{h+k}} \operatorname{sgn}(\pi) p^1\big(v_{\pi(1)}\big) \cdots p^h\big(v_{\pi(h)}\big)\, q^1\big(v_{\pi(h+1)}\big) \cdots q^s\big(v_{\pi(h+k)}\big) \\
&\quad = \sum_{\varrho \in \mathrm{Shuffle}(h,k)} \sum_{\sigma \in \mathfrak{S}_h} \sum_{\tau \in \mathfrak{S}'_k} \operatorname{sgn}(\varrho) \operatorname{sgn}(\sigma) \operatorname{sgn}(\tau) \qquad \text{(J.2)} \\
&\qquad\qquad p^1\big(v_{\varrho(\sigma(1))}\big) \cdots p^h\big(v_{\varrho(\sigma(h))}\big)\, q^1\big(v_{\varrho(\tau(h+1))}\big) \cdots q^k\big(v_{\varrho(\tau(h+k))}\big)
\end{aligned}
$$

[23] Equation (J.1) is easy to verify when ω and ϑ are decomposable, from which the general case follows.

Set $w_i := v_{\varrho(i)}$. Then $w_{\sigma(i)} = v_{\varrho(\sigma(i))}$ and hence

$$\begin{aligned}\sum_{\sigma\in\mathfrak{S}_h} p^1\big(v_{\varrho(\sigma(1))}\big)\cdots p^h\big(v_{\varrho(\sigma(h))}\big) &= \sum_{\sigma\in\mathfrak{S}_h} \operatorname{sgn}(\sigma) p^1\big(w_{\sigma(1)}\big)\cdots p^h\big(w_{\sigma(h)}\big)\\ &= \big(p^1\wedge\cdots\wedge p^h\big)\big(w_1,\ldots,w_h\big)\\ &= \big(p^1\wedge\cdots\wedge p^h\big)\big(v_{\varrho(1)},\ldots,v_{\varrho(h)}\big)\end{aligned}$$

It follows that the sum in (J.2) is

$$\begin{aligned}&\sum \operatorname{sgn}(\rho)\, \big(p^1\wedge\cdots\wedge p^h\big)\big(v_{\varrho(1)},\ldots,v_{\varrho(h)}\big)\, \big(q^1\wedge\cdots\wedge q^k\big)\big(v_{\varrho(h+1)},\ldots,v_{\varrho(h+k)}\big)\\ &\qquad = \sum \operatorname{sgn}(\varrho)\, \omega\big(v_{\varrho(1)},\ldots,v_{\varrho(h)}\big)\, \theta\big(v_{\varrho(h+1)},\ldots,v_{\varrho(k+s)}\big)\end{aligned}$$

where the sum is over $\varrho \in \mathrm{Shuffle}(h,k)$. This finishes the proof.

Problem J.4. Let $\omega \in \bigwedge^h V^*$ and $\theta \in \bigwedge^k V^*$. Prove that

$$i_v(\omega\wedge\theta) = i_v\omega\wedge\theta + (-1)^h\omega\wedge i_v\theta.$$

Solution. Either use Definition 23.7 and prove it directly (trivial) or use the second characterisation from problem J.3 (also trivial).

Problem J.5. Let $(M,\mathfrak{o})$ be an oriented smooth manifold with boundary. Let $\mu \in \Omega^m(M)$ be a volume form representing $\mathfrak{o}$. Let X be an outward pointing section.

(i) Prove that $i_X\mu$ restricts to define a volume form on ∂M.
(ii) Let $\partial\mathfrak{o}$ denote the orientation of ∂M determined by $i_X\mu$. Prove that (as the notation suggests) $\partial\mathfrak{o}$ only depends on $\mathfrak{o}$, and not on the particular choice of μ and X.

Solution. Let $X_1,\ldots,X_m$ be a local frame for TM on U (such that $U\cap\partial M\neq\emptyset$) with $X_1 = X$ and $X = \frac{\partial}{\partial x_1}$ in local coordinates (see Lemma 20.6 and proof of Lemma 24.37). As seen in the first part of Proposition 24.9, by changing the sign of X_2 (if needed) we can assume that $X, X_2,\ldots,X_m$ is a positively oriented frame. But then,

$$i_X\mu(X_2,\ldots,X_m) = \mu(X,X_2,\ldots,X_m)\neq 0,$$

by assumption on the vector fields and $\mu\in\mathfrak{o}$. This proves that $\iota_X\mu$ defines a volume form on ∂M.

For the second part, we pick another volume form τ belonging to the same equivalence class as μ, i.e. $\tau \in \mathfrak{o}$, and another outward pointing vector field Y on the boundary ∂M. Since $\dim M - \dim \partial M = 1$ (Proposition 24.30), we have that the space of

$$\{\text{strictly inward and outward pointing vector fields}\} \cup \{0\}$$

is 1-dimensional.[24] In particular, there exists a unique strictly positive function $f\colon \partial M \to \mathbb{R}_{>0}$ such that $X = fY$. By definition of $\mathfrak{o}$ we get, similar to f, another strictly positive function $g\colon M \to \mathbb{R}_{>0}$ such that $\mu = g\,\tau$. All in all:

$$i_Y\tau(X_2,\ldots,X_m) = \underbrace{(f\,g)}_{>0} \cdot i_X\mu(X_2,\ldots,X_m) \text{ on } \partial M.$$

Therefore Y and τ induce the same orientation on ∂M, proving the desired independence.

PROBLEM J.6.

(i) Prove that S^m is orientable.
(ii) Prove that any Lie group is orientable.
(iii) Prove that $\mathbb{R}P^m$ is orientable if and only if n is odd. *Hint*: Consider the antipodal map $p \mapsto -p$ on S^m.

SOLUTION.

(i) We will use a concrete model for S^m:

$$S^m := \{x = (x^0,\ldots,x^m) \in \mathbb{R}^{m+1} \mid |x| = 1\}.$$

On $\mathbb{R}^{m+1}$ we have the canonical differential $(m+1)$-form

$$\omega_{\mathbb{R}^{m+1}} := dx^0 \wedge \cdots \wedge dx^m,$$

which is clearly a volume form. On $\mathbb{R}^{m+1}\backslash\{0\}$ we have a canonical nowhere vanishing radial vector field $X \in \mathfrak{X}(\mathbb{R}^{m+1})$ which assigns to every point $p \neq 0$ the vector

$$X(p) = \mathcal{J}_p(p) \in T_p\mathbb{R}^{m+1}.$$

The interior product $i_X\omega_{\mathbb{R}^{m+1}}$ is a differential m-form $\mathbb{R}^{m+1}\backslash\{0\}$, so in particular the restriction $\omega_{S^m} := (i_X\omega_{\mathbb{R}^{m+1}})|_{S^m}$ is a differential m-form on S^m. We claim that ω_{S^m} is a volume form on S^m. To see this, recall that the differential of the inclusion $S^m \hookrightarrow \mathbb{R}^{m+1}$ identifies the tangent space of S^m at $p \in S^m$ with the orthogonal complement

[24] By strictly we mean $\xi \in T_p\partial M$ such that $dx_p^1(\xi) \neq 0$ and $dx_p^i(\xi) = 0$ for the remaining $i = 2,\ldots,m$.

$$p^{\perp} \subset \mathbb{R}^{m+1} \cong T_p\mathbb{R}^{m+1}$$

via $\mathcal{J}_p$, cf. Problem C.5. In particular, if $\xi_1, \ldots, \xi_m$ denotes a basis for T_pS^m, then $p, \mathcal{J}_p^{-1}\xi_1, \ldots, \mathcal{J}_p^{-1}\xi_m$ is a basis for $\mathbb{R}^{m+1} \cong T_p\mathbb{R}^{m+1}$. Hence,

$$\omega_{S^m}(\xi_1, \ldots \xi_m) = \omega_{\mathbb{R}^{m+1}}(X, \xi_1, \ldots, \xi_m) \neq 0.$$

Since this is true for every $p \in S^m$ we see that ω_{S^m} is indeed a volume form on S^m and thus S^m is orientable. Note that the above proof can be adapted to show the following more general statement: the boundary of an orientable manifold is orientable. In the above case, S^m is the boundary of the ball $\{x \mid |x| < 1\}$, which is clearly orientable, simply because it is an open subset of the orientable $\mathbb{R}^{m+1}$.

(ii) Let G denote an n-dimensional Lie group. The Lie algebra $\mathfrak{g} = T_eG$ is an n-dimensional real vector space and hence admits a volume form of the type $\omega_e := e_1^* \wedge \cdots \wedge e_n^*$, where $e_1, \ldots, e_n$ is some basis for $\mathfrak{g}$ with dual basis $e_1^*, \ldots, e_n^*$. At some point $g \in G$ we can define

$$\omega_g := (Dl_{g^{-1}}(g))^*\omega_e$$

which is the (linear) pull back of ω_e by the linear map $Dl_{g^{-1}}(g) \colon T_gG \to T_eG$. The map $g \mapsto \omega_g$ is clearly a smooth differential n-form which vanishes nowhere, because ω_e doesn't vanish and $Dl_{g^{-1}}(g)$ is a linear isomorphism. This shows that G is orientable.

(iii) Denote by $\pi \colon S^m \to \mathbb{R}P^m = S^m/(p \sim -p)$ the quotient map and by $a \colon S^m \to S^m$ the antipodal map

$$a(x) = -x.$$

$\mathbb{R}P^m$ is orientable if and only if there exists a nowhere vanishing differential m-form $\omega_{\mathbb{R}P^m}$ on $\mathbb{R}P^m$. Since the differential of π is an isomorphism (π is a covering map) at every point it follows that $\omega_{\mathbb{R}P^m}$ vanishes nowhere if and only if $\pi^*\omega_{\mathbb{R}P^m}$ vanishes nowhere. Since $\pi \circ a = \pi$ we also have

$$a^*\pi^*\omega_{\mathbb{R}P^m} = (\pi \circ a)^*\omega_{\mathbb{R}P^m} = \pi^*\omega_{\mathbb{R}P^m} \tag{J.3}$$

Putting these things together we conclude that, if $\mathbb{R}P^m$ is orientable, then $\pi^*\omega_{\mathbb{R}P^m}$ is a volume form on S^m, so it is a generator for the 1-dimensional real vector space $H^m_{\mathrm{dR}}(S^m)$, and (J.3) says that the induced linear map

$$a^* \colon H^m_{\mathrm{dR}}(S^m) \to H^m_{\mathrm{dR}}(S^m)$$

is the identity map (i.e. multiplication by 1). Hence, to see that $\mathbb{R}P^m$ is *not* orientable when m is even it suffices to check that a^* is *not* the identity in this case. In (i) we saw that

$$p \mapsto \omega_{S^m}|_p := \iota_x(dx^0 \wedge \cdots \wedge dx^m)|_{S^m}$$

defines a volume form on the sphere. Denote by $A \colon \mathbb{R}^{m+1} \to \mathbb{R}^{m+1}$ the antipodal map $A(x) = -x$, so that if $i \colon S^m \to \mathbb{R}^{m+1}$ denotes the inclusion then $i \circ a = A \circ i$

as a map $S^m \to \mathbb{R}^{m+1}$. Then

$$\begin{aligned}(a^*\omega_{S^m})|_x &= a^*(\omega_{S^m}|_{-p}) \\ &= a^*i^*(i_{X(-p)}[dx^0 \wedge \cdots \wedge dx^m]) \\ &= (i \circ a)^*(i_{X(-p)}[dx^0 \wedge \cdots \wedge dx^m]) \\ &= (A \circ i)^*(i_{X(-p)}[dx^0 \wedge \cdots \wedge dx^m]) \\ &= i^*A^*(i_{X(-p)}[dx^0 \wedge \cdots \wedge dx^m]) \\ &= i^*(i_{X(p)}[-dx^0 \wedge \cdots \wedge -dx^m]) \\ &= (-1)^{m+1}i^*(i_{X(p)}[dx^0 \wedge \cdots \wedge dx^m]) \\ &= (-1)^{m+1}\omega_{S^m}|_p\end{aligned}$$

Hence, we see that if m is even (so $(m+1)$ is odd) then

$$a^* \colon H^m_{\mathrm{dR}}(S^m) \to H^m_{\mathrm{dR}}(S^m)$$

is multiplication by -1 instead of 1, which concludes the proof that $\mathbb{R}P^m$ is *not* orientable when m is even.

To see that $\mathbb{R}P^m$ is orientable when m is odd, note that the above computation shows that in this case ω_{S^m} is an a^*-invariant volume form (in the sense that $a^*\omega_{S^m} = \omega_{S^m}$). This implies that ω_{S^m} descends to a volume form on $\mathbb{R}P^m$ when m is odd, finishing the proof that $\mathbb{R}P^m$ is orientable in this case.

Problem J.7. Let

$$\mathbb{R}^m_- := \mathbb{R}^m_{u^1 \le 0}, \qquad \mathbb{H}^m := \mathbb{R}^m_{u^m \ge 0}.$$

We can identify both $\partial\mathbb{R}^m_-$ and $\partial\mathbb{H}^m$ with $\mathbb{R}^{m-1}$. Endow both $\mathbb{R}^m_-$ and $\mathbb{H}^m$ with the standard orientation that they inherit from $\mathbb{R}^m$. Show that the induced orientation on $\partial\mathbb{R}^m_-$ is equal to the standard orientation on $\mathbb{R}^{m-1}$ for all n, but that the induced orientation on $\partial\mathbb{H}^m$ agrees with the standard orientation of $\mathbb{R}^{m-1}$ only when n is even.

Solution. Both on $\mathbb{R}^m_-$ and $\mathbb{H}^m$ the orientation is the one induced by the standard volume form $\mu = dx^1 \wedge \cdots \wedge dx^m$.

To determine the orientation induced on $\partial\mathbb{R}^m_-$, it is sufficient (by Problem J.6) to compute $i_X(\mu)$ for an arbitrarily chosen outward-pointing section $X(p) = X^i(p)\frac{\partial}{\partial x^i}|_p$ of $T\mathbb{R}^m_-|_{\partial\mathbb{R}^m_-}$. By definition X, is outward-pointing if and only if

$$dx^1(X)|_p = X^1(p) > 0 \quad \text{for every } p \in \partial\mathbb{R}^n_-,$$

consequently the obvious choice is the constant section $X = \frac{\partial}{\partial x^1}$, for which we get

$$i_X(\mu) = dx^2 \wedge \cdots \wedge dx^m.$$

This is a volume form on $\partial\mathbb{R}^n_-$ (cf. Problem J.5) which is exactly the standard one on $\mathbb{R}^{m-1} \simeq \{0\} \times \mathbb{R}^{m-1}$.

Similarly, on $\mathbb{H}^m$ a section $Y(p) = Y^i(p)\frac{\partial}{\partial x^i}|_p$ of $T\mathbb{H}^m|_{\partial\mathbb{H}^m}$ is outward-pointing if and only if

$$dx^m(Y)|_p = Y^m(p) < 0 \quad \text{for every } p \in \partial\mathbb{H}^n,$$

so choosing $Y = -\frac{\partial}{\partial x^m}|_p$ we compute[25]

$$i_Y(\mu) = (-1)^m dx^1 \wedge \cdots \wedge dx^{m-1},$$

and so this volume form induces the same orientation on $\partial\mathbb{H}^m$ as the standard one in $\mathbb{R}^{m-1} \simeq \mathbb{R}^{m-1} \times \{0\}$ if and only if m is even.

Problem J.8.

(i) Let V be a vector space of dimension n. A **symplectic form** on V is an element $\omega \in \mathrm{Alt}_2(V) \cong \bigwedge^2 V^*$ which is **non-degenerate** in the sense that $i_v(\omega) \equiv 0$ if and only if $v = 0$. Prove that if a symplectic form exists then $n = 2k$ is necessarily an even number.

(ii) A **symplectic manifold** is a smooth manifold M equipped with a *closed* differential 2-form ω such that ω_p is a symplectic form on T_pM for every $x \in M$. Prove that any symplectic manifold is orientable.

(iii) Let M be a smooth manifold. Define a 1-form $\Theta \in \Omega^1(T^*M)$ on the cotangent bundle via the formula:

$$\Theta_{p,\lambda}(\zeta) = \lambda\big(D\pi(p,\lambda)\zeta\big),$$

for $p \in M$, $\lambda \in T^*_pM$, and $\zeta \in T_{(p,\lambda)}T^*M$. Prove that $\omega := d\Theta$ is a symplectic form on T^*M. Thus every cotangent bundle is a symplectic manifold.

Solution. We will prove the first part by proving the following stronger statement:[26] there exists a basis $e_1, \ldots, e_k, f_1, \ldots, f_k$ of V satisfying

$$\forall i, j = 1, \ldots, r\colon\ \omega(e_i, e_j) = \omega(f_i, f_j) = 0, \text{ and } \omega(e_i, f_j) = \delta_{ij}.$$

We will refer to such a basis as a *symplectic basis* of (V, ω). If $\dim V < 1$ then there is nothing to prove, so we assume $\dim V \geq 1$. Pick a non-zero vector $e_1 \in V$. Non-degeneracy of ω and $e_1 \neq 0$ imply the existence of a non-zero $f_1 \in V$ such that

[25] Either compute this directly (easy) or use Problem J.4 with $\omega = dx^1 \wedge \cdots \wedge dx^{m-1}$, $\theta = dx^m$ and observe that there is another -1 coming from Y itself!

[26] This makes the proof of the remaining parts of this problem much more pleasant.

$$\omega(e_1, f_1) \neq 0.$$

By bilinearity of ω we can assume $\omega(e_1, f_1) = 1$ after rescaling f_1. Since ω is alternating and bilinear we know that f_1 cannot lie in the span of e_1 which in turn implies that the dimension of V is *strictly* greater than 1. If $\dim V = 2$ then we are done, otherwise we proceed as follows. For any subspace $W \subset V$ there exists the so called *symplectic complement*

$$W^{\omega} = \{v \in V \mid \omega(v, w) = 0 \text{ for all } w \in W\},$$

which is itself a subspace of V. Whenever $W \subseteq V$ is *symplectic*, i.e. ω restricts to a symplectic form on W, there is a splitting

$$V = W \oplus W^{\omega}.$$

Since $\dim V > 2$ we get for $W_1 := \operatorname{span}(e_1, f_1)$, which is symplectic,[27] a non-trivial symplectic complement W_1^{ω} with

$$V = W_1 \oplus W_1^{\omega}.$$

Again, pick a non-zero vector e_2, this time lying in W_1^{ω}, and another non-zero vector $f_2 \in V$ such that $\omega(e_2, f_2) = 1$ (works by non-degeneracy of ω). By definition of the symplectic complement, $e_2 \in W_1^{\omega}$ and $V = W_1 \oplus W_1^{\omega}$, it follows that $f_2 \in W_1^{\omega} \setminus \operatorname{span}(e_2)$ and hence

$$\dim \operatorname{span}(e_2, f_2) = 2,$$

which again implies

$$\dim V \geq 4 \text{ and } V = W_1 \oplus \underbrace{W_2 \oplus W_2^{\omega}}_{=W_1^{\omega}},$$

as W_2 is symplectic as well (same reasoning). If $W_2^{\omega} \neq 0$, then we proceed inductively (with $W_3 \subseteq W_2^{\omega}$ and so on) and observe that the procedure terminates at some point ($\dim V < +\infty$). This finishes the proof.

We continue with the second part of J.8. In Corollary 24.16 we have seen that a manifold M is orientable if and only if there exists a volume form, i.e. an everywhere non-vanishing top form

$$\mu \in \Omega^{2k}(M),$$

where $\dim M = 2k$ is even by the first part since ω_p is a symplectic form on $T_x M$ and $\dim M = \dim T_p M$. We claim that the k-wedge

$$\omega^k := \underbrace{\omega \wedge \cdots \wedge \omega}_{k\text{-times}} \in \Omega^{2k}(M)$$

defines a volume form. Indeed, by the first part we know that for each $p \in M$ there exists a symplectic basis $(e_{1,p}, \cdots, e_{r,p}, f_{1,p}, \cdots f_{k,p})$ for $(T_p M, \omega_p)$. By evaluating ω_p

[27] Indeed, $\omega|_{W_1} = e_1^* \wedge f_1^*$ which is clearly symplectic.

on this basis it is easy to see that

$$\omega_p = \sum_{i=1}^{k} e_{i,p}^* \wedge f_{i,p}^*,$$

where the $*$'s indicate the dual basis of $T_p^* M$. A straightforward combinatorial argument shows

$$\omega_p^k = k!\, e_{1,p}^* \wedge f_{1,p}^* \wedge \cdots \wedge e_{k,p}^* \wedge f_{k,p}^*,$$

thus proving $\omega_p^k \neq 0$ since

$$\omega_p^k \left(e_{1,p}, f_{1,p}, \ldots, e_{k,p}, f_{k,p}\right) = k! \neq 0.$$

Finally we move to the third and last part. The quickest way to see why $d\Theta$ defines a symplectic form on T^*M is by computing it in local coordinates. Let $(x^1, \cdots x^m)$ be the local coordinates on M associated to a chart $\sigma : U \to O$ and define, for every $i = 1, \ldots, m$,

$$y^i : T^*U \to \mathbb{R},\ y^i(p, \lambda) := \lambda \left(\left.\frac{\partial}{\partial x^i}\right|_p \right).$$

We can view $x^i : U \to \mathbb{R}$ also as smooth functions on T^*U by setting $x^i(p, \lambda) = x^i(p)$ which then leads us to local coordinates $(x^1, \ldots, x^m, y^1, \ldots, y^m)$ on T^*M (see Problem B.4 for more details). Let us show that on T^*U one has

$$\Theta = \sum_{i=1}^{m} y^i\, dx^i.$$

To prove this it suffices to compute Θ on the local frame field associated to our local coordinates, i.e.

$$\begin{aligned} \Theta_{(p,\lambda)} \left(\left.\frac{\partial}{\partial x^i}\right|_{(p,\lambda)} \right) &= \lambda \left(D\pi(p, \lambda) \left[\left.\frac{\partial}{\partial x^i}\right|_{(p,\lambda)} \right] \right) \\ &\overset{(1)}{=} \lambda \left(\left.\frac{\partial}{\partial x^i}\right|_p \right) \\ &= y^i(p, \lambda), \end{aligned}$$

where in (1) we used the fact that $\tilde{\sigma} \circ \pi \circ \tilde{\sigma}^{-1} = \mathrm{id}$, see solution of Problem B.4. On the other hand we have

$$\Theta_{(p,\lambda)} \left(\left.\frac{\partial}{\partial y^i}\right|_{(p,\lambda)} \right) = 0,$$

simply because the differential $D\pi$ maps all the $\frac{\partial}{\partial y^i}$'s to 0. All in all this proves the claim.

Using the Leibniz rule grants

$$d\Theta = \sum_{i=1}^{n} dy^i \wedge dx^i.$$

This 2-form is certainly closed[28] and non-degeneracy is a formal consequence of the first part of the proof.

(♣) PROBLEM J.9. Let M and N be smooth manifolds. Prove that if M has boundary and N does not, then $M \times N$ is a smooth manifold with boundary. Prove that if both M and N have non-empty boundary then $M \times N$ is not a smooth manifold with boundary,

SOLUTION. Suppose M has boundary and N does not. Let $m = \dim M$ and $n = \dim N$. We claim that $M \times N$ is a smooth manifold with boundary given by $\partial M \times N$.

We may always choose an atlas $\Sigma = \{\sigma_a : U_a \to Q_a \mid a \in A\}$ on M such that Q_a is an open subset of $\mathbb{R}^m_+$. Let $\mathcal{T} = \{\tau_b : V_b \to O_b \mid b \in B\}$ be an atlas on N. Then

$$\{\sigma_a \times \tau_b : U_a \times V_b \to Q_a \times O_b \mid (a, b) \in A \times B\}$$

is an atlas on $M \times N$ making it into a smooth manifold with boundary (since $Q_a \times O_b \subset \mathbb{R}^m_+ \times \mathbb{R}^n = \mathbb{R}^{m+n}_+$).

Showing that $\partial(M \times N) = \partial M \times N$ is equivalent to showing that

$$\operatorname{int}(M \times N) = \operatorname{int}(M) \times N. \tag{J.4}$$

Let $(p, q) \in M \times N$ and consider charts $\sigma : U \to \mathbb{R}^m_+$ and $\tau : V \to \mathbb{R}^n$ around p and q respectively. Then

$$\sigma \times \tau : U \times V \to \mathbb{R}^{m+n}_+$$

is a chart around (p, q) and

$$(\sigma \times \tau)(p, q) \in \operatorname{int}(\mathbb{R}^{m+n}_+)$$

if and only if $\sigma(p) \in \operatorname{int}(\mathbb{R}^m_+)$. It follows that $(p, q) \in \operatorname{int}(M \times N)$ if and only if $p \in \operatorname{int}(M)$. Thus (J.4) holds.

Now let M and N be smooth manifolds with non-empty boundary. Note that the boundary of a 0-dimensional manifold is empty, so our assumptions imply that $m = \dim M$ and $n = \dim N$ are both ≥ 1.

Suppose first that $m = n = 1$. It suffices to show that $\mathbb{R}^2_+$ and $\mathbb{R}_+ \times \mathbb{R}_+$ are not diffeomorphic. Suppose there exists a diffeomorphism

$$\phi : \mathbb{R}_+ \times \mathbb{R}_+ \xrightarrow{\sim} \mathbb{R}^2_+$$

[28] It is even exact!

and let $x_0 := (0,0)$. Since a diffeomorphism is also a homeomorphism, it follows that ϕ must send $\partial(\mathbb{R}_+ \times \mathbb{R}_+)$ to $\partial\mathbb{R}^2_+$, where we mean boundaries in the sense of topological manifolds with boundary. By assumption, the map ϕ extends on an open neighbourhood $U \subset \mathbb{R}^2$ of x_0 to a diffeomorphism $\tilde{\phi}\colon U \xrightarrow{\sim} V$, where V is open in $\mathbb{R}^2$. Consider the curves

$$\gamma_1\colon (-\epsilon_1, \epsilon_1) \to U, \quad t \mapsto (0,t)$$

and

$$\gamma_2\colon (-\epsilon_2, \epsilon_2) \to U, \quad t \mapsto (t,0).$$

Then the $\gamma_i'(0) \in T_{x_0}U$ are linearly independent for $i = 1, 2$. Since $\tilde{\phi}$ is a diffeomorphism, the vectors $D\tilde{\phi}(x_0)\big(\gamma_i'(0)\big)$ for $i = 1, 2$ must also be linearly independent. But for negative t, the curves $\tilde{\phi} \circ \gamma_i$ lie on the y-axis of $\mathbb{R}^2$, from which it follows that the tangent vectors $(\tilde{\phi} \circ \gamma_i)'(0)$ for $i = 1, 2$ are linearly dependent, a contradiction.

Note. One can generalise this argument to higher dimensions. In that case, one must instead take $n + m$ curves passing through a point $(p, q) \in \partial\mathbb{R}^m_+ \times \partial\mathbb{R}^n_+$ such that the tangent vectors are linearly independent, but whose images under the differential of a diffeomorphism are contained in an $(n + m - 1)$-dimensional subspace. We present a different argument of the same flavour below.

Now suppose that $m + n > 2$. Let $(p, q) \in \partial M \times \partial N$ and consider charts $\sigma\colon U \to Q \subset \mathbb{R}^m_+$ and $\tau\colon V \to O \subset \mathbb{R}^n$ around p and q respectively. As a consequence of the fact that the set of interior points and the set of boundary points on a manifold are disjoint, it follows that the image of (p, q) under $\sigma \times \tau$ is contained in $\partial\mathbb{R}^m_+ \times \partial\mathbb{R}^n_+$.

In order to obtain a contradiction, suppose that $M \times N$ is a smooth manifold with boundary. Then there exists a smooth chart $\rho\colon U' \to Q' \subset \mathbb{R}^{m+n}_+$ around (p, q). Without loss of generality, we may assume that $U' = U \times V$. We thus obtain a diffeomorphism

$$\rho \circ (\sigma \times \tau)^{-1}\colon Q \times O \xrightarrow{\sim} Q'.$$

Let

$$x = (x^i)_i := (\sigma \times \tau)^{-1}(p, q) \in \mathbb{R}^m \times \mathbb{R}^n.$$

By our assumptions, we have $x^1 = x^{m+1} = 0$. Let $S \subset \mathbb{R}^{m+n}$ be the $(n + m - 2)$-dimensional subspace consisting of points whose first and $(m + 1)$-st coordinates are zero. Then $\tilde{U} := (Q \times O) \cap S$ is open in S and $x \in \tilde{U}$. Consider the composite

$$\alpha\colon \tilde{U} \hookrightarrow Q \times O \to Q' \hookrightarrow \mathbb{R}^{m+n},$$

where the middle map is $\rho \circ (\sigma \times \tau)^{-1}$. By construction, the map α is smooth and the composite $(\sigma \times \tau) \circ \rho^{-1} \circ \alpha$ is the identity on $\tilde{U}$, so $D\alpha(x)$ is injective. Let

$$T := D\alpha(x)(T_x S) \subset \mathbb{R}^{m+n}.$$

Because T is $(n+m-2)$ dimensional, it must contain a vector v such that one of the first three components, v^1, v^2 or v^3 is non-zero. Renumbering coordinates and replacing v by $-v$ if necessary, we may assume that $v^1 < 0$. Let $\gamma\colon (-\epsilon, \epsilon) \to S$ be a smooth curve such that $\gamma(0) = x$ and $D\alpha(\gamma'(0)) = v$. Then $\alpha \circ \gamma(t)$ has negative x^1 coordinate for all small $t > 0$, which contradicts the fact that α takes values in $\mathbb{R}^{m+n}_+$.

Problem Sheet K

PROBLEM K.1. A singular k-cube $c\colon C^k \to M$ is said to be **degenerate** if there exists $1 \le i \le k$ such that c does not depend on u^i. Prove that if $c\colon C^k \to M$ is a degenerate singular k-cube then $\int_c \omega = 0$ for any $\omega \in \Omega^k(M)$.

SOLUTION. Let $c\colon C^k \to M$ be a degenerate singular k-cube and let $\omega \in \Omega^k(M)$. Recall that we can write $c^*\omega$ as $h\, dx^1 \wedge \cdots \wedge dx^k$, where $h \in C^\infty(C^k)$ is the function

$$h = c^*(\omega)\Big(\frac{\partial}{\partial x^1}, \dots, \frac{\partial}{\partial x^k}\Big).$$

Fix $p \in \mathrm{Int}(\mathrm{C}^k)$. By definition,

$$c^*(\omega)_p\Big(\frac{\partial}{\partial x^1}, \dots, \frac{\partial}{\partial x^k}\Big) = \omega_{c(p)}\Big(Dc(p)\Big[\frac{\partial}{\partial x^1}\Big], \dots, Dc(p)\Big[\frac{\partial}{\partial x^k}\Big]\Big) \tag{K.1}$$

Let $1 \le i \le k$ be such that c does not depend on x^i. Consider the map

$$j^i_p\colon C^1 \to C^k, \quad t \mapsto (p^1, \dots, p^{i-1}, t, p^{i+1}, \dots, p^k).$$

Since $c \circ j^i_p$ is constant, it follows that

$$Dc(p)\Big[\frac{\partial}{\partial x^i}\Big] = D(c \circ j^i_p)(p^i)\Big[\frac{\partial}{\partial t}\Big] = 0.$$

As $\omega_{c(p)}$ is alternating, we conclude that (K.1) is zero. Since this holds for all $p \in \mathrm{Int}(\mathrm{C}^k)$, this implies that $\int_c \omega = 0$.

PROBLEM K.2. Let $c\colon C^k \to M$ be a smooth singular k-cube in M and let $\varphi\colon C^k \to C^k$ be an orientation preserving diffeomorphism. Let $\tilde{c} := c \circ \varphi$. Prove that for any $\omega \in \Omega^k(M)$, one has

$$\int_c \omega = \int_{\tilde{c}} \omega.$$

SOLUTION. We write $c^*\omega = h\, dx^1 \wedge \cdots \wedge dx^k$ for $h \in C^\infty(C^k)$. We have

$$\begin{aligned}\int_c \omega := \int_{C^k} c^*\omega &= \int_{C^k} h \\ &\overset{(*)}{=} \int_{C^k} (h \circ \varphi)|\det D\varphi| \\ &\overset{(**)}{=} \int_{C^k} (h \circ \varphi)(\det D\varphi) \\ &\overset{(***)}{=} \int_{C^k} \varphi^* c^* \omega \\ &= \int_{\tilde{c}} \omega,\end{aligned}$$

where $D\varphi$ is the Jacobian matrix of φ. Here $(*)$ is just the classical change of variable formula from multivariable calculus, and $(**)$ follows from the assumption that φ is orientation preserving. Finally, equality $(***)$ comes from pullback formula

$$\varphi^*(h\,dx^1 \wedge \cdots \wedge dx^n) = (h \circ \varphi)(\det D\varphi)\,dx^1 \wedge \cdots \wedge dx^n.$$

PROBLEM K.3. Prove that there does not exist a compact symplectic manifold (M, ω) without boundary with the property that ω is exact.

SOLUTION. Suppose for contradiction that there exists a closed exact symplectic manifold

$$(M^{2n}, \omega = d\lambda).$$

By the second part of problem J.8, or rather its proof, we do know that ω^n defines a volume form on M. Without loss of generality we therefore assume

$$\omega^n > 0,$$

which readily implies

$$\int_M \omega^n > 0.$$

On the other hand exactness of ω implies exactness of ω^n as one can see by the computation

$$\omega^n = (d\lambda)^n = \overbrace{d\lambda \wedge \cdots \wedge d\lambda}^{n} = d(\lambda \cdot \overbrace{d\lambda \wedge \cdots \wedge d\lambda}^{n-1}).$$

The last equality is a consequence of $d^2 = 0$ and the Leibniz-rule. Thus, for $\alpha := \lambda \cdot (d\lambda \wedge \cdots \wedge d\lambda)$ we have

$$\omega^n = d\alpha.$$

We finally obtain our desired contradiction via Stokes' Theorem and the fact that M has empty boundary, i.e. $\partial M = \emptyset$, as it is a *closed* manifold by assumption. We have

$$0 < \int_M \omega^n = \int_M d\alpha = \int_{\partial M} \alpha = 0,$$

which is a contradiction and concludes the proof.

PROBLEM K.4. Find a closed $(m-1)$-form on $\mathbb{R}^m \setminus \{0\}$ that is not exact.

SOLUTION. Our strategy consists in pulling back a volume form on S^{m-1} to $\mathbb{R}^m \setminus \{0\}$. Pick a volume form ω on S^{m-1} inducing the standard orientation and define the smooth function

$$r \colon \mathbb{R}^m \setminus \{0\} \to S^{m-1},\ r(x) := \frac{x}{\|x\|}.$$

Observe that r is a left inverse of the inclusion $i \colon S^{m-1} \hookrightarrow \mathbb{R}^m \setminus \{0\}$. Now the claim is that the pullback

$$r^*\omega \in \Omega^{m-1}\left(\mathbb{R}^m \setminus \{0\}\right)$$

defines a closed non-exact $(m-1)$-form. Showing that it is closed is straightforward by dimension reasons of the sphere, more precisely

$$dr^*\omega = r^* \underbrace{d\omega}_{=0} = 0.$$

To see why $r^*\omega$ is non-exact we argue by contradiction. Assume for contradiction that there exists $\lambda \in \Omega^{m-2}\left(\mathbb{R}^m \setminus \{0\}\right)$ such that

$$d\lambda = r^*\omega.$$

From this we deduce that ω is also exact, as

$$\omega = (r \circ i)^*\omega = i^*\left(r^*\omega\right) = i^* d\lambda = di^*\lambda.$$

But this leads to a contradiction using the fact that ω is a volume form and from Stokes' Theorem we have

$$0 < \int_{S^{m-1}} \omega = \int_{S^{m-1}} di^*\lambda = \int_{\partial S^{m-1}} i^*\lambda = 0,$$

which finishes the proof.

PROBLEM K.5. Let M be a smooth manifold, let $X \in \mathfrak{X}(M)$, and let A be a tensor field. Let Φ_t denote the flow of X. Prove that

$$\frac{d}{dt}\bigg|_{t=t_0} \Phi_t^* A = \Phi_{t_0}^*(\mathcal{L}_X A).$$

SOLUTION. Changing variable in the limit of the difference quotient, we can write

$$\frac{d}{dt}\bigg|_{t=t_0} \Phi_t^*(A) = \frac{d}{dt}\bigg|_{t=0} \Phi_{t_0+t}^*(A).$$

By the properties of the flow, we then have $\Phi_{t_0+t} = \Phi_{t_0} \circ \Phi_t$, and consequently by the properties of the pull-back, this implies $\Phi_{t_0+t}^*(A) = \Phi_{t_0}^*(\Phi_t^*(A))$. Moreover, since the pull-back is a linear operation, it commutes with derivatives and thus we conclude that

$$\frac{d}{dt}\bigg|_{t=0} \Phi_{t_0+t}^*(A) = \frac{d}{dt}\bigg|_{t=0} \Phi_{t_0}^*(\Phi_t^*(A)) = \Phi_{t_0}^*\left(\frac{d}{dt}\bigg|_{t=0} \Phi_t^*(A)\right) = \Phi_{t_0}^*(\mathcal{L}_X(A)),$$

as desired.

PROBLEM K.6. Let $\varphi\colon M \to N$ be a diffeomorphism of connected oriented manifolds and let $\omega \in \Omega_c^m(N)$. Prove that

$$\int_M \varphi^*(\omega) = \pm \int_N \omega,$$

where the + signs occurs if and only if φ is orientation preserving.

SOLUTION. Assume first that ω and $\varphi^\star(\omega)$ are compactly supported in the domains of single charts whose coordinates we denote $y = (y^1, \ldots, y^m)$ and $x = (x^1, \ldots, x^m)$ respectively. We may then write $\omega(y) = f(y)\, dx^1 \wedge \cdots \wedge dx^m$ for some smooth function f, and consequently (by the properties of the exterior derivative and of the wedge product)

$$\begin{aligned}\varphi^\star(\omega)(x) &= f(\varphi(x))\, d\varphi^1(x) \wedge \cdots \wedge \varphi^m(x)\\ &= f(\varphi(x)) \det(D\varphi(x))\, dx^1 \wedge \cdots \wedge dx^m,\end{aligned}$$

and so, with the standard change of variable formula for the integrals and the rule on the derivative of the inverse function, setting $y = \varphi(x)$, we see that

$$
\begin{aligned}
\int \varphi^\star(\omega) &= \int f(\varphi(x))\det(D\varphi(x))\,dx^1\cdots dx^m \\
&= \int f(y)\det D\varphi(\varphi^{-1}(y))\,|\det D(\varphi^{-1})(y)|\,dy^1\cdots dy^m \\
&= \int f(y)\frac{|\det D(\varphi^{-1})(y)|}{\det D(\varphi^{-1})(y)}\,dy^1\cdots dy^m \\
&= \pm\int f(y)\,dy^1\cdots dy^m \\
&= \pm\int \omega,
\end{aligned}
$$

where the plus sign occurs if φ (and so φ^{-1}) is orientation-preserving and the minus sign if if it is orientation-reversing. This proves the result in this special single-chart case.

For the general case, we choose a positively oriented atlas $(U_a, x_a)_{a\in A}$ for M so that, for each U_a, $\varphi(U_a)$ is contained in a single coordinate chart y_b of N (this is possible because φ is smooth and in particular continuous), and $(\psi_a)_{a\in A}$ a correspondent partition of unity. Since φ is a diffeomorphism, $(\varphi(U_a), x_a\circ\varphi^{-1})_{a\in A}$ will then constitute a (positively or negatively, depending on φ) oriented atlas for N, and $(\psi_a\circ\varphi^{-1})_{a\in A}$ a corresponding partition of unity. We can then compute, thanks to the special case above:

$$
\int_N \varphi^\star(\omega) = \sum_a \int_{\varphi(U_a)} (\psi_a\circ\varphi^{-1})\,\varphi^\star(\omega) = \sum_a \pm\int_{U_a}\psi_a\,\omega = \pm\int_M \omega,
$$

where again, the plus sign occurs if φ is orientation-preserving and the minus sign if if it is orientation-reversing.

Problem K.7. Let G be a compact connected Lie group.

(i) Prove there exists a unique **normalised left-invariant** volume form μ on G, i.e. a volume form μ such that $\int_G \mu = 1$ and $l_g^*\mu = \mu$ for all $g\in G$.
(ii) This allows us to define the integral of a *function* on G via:

$$
\int_G f := \int_G f\mu, \qquad f\in C^\infty(G).
$$

Prove that for all $f\in C^\infty(G)$ and $g\in G$, one has

$$
\int_G f = \int_G (f\circ l_g) = \int_G (f\circ r_g), \qquad \forall f\in C^\infty(G),\ g\in G.
$$

Solution. Part (i): observe that two volume forms on G which are left-invariant and coincide at the identity $e \in G$, coincide everywhere. This is because, if μ is such a form then $\mu|_{T_gG} \equiv Dl_{g^{-1}}(g)^\star \mu|_{T_eG}$ for all $g \in G$, so $\mu|_{T_gG}$ is uniquely determined by $\mu|_{T_eG}$. Hence, it suffices to show that two left-invariant volume forms on G coincide on T_eG up to scaling, but this follows from basic linear algebra: we know that

$$\bigwedge^{\dim G} T_e^*G$$

is 1-dimensional.

Part (ii): by definition we have

$$\begin{aligned}\int_G (f \circ l_g) &= \int_G (f \circ l_g)\mu \\ &= \int_G (f \circ l_g) l_g^* \mu \\ &= \int_G l_g^*(f\mu) \\ &= \int_G f\mu \\ &= \int_G f,\end{aligned}$$

where at the second to last equality we make use of problem K.6. and the fact that $l_g : G \to G$ is orientation preserving (since $l_g^*\mu = \mu$). The equality $\int f = \int (f \circ r_g)$ follows as before once we verify that

$$r_g^*\mu = \mu.$$

To see this, first observe that $r_g^*\mu$ is itself a left-invariant volume form since r_g is a diffeomorphism and

$$l_h^* r_g^* \mu = (r_g \circ l_h)^* \mu = (l_h \circ r_g)^* \mu = r_g^*(l_h^*\mu) = r_g^*\mu,$$

for all $h \in G$. As observed above, $r_g^*\mu$ has to be a non-zero multiple of μ by dimension reasons, and since $r_g : G \to G$ is orientation preserving, the multiple is strictly positive. Thus for every $g \in G$ there exists a $\kappa(g) \in \mathbb{R}$ such that

$$r_g^*\mu = e^{\kappa(g)}\mu.$$

Using $r_h \circ r_g = r_{gh}$ and the equation above it is easy to show that $\kappa : (G, \cdot) \to (\mathbb{R}, +)$ is a Lie group homomorphism. But then $\kappa \equiv 0$—indeed, assume for contradiction that there is some $g \in G$ with $\kappa(g) \neq 0$. Then $\kappa(g^n) = n \cdot \kappa(g)$ diverges as $n \in \mathbb{N}$ goes to infinity, however $(g^n) \subseteq G$ admits a convergent subsequence since G is compact—contradiction! Therefore we have shown

$$r_g^*\mu = e^{\kappa(g)}\mu = e^0\mu = \mu.$$

This finishes the proof.

(♣) Problem K.8. For this problem you may assume that for any compact connected orientable smooth manifold M^m, one has $H^m_{\mathrm{dR}}(M) \cong \mathbb{R}$, and that an explicit isomorphism is given by

$$\int : H^m_{\mathrm{dR}}(M) \to \mathbb{R}, \qquad [\omega] \mapsto \int_M \omega.$$

Let $\varphi\colon M \to N$ be a smooth map between compact connected orientable smooth manifolds of dimension m. Then $\varphi^*\colon H^m_{\mathrm{dR}}(N) \to H^m_{\mathrm{dR}}(M)$ is a linear map between one-dimensional vector spaces, and hence is multiplication by a number. We call this number the **degree** of φ. Explicitly,

$$\int_M \varphi^*\omega = \deg(\varphi) \int_N \omega, \qquad \omega \in \Omega^m(N).$$

(i) Let $q \in N$ denote a regular value of φ. Given $p \in \varphi^{-1}(q)$, let

$$\operatorname{sgn}_p(\varphi) := \begin{cases} +1, & \text{if } D\varphi(p) \text{ is orientation preserving,} \\ -1, & \text{if } D\varphi(p) \text{ is not orientation preserving.} \end{cases}$$

Prove that

$$\deg(\varphi) = \sum_{p \in \varphi^{-1}(q)} \operatorname{sgn}_p(\varphi).$$

Thus $\deg(\varphi)$ is an integer.

(ii) Prove the **Hairy Ball Theorem**: if m is even then any vector field on S^m has at least one zero.

Solution. Part (i): by Sard's Theorem there exists a regular value $q \in N$ of $\varphi\colon M \to N$. The Implicit Function Theorem then implies that $\varphi^{-1}(q) \subset M$ is a 0-dimensional manifold, i.e. a collection of isolated points in M. Since M is compact and $\varphi^{-1}(q)$ does not have any accumulation points, it follows that $\varphi^{-1}(q)$ is finite. We enumerate

$$\varphi^{-1}(q) = \{p_1, \dots, p_l\} \subset M.$$

By definition each p_i defines a regular point of φ which then readily implies that each differential

$$d\varphi(p_i)\colon T_{p_i}M \to T_qM$$

is a linear isomorphism (remember $\dim M = \dim N$!). Therefore the Inverse Function Theorem grants the existence of pairwise disjoint charts

$$U_1, \dots, U_l \subset M, \; x_i\colon U_i \to \mathbb{R}^m$$

such that φ restricted to each one of these is a diffeomorphism onto its image, i.e.

$$\varphi\big|_{U_i}\colon U_i \xrightarrow{\sim} \varphi(U_i), \text{ for all } i = 1, \dots, l.$$

Now we define the open sets

$$V = \bigcap_{i=1}^{l} \varphi(U_i) \text{ and } U_i' := U_i \cap \varphi^{-1}(V).$$

By shrinking U_i (and hence V) if necessary, we can assume that V is the domain of a chart on N, say (V, y). Let

$$g \in C^\infty(N)$$

be a smooth function with support in V and define the top form

$$\omega := g\, dy^1 \wedge \cdots \wedge dy^n$$

on N, where the y^i's are the local coordinates coming from (V, y). Due to our choices, the pullback form $\varphi^*\omega$ on M has support lying inside the union $\bigcup_i U_i'$ and

$$\varphi(U_i') = V,$$

hence

$$\int_M \varphi^*\omega = \sum_{i=1}^{l} \int_{U_i'} \varphi^*\omega \overset{(1)}{=} \sum_{i=1}^{l} \operatorname{sgn}_{p_i}(\varphi) \int_V \omega = \sum_{i=1}^{l} \operatorname{sgn}_{p_i}(\varphi) \int_N \omega,$$

where step (1) follows from Problem K.6.. This proves

$$\deg(\varphi) = \sum_{p \in \varphi^{-1}(q)} \operatorname{sgn}_p(\varphi)$$

and finishes the first part.

For part (ii) we make the crucial observation that the degree is a homotopy invariant, i.e. two smoothly homotopic maps $\varphi, \psi\colon M \longrightarrow N$ have the same degree

$$\deg(\varphi) = \deg(\psi).$$

This follows immediately from the definition of $\deg(\varphi)$ as the scalar of the induced map $\varphi^*\colon H^m_{\mathrm{dR}}(M) \cong \mathbb{R} \longrightarrow H^m_{\mathrm{dR}}(N) \cong \mathbb{R}$, and Theorem 27.5, which asserts that two smoothly homotopic maps induce the same linear map on the de Rham cohomology.

We proceed by contradiction and assume that for m even there exists a vector field X on S^m with no zeros. The strategy is to build a homotopy between the antipodal map and the identity, which would lead to a contradiction as the former has degree -1 whereas $\deg(\mathrm{id}_{S^n}) = 1$.

For every $p \in S^m$, there exists a *unique* semicircle $\gamma_p \approx S^1$ on S^m determined by the direction $X(p) \in T_pS^m \setminus \{0\}$. Each such semicircle can be viewed as a smooth loop

$$\gamma_p\colon [0, 2] \longrightarrow S^m$$

satisfying

$$\gamma_p(0) = \gamma_p(2) = p \text{ and } \gamma_p(1) = -p.$$

With this we are able to define a smooth homotopy

$$H\colon [0,1] \times S^m \to S^m,\ H(t,p) := \gamma_p(t)$$

that connects the identity

$$H(0,p) = \gamma_p(0) = p$$

to the antipodal map

$$H(1,p) = \gamma_p(1) = -p.$$

The fact that H is smooth follows from γ_p being uniquely defined by $X(p)$, where the later is smooth by definition. This finishes the proof.

Appendix B
Contents of Volume II

W. J. Merry, *Lectures on Differential Geometry I*,
https://doi.org/10.1007/978-3-032-03733-6

References

1. Merry, W.J.: Lectures on Differential Geometry II: Connections & Riemannian Geometry. Springer, Berlin (2025)
2. Milnor, J.W.: Topology from the Differentiable Viewpoint. Princeton Landmarks in Mathematics. Princeton University Press, Princeton (1998)
3. Robbin, J., Salamon, D.: Introduction to Differential Geometry. Springer Studium Mathematik (Master). Springer, Berlin (2022)
4. Stoll, C.: Klein Bottle. https://www.kleinbottle.com/. Accessed 13 March 2024
5. Wikipedia. Abstract nonsense. https://en.wikipedia.org/wiki/Abstract_nonsense
6. Wikipedia. Contact geometry. https://en.wikipedia.org/wiki/Contact_geometry
7. Wikipedia. Heine-Borel theorem. https://en.wikipedia.org/wiki/Heine%E2%80%93Borel_theorem
8. Wikipedia. Hilbert's fifth problem. https://en.wikipedia.org/wiki/Hilbert%27s_fifth_problem
9. Wikipedia. Leibniz integral rule. https://en.wikipedia.org/wiki/Leibniz_integral_rule
10. Wikipedia. Noether's Theorem. https://en.wikipedia.org/wiki/Noether%27s_theorem
11. Wikipedia. Quotient metric spaces. https://en.wikipedia.org/wiki/Metric_space#Quotient_metric_spaces
12. Wikipedia. Russell's paradox. https://en.wikipedia.org/wiki/Russell%27s_paradox
13. Wikipedia. Sophus Lie. https://en.wikipedia.org/wiki/Sophus_Lie

W. J. Merry, *Lectures on Differential Geometry I*,
https://doi.org/10.1007/978-3-032-03733-6

Index

W. J. Merry, *Lectures on Differential Geometry I*,
https://doi.org/10.1007/978-3-032-03733-6

MIX
Papier aus verantwortungsvollen Quellen
Paper from responsible sources
FSC® C105338

If you have any concerns about our products,
you can contact us on
ProductSafety@springernature.com

In case Publisher is established outside the EU,
the EU authorized representative is:
Springer Nature Customer Service Center GmbH
Europaplatz 3, 69115 Heidelberg, Germany

Printed by Libri Plureos GmbH
in Hamburg, Germany